PERKINS' SERIES.

# PLANE AND SOLID

# GEOMETRY.

TO WHICH IS ADDED

PLANE AND SPHERICAL

# TRIGONOMETRY

AND

# MENSURATION.

ACCOMPANIED WITH ALL THE NECESSARY

LOGARITHMIC AND TRIGONOMETRIC

TABLES.

BY GEORGE R. PERKINS, LL.D.

NEW YORK:
D. APPLETON & CO., 346 & 348 BROADWAY,
AND 16 LITTLE BRITAIN, LONDON.
M DCCC LVI.

Entered according to Act of Congress, in the year 1854,
By GEORGE R. PERKINS,
In the Clerk's Office of the District Court for the Northern District of New York.

# PREFACE.

This work is not designed to take the place of the "Elements of Geometry," but rather to give a more extended and com prehensive course, better adapted to the use of Advanced Schools and Colleges.

In the *First Part* of the work, which is confined to the theoretical principles of Geometry, great care has been taken to classify and arrange the various Theorems, by bringing together such as correspond to analogous subjects. We have also arranged the Problems by themselves, and have in no case mixed them up with the Theorems.

In the preparation of the Geometrical portion of this work, we have made free use of the admirable work of Vincent. The edition used is the *fifth*, revised by its author, with the assistance of the distinguished Bourdon. This is a most excellent and valuable French work, but it is not in all respects well adapted to the wants of our Higher Institutions; we have, therefore, preferred to make selections from it, rather than to give a translation of the whole. We have endeavored to arrange and present the different Propositions in the manner which to us appeared best calculated to impart a thorough knowledge of the most important principles of the science of Geometry. As a general thing, we have adhered to the ordinary method of conducting our demonstrations; still we have not hesitated to introduce the algebraic notation, when by so doing we could present the subject in a clearer and more satisfactory manner.

Throughout the *Second Part*, which is of a more mixed or practical nature, we have studied to present the whole in a distinct and clear manner, giving in all cases the full and complete work of all examples necessary to illustrate each principle. We would direct the attention in particular to the chapters on Spherical Trigonometry. This subject is generally regarded as intricate and difficult. It is believed the student will find the arrangement and full development of the different cases as here given, nearly as simple and of as easy comprehension as those of Plane Trigonometry.

We might particularize other portions of the work, but will content ourselves by remarking that we have spared no pains to make the whole acceptable to the mathematical student of the present age. How well we have succeeded remains to be shown.

GEO. R. PERKINS.

UTICA, September, 1854.

# CONTENTS.

## PART I.

## FOURTH BOOK.



## FIFTH BOOK.



## SIXTH BOOK.



## SEVENTH BOOK.



## EIGHTH BOOK.

# PART II.

## CHAPTER I.



## CHAPTER II.



## CHAPTER III.



## CHAPTER IV.



## CHAPTER V.

### SPHERICAL TRIGONOMETRY.



## CHAPTER VI.



## CHAPTER VII.



## CHAPTER VIII.

PART FIRST.

---

# PLANE AND SOLID GEOMETRY.

# INTRODUCTION.

1. All bodies occupy, in the *indefinite space* which embraces the material universe, a *determinate*, or *finite place* which we properly call a *space*.

This finite space occupied by the body, has *limits* or *boundaries*, which separate it from the rest of space. These boundaries constitute what we call the *Surface* of the body. Surface being then the separation of a body from the rest of space, belongs as well to the one as the other; and as an infinite number of bodies may exist, in indefinite space, each having its proper surface, it follows that

*In space we may conceive of an infinite number of surfaces.*

When one surface is cut by another surface, the place of their mutual intersection is called a *Line*. This line belongs to each of the surfaces. Since the intersection of two surfaces gives a line, and any surface may be intersected by an infinite number of other distinct surfaces, it follows that

*On any surface whatever, we may conceive an infinite number of lines.*

The place of meeting, or of intersection of two lines, is called a *Point*. This point is common to the two lines. Since a point results from the meeting of two lines, and any line may be met by an infinite number of other distinct lines, it follows that

*Any line may be regarded as having an infinite number of points.*

2. Although we acquire the notion of a point by the consideration of lines, the notion of a line by the consideration of surfaces, and that of the surface by the consideration of a body; that is to say, of *material* things, we must not conclude from this, that points, lines, and surfaces are themselves *material* objects. In virtue of an inherent faculty of the mind, we may readily conceive of a point without the lines which determine it,

of a line without the surfaces of which it is the intersection, of a surface independently of the body, or of the space of which it is the limit; in short, we may conceive of space itself, as absolutely immaterial. It is the result of these different abstractions which we call *a point*, *a line*, *a surface*, or *a space*. It is the same when we speak of *the points of a line*, *of a surface*, *of a space; the lines of a surface*, &c.

3. A space, a surface, and a line, may be considered in two distinct ways. We may consider them in regard to their different *forms*, which we name in general their *Figures*. Or, we may consider them in regard to their relative *magnitudes*, which is comprehended under the name of *Extension*.

Extension takes the particular name of *Volume*, of *Area*, or of *Length*, according as it is applied to a space, a surface, or a line. Thus, the length of a line, or its *linear extension*, is the magnitude of this line, estimated or measured in units of a line. In the same way the *area* of a surface, or its *superficial extension*, is the magnitude of this surface, estimated or measured in units of a surface. In short, the *volume* or the *extension of space* is the magnitude of this space, estimated or measured in units of space.

GEOMETRY is the science which treats of these two principal objects: The *properties* of different kinds of *figures*, and the *measure of extension*, considered under the different circumstances, as already noticed.

## THE POINT.

4. *A point has neither figure nor extension*; it is this, above all, which distinguishes it from all other objects of Geometry, which are all capable of being *described* and *measured*.

Nevertheless as it is often necessary to consider one or more isolated points, we represent their position by a dot or distinct mark made on a surface with the pen, pencil, or crayon, and we distinguish them from each other, usually, by letters. Thus we say the point A, the point B, the point C, &c.

.A .B .C .D

The position of a point in reference to any other point is determined by its direction and distance from that point. This *distance*, which is the length of the shortest line which can unite these points, is obviously mutual to the two points; that is, the distance of the point B from the point A is the same as the distance of the point A from the point B.

The *direction* of the point A from the point B is *opposite* to the *direction* of the point B from the point A.

## THE STRAIGHT LINE.

5. Of all geometrical lines, the simplest is the *straight line.*

Although the idea of a straight line, is the first to which we are conducted by our experience, and the use of our senses, still it is very difficult to define it. The definition usually given is as follows: *A straight line is the shortest distance between two points.*

A more general definition is as follows: A straight line is an indefinite line, such, that any limited portion whatever, is the shortest distance between the points which fix this limit.

In any line the direction of one of its points, B, from another point, A, as has already been noticed, is opposite the direction of A from B; so that a line has *two different directions exactly opposite*, either of which may be regarded as the direction of the line.

*A Straight line* is one which has the same direction throughout its whole extent.

*A Curved line* is one which changes its direction at every point.

Two straight lines are evidently capable of *superposition*, that is, of being placed the one on the other, so as to coincide. Hence, two straight lines coincide throughout their whole extent when they have two points common. Or, in other words, two points determine the position of a straight line. We also infer that two distinct straight lines can intersect or meet each other in only one point.

## THE PLANE.

6. The plane surface, or, as usually expressed, the *Plane*, is the simplest of all surfaces. It may be defined as follows: A plane is an indefinite surface, on which we may conceive that through each of its points a straight line may be made exactly to coincide with it throughout its whole extent.

It immediately results from this definition, and the nature of a straight line, That any line, having two of its points common with the plane, lies wholly in this plane. Hence, a straight line cannot be partly in a plane and partly out of it.

When a straight line has only one point in common with a plane, it is said to meet or pierce the plane, and the plane is said to cut the line, and the segments or portions of the line thus separated will be on different sides of this plane.

## THE CIRCLE.

7. When a line is not a straight line, or made up of finite portions of straight lines, it is called a *curved line.*

The simplest of all *curved lines* is the *circumference of a circle*, which may be thus defined: *The circumference of a circle* is a plane curve returning into itself, every point of which is equally distant from a certain point in its plane, which point is called the *centre* of the circumference. The portion of the plane limited by this circumference is called a *circle.* Any portion of the circumference of a circle is called an *arc.* When the arc is equal to one-fourth of the circumference it is called *a quadrant.* Each of the straight lines drawn from the centre to the circumference is called *a radius.* Hence, *all radii of the same circle are equal.*

A line passing through the centre and terminating in both directions by the circumference, is called *a diameter. All diameters of the same circle are equal,* since each is twice the radius.

## THE ANGLE.

8. When two straight lines meet, the opening between them is called a *plane angle*, or simply *an Angle.* The magnitude of the angle does not depend upon the lengths of these lines, but only upon the difference of their directions.

If in any circle we draw two radii, the distance between their extremities which terminate in the circumference, will embrace an arc. When the arc between the two radii is equal to a quadrant, these radii form with each

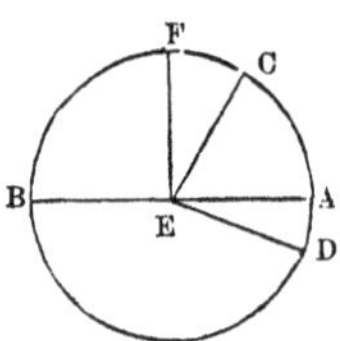

other an angle, which we call a *right angle*. And the radii are said to be the one *perpendicular* to the other. When the arc between the radii is less than a quadrant, the angle is called *acute*. When the arc is greater than a quadrant, the angle is called *obtuse*.

The magnitude of an angle may be estimated or measured by means of any particular angle, taken as the unit angle. The right angle is generally the angle chosen as the unit angle.

## THE RULER AND THE COMPASS.

9. The straight line and the circumference of the circle, which are the only lines treated of in Elementary Geometry, are respectively traced or drawn upon a plane, by the aid of the *Ruler* and of the *Compass*.

These instruments are so simple, and of such general use, as to need no description in this place. With the Ruler we can draw a straight line on a plane from any one point to any other point. With the Compass we can describe on a plane the circumference of a circle having any given point as a centre, and for its radius any given line.

## METHODS OF DEMONSTRATION.

10. There are two distinct methods employed in Geometrical demonstration: The *Direct Method*, and the *Indirect Method*.

The most simple process of direct demonstration is the principle of *superposition*, which consists in being able to make two figures exactly coincide, by applying the one upon the other.

The demonstration is also direct when we employ, by a direct course of reasoning, axioms, definitions, and principles already established.

The indirect method, known under the name of *Reducing to an absurdity*, consists in first supposing the proposition not to be true; afterwards, by certain deductions, to draw, from truths already recognized as rigorously exact, a result contradictory to some one of these truths, or to the proposition itself.

We will terminate this subject by noticing two kinds of *false reasoning*, very common with beginners, and against which they should be constantly on their guard.

The first is called, *Reasoning in a circle.*

The second is called, *Begging the question.*

We are said to reason in a circle when, in the demonstration of a proposition, we employ, either implicitly or explicitly, a second proposition, which cannot, itself, be established without the aid of the first.

We are said to beg the question, when, in order to establish a proposition, we employ the proposition itself.

# GEOMETRY.

## FIRST BOOK.

## THE PRINCIPLES.

### DEFINITIONS.

I. GEOMETRY is the science of *Position* and *Extension*.

II. A *Point* has merely position, without any extension.

III. *Extension* has three dimensions; *Length*, *Breadth*, and *Thickness*.

IV. A *Line* has only one dimension; length.

V. A *Surface* has two dimensions; length and breadth.

VI. A *Solid* has three dimensions; length, breadth, and thickness.

VII. A *Straight line* is one which has the same direction through its whole extent. In reality a line has two directions, the one exactly opposite the other; either of which may be considered as its direction.

VIII. A *Broken line* is one which is made up of two or more straight lines.

IX. A *Curved line* is one which changes its direction at every point.

X. *Parallel lines* are those which have the same direction.

XI. An *Angle* is the difference in direction of *two* straight lines meeting or crossing each other.

The *Vertex* of the angle is the point where its *sides* meet.

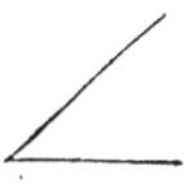

XII. When one straight line meets or crosses another, so as to make the adjacent angles equal, each of these angles is called a *Right angle*, and the lines are said to be *perpendicular* to each other.

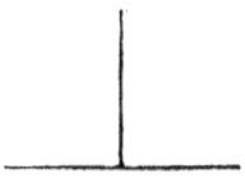

XIII. An *Acute angle* is one which is less than a right angle.

XIV. An *Obtuse angle* is one which is greater than a right angle.

XV. When the sum of two angles is equal to a right angle, they are called *Complementary* angles; each being the complement of the other.

XVI. When the sum of two angles is equal to two right angles, they are called *Supplementary* angles; each being the supplement of the other.

XVII. A *Plane* is a surface which is straight in every direction, or one with which a straight line, joining any two of its points, will coincide.

XVIII. When a surface is not a plane surface it is called a *Curved surface.*

XIX. A *plane figure* is a limited portion of a plane. When it is limited by straight lines, the figure is called a *rectilineal figure*, or a *polygon;* and the limiting lines, taken together, form the *contour* or *perimeter* of the polygon.

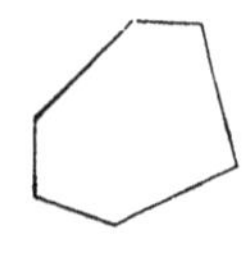

XX. The simplest kind of polygon is one having only three sides, and is called a *triangle.* A polygon of four sides is called a *quadrilateral;* that of five sides is called a *pentagon;* that of six sides is called a *hexagon;* one of seven sides is called a *heptagon;* one of eight sides an *octagon;* one of nine sides a *nonagon*, and so on for figures of a greater number of sides.

XXI. A triangle having the three sides equal, is called an *equilateral triangle;* one having two sides equal, is called an *isosceles triangle;* and one having no two sides equal, is called a *scalene triangle.*

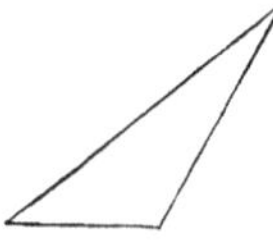

XXII. A triangle having a right angle, is called a *right-angled triangle.* The side opposite the right angle is called the *hypotenuse.*

XXIII. A triangle having its three angles acute, is called an *acute-angled triangle.*

XXIV. A triangle having an obtuse angle, is called an *obtuse-angled triangle.*

XXV. When the opposite sides of a quadrilateral are parallel, the figure is called a *parallelogram.*

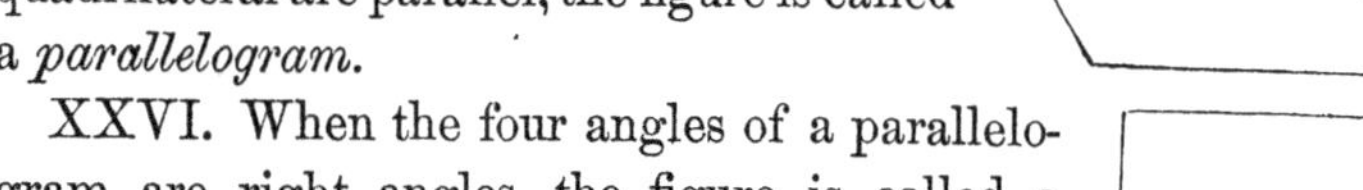

XXVI. When the four angles of a parallelogram are right angles, the figure is called a *rectangle.*

XXVII. When the four sides of a rectangle are equal, the figure is called a *square.*

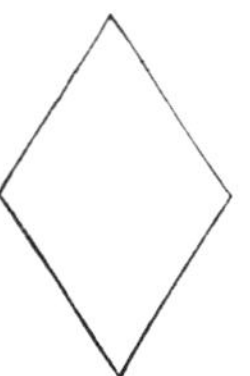

XXVIII. When the four sides of a parallelogram are equal, and the angles not right, the figure is called a *rhombus.*

XXIX. When only two sides of a quadrilateral are parallel, the figure is called a *trapezoid.*

XXX. A *diagonal* of a polygon is a line joining the vertices of two angles, not adjacent.

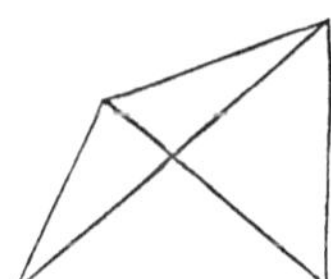

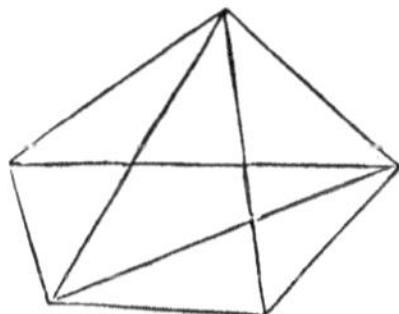

### DEFINITIONS OF TERMS.

I. A *demonstration* is a logical train of reasoning employed to establish an asserted truth.

II. A *proposition* is a statement of a truth to be demonstrated, or of an operation to be performed.

III. An *axiom* is a self-evident proposition, of which the simple mention carries a conviction of its truth.

IV. A *postulate* is a proposition, the truth of which is required to be admitted without demonstration, notwithstanding it does not present the same degree of evidence as in the case of the axiom.

V. A *theorem* is a proposition which requires a demonstration.

VI. A *corollary* is an immediate consequence of one or more propositions. If any new course of reasoning is required to establish it, this reasoning is so simple that it may be supplied without much inconvenience.

VII. A *problem* is a question proposed, which requires a solution.

VIII. A *lemma* is a subsidiary truth, employed for the demonstration of a theorem, or the solution of a problem.

IX. A *scholium* is a remark on one or several preceding propositions, which tends to point out their connection, their use, their restriction, or their extension.

X. An *hypothesis* is a supposition, made in the statement of a proposition, or in the course of its demonstration.

## REMARKS IN REFERENCE TO SIGNS, SYMBOLS, ETC.

The signs and symbols which we shall employ, have the same signification in Geometry as in Algebra, to which, for their full explanation, we shall refer the student.

For the doctrine of ratios and proportions, we will also refer the student to the method explained in the Algebra.

There is this difference between geometrical ratios of magnitudes, and ratios of numbers: All numbers are *commensurable;* that is, their ratio can be accurately expressed: but many magnitudes are *incommensurable;* that is, their ratio can be expressed only by approximation; which approximation may, however, be carried to any extent we desire. Such is the ratio of the circumference of a circle to its diameter, the diagonal of a square to its sides, etc. Hence many have deemed the arithmetical method not sufficiently general to apply to geometry. This would be a safe inference, were it necessary in all cases to assign the specific ratio between the two terms compared. But this is not the case. Such ratios themselves may be unknown, indeterminate, or irrational, and still their equality or inequality may be as completely determined by the arithmetical method as by the more lengthy method of the Greek geometers.

## AXIOMS.

I. Things which are equal to the same thing, are equal to each other.

II. When equals are added to equals, the wholes are equal.

III. When equals are taken from equals, the remainders are equal.

IV. When equals are added to unequals, the wholes are unequal.

V. When equals are taken from unequals, the remainders are unequal.

VI. Things which are double of the same or equal things, are equal.

VII. Things which are halves of the same thing, are equal.

VIII. Every whole is equal to all its parts taken together, and greater than any of them.

IX. Things which coincide, or fill the same space, are identical.

X. All right angles are equal to one another.

XI. A straight line is the shortest distance between two points.

XII. Through the same point only one straight line can be drawn parallel to another.

XIII. Only one straight line can be drawn joining two given points.

XIV. Straight lines which are parallel to the same line are parallel to each other.

## POSTULATES.

I. Let it be granted that a straight line may be drawn from any point to any other point.

II. That a terminated straight line may be produced, in either direction, to any extent.

III. That a straight line may be bisected or halved.

IV. That from a point, either within or without a straight line, a perpendicular may be drawn to the line.

V. That a line may be drawn, making any given angle with another line.

VI. That a line may be drawn from the vertex of an angle, bisecting it.

---

## OF ANGLES.

### THEOREM I.

*When a straight line meets or crosses another, the adjacent angles are supplements; and the opposite angles are equal.*

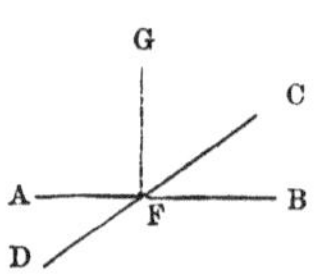


For, drawing FG perpendicular to AB, we see that the angle AFC exceeds the right angle AFG, by the angle CFG; while the angle BFC is less than the right angle BFG by the same angle CFG. Hence, the sum of AFC, BFC is equal to two right angles, and they are therefore supplements of each other (D. XVI.).*

In the same way by drawing a line perpendicular to CD, it may be shown that the angles CFA and AFD are supplements; consequently, BFC + CFA = CFA + AFD (A. I.): from each taking CFA, we have BFC equal to its opposite angle AFD. In a similar manner we have CFA=DFB.

*Cor.* I. If either of the four angles, formed by the intersection of two straight lines, is a right angle, the remaining three angles will each be right, and the two lines will be mutually perpendicular.

When the angles are not right there will be two equal acute angles, and two equal obtuse angles. The acute angle and the obtuse angle will be supplementary.

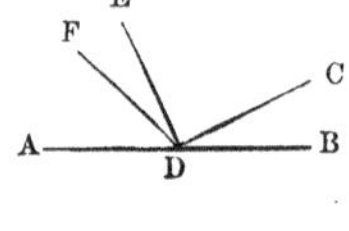


*Cor.* II. All the angles which can be made at any point D, by any number of lines on the same side of AB, are together equal to two right angles.

---

* In the references we shall use the following abbreviations: A. for Axiom, B. for Book, C. for Corollary, D. for Definition, P. for Problem, S. for Scholium, T. for Theorem.

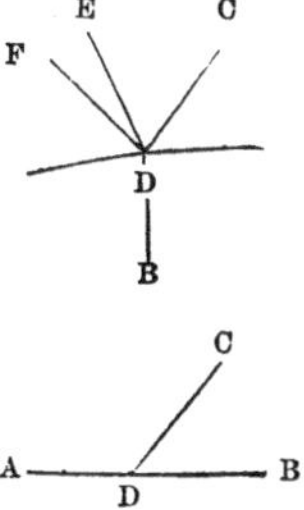


*Cor.* III. And as all the angles that can be made on the other side of AB are also equal to two right angles; therefore all the angles that can be made around the point D, by any number of lines, are together equal to four right angles.

*Cor.* IV. If two adjacent angles, ADC and BDC, are supplementary, their exterior sides AD and BD will form one and the same line.

*Scholium.* Since a straight line has two different directions, exactly opposite each other, it is sometimes considered as making with itself an angle equal to two right angles.

THEOREM II.

*If through the vertex of any angle, lines are drawn perpendicular respectively to its sides, they will form a new angle, either equal to the first, or supplementary to it.*

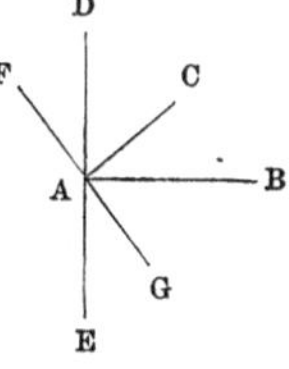


Let BAC be the given angle, DE perpendicular to AB, and FG perpendicular to AC.

Then we shall have the angle DAF equal to BAC, since each is the complement of CAD (D. XV.). The angle EAG being opposite DAF must also equal BAC (T. I). But the angle DAG, or its opposite angle FAE, is supplementary with DAF, and consequently supplementary with its equal angle BAC.

THEOREM III.

*Two angles having their corresponding sides parallel are either equal or supplementary.*

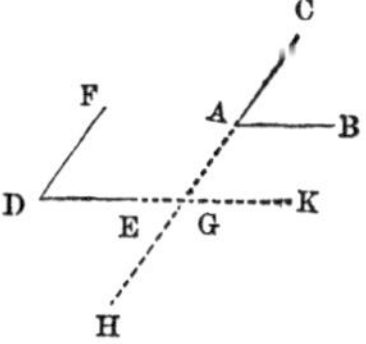


*First.* When the sides AB and AC, forming the angle at A, have respectively the same directions as the sides DE and DF forming the angle at D.

Produce CA and DE until they cross each other at G. Now, since AB and DE are parallel, they have the same direction in reference to the line CG (D. X.), consequently the angle BAC is equal to KGC. And since AC and DF are parallel they have the same direction in reference to the line DG; consequently, the angle KGC is equal to EDF; therefore (A. I.), the angle BAC is equal to EDF.

*Secondly.* When the sides AB and AC are respectively in opposite directions to DE and DF.

As before, we have the angle BAC equal to KGC, and the angle EDF equal to EGH. But KGC is equal to its opposite angle EGH (T. I.), consequently, the angle BAC is equal to EDF.

*Thirdly.* When one of the sides, as AB, has an opposite direction to DE, its corresponding side.

As before, the angle BAC is equal to KGC, and EDF is equal to EGC. But KGC and EGC are supplementary (T. I.), consequently BAC and EDF are supplementary.

*Cor.* Two angles having their corresponding sides perpendicular, are either equal or supplementary.

For, if we draw through the vertex of the first angle lines respectively parallel to the sides of the second angle, we should thus form a third angle, which will be equal to the second. But this third angle is either equal to the first or supplementary to it (T. II.). Consequently the second angle is either equal to the first or supplementary to it (A. I.).

### THEOREM IV.

*When all the sides of a polygonal figure are produced, in the same direction, the sum of all the exterior angles will be equal to four right angles.*

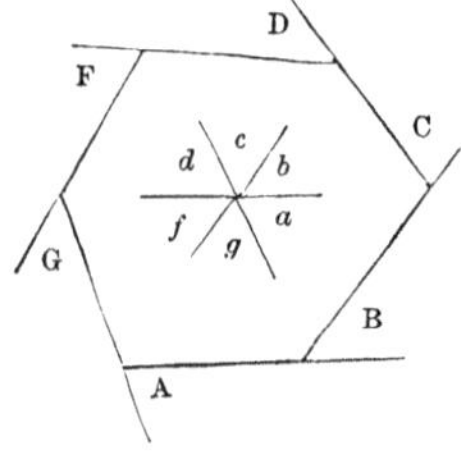


For, if from any point in the same plane, straight lines be drawn parallel respectively to the sides of the figure, the angles contained by the straight lines about that point will be equal to the exterior angles of the figure (T. III.), each to each. Thus the angles $a$, $b$, $c$, etc., are respectively equal to the exterior angles A, B, C, etc.; but the former angles are together equal to four right angles (T. I., C. III.); therefore all the exterior angles of the figure are together equal to four right angles.

*Scholium.* This proposition must be restricted to the case in which the polygon is convex. A convex polygon may be defined to be one, such that no side, by being produced in either direction, can divide the polygon. The polygon ABCDFG is not convex, since it may be divided by producing either of the sides CD or FD. This polygon is said to have a *re-entering* angle at D.

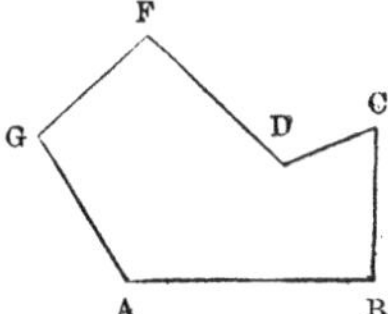


THEOREM V.

*In any convex polygon, the sum of all the interior angles, taken together, is equal to twice as many right angles as the polygon has sides, wanting four right angles.*

Let ABCDFG be a convex polygon. Conceive the sides to be produced all in the same direction, forming exterior angles, which we will denote by the capital letters A, B, C, etc., while their corresponding interior angles are denoted by the small letters $a$, $b$, $c$, etc. Now any exterior angle, together with its adjacent interior angle, as $A + a$, is equal to two right angles (T. I.), therefore the sum of all the interior angles, together with all the exterior angles, is equal to twice as many right angles as the polygon has sides; but the sum of all the exterior angles is equal to four right angles (T. IV.); therefore the sum of all the interior angles is equal to twice as many right angles as the polygon has sides, wanting four right angles.

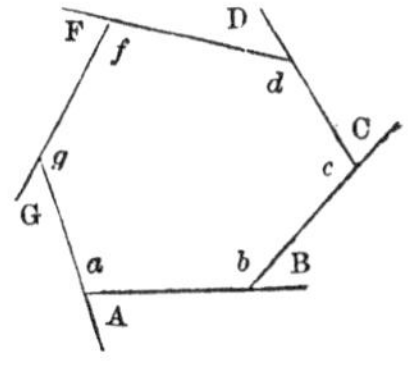


*Cor.* I. In any triangle, the sum of the three angles is equal to two right angles.

*Cor.* II. If one angle of a triangle is a right angle, as in the case of a right-angled triangle (D. XXII.), each of the other angles must be acute, and they are complementary, since their sum must equal a right angle. Each of the angles of a triangle may be acute, but only one of the angles can be obtuse.

*Cor.* III. If two triangles have two angles of the one respectively equal to two angles of the other, their third angles will be equal, and the triangles will be mutually equiangular.

*Cor.* IV. In any quadrilateral, the sum of the four interior angles is equal to four right angles.

*Cor.* V. If two angles of a quadrilateral are right, the other two angles will be supplementary.

*Cor.* VI. If from a point D within a triangle ABC, we draw two lines to the extremities of one of the sides AB, the angle ADB formed by these lines will be greater than the angle ACB, opposite this side.

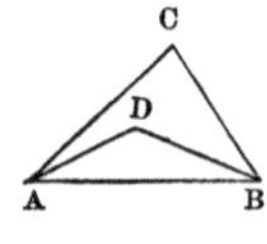


For, the sum of the two angles DAB, DBA is obviously less than the sum of the two angles CAB, CBA. And since the sum of the three angles of every triangle is equal to two right angles, it follows that the remaining angle ADB, of the triangle ADB, must be greater than the remaining angle ACB of the triangle ACB.

THEOREM VI.

*The exterior angle, formed by producing one of the sides of a triangle, is equal to the sum of the two interior and opposite angles of the triangle.*

In the triangle ABC, if the side AB be produced to D, the exterior angle CBD will be equal to the sum of the two angles at A and at C.

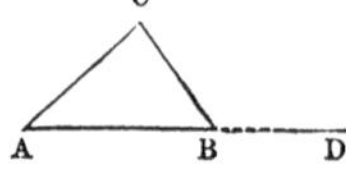


For, the angle CBD, together with its adjacent angle, CBA, is equal to two right angles (T. I.), and the sum of A and C, together with the same angle CBA, is also equal to two right angles (T. V., C. I.); hence, the exterior angle is equal to the sum of the two interior and opposite angles.

*Cor.* The exterior angle is greater than either of the interior and opposite angles.

## OF THE SIDES OF TRIANGLES.

### THEOREM VII.

*Either side of any triangle is less than the sum of the other two sides, and greater than their difference.*

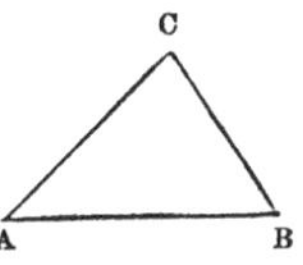


*First.* The straight line AB is the shortest distance between A and B (A. XI.), and therefore shorter than the broken line AC + CB. The same reasoning applies to each of the sides. Hence, we have $AB < AC + CB$; $AC < AB + CB$; $BC < AB + AC$.

*Secondly.* Since $AC < AB + CB$, we have $AC - CB < AB$ (A. V.). That is, $AB > AC - CB$. In a similar manner we deduce $AC > AB - CB$; $CB > AB - AC$.

### THEOREM VIII.

*If from any point within a triangle two lines be drawn to the extremities of either side, the sum of these two lines will be less than the sum of the other two sides.*

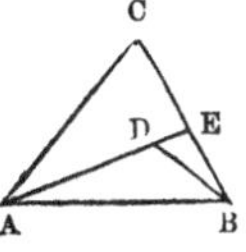


From the point D, suppose the lines DA and DB to be drawn to the extremities of the side AB; then will $DA + DB < AC + BC$.

For, producing AD until it meets BC at E, we have (T. VII.)

$DB < DE + EB$, consequently (A. IV.),
$AD + DB < AD + DE + EB$, that is,
$AD + DB < AE + EB$.

Again, we have

$AE < AC + CE$, consequently,
$AE + EB < AC + CE + EB$, that is,
$AE + EB < AC + CB$.

Comparing these conditions, we have

$$AD + DB < AC + CB.$$

## OF PERPENDICULAR AND OBLIQUE LINES.

### THEOREM IX.

*If two straight lines have two points common they will coincide throughout their whole extent.*

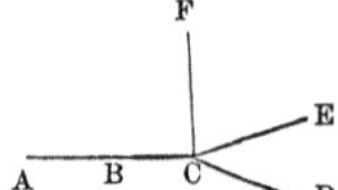


If A and B are points common to two straight lines, they will coincide between A and B (A. XIII.).

If possible, we will suppose that when produced, they begin to separate at C, the one taking the direction CD, and the other the direction CE. Draw CF perpendicular to AC; then since the lines ACE and ACD are each straight, we have the angle FCD and FCE equal, each being equal to a right angle (T. I., C. I.); that is, the whole is equal to one of its parts, which is impossible (A. VIII.). It is therefore absurd to suppose these lines can separate when produced. Hence, if two straight lines have two points common they will coincide throughout their whole extent.

### THEOREM X.

*Through a given point in a straight line only one perpendicular can be drawn to this line.*

D E A C B

If there could be two perpendiculars, as CD and CE, the angles BCD and BCE would be equal, each being a right angle; that is, the whole would be equal to its part, which is impossible (A. VIII.). Hence, it is absurd to suppose that more than one perpendicular can be drawn to a given line through any one of its points.

### THEOREM XI.

*From a given point without a straight line, only one perpendicular can be drawn.*

Let C be the point, and AB the given line. If possible, suppose we can draw the two perpendiculars CD and CE. Revolve the figure CDE about DE as a hinge, until it returns into its primitive plane on the opposite side of AB, having the position FDE. Since each of the angles FDE and CDE is right, CDF is a straight line (T. I., C. IV.). By the same reasoning we have CEF a straight line. That is, we have two straight lines joining the points C and F, which is impossible (A. XIII.). Hence, from a point without a straight line only one perpendicular can be drawn to this line.

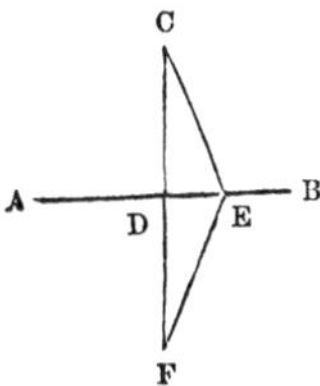


THEOREM XII.

*If from a point without a line, a perpendicular be drawn, and several oblique lines:*

I. *The perpendicular will be shorter than any oblique line.*

II. *Any two oblique lines which terminate at equal distances from the foot of the perpendicular, will be equal.*

III. *Of two oblique lines terminating at unequal distances from the foot of the perpendicular, the one at the greater distance will be the longer.*

Let A be the given point, BC the given line, AD perpendicular, and AB, AE, and AC oblique lines.

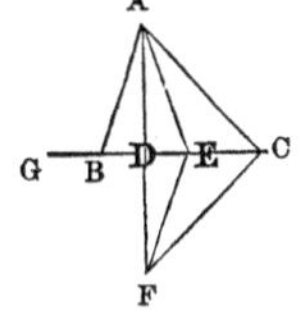


*First.* As in the last Theorem, suppose ADC to revolve about BC into the position FDC, on the opposite side of BC in its primitive plane. We have FD=AD; FE=AE; and FC=AC. But since each of the angles FDE and ADE is a right angle, ADF is a straight line (T. I., C. IV.), and therefore shorter than the broken line AE+EF; hence, AD, the half of AF, is less than AE, the half of AE+EF. That is, the perpendicular is shorter than any oblique line.

*Secondly.* If we suppose the figure ADB to revolve about AD, the point B will coincide with E, since DB is equal to DE, and the angle ADB is equal to ADE, each being a right angle. Hence, the oblique line AB will coincide with AE. That is, two oblique lines which terminate at equal distances from the foot of the perpendicular are equal.

*Thirdly.* Returning to the figure as first revolved, we have, since E is a point within the triangle ACF, AC + CF > AE + EF (T. VIII.); hence, AC, the half of AC + CF, is greater than AE, the half of AE + EF. That is, the oblique line terminating farther from the foot of the perpendicular is the longer.

*Cor.* I. The perpendicular measures the shortest distance from a point to a line.

*Cor.* II. From the same point without a straight line, only two equal oblique lines can be drawn, one on each side of the perpendicular.

THEOREM XIII.

*If through the middle point of a straight line, a perpendicular be drawn:*

I. *Any point in this perpendicular will be equally distant from the extremities of this line.*

II. *Any point without the perpendicular will be unequally distant from the extremities of this line.*

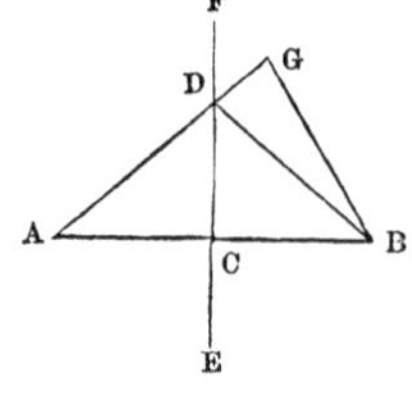


Let AB be the given line, DE a perpendicular through C, its middle point.

*First.* Let D be any point in this perpendicular. Drawing DA and DB, we know these lines to be equal, since they terminate at equal distances from C, the foot of the perpendicular DC (T. XII.). That is, D is equidistant from A and B. In the same manner we may show that any other point of this perpendicular, as E or F, is equidistant from the extremities of the given line.

*Secondly.* Suppose G to be a point without the perpendicular. If we draw GA and GB, one of these lines must cut the perpendicular. We will suppose GA to cut it at D. Draw DB, then GB will be less than GD + DB (T. VII.). But, we already have DB = DA, hence GB is less than GD + DA, or less than GA. That is, the point G is unequally distant from the extremities of the given line.

In the same manner we can show that any other point without the perpendicular is unequally distant from the extremities of this line.

*Scholium.* The most distant extremity A of this line will be on the side of the perpendicular opposite the point G.

*Cor.* I. Any point equidistant from the extremities of a straight line is situated in the perpendicular which bisects this line.

*Cor.* II. If a straight line have any two of its points equidistant from the extremities of a second line, it will be perpendicular to the second line and bisect it.

### THEOREM XIV.

*If a line be drawn bisecting a given angle, that is, dividing it into two equal angles:*

I. *Any point in this bisecting line will be equidistant from the sides of the angle.*

II. *Any point without this bisecting line will be unequally distant from the sides of the angle.*

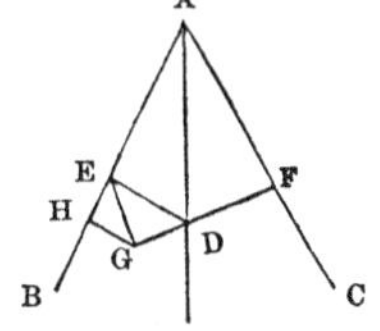


*First.* Let the given angle BAC be bisected by the line AD, then will any point as D in this line be equidistant from the lines AB and AC.

The distance of D from these lines will be measured by the perpendiculars DE and DF (T. XII., C. I.).

If we conceive the figure folded about the line AD, so that the portion on the left of AD may be superposed upon the portion on the right, the line AB will then coincide with AC, since the angle BAD = CAD. Consequently the perpendiculars DE and DF will coincide (T. XI.), and are therefore equal.

*Secondly.* Suppose G to be a point without the bisecting line. If we draw the perpendiculars GF and GH, one of these must cut the bisecting line. Let GF cut it at D. Draw DE perpendicular to AB, and join G and E: then will GH be less than GE (T. XII.). But GE is less than GD + DE (T. VII.); therefore, GH is less than GD + DE, or less than GD + DF, since by the first case of this Theorem DF = DE. Hence, finally, GH is less than GF.

*Cor.* I. Any point equidistant from the sides of a given angle is situated on the line bisecting this angle.

*Cor.* II. If a straight line have two of its points equidistant from the sides of a given angle, it will bisect this angle.

## OF PARALLEL LINES.

### THEOREM XV.

*When two lines are perpendicular to the same line, they are parallel.*

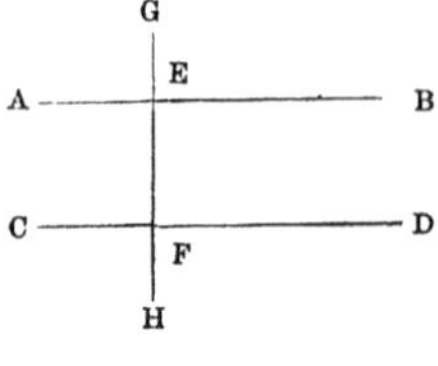


If the lines AB and CD are perpendicular to GH they will be parallel.

For, since they are perpendicular to GH, we have the angles BEG and DFE equal, each being a right angle; hence, these lines have the same direction, and are therefore parallel (D. X.).

*Cor.* A line which is perpendicular to one of two parallels, is also perpendicular to the other.

### THEOREM XVI.

*Two parallel lines are throughout their whole extent equally distant.*

Suppose AB and CD to be parallel.

Through any two points, as E and G, of the line CD, draw EF and GH perpendicular to CD, they will also be perpendicular to AB (T. XV.), and they will measure the distance of E and G respectively from the line AB (T. XII., C. I.).

We are now to prove these lines equal. Through K, the middle of EG, draw KL perpendicular to CD, and it will also be perpendicular to AB. Now, suppose the portion of the figure on the left of KL to revolve about KL as a hinge, until it returns into its primitive plane, on the right of KL. The angles at K and L being right, KE will take the direction of KG, and LF the direction of LH, and E will coincide with G, since KE = KG, and since EF and GH are each perpendicular to CD, they must coincide, which proves them equal.

*Cor.* Two parallels cannot meet, however far they are produced.

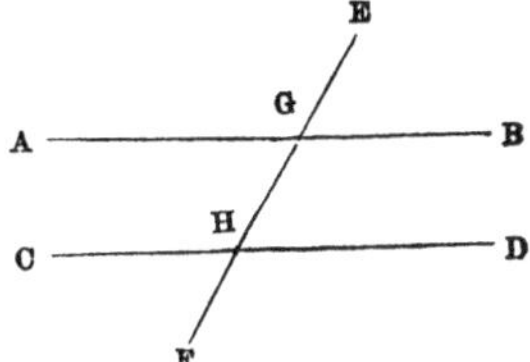


*Scholium.* When two parallels are cut by a third line, the angles formed receive particular names.

WHEN TAKEN SEPARATELY:

1. The four angles AGF, CHE, BGF, and DHE, situated within the parallels, are called *interior angles.*

2. The four angles AGE, CHF, BGE, and DHF, situated without the parallels, are called *exterior angles.*

WHEN COMPARED TWO AND TWO:

1. The two angles AGF and CHE, as well as BGF and DHE, are called *interior angles on the same side;* that is, on the same side of the *secant* or cutting-line.

2. The two angles AGE and CHF, as well as BGE and DHF, are called *exterior angles on the same side.*

3. The two angles AGF and DHE, as well as CHE and BGH, being interior angles on different sides of the secant, are called *alternate interior angles.*

4. The two angles AGE and DHF, as well as CHF and BGE, are called *alternate exterior angles.*

We have noticed two pairs of each of the four distinct kinds of angles.

5. There are still four pairs of angles, referred to as *Corresponding angles*, namely: AGE and CHE, AGF and CHF, BGE and DHE, BGH and DHF. These last angles, which establish the direction of the lines, are sometimes called *opposite exterior* and *interior angles.*

THEOREM XVII.

*If two straight lines are cut by a third line, making the sum of the two interior angles on the same side equal to two right angles, the two lines will be parallel.*

If the two lines AB and CD are cut by the line EF, making BGH + DHG = 2 right angles. these lines will be parallel.

For we have BGH + BGE = 2 right angles (T. I.), therefore BGH + DHG = BGH + BGE; taking BGH from each we have DHG = BGE, hence the lines AB and CD have the same direction, and are consequently parallel (D. X.).

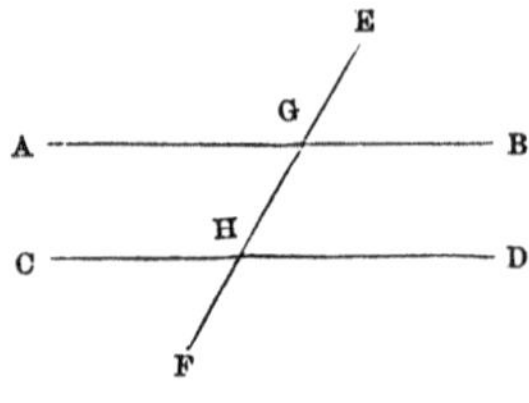


*Cor. If two straight lines are cut by a third line, they will be parallel:*

I. *When the sum of the exterior angles on the same side is equal to two right angles.*

II. *When the alternate interior angles are equal.*

III. *When the alternate exterior angles are equal.*

IV. *When the corresponding angles are equal.*

*First.* Suppose EGB + FHD = 2 right angles.

We have EGB = AGH, and FHD = CHG (T. I.), hence AGH + CHG = 2 right angles, which agrees with the Theorem itself, hence AB and CD are parallel.

*Secondly.* Suppose AGH = GHD.

We have AGH = EGB (T. I.), consequently EGB = GHD; that is, the lines AB and CD have the same direction, and are therefore parallel.

*Thirdly.* Suppose EGB = CHF.

We have CHF = GHD, consequently EGB = GHD, and the lines have the same direction, and are therefore parallel.

*Fourthly.* Since the corresponding angles are equal, the lines have the same direction, and are therefore parallel.

## THEOREM XVIII.

*If two parallels are cut by a third line, the sum of the two interior angles on the same side is equal to two right angles.*

Since the lines are parallel they have the same direction, and their corresponding angles must be equal. We therefore have EGB = GHD; to each add BGH, and we shall have EGB + BGH = BGH + GHD.

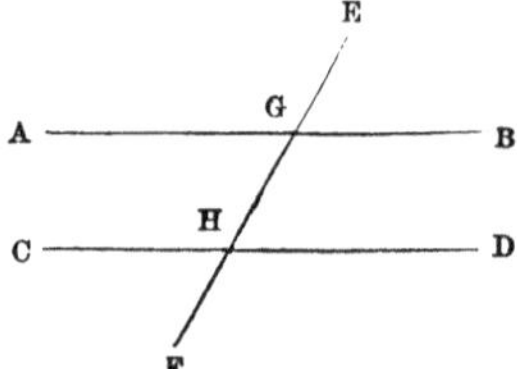

But EGB + BGH = 2 right angles (T. I.), consequently BGH + GHD = 2 right angles.

*Cor.* It follows immediately from D. X., in connection with T. I., C. I., that if either one of the eight angles formed by the intersection of a straight line with two parallels is a right angle, the other seven will also be right. When these angles are not right, there will be four equal acute angles, and four equal obtuse angles. These angles will be supplementary. Any two of the acute angles, or any two of the obtuse angles taken together, constitute a pair of corresponding angles, and fix the direction of the parallels. Hence it readily follows, that—

*If two parallels are cut by a third line, we shall have:*

I. *The sum of the exterior angles on the same side equal two right angles.*

II. *The alternate interior angles equal.*

III. *The alternate exterior angles equal.*

IV. *The corresponding angles equal.*

### THEOREM XIX.

*If two straight lines are cut by a third line, and the sum of the interior angles on the same side is not equal to two right angles, the two lines will meet, if sufficiently produced.*

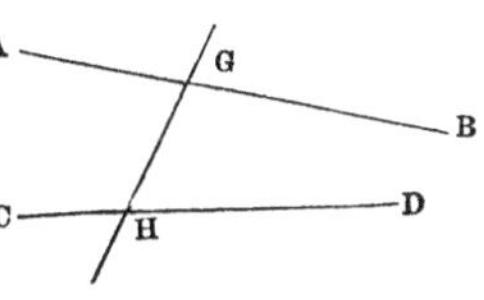


If we suppose the angle AGH + CHG > 2 right angles, then, obviously, will the angle BGH + DHG < 2 right angles, since these angles are respectively supplementary angles. This being supposed, these lines will meet. For if they do not meet they are parallel; but they are not parallel, since the sum of the interior angles on the same side is not equal to two right angles. Hence these lines will meet if produced sufficiently far.

*Scholium.* It is evident they will meet in the direction of B and D, on that side of the secant line which has the sum of the interior angles less than two right angles.

## OF TRIANGLES.

### THEOREM XX.

*If two triangles have two sides and the included angle of the one, equal to the two sides and the included angle of the other, the triangles will be identical, or equal in all respects.*

In the two triangles ABC, DFG, if the side CA be equal to the side DG, and the side CB equal to the side GF, and the angle C equal to the angle G, then will the two triangles be identical, or equal in all respects.

For, conceive the triangle ABC to be placed upon the triangle DFG in such a manner that the point C may coincide with the point G, and the side CA with the equal side GD. Then, since the angle G is equal to the angle C, the side CB will take the direction of the side GF. Also CB being equal to GF, the point B will coincide with the point F; consequently the side AB will coincide with DF. Therefore the two triangles are identical, and have all their other corresponding parts equal (A. IX.), namely, the side AB equal to the side DF, the angle A equal to the angle D, and the angle B equal to the angle F.

*Cor.* Two right-angled triangles are equal, when the two sides containing the right angle of the one are respectively equal to the two sides containing the right angle of the other.

### THEOREM XXI.

*If two triangles have two sides of the one respectively equal to two sides of the other, and the included angle of the first greater than the included angle of the second, the third side of the first will be greater than the third side of the second.*

In the two triangles ABC and DEF, suppose CA and CB of the first to be respectively equal to FD and FE of the second, and the included angle ACB > DFE, then will AB > DE.

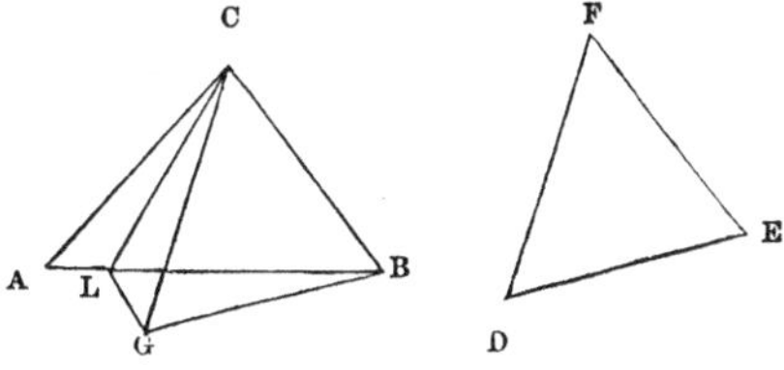


Apply the triangle DEF upon the triangle ABC, making EF coincide with BC, so that the triangle may, in its new position, be represented by GBC. Draw CL bisecting the angle ACG (Post. VI.), and meeting AB at L, draw GL. The two triangles ACL, GCL, are equal (T. XX.), and LG = LA. Therefore, AB = AL + LB = LG + LB, which is greater than BG (T. VII.). That is, AB > DE.

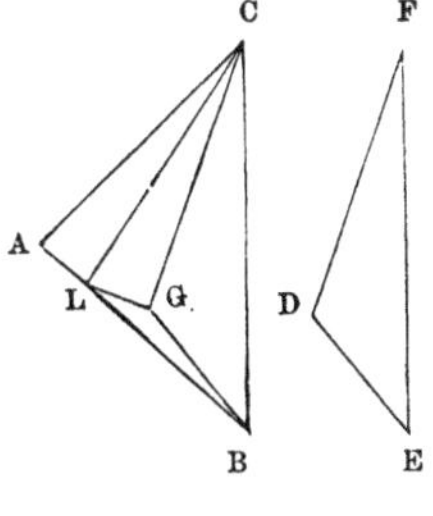


Although the triangle DEF, when applied to ABC, may have three different positions, as here represented, still our demonstration is alike applicable to either case.

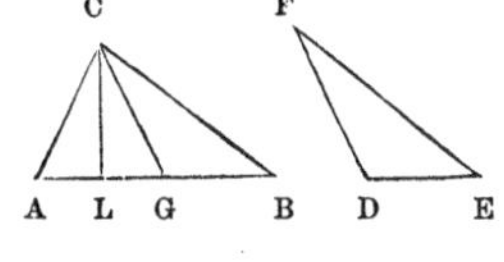


*Cor. Conversely.* If two triangles have two sides of the one respectively equal to two sides of the other, and the third side of the first greater than the third side of the second, the included angle of the first will be greater than the included angle of the second.

For, if the included angle of the first is not greater than the included angle of the second, it must be either equal or less. If it were equal, the third sides would be equal (T. XX.). If it were less, the third side of the first would be less than the third side of the second, by the Theorem itself. Both these results are contrary to the hypothesis. Consequently, the included angle of the first is greater than the included angle of the second.

### THEOREM XXII.

*If two triangles have two angles and the interjacent side of the one equal to two angles and the interjacent side of the other, the triangles will be identical, or equal in all respects.*

In the two triangles ABC, DFG, if the angle A is equal to the angle D, the angle B equal to the angle F, and the side AB equal to the side DF, then will the triangles be identical, or equal in all respects.

For, conceive the triangle ABC to be placed on the triangle DFG, in such a manner that the side AB may coincide with the equal side DF. Then, since the angle D is equal to the angle A, the side AC will take the direction of the side DG; also, since the angle F is equal to the angle B, the side BC will take the direction of the side FG; consequently the point C must coincide with the point G. Therefore the two triangles are identical (A. IX.), having the two sides AC and BC respectively equal to DG and FG, and the remaining angle C equal to the remaining angle G.

### THEOREM XXIII.

*In an isosceles triangle, the angles at the base are equal; or, if a triangle have two sides equal, the angles opposite those sides will be equal.*

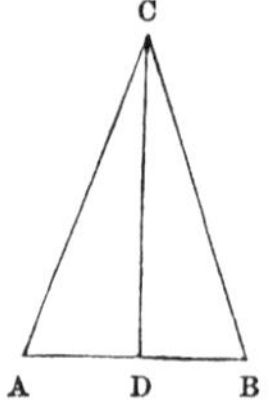


If the triangle ABC have the side AC equal to the side BC, then will the angle B be equal to the angle A.

For, conceive the angle C to be bisected, or divided into two equal parts, by the line CD, making the angle ACD equal to the angle BCD. Then the two triangles ADC and BDC will be identical, or equal in all respects (T. XX.), and consequently the angle B is equal to the angle A.

*Cor.* I. The line which bisects the vertical angle of an isosceles triangle, bisects the base perpendicularly, and divides the triangle itself into two equal parts.

*Cor.* II. An equilateral triangle is equi-angular, that is, has all its angles equal.

### THEOREM XXIV.

*If a triangle have two angles equal, the sides opposite those angles will be equal, and the triangle will be isosceles.*

In the triangle ABC, if the angles CAB, CBA, are equal, the opposite sides BC, AC, will be equal.

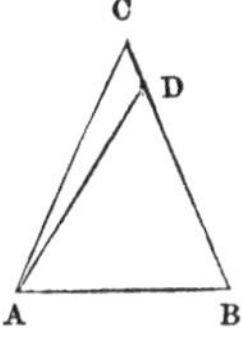


For, if they are not equal, suppose BC > AC, and take BD = AC. Then in the two triangles BAD and ABC we have the side BD = AC by supposition, the side AB common, and the included angle ABD of the first triangle equal to BAC, the included angle of the second triangle. Hence (T. XX.), the triangle BAD is equal to ABC, that is, a part is equal to the whole, which is impossible (A. VIII.).

There is, therefore, no inequality of the sides CB and CA, that is, they are equal, and the triangle is isosceles.

### THEOREM XXV.

*When two triangles have the three sides of the one respectively equal to the three sides of the other, the triangles will be identical, and equal in all respects.*

Let the two triangles ABC, ABD, have their sides respectively equal, namely, AB equal to AB, AC equal to AD, and BC equal to BD, then will these triangles be identical.

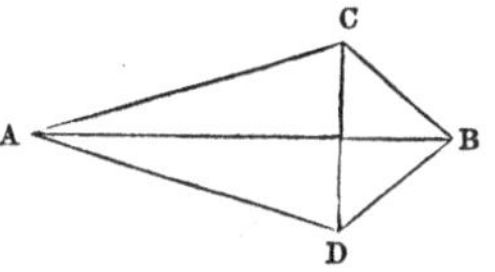


For, conceive the two triangles to be joined together by their longest equal sides, and draw the line CD; then in the triangle ACD, since AC is equal to AD, we have the angle ACD equal to the angle ADC (T. XXIII.). In like manner, in the triangle BCD, since BC is equal to BD, we have the angle BCD equal to the angle BDC. Hence the angle ACB, which is the sum of ACD and BCD, is equal to the angle ADB, which is the sum of ADC and BDC. Since, then, in the triangle ACB, we have the two sides AC and BC, and their included angle ACB, equal respectively to the two sides AD and BD, and their included angle ADB, of the triangle ADB, it therefore follows that these triangles are identical (T. XX.).

## THEOREM XXVI.

*Two right-angled triangles are equal, when the hypotenuse and a side of the one, are respectively equal to the hypotenuse and a side of the other.*

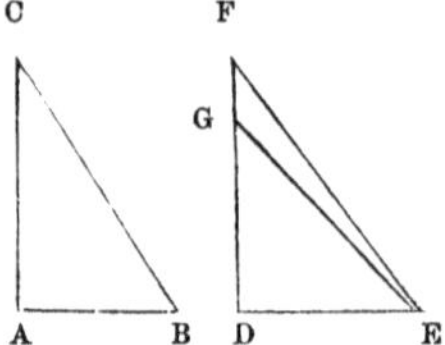


Suppose BC = EF and AB = DE, then will the two right-angled triangles, ABC and DEF, be equal.

For, if we apply the triangle ABC upon the triangle DEF, so that AB may coincide with its equal DE, the side AC will take the direction of DF, since the angle at A is equal to the angle at D, each being a right angle. If, now, BC coincides with EF, there will be a complete coincidence, and the two triangles will be equal. If possible, suppose BC to take the position EG, so that the triangle ABC may be represented by DEG; then, since the hypotenuses of the two triangles are equal, we shall have EG = EF, that is, two oblique lines are drawn from E, terminating at unequal distances from D, the foot of the perpendicular ED, which is impossible (T. XII.). It is, therefore, absurd to suppose BC to take any position different from EF. Hence, the triangles coincide throughout, and are in all respects equal.

*Cor.* Two right-angled triangles will also be equal, when the hypotenuse and an acute angle of the one are respectively equal to the hypotenuse and an acute angle of the other. For, since the sum of the two acute angles of any right-angled triangle is equal to a right angle (T. V., C. II.), it follows that the remaining acute angles of the two triangles will be equal. Hence, the two triangles will have two angles and the interjacent side of the one equal, respectively, to two angles and the interjacent side of the other. Consequently, they will be equal (T. XXII.).

## THEOREM XXVII.

*The greater side of every triangle is opposite the greater angle, and, conversely, the greater angle is opposite the greater side.*

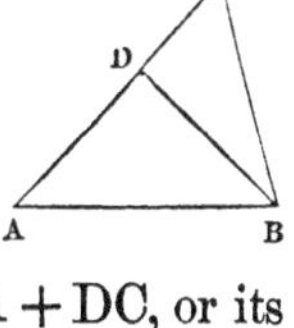


*First.* Suppose, in the triangle ABC, the angle CBA > CAB, then will the side CA > CB.

For, draw BD, making the angle DBA = DAB, and DA will equal DB (T. XXIV.). To each adding DC, we have DA + DC = DB + DC, but DB + DC is greater than CB (T. VII.); hence, DA + DC, or its equal CA, is greater than CB.

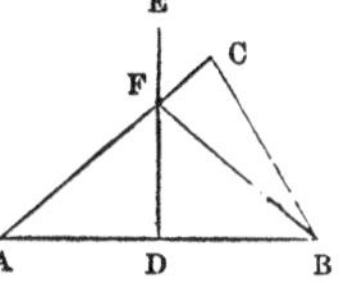


*Secondly.* Suppose, in the triangle ABC, the side CA > CB, then will the angle CBA > CAB.

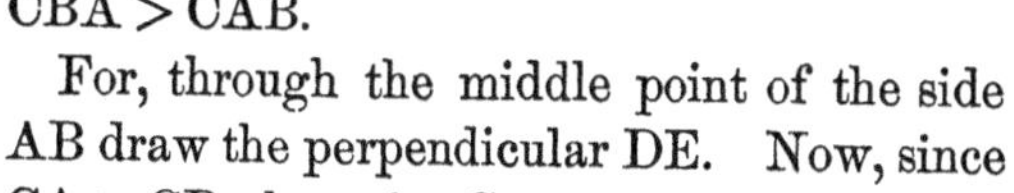

For, through the middle point of the side AB draw the perpendicular DE. Now, since CA > CB, the point C must be on the side of DE opposite A (T. XIII., S.); consequently, the perpendicular must cut CA at some point, as F. Draw FB, and we shall have FB = FA (T. XIII.), and angle FBA = FAB (T. XXIII.); consequently, FBA + FBC, which equals CBA. is greater than CAB.

---

## OF QUADRILATERALS.

### THEOREM XXVIII.

*If the opposite angles of a quadrilateral are equal, the figure is a parallelogram.*

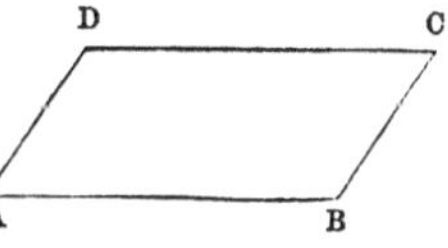


We have the angles A + B + C + D = 4 right angles (T. V., C. IV.), true for all quadrilaterals. Now, if we suppose A = C, and B = D, we shall find 2A + 2D = 4 right angles, or A + D = 2 right angles, hence AB and DC are parallel (T. XVII.). In a similar manner, we find 2A + 2B = 4 right angles, or A + B = 2 right angles, and, as before, AD and BC are parallel. Consequently the figure is a parallelogram (D. XXV.).

### THEOREM XXIX.

*If the opposite sides of a quadrilateral are equal, the figure is a parallelogram.*

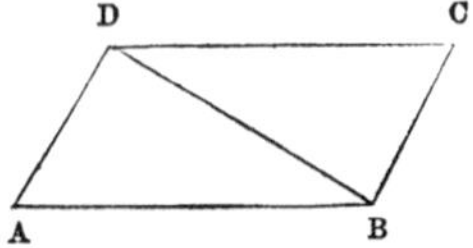


If we suppose AB = DC, and AD = BC, and draw the diagonal BD, we shall have the three sides of the triangle ABD respectively equal to the three sides of the triangle CDB, consequently they are equal, (T. XXV.), and the angle ABD = CDB, which are alternate interior angles in reference to the sides AB and DC, hence AB and DC are parallel (T. XVII., C.).

The equality of the two triangles also gives the angle ADB = CBD, which are alternate interior angles in reference to the sides AD and BC, hence these sides are parallel. Consequently the figure is a parallelogram (D. XXV.).

### THEOREM XXX.

*If two opposite sides of a quadrilateral are equal and parallel, the figure is a parallelogram.*

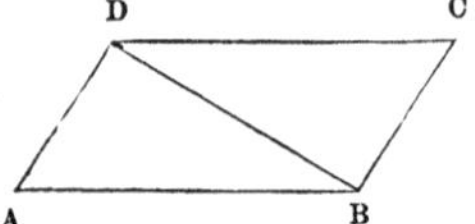


In the quadrilateral ABCD, suppose AB equal and parallel to DC, then will the figure be a parallelogram. For, drawing BD, the two triangles ABD, CDB, have the common side BD, and AB of the first equal to CD of the second; also, the included angles ABD, CDB, equal, being alternate angles with reference to the parallels AB, DC. Hence these triangles are equal (T. XX.), and the angle ADB is equal to CBD, but these are alternate angles with reference to AD and BC, consequently AD and BC are parallel, and the figure is a parallelogram.

### THEOREM XXXI.

*The opposite angles of a parallelogram are equal; so also are the opposite sides equal.*

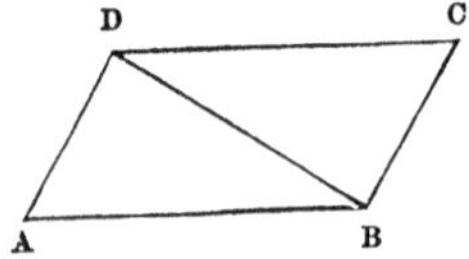


Since the figure is a parallelogram, we have AB and DC parallel, also AD and BC parallel; hence, if we draw the diagonal BD, we shall have the angle ABD = CDB, being alternate interior angles in reference to the parallels AB and DC (T. XVIII., C.). For a similar reason the angle ADB = CBD. Therefore, the two triangles ABD and CDB have a common side BD, and the adjacent angles of the one equal respectively to the adjacent angles of the other, consequently they are equal (T. XXII.). Hence the angle A = C, and ABD + CBD = CDB + ADB, or ABC = CDA. That is, the opposite angles of a parallelogram are equal.

The equality of the triangles also gives AB = DC, and AD = BC; that is, the opposite sides of a parallelogram are also equal.

*Cor.* I. The diagonal of a parallelogram divides it into two equal parts.

*Cor.* II. Two parallels included between two other parallels are equal.

*Cor.* III. If one angle of a parallelogram is right, the other three will be right also.

### THEOREM XXXII.

*The diagonals of a parallelogram mutually bisect each other, that is, they divide each other into halves.*

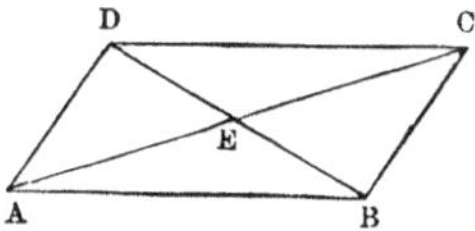


In the two triangles EAB and ECD, we have the angle EAB of the first equal to ECD of the second, being alternate angles in reference to the parallels AB and DC. For a like reason the angle EBA is equal to EDC. We also have the interjacent sides AB and DC equal, consequently these triangles are equal (T. XXII.), and AE = EC, BE = ED.

*Cor.* If the four sides of the parallelogram are equal, as in the case of a *rhombus*, we have AB = AD, and the two triangles AEB and AED will have the three sides of the one equal to the three sides of the other respectively, consequently they will be equal (T. XXV.), and the angle AEB = AED, that is, in a rhombus the diagonals bisect each other at right angles.

*Scholium.* The point E, where the diagonals intersect, is called the centre of the parallelogram.

### THEOREM XXXIII.

*The straight line joining the middle points of the oblique sides of a trapezoid, will be parallel to the other sides, and equal to half their sum.*

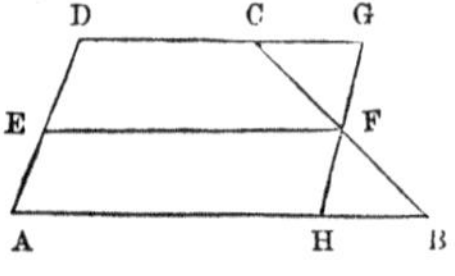


Suppose EF to join the middle points E and F, of the oblique sides AD and BC of the trapezoid ABCD. Through F draw GH parallel to AD, and meeting DC produced. In the two triangles, FBH and FCG, we have the side FB = FC, the angle FBH equal to its alternate angle FCG, and the angle BFH equal its opposite angle CFG, hence these triangles have two angles and the interjacent side of the one, respectively equal to the two angles and the interjacent sides of the other, and are therefore equal (T. XXII.). Consequently, FG = FH; that is, FG equals half of HG. But HG is equal to AD, since the figure AHGD is a parallelogram. Therefore, FG is half of AD, but ED is also half of AD, hence FG and ED are equal and parallel, and the figure EFGD is a parallelogram, and EF is parallel to DC, and also to AB, since DC and AB are parallel.

Again, since EF = DG = AH; and CG = HB, by reason of the equality of the two triangles FGC and FHB, we have EF as much greater than DC as it is less than AB. Hence, EF is one half of the sum of DC and AB.

*Cor.* If we suppose the side DC of the trapezoid to be reduced to a point, the trapezoid will then change to a triangle. Hence, the straight line joining the middle points of any two sides of a triangle, will be parallel to the third side, and equal to half of it.

### THEOREM XXXIV.

*The four lines joining the middle points of the adjacent sides of a quadrilateral form a parallelogram.*

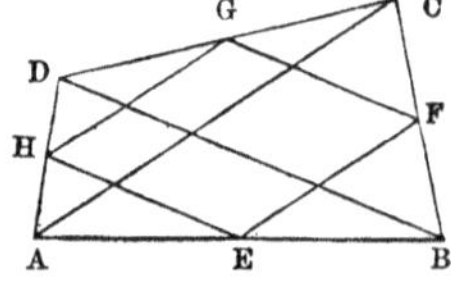


If we draw the diagonals of the quadrilateral, we shall have EF and HG, each parallel to AC, and equal to half of it (T. XXXIII., C.); consequently they are equal and parallel. For the same reason EH and FG are each parallel to BD, and equal to half of it. Therefore the figure EFGH is a parallelogram.

*Cor.* The two lines joining the middle points of the opposite sides of a quadrilateral mutually bisect each other, since they are the diagonals of a parallelogram.

## ADDITIONAL THEOREMS OF TRIANGLES.

### THEOREM XXXV.

*In any triangle if a line be drawn from either angle to the middle of the opposite side:*

I. *When this line is equal to the half side, the angle will be a right angle.*

II. *When this line is greater than the half side, the angle will be acute.*

III. *When this line is less than the half side, the angle will be obtuse.*

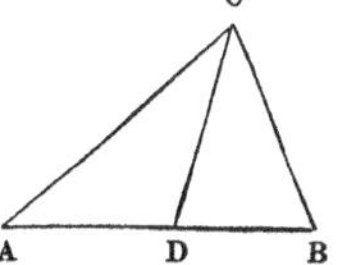


In the triangle ABC, suppose the line CD to be drawn from the angle C to the middle of the side AB.

*First.* When CD = AD = DB, we have, angle A = DCA, angle B = DCB (T. XXIII.), hence A + B = DCA + DCB; that is, the angle ACB is one half the sum of the three angles of the triangle, and is therefore a right angle (T. V., C. I.).

*Secondly.* When CD > AD or DB, we have angle A > DCA, angle B > DCB (T. XXVII.), hence A + B > DCA + DCB; that is, the angle ACB is less than half the sum of the three angles of the triangle, and it is therefore an acute angle.

*Thirdly.* When CD < AD or DB, we have angle A < DCA, angle B < DCB (T. XXVII.); hence, A + B < DCA + DCB; that is, the angle ACB is greater than half the sum of the three angles of the triangle, and it is therefore an obtuse angle.

*Scholium.* This Theorem affords a very simple method of determining the kind of angle in any given triangle.

### HEOREM XXXVI.

*The three lines bisecting the three angles of a triangle, intersect each other in the same point.*

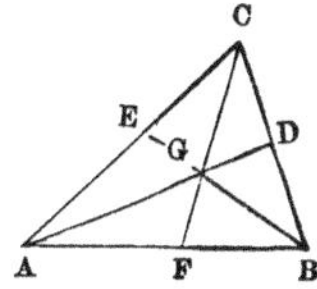


Every point in the bisecting line AD is equally distant from AB and AC (T. XIV.). For the same reason every point of the bisecting line BE is equally distant from AB and BC. Hence, the point G, where these two lines intersect, is equally distant from AC and BC; consequently, it is in the line CF which bisects the angle ACB (T. XIV., C. I.). That is, this point is common to the three bisecting lines.

*Cor.* If the sides AB and AC are produced, and the exterior angles are bisected by the lines BF and CF, the point F, where they intersect, will be on the line AF, which bisects the angle at A.

*Scholium.* This common point, equidistant from the three sides of a triangle, is always within the triangle.

### THEOREM XXXVII.

*The three perpendiculars bisecting the three sides of a triangle intersect each other in the same point.*

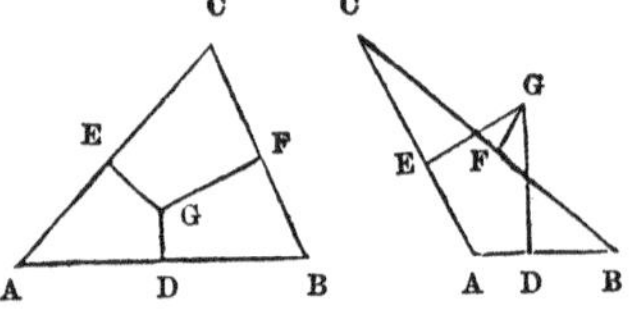


Every point in the perpendicular DG is equidistant from A and B (T. XIII.). For the same reason every point in the perpendicular EG is equidistant from A and C. Hence, the point G, where these perpendiculars intersect, is equidistant from B and C; consequently, it is in the perpendicular which bisects the side BC (T. XIII., C. I.). That is, this point is common to the three perpendiculars.

*Scholium.* This common point, equidistant from the three angles of a triangle, may be either within or without the triangle; or, in the case of a right-angled triangle, it will evidently be at the middle point of the hypotenuse.

### THEOREM XXXVIII.

*The three lines passing through the three angles of a triangle, and bisecting the opposite sides, will intersect each other in the same point.*

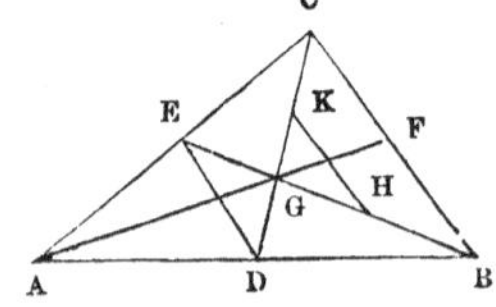


We will first consider the two lines CD, BE, which intersect at G. Draw HK joining the middle points of GB, GC, and it will be parallel to BC and equal to half of it (T. XXXIII., C.). For the same reason DE, which joins the middle points of AB, AC, will be parallel to BC and equal to half of it; consequently, HK and DE are equal and parallel. Now in the two triangles GHK, GED, we have the angle GHK equal to its alternate angle GED, and GKH equal to its alternate angle GDE, also the side HK = DE; hence, these triangles are equal (T. XXII.), and GH = GE, but GH was taken one half of BG, therefore we have EG one third of EB. For the same reason, DG is one third of DC.

If we now consider the two lines CD, AF, we can, by the same kind of demonstration already employed, show that the point where they intersect is one third of FA, measured from F on FA; and one third of DC, measured from D on DC, as before. Hence, the three lines intersect at the same point.

*Scholium.* This common point, situated at one third the distance of the middle points of the sides of a triangle from their opposite angles, is always within the triangle.

### THEOREM XXXIX.

*The three lines passing through the three angles of a triangle, perpendicular to the opposite sides, will intersect each other in the same point.*

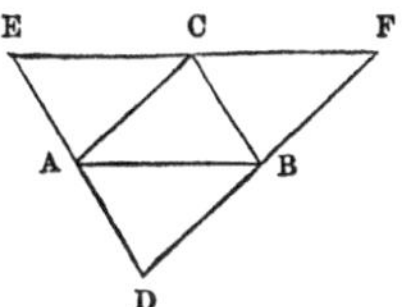


Let ABC be the given triangle. Through the vertices draw lines parallel to the opposite sides, thus forming a second triangle DEF.

Then will the figures ABCE, ABFC, and ADBC be parallelograms, and we have AE and AD each equal to BC; hence, A is the middle point of ED. For a similar reason, B is the middle point of DF, and C the middle point of EF. From this we see that lines drawn from A, B, C, perpendicularly to BC, AC, AB, will be the same as the three perpendiculars bisecting the sides of the triangle DEF, which perpendiculars we already know must intersect each other in the same point (T. XXXVII.).

Hence, three lines passing through the three angles of a triangle perpendicularly to the opposite sides, will intersect each other in the same point.

*Scholium.* This common point may be either within or without the triangle (T. XXXVII.).

# SECOND BOOK.

## THE CIRCLE, AND ITS COMBINATION WITH THE STRAIGHT LINE.

### DEFINITIONS.

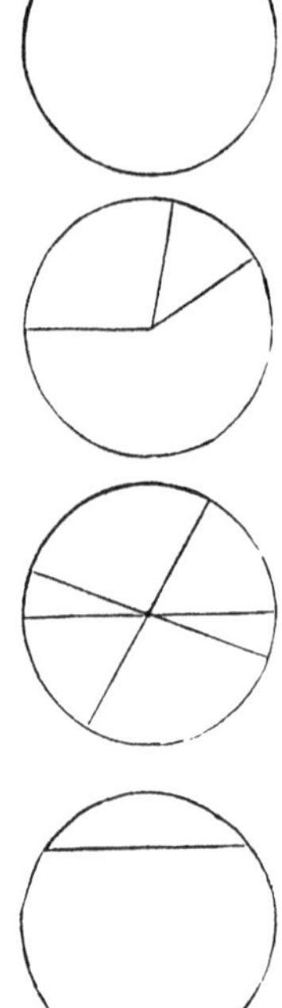

I. The *circumference* of a circle is a curved line, all the points of which are equally distant from a point within called the *centre*. The *circle* is the space bounded by the circumference.

II. Any straight line drawn from the centre to the circumference is called a *radius*. Hence, all radii of the same circle are equal.

A line passing through the centre and terminating in both directions by the circumference, is called a *diameter*. Hence, all diameters of the same circle are equal, each being made up of two radii.

III. Any portion of the circumference of a circle is called an *arc*. One fourth of the entire circumference is called a *quadrant*.

The straight line joining the extremities of an arc is called a *chord*. The chord is said to *subtend* the arc.

Every chord corresponds always to two arcs, which together make up the entire circumference.

It is the smaller arc which is referred to as the subtended arc, unless otherwise expressed.

The portion of a circle included by an arc and its chord is called a *segment.*

The portion included between two radii and the intercepted arc is called a *sector.*

IV. When a straight line cuts the circumference of a circle it is called a *secant.*

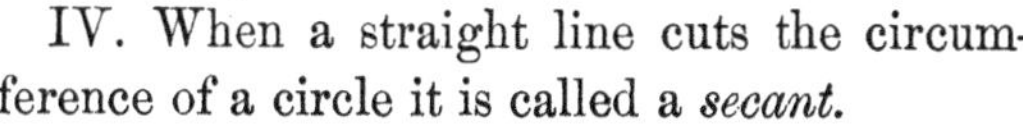

When a straight line touches the circumference in only one point it is called a *tangent;* and the common point of the line and circumference is called the *point of contact.*

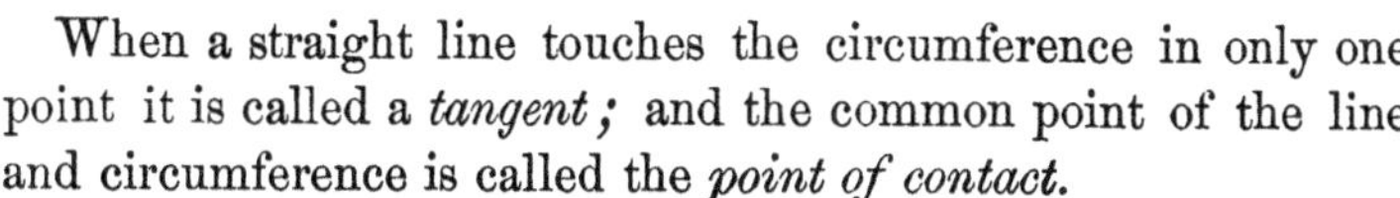

Two circumferences are tangent to each other when they have only one point in common.

Two circumferences are *concentric* when they have the same centre.

V. A line is *inscribed* in a circle when its extremities are in the circumference.

An angle is inscribed in a circle when its sides are inscribed.

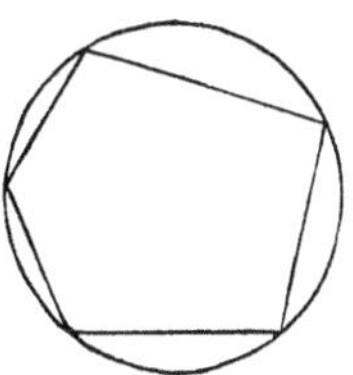

A polygon is inscribed in a circle when its sides are inscribed; and under the same circumstances, the circle is said to *circumscribe* the polygon.

A circle is inscribed in a polygon when its circumference touches each side, and the polygon is said to be *circumscribed about the circle.*

By *an angle in a segment* of a circle, is to be understood an angle whose vertex is in the arc, and whose sides intercept the chord of said arc; and by *an angle at the centre*, is meant one whose vertex is at the centre. In both cases the angles are said to be *subtended* by the chords or arcs which their sides include.

VI. Any polygonal figure is said to be *equilateral* when all its sides are equal; and it is *equiangular* when all its angles are equal.

Two polygons are said to be *mutually equilateral* when their corresponding sides, taken in the same order, are equal. When this is the case with the corresponding angles, the polygons are said to be *mutually equiangular.*

A *regular polygon* has all its sides equal, and all its angles equal. If all the sides are not equal, or all the angles are not equal, the polygon is *irregular.*

A regular polygon may have any number of sides not less than three. The *equilateral triangle* is a regular polygon of three sides. The *square* is also a regular polygon of four sides.

---

## OF CHORDS, SECANTS, AND TANGENTS.

### THEOREM I.

*Every diameter divides the circle and its circumference into two equal parts.*

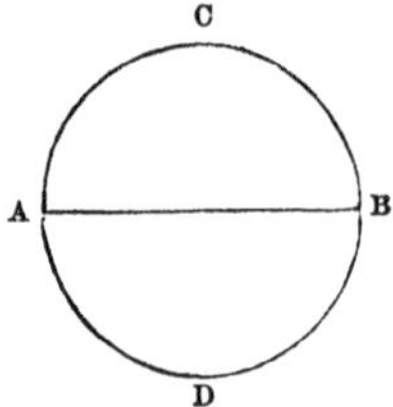


Revolve the portion ACB about the diameter AB as a hinge, until it returns to its primitive plane, on the opposite side of AB; then will the portion of the circumference ACB wholly coincide with ADB. For, if not, there would be points in the circumference unequally distant from the centre, which is impossible (D. II.). Hence, the diameter divides the circle and its circumference into two equal parts.

### THEOREM II.

*A straight line cannot meet the circumference in more than two points.*

For if it could, drawing radii to these points, we should have more than two equal lines drawn from the same point to a straight line, which is impossible (B. I., T. XII., C. II.).*

---

* When a reference is made from one Book to another Book, the Book referred to will be given as above; but when the Book is not given, the reference is confined to the Book in which the proposition occurs.

### THEOREM III.

*The diameter of a circle is greater than any chord.*

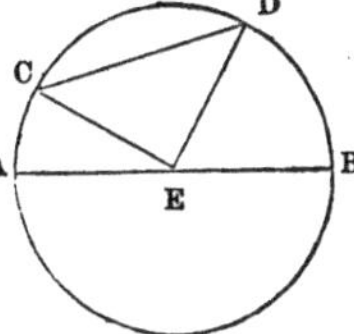


The diameter AB is greater than any chord, as CD. For, drawing the radii EC and ED, we have EC + ED > CD (B. I., T. VII.). But the diameter AB = EC + ED (D. II.); hence, AB > CD.

### THEOREM IV.

*The radius drawn perpendicular to a chord, bisects the chord, and also bisects its subtended arc.*

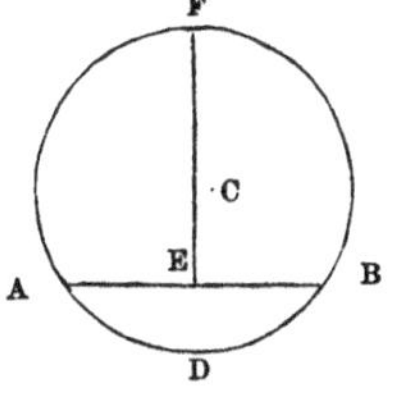


Let the radius CD be drawn perpendicular to the chord AB.

Produce DC to F, and apply the figure DAF to DBF by revolving it about DF. Then the arc DAF will coincide with DBF (T. I.). And since the angles DEA and DEB are right, the line EA will coincide with EB. Hence we have AE = EB, arc AD = arc DB. Also we have arc AF = arc FB.

*Scholium.* The straight line CD fulfils four different conditions: 1st, it passes through the centre; 2d, it passes through the middle of the chord; 3d, it passes through the middle of the subtended arc; 4th, it is perpendicular to the chord. Any two of these conditions are sufficient to determine the direction of the line. Hence we also have the following propositions:

A radius drawn bisecting a chord is perpendicular to it, and bisects the subtended arc.

A radius drawn bisecting an arc, will also bisect its chord perpendicularly.

A line drawn bisecting a chord and its subtended arc, will pass through the centre, and be perpendicular to the chord.

A perpendicular bisecting a chord, will also bisect its subtended arc, and pass through the centre.

A line drawn from the middle of an arc perpendicular to its chord, will bisect the chord, and pass through the centre.

### THEOREM V.

*A line perpendicular to a radius at its extremity, is tangent to the circumference.*

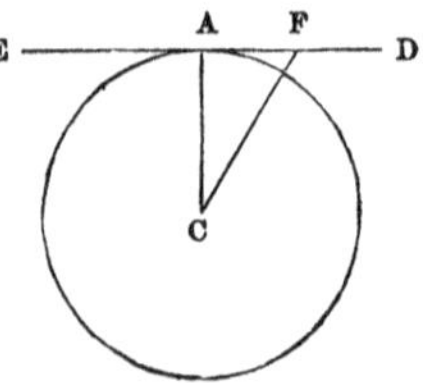


If ED is drawn perpendicular to the radius CA, at its extremity, it will be tangent to the circumference; that is, it will have only the point A in common with the circumference. For any other point, as F, being joined with the centre, gives an oblique line greater than the radius (B. I., T. XII.). Hence, A is the only point common to the straight line and the circumference; consequently, this line is tangent to the circumference (D. IV.).

*Cor.* I. If a straight line is tangent to the circumference of a circle, it will be perpendicular to the radius drawn to the point of contact.

For all other points of the tangent line, except that of contact, are situated without the circumference, and therefore at a greater distance from the centre than the radius. Hence the radius, being the shortest line which can be drawn from the centre to the tangent, is perpendicular to it (B. I., T. XII.).

*Cor.* II. Only one tangent can be drawn through the same point of the circumference.

*Cor.* III. From a point within the circumference no tangent can be drawn.

*Cor.* IV. The perpendiculars drawn at the extremities of a diameter will be parallel tangents, and conversely two parallel tangents will have their points of contact situated at the extremities of the same diameter.

### THEOREM VI.

*Parallel secants or tangents intercept equal arcs of the circumference.*

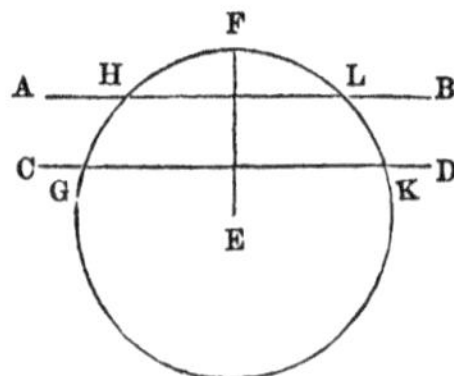


*First.* When the parallels are both secants. Draw the radius EF perpendicular to AB, and it will also be perpendicular to its parallel CD, and we shall have (T. IV.) *arc* GF = *arc* KF and *arc* HF = *arc* LF. Consequently, *arc* GF — *arc* HF

= *arc* KF − *arc* LF; that is, the intercepted arcs GH and KL are equal.

*Secondly*. When one of the parallels is a secant and the other a tangent. Draw the radius EF to the point of contact, and it will be perpendicular to the tangent (T. V., C. I.), and consequently perpendicular to its parallel CD. Hence the *arc* GF = *arc* KF (T. IV.).

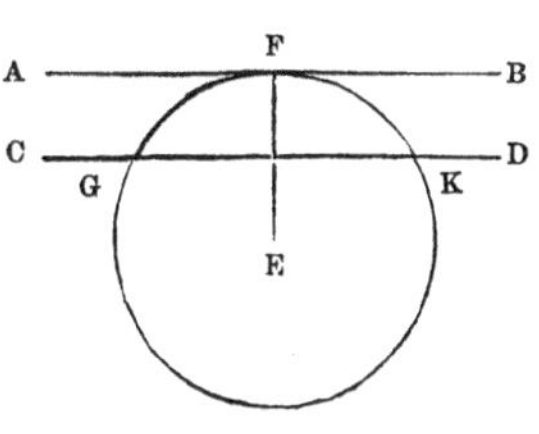


*Thirdly*. When both parallels are tangents. The line FG, joining the points of contact, will be a diameter (T. V., C. IV.), consequently the intercepted arcs will be semi-circumferences, and therefore equal.

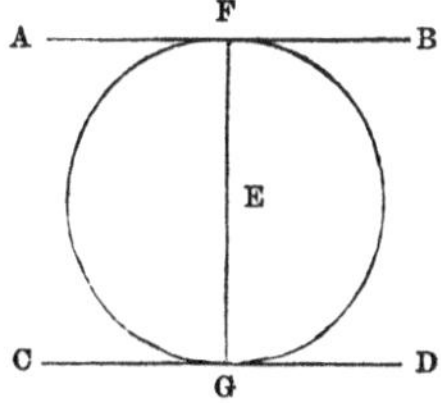


*Cor*. Conversely, if two secants intercept equal arcs of the same circumference they will be parallel, provided they do not intersect, within the circumference.

If a secant and a tangent intercept equal arcs of the same circumference, they will be parallel.

If two tangents intercept equal arcs of the same circumference they will be parallel.

### THEOREM VII.

*In the same circle, or in equal circles, if two arcs are equal, they will have equal chords, which will be equally distant from the centre.*

*If the arcs are unequal and each less than the semi-circumference, the greater arc will have the greater chord, which will be nearer the centre.*

Two circles having equal radii, by superposition, may be made exactly to coincide. Hence we may confine our demonstration to the case of chords in the same circle.

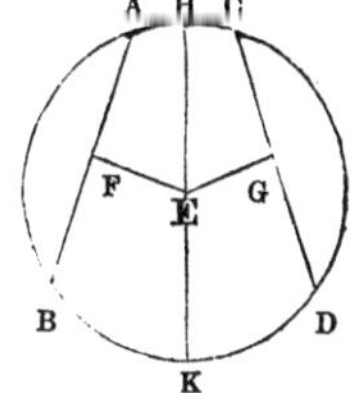


*First*. Suppose the arc AB = CD. From the centre E draw EF and EG perpendicular to the chords AB and CD. Also through H, the middle point of the arc AC, draw the diameter HK. If we revolve the portion of the figure on the left of HK over upon the portion on the right, the arcs HA and HAB will coincide with their equal arcs HC and HCD, consequently the chords AB and CD will be equal, and they will have the same perpendicular; that is, EF will equal EG, and the chords will be equally distant from the centre.

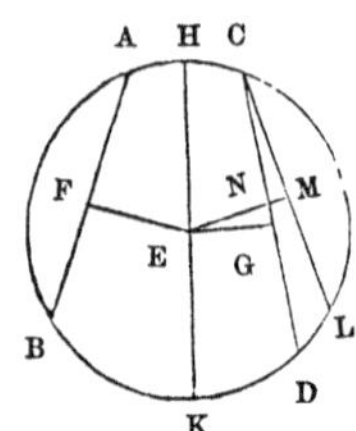


*Secondly.* Suppose the arc AB less than CD. Applying the portion of the figure on the left of HK upon the portion on the right, as before, we see that the point B will fall at some point L, between C and D, since the arc AB is less than CD. Hence, the middle of the arc CL, which is in the prolongation of the perpendicular EM, will be nearer C than the middle of the arc CD which is in the prolongation of the perpendicular EG; consequently the point G is between D and N, the point where the perpendicular EM intersects CD. Therefore CG, the half of CD, is greater than CN, but CN being an oblique line in reference to EM is greater than the perpendicular CM; consequently, CD > CL, or CD > AB. That is, the greater arc has the greater chord.

Again, we have EG < EN, and EN < EM, consequently EG < EM, or EG < EF; that is, the chord corresponding to the greater arc is nearer the centre.

*Cor.* Conversely, in the same circle, or in equal circles:

1. If the chords are equal, the arcs will be equal. And these chords will be equally distant from the centre.

2. The greater chord will correspond to the greater arc. And the greater chord will be nearer the centre.

We may also add:

3. Chords equally distant from the centre are equal, and subtend equal arcs.

4. Of two chords unequally distant from the centre, the one nearer the centre is the greater, and it subtends the greater arc.

*Scholium.* We must keep in mind that the arcs considered in this Theorem are in each case less than a semi-circumference.

## OF THE MEASURE OF ANGLES.

### THEOREM VIII.

*In the same circle, or equal circles, equal angles at the centre correspond to equal arcs. Conversely, equal arcs correspond to equal angles.*

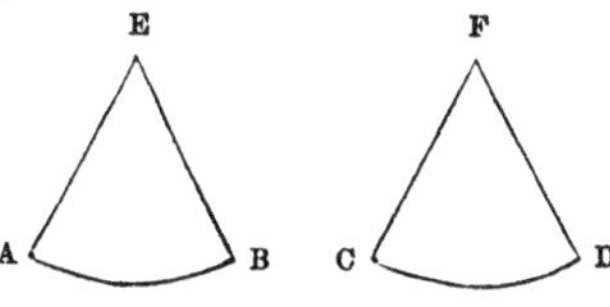


*First.* Suppose the angle at E to equal the angle at F. Apply the angle AEB upon the angle CFD; EA and EB will take, respectively, the directions of FC and FD; and since the radii are equal, the points A and B will coincide respectively with the points C and D. Hence the *arc* AB = *arc* CD.

*Secondly.* Suppose the arcs AB and CD equal. Then, as before, applying the figure ABE upon CDF, so that the radius AE may coincide with its equal radius CF, then will the arcs AB and CD coincide, since they are equal, and the point B will coincide with the point D; consequently the angle AEB is equal to the angle CFD.

*Cor.* In the same circle, or equal circles, the greater angle at the centre corresponds to the greater arc, and, conversely, the greater arc corresponds to the greater angle.

### THEOREM IX.

*In the same circle, or in equal circles, angles at the centre are to each other as their included arcs.*

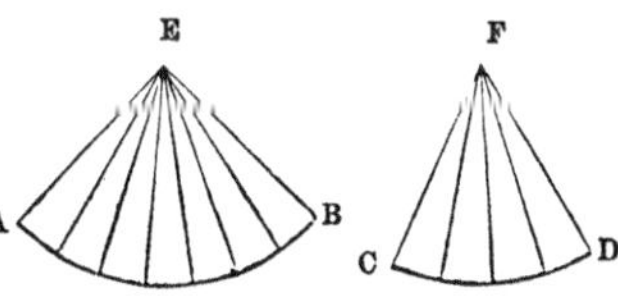


*First.* When the two arcs AB and CD are commensurable that is, when they are to each other as two whole numbers, for instance, as 7 to 4. If we divide the arc AB into seven equal portions, and the arc CD into four, and draw radii to these points of division, we shall divide the angle AEB into seven angles and CFD into four, and these angles will be equal, since their arcs are (T. VIII.). Hence the angle AEB will be to the angle CFD as 7 to 4; that is, as the arc AB to the arc CD.

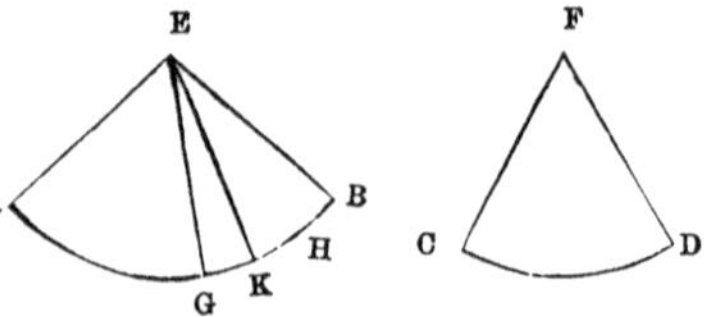


*Secondly*. When the arcs AB and CD are not commensurable. Apply the smaller angle CFD upon the larger AEB, so that it may take the position AEG. If now, these angles are not to each other as their arcs, suppose we have

*angle* AEB : *angle* AEG : : *arc* AB : *arc* AH greater than arc AG.

If we conceive the arc AB to be divided into equal portions, each of which shall be less than arc GH, there will be at least one point of division between G and H, as at K. Draw the radius EK, and then since the arcs AB and AK are commensurable, we have by the first part of this Theorem,

*angle* AEB : *angle* AEK : : *arc* AB : *arc* AK.

From these two proportions, since their antecedents are the same, we deduce

*angle* AEG : *angle* AEK : : *arc* AH : *arc* AK,

which cannot be, for the antecedent of the first couplet is less than its consequent, while in the second couplet the antecedent is greater than its consequent.

Hence the angle AEB cannot be to the angle AEG as the arc AB is to an arc greater than AG. And by a similar process we may show that the angle AEB cannot be to the angle AEG as the arc AB is to an arc less than AG. Consequently we must have in the same circle, or in equal circles, the angles at the centre to each other as their corresponding arcs.

*Scholium*. Since, in the same circle, or in equal circles, the angles at the centre are to each other as their corresponding arcs, we may use the arcs as the measure of their corresponding angles. In the case of a right angle, its measuring arc is a quadrant; in the case of an angle equal to two right angles, the measuring arc is a semi-circumference. We sometimes regard the right angle as the unit angle, so that acute angles would all be less than 1, and obtuse angles would exceed 1 and be less than 2.

THEOREM X.

*An inscribed angle is measured by half the arc included between its sides.*

*First.* When the centre of the circle is within the angle.

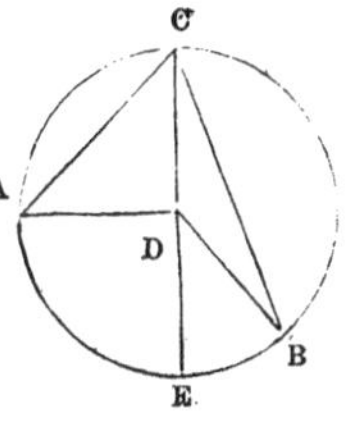


Draw the diameter CE and the radii DA and DB. Now, since the triangle DAC is isosceles, the exterior angle ADE is double the angle DCA. The angle ADE is measured by the arc AE (T. IX., S.); hence the angle ACE is measured by one half of the arc AE. In the same way we may show that the angle ECB is measured by one half of the arc EB; consequently the angle ACB is measured by one half its included arc AEB.

*Secondly.* When the centre of the circle is without the angle.

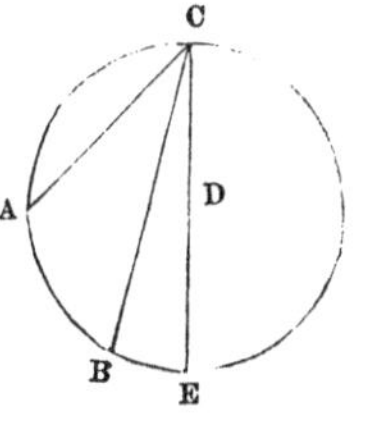


As before, draw the diameter CE, and we shall have the angle ACE, measured by one half of the arc ABE, also the angle BCE measured by half the arc BE; consequently, their difference, the angle ACB, is measured by half the difference of these arcs; that is, by half the included arc AB.

*Cor.* I. Each angle inscribed in a semicircle is a right angle, since it is measured by half the semi-circumference, or by a quadrant.

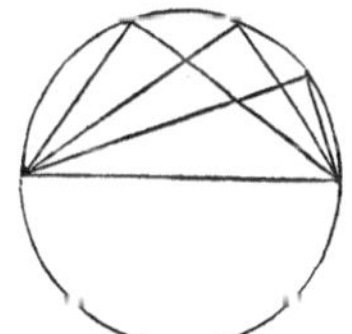

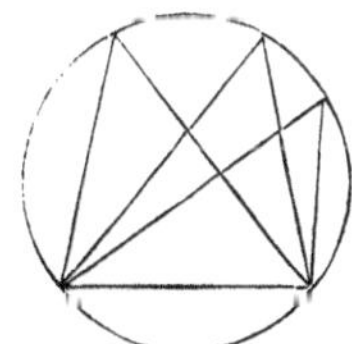

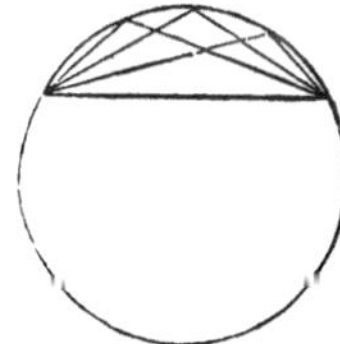

Each angle is a segment greater than a semicircle is acute, since half the included arc, which measures it, is less than a quadrant.

Each angle in a segment less than a semicircle is obtuse, since half the included arc, which measures it, is greater than a quadrant.

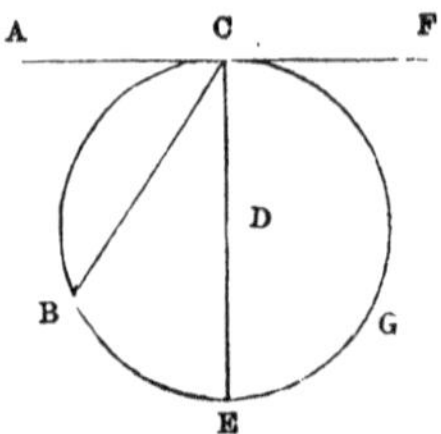


*Cor.* II. An angle, ACB, formed by a tangent and chord, has also for its measure one half the arc CB included between its sides. For, drawing the diameter CE, we have ACE a right angle (T. V., C. I.), and it has for its measure one half of the semi-circumference CBE. The angle BCE has for its measure one half the arc BE; hence, the difference of these angles; that is, the angle ACB has for its measure one half the arc CB.

If we consider the supplementary angle FCB, we have the right angle FCE measured by one half the arc CGE; hence, the sum of the angles FCE and ECB; that is, the angle FCB has for its measure one half the included arc CGEB.

### THEOREM XI.

*The angle formed by the intersection of two chords is measured by half the sum of the included arcs. And the angle formed by the intersection of two secants is measured by half the difference of the included arcs.*

Drawing the chord AF parallel to CD, we have the arc FD= AC (T. VI.), and the angle BAF=BED (B. I., T. XVIII., C.), but the angle BAF is measured, in the case of the chords, by

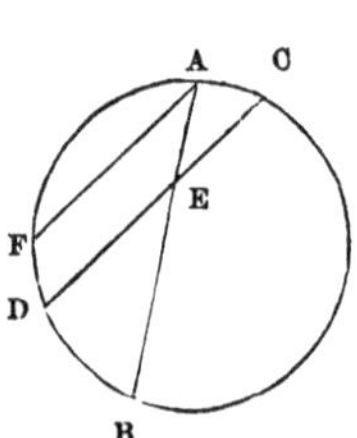


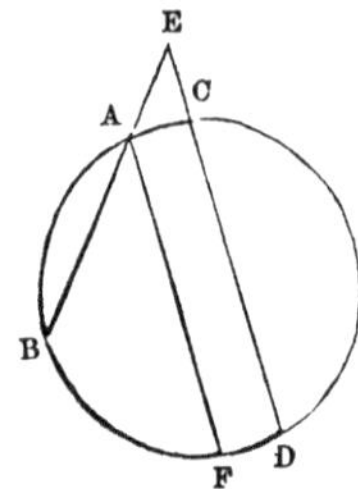


half the sum of the arcs BD and DF; that is, by half the sum of BD and AC. But in the case of the secants, the angle BAF is measured by half the difference of BD and DF; that is, by half the difference of BD and AC. Consequently, the angle formed by the intersection of two chords is measured by half the sum of the included arcs; and the angle formed by the intersection of the two secants is measured by half the difference of the included arcs.

*Cor.* The angle AEC formed by the intersection of two tangents, is measured by half the difference of the concave and convex arcs AFC and AC, comprehended between the points of contact.

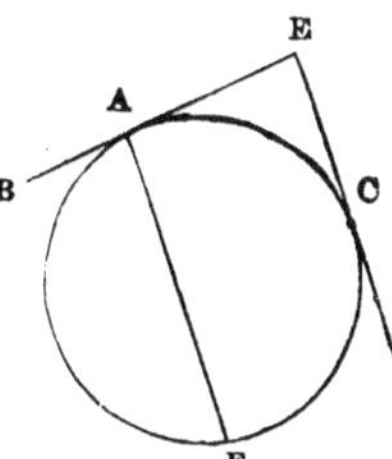


For, drawing AF parallel to CE, we have the arc FC=AC (T. VI.), and the angle BAF=BEC, but the angle BAF is measured by half the arc AF (T. X., C. II.); that is, by half the difference of AFC and FC, or which is the same, by half the difference of the concave arc AFC and the convex arc AC.

---

## OF INSCRIBED AND CIRCUMSCRIBED POLYGONS.

### THEOREM XII.

*All triangles are capable of being inscribed in a circle and of circumscribing a circle.*

*First.* The three perpendiculars bisecting the three sides of the triangle ABC, meet in the same point G, which is equally distant from A, B, and C (B. I., T. XXXVII.). Hence, if with G as a centre, a circumference be described with a radius equal to the distance from G to either angle of the triangle, it will circumscribe the triangle, and consequently the triangle will be inscribed in the circle.

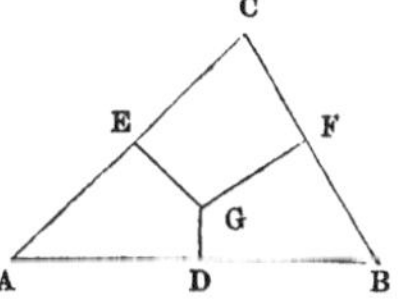


*Secondly.* The three lines bisecting the three angles of the triangle ABC meet in the same point G, which is equally distant from the three sides of the triangle (B. I., T. XXXVI.). Hence, if with G as a centre, a circumference be described with a radius equal to the distance from G to either side of the triangle, it will be inscribed in the triangle, and consequently the triangle will circumscribe the circle.

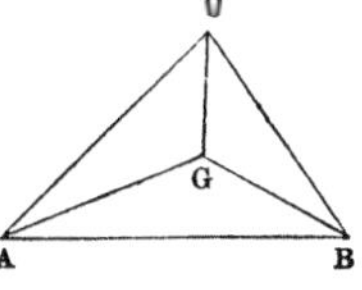

*Scholium* I. Since, in the first case, the perpendiculars can meet in only one point, there can be only one circumference circumscribing a triangle. That is, through three points not in the same straight line, only one circumference can be made to pass. Hence, when two circumferences have three points common, they must coincide.

If the three points A, B, and C, are all in the same straight line, the perpendiculars drawn bisecting AB, BC and CA will be parallel, and cannot meet, in which case there can be no centre, unless we regard the centre as at an *infinite* distance.

*Scholium* II. If, in the second case, the sides AB and AC are produced, the point F, where the lines bisecting the exterior angles meet, is also equally distant from the three lines forming the triangle. Therefore, with F as a centre, a new circumference may be described tangent to the three sides. This circle is called the *escribed* circle. Hence, if lines be drawn bisecting the angles, and the exterior angles of a triangle, they will intersect each other by threes at the centres of the *inscribed* and *escribed* circles; thus the three lines which bisect the angles will meet at the same point, giving the centre of the inscribed circle. Any one of the lines bisecting an angle of the triangle will intersect, at the same point, two of the lines which bisect the exterior angles, giving the centre of an escribed circle.

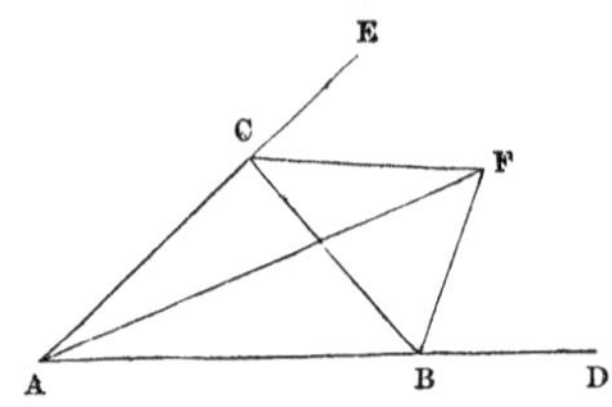


*Scholium* III. This Theorem, taken in connection with Theorem XXXV., Book First, shows that, when the triangle is right-angled at C, the centre of the circumscribed circle will be at D, the middle point of the hypotenuse, for in this case the three lines DA, DB, and DC are all equal.

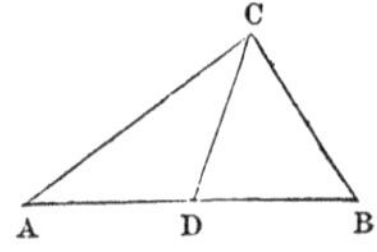


If the triangle is isosceles, the centre of the circumscribing circle as well as that of the inscribed circle will be in the line bisecting the angle formed by the equal sides.

If the triangle is equilateral, the centres of the two circles will coincide, and the circumscribed and inscribed circles will then be concentric.

### THEOREM XIII.

*In any inscribed quadrilateral, the sum of the opposite angles is equal to two right angles.*

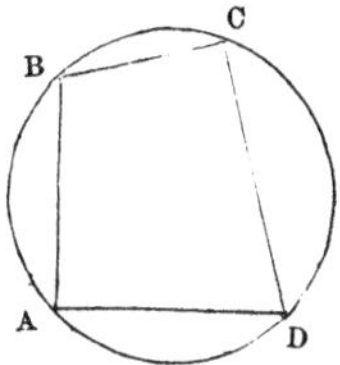


The angle at B is measured by half the arc CDA, and the opposite angle at D is measured by half the arc ABC; hence the sum of the angles at B and D is measured by half the entire circumference, which is the measure of two right angles.

*Cor.* Conversely, a quadrilateral may be inscribed when the sum of its opposite angles is equal to two right angles.

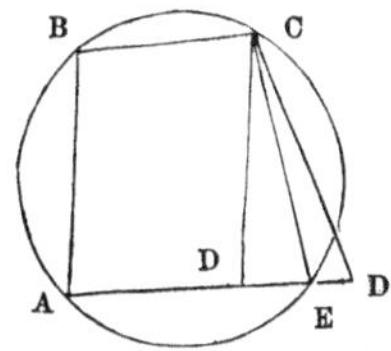


If we describe a circumference through the three points A, B, and C, which can always be done (T. XII., S. I.), and it does not pass through D, then the point D must be either within or without this circumference. First suppose it within. Produce AD until it meet the circumference at E, then; by the Theorem itself, we shall have the sum of ABC and CEA = 2 right angles: but by hypothesis ABC + CDA = 2 right angles; consequently CEA = CDA, which is absurd (B. I., T. VI., C.). Hence, the point D cannot be within the circumference. By a similar process, we can show that it cannot be without the circumference. It must therefore be in the circumference.

*Scholium.* The *rectangle*, which includes the *square*, is the only parallelogram which can be inscribed in a circle. The diagonals of the rectangle are diameters of the circumscribing circle.

### THEOREM XIV.

*When a quadrilateral circumscribes a circle, the sums of the opposite sides are equal.*

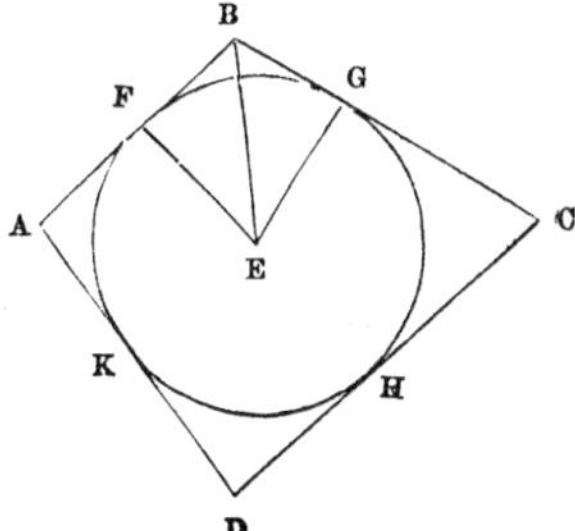


From the centre draw EF and EG to the points of contact of the sides AB and BC, and they will be perpendicular to those sides (T. V., C. I.). The two right-angled triangles EFB,

EGB, will be equal, since they have the hypotenuse and a side of the one respectively equal to the hypotenuse and a side of the other (B. I., T. XXVI.), and BF = BG. In a similar manner we may show that AF = AK; CH = CG; DH = DK. Consequently by adding, we have AB + CD = BC + AD.

*Cor.* When the sums of the opposite sides of a quadrilateral are equal, it is capable of circumscribing a circle.

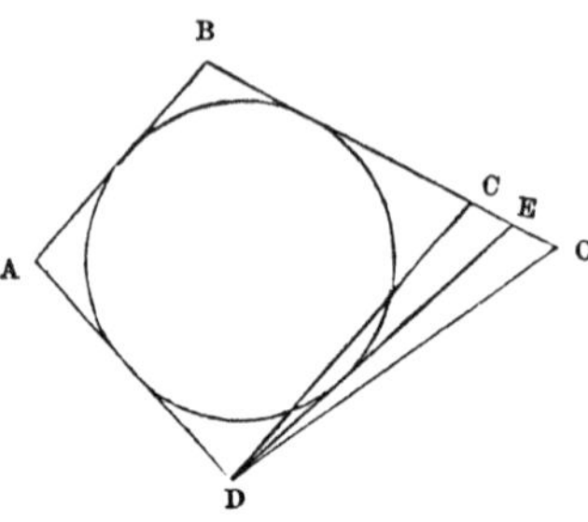


For, if we describe a circumference tangent to AD, AB, and BC, which can always be done (T. XII., S. I. or II.), it will also be tangent to DC. For, if the side DC is not tangent to the circumference, it must be either a secant or lie wholly without the circumference. We will first suppose it to be a secant. If DE is drawn tangent to this circumference, we shall have AD + BE = AB + DE, but by hypothesis we have AD + BC = AB + DC; hence, by subtraction, we obtain EC = DE − DC; that is, one side of a triangle is equal to the difference of the other two sides, which is impossible (B. I., T. VII.). Hence the side DC cannot be secant. In a similar way we can show that it cannot be wholly without the circumference. It must therefore be tangent.

*Scholium.* The *rhombus* and the *square* are the only quadrilaterals capable of circumscribing a circle.

## THEOREM XV.

*All regular polygons are capable of being inscribed in a circle, and of circumscribing a circle.*

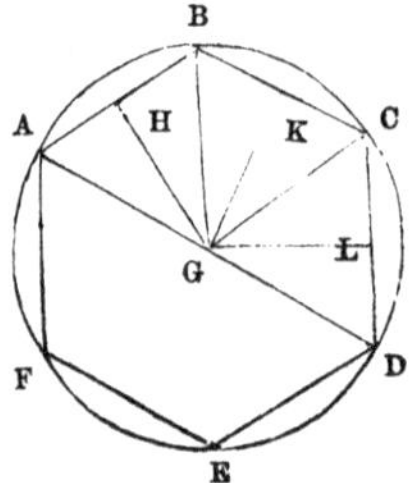


Let ABCDEF be any regular polygon; and through three consecutive angles A, B, and C, describe the circumference of a circle (T. XII.). This circumference will also pass through all the other angles of the polygon. For, drawing from the centre G, the lines GA, GB, GC, and GD, we shall form three triangles, GAB, GBC,

GCD. In the two triangles GAB and GBC, we have GB common, AB = BC and GA = GC; hence the triangles are equal (B. I., T. XXV.), and being isosceles since GA = GB = GC, we have the angles GAB = GBA = GBC = GCB. That is, each of these angles is one half the angle of the polygon; hence the angle GCB = GCD. Now, comparing the two triangles GBC and GCD, we have the side GC common, BC = CD, and the angle GCB = GCD; these triangles are therefore equal (B. I., T. XX.); and we have GD = GB = GC = GA; hence, the circumference which passes through A, B, and C, will also pass through D.

In a similar manner we can show that the circumference which passes through B, C, and D, will also pass through E, and so on, for all the angles of the polygon, which demonstrates the first part of the Theorem.

As to the second part, we observe, that the sides AB, BC, CD, etc., are equal chords, in reference to the circumscribed circle, and therefore they are equally distant from the centre G (T. VII., C. I.). If then, with G as a centre, and with the perpendicular GH as a radius, we describe a circumference, it will be tangent to all the sides of the polygon. This circle will be inscribed in the polygon, and the polygon will circumscribe the circle.

*Scholium.* The point F, the common centre of the inscribed and circumscribed circles, is called the *centre of the regular polygon.*

The radius of the circumscribing circle is called the *radius of the polygon*, and the radius of the inscribed circle is called the *apothem.*

The angles AGB, BGC, CGD, etc., are called *angles at the centre.*

The lines GA, GB, GC, etc., which are the radii of the polygon, bisect the angles of the polygon. The apothems GH, GK, GL, etc., which are the radii of the inscribed circle, bisect the sides of the polygon perpendicularly, and also bisect the angles at the centre, since they would, if produced, bisect the arcs which measure the angles at the centre.

THEOREM XVI.

*If through the corners of a regular polygon, already inscribed in a circle, we draw tangents, we shall thus circumscribe a regular polygon of the same number of sides.*

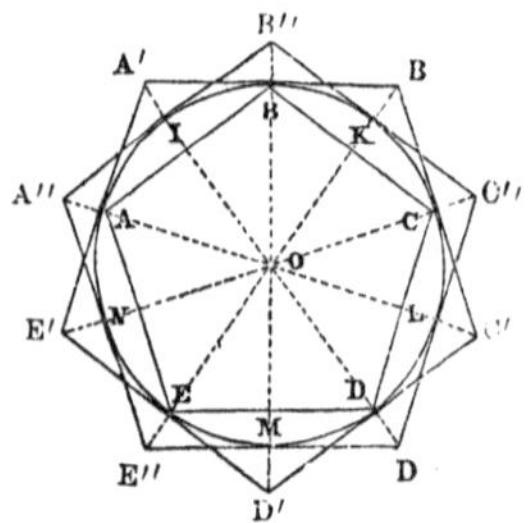


The tangents thus drawn will form with the sides of the inscribed polygon, AB, BC, CD, etc., a series of triangles, AA′B, BB′C, CC′D, etc., all isosceles and equal, since AB = BC = CD = etc., and the angles A′AB, A′BA, B′BC, B′CB, C′CD, C′DC, etc., are all equal, having for their measure half of the equal arcs AB, BC, CD, etc. (T. X., C. II.). Hence, all the angles, A′, B′, C′, etc., of the circumscribed polygon are equal. And since AA′ = A′B = BB′ = B′C = CC′ = etc., we have A′B′ = B′C′ = C′D′ = etc. That is, this circumscribed polygon has all its angles equal, and all its sides equal; it is therefore regular (D. VI.).

*Scholium.* Since the radius OA′ of the circumscribed polygon bisects the angle E′A′B′ (T. XV., S.), we have, in the two right-angled triangles OAA′ and OBA′, the acute angle OA′A of the one, equal to the acute angle OA′B of the other; consequently, the remaining acute angles are equal; that is, the angle A′OA = A′OB. In the same way we can show that the angle B′OB = B′OC. But the angle A′OB = B′OB, since the apothem FB bisects the angle A′FB′ at the centre (T. XV., S.). Hence, all the angles AOA′, A′OB, BOB′, B′OC, COC′, etc., are equal; and their measuring arcs AI, IB, BK, KC, CL, etc., are all equal.

If then we suppose the circumscribed polygon to revolve in its own plane about its centre O, as a pivot, so that A may reach the point I, B will then coincide with K, C with L, etc. The circumscribed polygon will by this means, without having its relative parts in the least changed, assume a new position, having its sides parallel with the sides of the inscribed polygon.

## OF SECANT AND TANGENT CIRCLES.

### THEOREM XVII.

*If two circumferences of circles have two points in common, the line joining their centres will bisect their common chord perpendicularly.*

For, the line bisecting a chord at right angles passes through the centre (T. IV., S.); and as this chord is common to both circles, this bisecting line must pass through both centres.

### THEOREM XVIII.

*When two circumferences have only one point in common, this point will be situated on the line joining their centres.*

If A and B are the centres of two circles having only one point in common, that point will be situated on the line AB joining their centres. For, if not, suppose it to be situated without this line, as at C. Draw CE perpendicular to AB, and prolong it until ED = CE; then we evidently have AC = AD, also BC = BD, and since C was a common point of the circumferences, D will also be a common point. These circumferences will then have two points in common, which is contrary to the hypothesis. Hence, the point in common to the two circumferences must be in the line joining their centres.

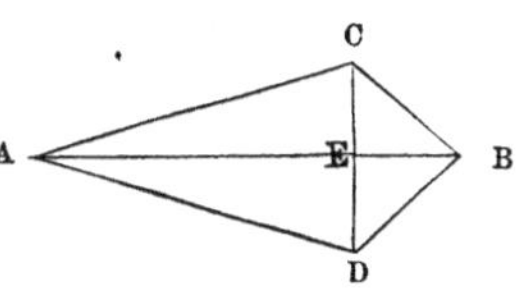


*Scholium.* Suppose we consider two circles whose centres are at A and B, the one at B being the smaller. Draw CF passing through their centres, and at the extremities of their diameters, CD and EF, draw the perpendiculars GG′, HH′, KK′, LL′. If we suppose the circle, whose centre is at A, fixed, and the other to move towards it, so that the centre B may move in the line AB, we may notice *five distinct positions* of these circles.

*First.* When the tangent KK′ is on the right of HH′. In this case, each circle will be wholly without the other, and they have no common point.

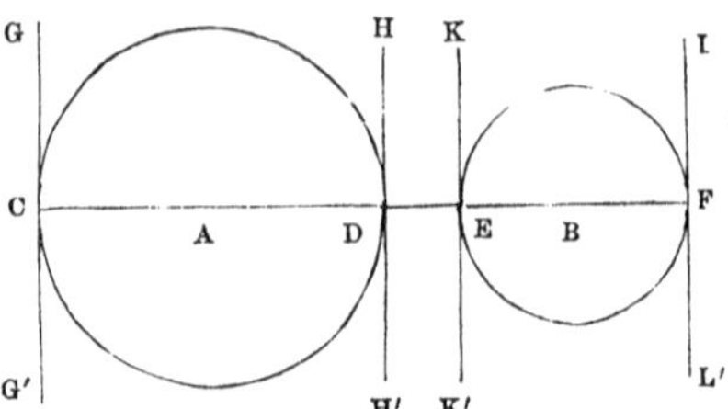


*Secondly.* When the tangent KK′ coincides with HH′, and becomes a common tangent to the two circumferences. In this case, the point D is the only point common to the two circumferences, and the circles are said to touch *externally*.

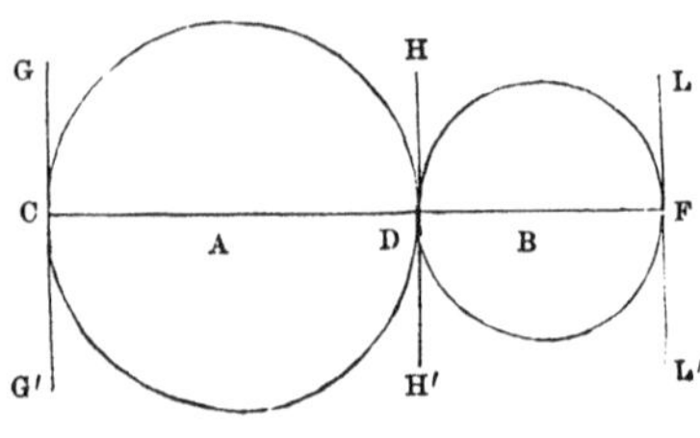


*Thirdly.* When the limiting tangents of the smaller circle are on opposite sides of HH′. In this case, there will be a portion of space common to the two circles; hence their circumferences must necessarily cut each other in two points.

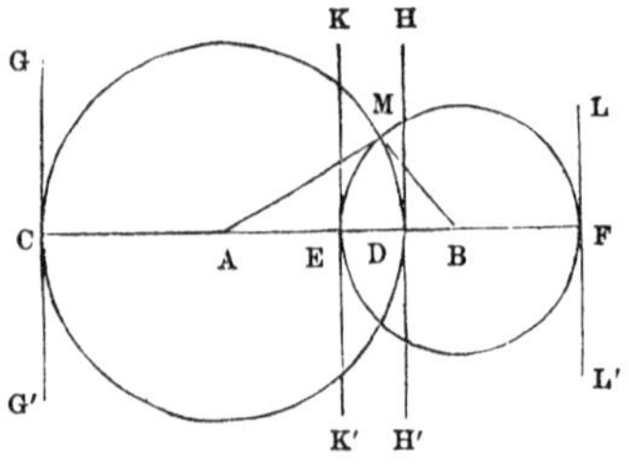


*Fourthly.* When the tangent LL′ coincides with HH′, and becomes a common tangent to the two circumferences. In this case, D is the only point common to the two circumferences. For, drawing any radius, AM, of the larger circle, cutting the circumference of the smaller circle at N, and we shall have

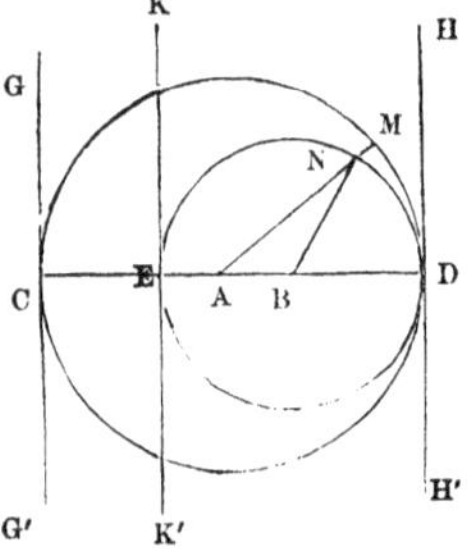


$$AN < AB + BN, \text{ or } AN < AB + BD, \text{ or } AN < AD, \text{ or } AM.$$

Hence, every point in the circumference of the smaller circle, except the point D, is within the circumference of the larger circle. The circles are said to touch *internally*.

*Fifthly.* When the limiting tangents of the smaller circle are both situated between those of the larger circle. In this case the two circumferences have no common point, and the smaller is wholly within the larger. For, joining any point of its circumference, as N, with the centres, we have

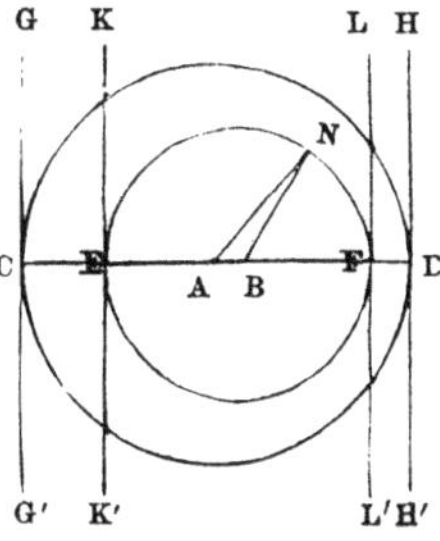


$$AN < AB + BN, \text{ or } AN < AB + BF,$$

but $AB + BF = AF$, which is less than AD; hence, $AN < AD$.

### THEOREM XIX.

*When two circumferences have no point in common, the distance between their centres is either greater than the sum of their radii, or less than their difference.*

For by the Scholium under the last Theorem we have two positions of the circumferences in which they have no common point.

*First.* When they are, the one wholly without the other, in which case the distance between their centres is greater than the sum of their radii.

*Secondly.* When the smaller circumference lies wholly within the larger; in which case the distance between their centres is less than the difference of their radii.

If we denote the greater radius by R, the less radius by R′, and the distance between their centres by D, we shall have, when the circles are *exterior*,

$$D > R + R',$$

and when they are *interior*—that is, when the one is wholly within the other—we shall have

$$D < R - R'$$

### THEOREM XX.

*When two circumferences are tangent to each other, the distance between their centres is equal to the sum of their radii, or equal to their difference.*

There are two positions of the circumference in which

they are tangent to each other (T. XVIII., S.). In the first case,

$$D = R + R',$$

and in the second case,

$$D = R - R'.$$

THEOREM XXI.

*When two circumferences cut each other, the distance between their centres is less than the sum of their radii, and greater than their difference.*

By the third case under the Scholium of Theorem XVIII. (see figure of the same), we see that the point M, common to the two circumferences, must be necessarily without the line AB which joins their centres; hence the three points A, B, and M must form a triangle, consequently we must have

$$D < R + R' \text{ and } D > R - R'.$$

*Scholium.* The reciprocals of the foregoing Theorems may readily be demonstrated by the method of reducing to an absurdity. They may be thus stated:

When two circumferences are in the same plane:

1. If $D > R + R'$, the two circles are exterior the one to the other.

2. If $D = R + R'$, the two circles will touch externally.

3. If $D < R + R'$, and at the same time $D > R - R'$, the two circles will cut each other in two points.

4. If $D = R - R'$, the two circles will touch internally.

5. If $D < R - R'$, the smaller circle will lie wholly within the other.

# PROBLEMS,

## WHICH REFER TO THE FIRST AND SECOND BOOKS.

---

## OF PERPENDICULARS, ANGLES, AND PARALLELS.

### PROBLEM I

*From any given point of an indefinite straight line, to draw a perpendicular to this line.*

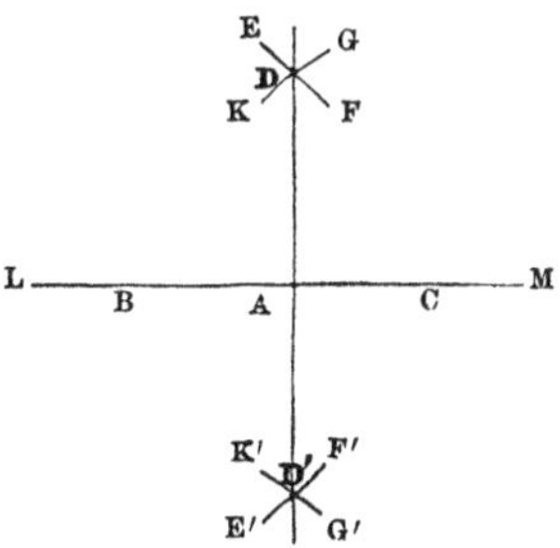


Let A be the given point of the indefinite line LM. With the compass, take two equal distances, AB and AC. With A and B as centres, and with a radius greater than AB or AC, describe two circumferences intersecting each other at D. Draw DA, and it will be the perpendicular required.

For, by construction, the two circumferences described are such that the distance of their centres BC is less than the sum of their radii, since each radius is greater than half this distance, and it is at the same time greater than the difference of their radii, which is zero; hence these circumferences cut each other in some point, as D (T. XXI., S.); and if we draw DA, it must be perpendicular to BC (B. I., T. XIII., C. II.).

*Scholium.* The two circumferences cut each other in a second point D′, on the other side of the line LM, which point may be used to verify our work, since the three points D, A, D′ must be in the same straight line.

In practice, instead of describing the entire circumferences, we trace only the arcs EF, GK, E′F′, G′K′, in the immediate vicinity of the points of intersection.

### PROBLEM II.

*From a given point without a straight line, to draw a perpendicular to this line.*

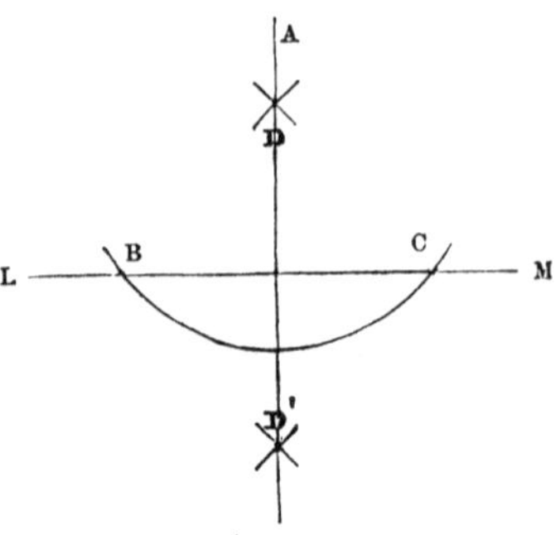


Let A be the given point without the line LM. With A as a centre and with a radius sufficiently great, describe an arc cutting LM in two points B and C; and with B and C as centres, and with a new radius greater than half of BC, describe two new arcs cutting each other in D and D′; draw ADD′, and it will be the perpendicular required. For the points A, D, and D′ are, by construction, equally distant from B and C, consequently the line ADD′ is perpendicular to BC or LM (B. I., T. XIII., C. II.).

### PROBLEM III.

*To divide a given straight line of a determinate length into two equal parts; that is, to bisect a given line.*

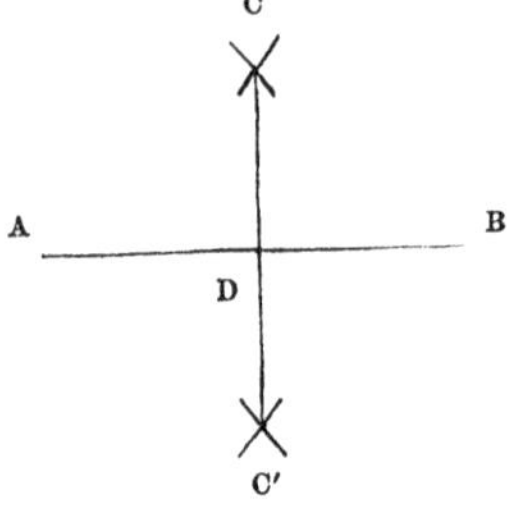


Let AB be the line which we wish to bisect.

With A and B as centres, and with a radius greater than half of AB, describe arcs cutting each other in C and C′, draw CC′, and it will bisect AB perpendicularly at D. The same demonstration as in preceding problems applies.

*Cor.* The following problems follow immediately from the foregoing:

*First.* To describe on a given straight line as a diameter a semi-circumference, or the entire circumference. For this question is reduced to finding the middle point of this line.

*Secondly.* To describe a circumference through three given

points. For, having joined these points, two and two, the whole is reduced to drawing perpendiculars through the middle points of these lines. Where these perpendiculars meet, will be the centre of the circle. The radius will be found by joining the centre and either of the given points.

*Thirdly.* To find the centre of a circle, or of an arc already described. For, taking at pleasure any three points on the arc and proceeding as in the last case, we shall find the centre.

*Fourthly.* To divide an arc into two equal parts. For, draw the chord, and then draw a perpendicular through its middle point (T. IV., S.).

### PROBLEM IV.

*To draw from a given point a perpendicular to a given line which can be produced only in one direction.*

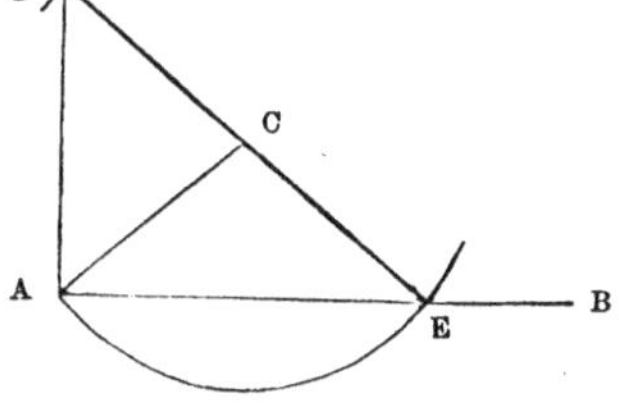


*First.* When the given point A is in the given line AB, which line we will suppose cannot be prolonged to the left of A.

Take any point, as C, for a centre, and with CA as a radius describe an arc cutting AB at E. Draw EC, and produce it until CD is equal to CE. Then drawing AD, it will be perpendicular to AB. For, by construction, CA = CE = CD, and the angle DAB is right (B. I., T. XXXV.).

*Secondly.* When the given point D is without the given line AB.

Draw any line, as DE, meeting AB and with C its middle point as a centre, and with CE = CD as a radius, describe an arc cutting AB in A. Join A and D, and it will be the perpendicular required. For, by construction we have, as before, CA = CE = CD ; consequently, DAE is a right angle

### PROBLEM V.

*Through a given point in an indefinite straight line to draw a second line, forming with the first a given angle.*

Let M be the given angle, and A the given point in the indefinite line AX. With M and A as centres with a radius MN = AB, describe the arc NP comprised between the sides of the given angle and the indefinite arc BB′. Then with B as a centre, and with a radius equal to the chord of the arc NP, describe a new arc cutting the indefinite arc BB′ in C. Draw AC, and BAC will be the angle required. For, by construction, the chords of the two arcs BC and NP are equal; hence the two arcs are equal, and consequently the angles are equal (T. VIII.).

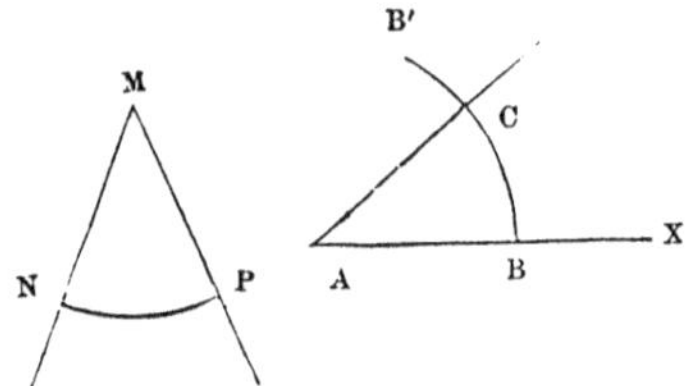


*Scholium* I. To divide an angle into two, four, eight, etc., equal parts, it is sufficient to divide the corresponding arc first into two equal parts, and afterwards to divide each half into two equal parts, and so on.

*Scholium* II. The same problem furnishes the means of determining the supplement of the sum of two given angles, or, in other words: Knowing two angles of a triangle, to determine the third.

After forming the angle BAC equal to the first of the given angles, draw AD, making the angle CAD equal to the second of the given angles. Then produce BA to B′ and DAB′ will be the supplement required.

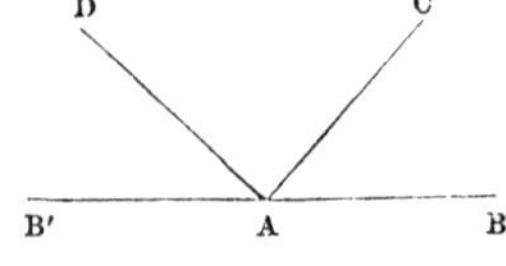


PROBLEM VI.

*From a point without a straight line to draw a parallel to this line.*

Let C be the given point, and AB the given line.

Draw any line from C meeting AB, as CD; and with C and D as centres with CD for a radius, describe the indefinite arc DD′, and the definite arc CE terminating at E in the given line AB. Take the arc DF equal to CE, and draw CF, which will be the parallel required.

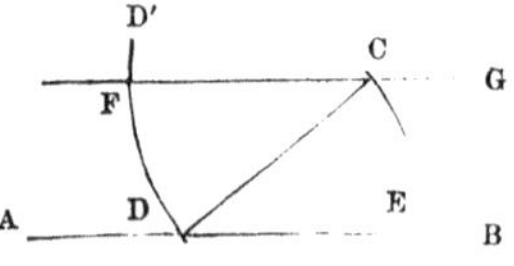


For, by construction, we have angle FCD = angle CDE (T. VIII.), hence FG and AB are parallel (B. I., T. XVII., C.).

*Cor.* Problems V. and VI. afford the means of solving the following:

*Through a given point*, C, *without a straight line* AB, *to draw a second line meeting the first under a given angle* M.

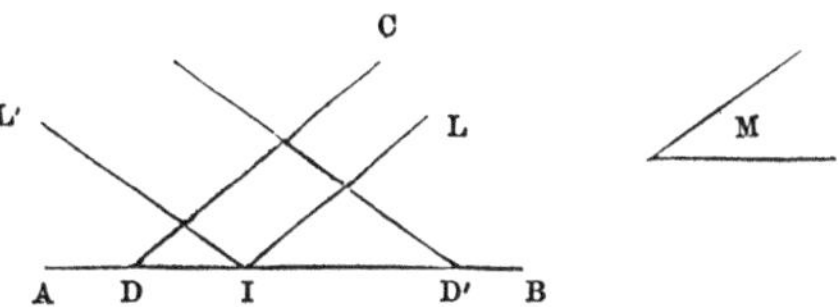


For, at any point, I, of the line AB, draw IL, making the angle LIB = angle M; then through C draw CD parallel to LI, which is evidently the line required. Since from any point, I, two lines IL and IL′ may be drawn, making with AC an angle equal to M, if M is not a right angle, it follows that the problem admits of two solutions, unless M is a right angle.

### PROBLEM VII.

*On a given straight line, as a chord, to describe an arc of a circle capable of containing a given angle.*

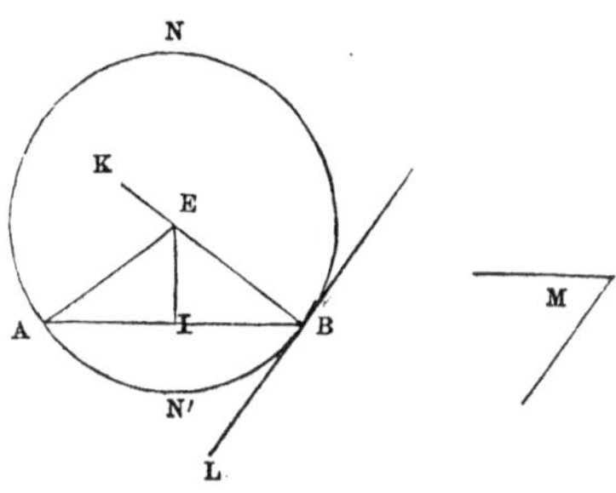


Let AB be the given line, and M the given angle.

Draw BL, making the angle ABL = M, and draw BK perpendicular to BL. Through I the middle point of AB draw the perpendicular IE meeting BK in E. From the point E with a radius EA = EB, describe a circle. The portion ANB is the arc required. For, by construction, BL is a tangent, being perpendicular to the radius EB, and the angle ABL = M, is measured by half the arc AN′B (T. X., C. II.). But each angle in the segment ANB is also measured by half the same arc (T. X.).

## CONSTRUCTION OF POLYGONS.

### PROBLEM VIII.

*To construct a triangle when we have given:*

I. *One side and the two adjacent angles.*

II. *Two sides and the included angle.*

III. *The three sides.*

It is not necessary to give at length the construction of the first and second cases, since they are so readily deduced from the Corollary to Problem VI. We remark, however, that in the first, in order that the construction may be possible, the sum of the two given angles must be less than two right angles. And in the second case, the construction is always possible, since, after having drawn two lines, forming with each other the required angle, we can always take on these lines portions equal to the given sides.

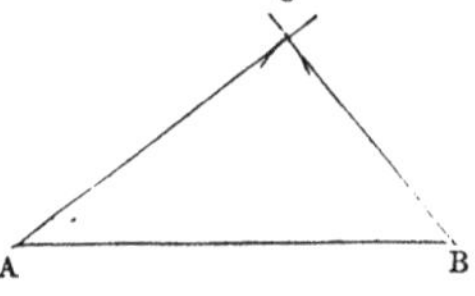


*In the third case.* Draw AB equal to one of the given sides, and with A as a centre with a radius equal the second of the given sides, describe an arc; also, with B as a centre and with a radius equal to the third side, describe an arc intersecting the former arc in C. Draw CA and CB, and the triangle ABC will evidently satisfy the conditions announced.

That this construction may be possible, it is necessary that the first side may be less than the sum of the other two sides, and greater than their difference (B. I., T. VII.).

*Scholium.* If, in the first case, instead of having the two adjacent angles given, we had one of the adjacent angles and the opposite angle, we could, by Prob. V., Scholium II., find the third angle, and then the two adjacent angles would be known.

### PROBLEM IX.

*To construct a triangle when two sides are given and an angle opposite one of these sides.*

*First.* When the side opposite the given angle is greater

than the adjacent side, the given angle being of any magnitude whatever.

Draw the two lines AB and AC, forming the angle BAC equal to the given angle; take AD equal to the adjacent side, and with D as a centre, and with a radius equal to the opposite side, describe an arc cutting AB in E. Draw DE, and the triangle DAE will fulfil the conditions announced.

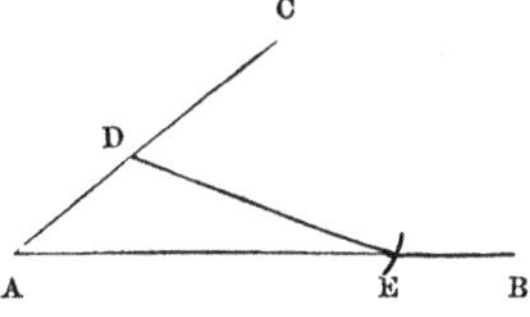


*Secondly.* When the side opposite the given angle is less than the adjacent side, the given angle being acute.

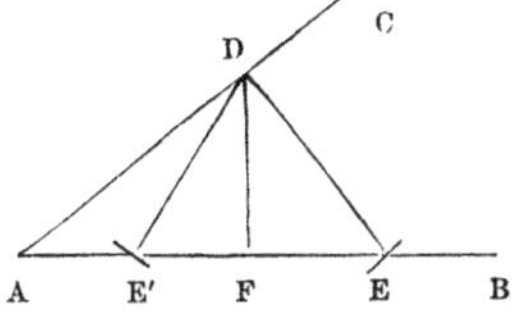


As before, make the angle BAC equal to the given angle, take AD equal to the adjacent side, and with D as a centre, and with a radius equal to the opposite side, describe an arc cutting AB in two points E and E′: drawing DE and DE′, we have two triangles DAE and DAE′ fulfilling the required conditions. If, however, the opposite side is equal to the perpendicular DF, the arc will touch AB, and the two points E and E′ will unite with F, and there will be only one triangle, which will be right-angled at F. But if it is less than this perpendicular, the arc cannot either intersect or touch AB, and the construction will be impossible. It is obvious that the construction would be impossible if the given angle were either right or obtuse, so long as the opposite side is less than the adjacent side.

*Scholium.* It is not necessary to consider the case when the two sides are equal, for the triangle would then be isosceles; and having one angle given, the other angles could readily be found, and then it would be included in Problem VIII.

### PROBLEM X.

*A regular inscribed polygon being given, to construct a regular circumscribed polygon of the same number of sides; and conversely.*

*First construction.* Through the vertices A, B, C, etc., of the inscribed polygon, draw tangents to the circumference, and they will form the polygon required

*Second construction.* Through the extremities I, K, L, etc., of the radii drawn perpendicular to the sides AB, BC, CD, etc., of the inscribed polygon, draw tangents which will form the polygon required.

We have already shown that the two polygons thus obtained are equal (T. XVI., S.).

*Reciprocally.* To obtain the inscribed polygon, when we already know the circumscribed polygon A′B′C′D′E′, we join the consecutive points of contact A, B, C, etc., of the circumscribed polygon. These chords, AB, BC, CD, etc., are equal, since their arcs AIB, BKC, CLD, etc., are equal; and they form with each other equal angles, since the arcs which measure them are equal.

Or we could equally as well use the circumscribed polygon A″B″C″D″E″, by joining the consecutive points A, B, C, etc., where the straight lines OA″, OB″, OC″, etc., drawn from the centre to the vertices of the circumscribed polygon, meet the circumference.

The polygon ABCDE is regular, since the angles at the centre A″OB″, B″OC″, etc., being equal, their arcs which measure them are equal, and consequently the chords of these arcs are equal.

PROBLEM XI.

*Having given two regular polygons, the one inscribed and the other circumscribed, each of the same number of sides, to inscribe and circumscribe regular polygons of double the number of sides, and of half the number of sides.*

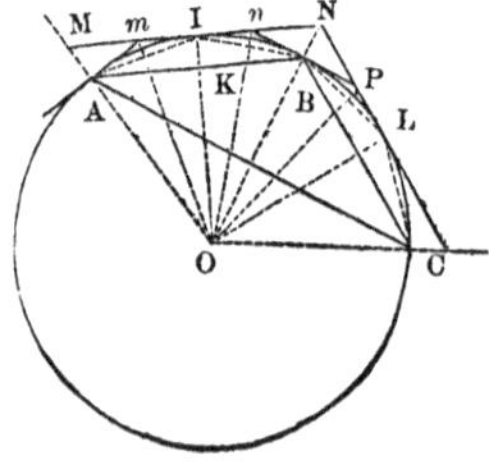


Let AB and MN be sides of the given polygons.

*First.* To obtain an inscribed polygon of double the number of sides, it is evidently sufficient to join the points A, B, with the point I, the middle of the arc AB, and to repeat the operation for each of the arcs BC, CD, etc.

The chords AI, IB, BL, etc., are equal, since they are subtended by equal arcs.

*Secondly.* We obtain the corresponding circumscribed polygon, by drawing through the points A, B, two tangents which terminate at the points $m$, $n$, on the tangent MN, and repeating the operation at the points C, D, etc.

The line $mn$ is the side of a circumscribed polygon of double the number of sides; and A$m$, $n$B are half sides.

In effect, the right-angled triangles OA$m$, OI$m$ are equal, having the common hypotenuse O$m$ and the side OA = OI; consequently A$m$ = I$m$, and the angle AO$m$ = angle IO$m$.

We see also that the angle $m$O$n$ is half the angle at the centre of each of the given polygons, and it is therefore the angle at the centre of the polygon sought.

*Thirdly.* As to the inscribed and circumscribed polygons of half the number of sides, we may obtain them by drawing through the alternate points A, C, E, etc., chords AC, CE, etc. And through the same points drawing tangents, having no regard to the intermediate points B, D, F, etc.

*Scholium.* It is evidently necessary in the third case, in order that the problem may be possible, to admit that the given polygons have an even number of sides.

---

## OF CONTACT

### PROBLEM XII.

*Through a given point without the circumference of a circle to draw a tangent to this circumference.*

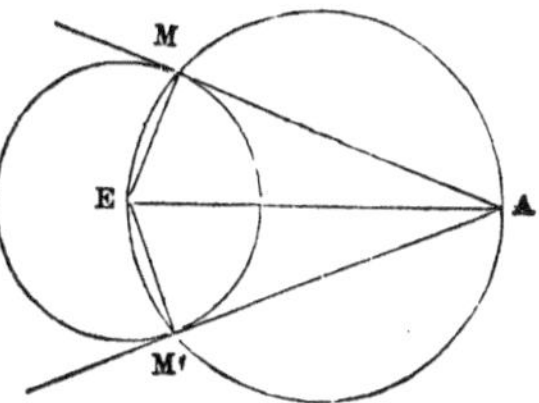


Let E be the centre of the given circle, and A the given point. Draw AE, and upon it, as a diameter, describe a circumference which will cut the given circumference in two points M, M′, AM, and AM′, being drawn, will be the tangents required. For, the angles EMA, EM'A being in a semicircle, are right (T., X. C. I.). Consequently, AM and AM' are respectively perpendicular to the radii EM and EM', and therefore tangent to the circumference (T. V.).

*Scholium.* The case when the given point is in the circumference offers no difficulty, since we then draw a radius to the given point, and through its extremity draw a perpendicular.

---

## OF COMMON MEASURE.

### PROBLEM XIII.

*To find the common measure of two given lines, provided they have one, and consequently their numerical ratio.*

Let AB and CD be the given lines.

From the greater AB cut off parts equal to the less line CD, as many times as possible; for example, twice, with the remainder FB.

From the line CD cut off parts equal to FB, as many times as possible; for example, once, with the remainder GD.

From the first remainder FB cut off parts equal to the remainder GD, as many times as possible; for example, once, with the remainder HB.

From the second remainder GD cut off parts equal to the third remainder HB, as many times as possible; for example, twice, without a remainder.

The last remainder, HB, will be a common measure of the given lines.

If we regard HB as a unit, GD will be 2, and

$$FB = FH + HB = GD + HB = 3;$$
$$CD = CG + GD = FB + GD = 3 + 2 = 5;$$
$$AB = AF + FB = 2\,CD + FB = 10 + 3 = 13.$$

Therefore the line AB is to the line CD as 13 to 5.

If AB is taken for the unit, CD will be $\frac{5}{13}$; but if CD be taken as the unit, AB will be $\frac{13}{5}$.

If AB is $\frac{1}{2}$ of a yard, then CD will be $\frac{5}{13}$ of $\frac{1}{2}$ a yard, or $\frac{5}{26}$ of a yard.

Again, if CD is $\frac{2}{3}$ of a foot, then AB will be $\frac{13}{5}$ of $\frac{2}{3}$ of a foot $= \frac{26}{15}$ of a foot; and so on for other comparisons.

*Cor.* Since arcs of the same circle, or of equal circles, are capable of superposition, this same method may be employed when we wish to find a *common measure of two arcs of the same circle, or of equal circles.*

*Scholium.* Magnitudes frequently have no common measure; that is, they are *incommensurable;* in which case, we shall always, by the foregoing method, have a remainder, however far we carry our successive operations. But these successive remainders would finally become so small that they might practically be neglected. Consequently, we can always find two numbers whose ratio shall approximate as close as we please to the ratio of the given incommensurable magnitudes.

This being admitted, we say that two incommensurable magnitudes A and B are proportional to two other incommensurable magnitudes A′ and B′, or that the ratio of A to B is equal to the ratio of A′ to B′, when we obtain for each the same numerical ratio at each successive step of the approximation.

### PROBLEM XIV.

*Show, geometrically, that the side of a square and its diagonal are incommensurable.*

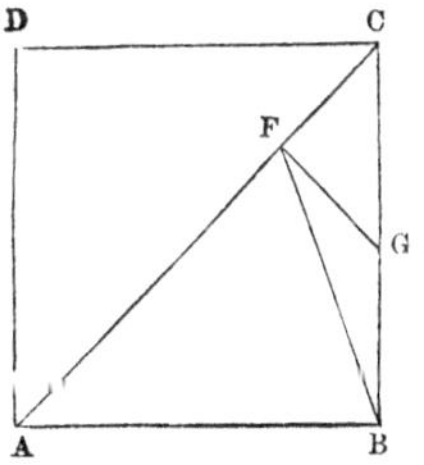

Let ABCD be the square, and AC its diagonal.

Cutting off AF from the diagonal equal to AB, a side of the square, we have the remainder CF, which must be compared with CB.

If we join FB, and draw FG perpendicular to AC, the triangle BGF will be isosceles.

For, the angle ABG = AFG, each being a right angle; and since the triangle ABF is isosceles, the angle ABF=AFB. Therefore, subtracting the angle ABF from ABG, the remainder FBG will equal the angle BFG, found by subtracting the angle AFB from AFG. Consequently the triangle BGF is isosceles, and BG = FG; but, since AC is the diagonal of a square, the angle FCG is half a right angle; but CFG is a right angle, and consequently FGC is also half a right angle, and CG is the diagonal of a square whose side is CF.

Hence, after CF = FG = BG has been taken once from CB, it remains to take CF from CG; that is, to compare the side of a square with its diagonal, which is the very question we set out with, and of course we shall find precisely the same difficulty in the next step of the process; so that, continue as far as we please, we shall never arrive at a term in which there will be no remainder. Therefore there is no common measure of the diagonal and side of a square.

# THIRD BOOK.

---

## THE PROPORTIONS OF STRAIGHT LINES AND THE AREAS OF RECTILINEAL FIGURES.

### DEFINITIONS.

I. *Two figures are similar* when they resemble each other in all their parts, differing only in their comparative magnitudes, so that every point of the one has a corresponding point in the other. If, in two similar figures, we conceive a line joining any two points of the one, and another line joining the corresponding points of the other, the numerical ratio of these lines, which is always the same, is called the *ratio of similitude.*

When the ratio of similitude is equal to unity, the two similar figures become equal in all respects.

II. *Two triangles are similar* when their corresponding sides are proportional.

III. *Two polygons are similar* when they are capable of being decomposed into the same number of similar triangles, each to each, having the same order of arrangement.

IV. In similar figures *homologous points, lines, or angles,* are those which have like positions, in regard to the two figures.

V. The *area* of a surface, or its superficial extent, is the numerical ratio of this surface referred to its *unit surface.* We readily see that figures of very different forms may have the same superficial extent. When we wish to express the property that two surfaces have the same area, without, however, being equal, or capable of superposition, we say they are *equivalent.*

VI. In determining the area of surfaces, it is found convenient to call the *bases* of a parallelogram the sides which are parallel. The *altitude* or *height* is the common perpendicular to these bases, of which one is called the *inferior* or *lower* base, and the other the *superior* or *upper* base.

In the rectangle, two consecutive sides form the *base* and the *height* respectively. In the square, the base and height are equal.

In the case of a triangle, either side may be regarded as the base. The height will be the perpendicular, drawn from the opposite angle to the base, or to the base produced.

---

## PROPORTIONAL LINES.

### THEOREM I.

*If equal portions are taken on a straight line, and through the points of division parallel lines are drawn meeting a second line, the intercepts of this second line will be equal also.*

If the portions CD, DE, and EF, of the straight line AB, are equal; and if through the points C, D, E, and F, the parallels CC′, DD′, EE′, and FF′, are drawn, meeting the line A′B′, then will the intercepts C′D′, D′E′, and EF′ be equal.

For, drawing C*c*, D*d*, and E*e*, parallel to A′B′, the triangles CD*c*, DE*d*, and EF*e* will be equal, since the sides CD, DE, and EF are equal, and their corresponding angles are obviously equal (B. I., T. III.). Hence, C*c* = D*d* = E*e*, and consequently C′D′= D′E′= E′F′ (B. I., T. XXXI., C. II.).

### THEOREM II.

*If a line is drawn parallel to the bases of a trapezoid, intersecting the sides, it will divide them into proportional parts.*

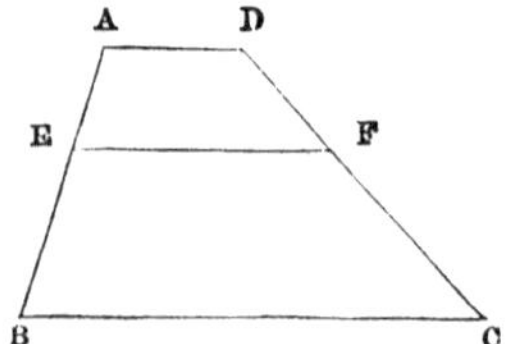


If EF is parallel to AD and BC, we shall have

AE : EB : : DF : FC.

*First.* When AE and EB are commensurable; that is, when they are to each other as two whole numbers, as 7 to 11 for instance, so that

$$AE : EB :: 7 : 11,$$

then will DF and FC be to each other in the same ratio.

For, if we conceive the straight line AB divided into $7+11=18$ equal parts, and through the points of division parallels to BC to be drawn, they will divide DC into 18 equal parts (T. I.), of which 7 will belong to DF, and 11 to FC. Hence,

$$DF : FC :: 7 : 11.$$

*Secondly.* When AE and EB are incommensurable.

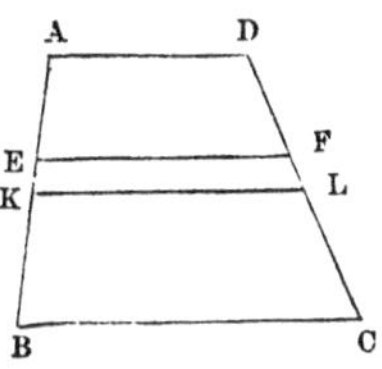


We may conceive AB divided into equal parts so small as to bring K, one of the points of division, as near as we please to E. Then drawing KL, we have, by the first case,

$$AK : KB :: DL : LC.$$

And as this will continue, however small the divisions are, we shall have (B. II., P. XIII., S.)

$$AE : EB :: DF : FC.$$

### THEOREM III.

*If a line be drawn parallel to either side of a triangle, cutting the other sides, it will divide them into proportional parts.*

Although this Theorem might have been made a corollary to the last, we have preferred making it one of our principal Theorems, not only on account of its own importance, but also by reason of the important corollaries which it is capable of furnishing.

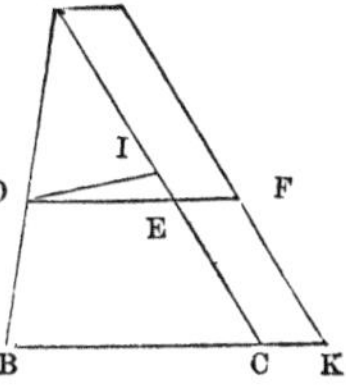


Let DE be drawn parallel to the side BC of the triangle ABC, cutting the other sides at D and E; then we shall have

$$AD : DB :: AE : EC.$$

Through A draw AG parallel to BC, and through any point, as G, of this parallel, draw

GK parallel to AC meeting BC produced. Also produce DE to F. We have (T. II.)

$$AD : DB :: GF : FK;$$

but, by reason of the parallels AC, GK, we have $GF = AE$, $FK = EC$; consequently,

$$AD : DB :: AE : EC.$$

*Cor.* I. *Conversely.* If a straight line DE divide two sides of a triangle into proportional parts, it will be parallel to the third side. For, if this is not the case, we might draw through D another straight line DI parallel to BC, and we should then have

$$AD : DB :: AI : IC;$$

but by hypothesis,

$$AD : DB :: AE : EC;$$

hence, by equality of ratios, we have

$$AI : IC :: AE : EC,$$

which is absurd, since $AI < AE$, and $IC > EC$.

*Cor.* II. From the proportion

$$AD : DB :: AE : EC$$

we have, by *composition*,

$$AD + DB : AD :: AE + EC : AE,$$
$$AD + DB : DB :: AE + EC : EC,$$

which gives these new propositions:

$$AB : AD :: AC : AE \text{ and } AB : DB :: AC : EC.$$

*Conversely.* If a straight line DE is so drawn as to give

$$AB : AD :: AC : AE,$$

we shall have, by *division*,

$$AB - AD : AD :: AC - AE : AE,$$

that is, $DB : AD :: EC : AE,$

which is the same as $AD : DB :: AE : EC;$

consequently, the straight line DE is parallel to BC (C. I.).

*Scholium.* Since, in every proportion, the product of the extremes is equal to the product of the means, we derive from the above proportion, $AD : DB :: AE : EC$, this equation, $AD \times EC = DB \times AE$.

Now, in order to comprehend the sense which we ought to attribute to the expression, *the product of two lines*, AD × EC, or DB × AE, it is necessary to suppose that the straight lines AD, EC, DB, and AE, have been referred to the same *linear unit;* and instead of lines, properly speaking, we have only to consider, in reality, *abstract numbers* expressing the ratio of these lines to their unit. We ought then, as in Arithmetic, to attach to the word *product* the idea of the multiplication of two *numbers.*

Hereafter, we may be required to multiply *a surface by a line*, *a surface by a surface*, and even *a volume by a volume*, etc. But in all these cases it must be understood that it is the ratios of these geometrical magnitudes to their respective units, which are thus multiplied.

---

## SIMILAR TRIANGLES.

### THEOREM IV.

*A straight line drawn parallel to either side of a triangle, intersecting the other two sides, will form a new triangle similar to the first.*

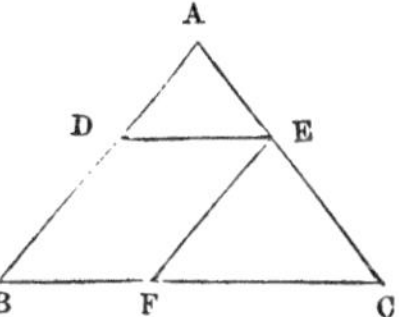


If DE is parallel to BC, we shall have

AB : AD : : AC : AE : : BC : DE.

For, since DE is parallel to BC, we have (T. III., C. II.)

AB : AD : : AC : AE.

Drawing EF parallel to AB, we also have

AC : AE : : BC : BF or DE;

hence, we have, by equality of ratios,

AB : AD : : AC : AE : : BC : DE.

Consequently, these triangles are similar (D. II.).

### THEOREM V.

*Two similar triangles, that is to say, which have their sides proportional, have their homologous angles equal, each to each.*

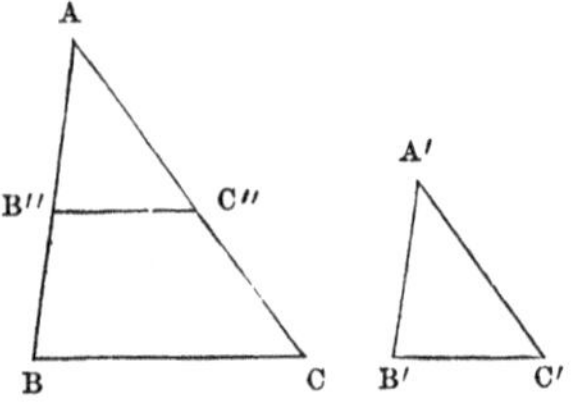


Let ABC and A′B′C′ be the two similar triangles.

Take AB″=A′B′, and draw B″C″ parallel to BC; the two triangles A′B′C′, AB″C″, being each similar to ABC, the one by hypothesis, and the other by the preceding Theorem, are similar to each other, and as we have, by construction, AB″= A′B′, these triangles must be equal, since the ratio of similitude is unity (D. I.). Now, since B″C″ is parallel to BC, the angles of the triangle AB″C″ are respectively equal to the angles of the triangle ABC; that is to say, we have the angle at A common, B″=B, C″=C. Then, also, we must have A′=A, B′=B, C′=C.

We see that the *equal angles* are opposite the *homologous sides.*

### THEOREM VI.

*Two triangles, which have the angles of the one respectively equal to the angles of the other, are similar, and the sides opposite the equal angles are homologous.*

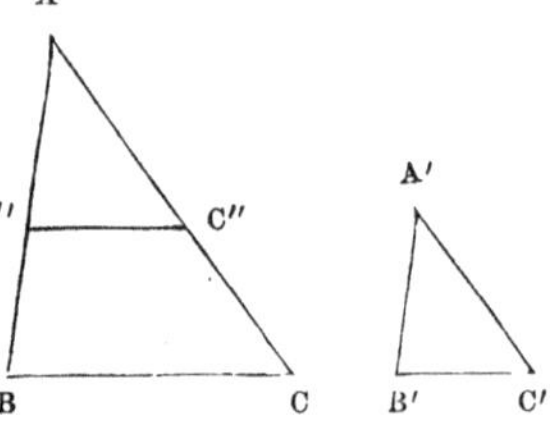


In the two triangles ABC, A′B′C′, suppose A=A′, B=B′, C=C′.

As in the preceding Theorem, take AB″=A′B′ and draw B″C″ parallel to BC. The two triangles A′B′C, AB″C″, are equal, having a side equal, A′B′=AB″, and the adjacent angles equal, A′=A, B′=B=B″. Now the triangle AB″C″ is similar to the triangle ABC (T. IV.); hence its equal A′B′C′ will be similar to ABC. Consequently we have this series of equal ratios:

$$AB : A'B' :: AC : A'C' :: BC : B'C'.$$

The *homologous* sides, AB and A′B′, AC and A′C′, BC and B′C′, as we see, are *opposite* the equal angles $C' = C$, $B' = B$, $A' = A$.

THEOREM VII.

*Two triangles are similar, when the sides of the one are respectively parallel or perpendicular to the sides of the other, and the homologous sides are the parallel or the perpendicular sides.*

From the preceding Theorem we see that the demonstration will be made out if we can show that these triangles have their angles respectively equal.

Let AB and A′B′, AC and A′C′, BC and B′C′, represent the respective sides, either parallel or perpendicular, we are to show that $C = C'$, $B = B'$, $A = A'$.

In effect, we know (B. I., T. III., C.) that the angles C and C′, B and B′, A and A′, must be either equal or supplementary.

Now, in the first place, it is impossible that three set of angles should be supplementary, since we should then have

$$A + A' + B + B' + C + C' = 6 \text{ right angles},$$

which is absurd (B. I., T. V., C. I.).

Neither can two set of these angles be supplementary; as, for example, A and A′, B and B′, for then we should have

$$A + A' + B + B' = 4 \text{ right angles},$$

which is also absurd.

It is then necessary that two angles, at least, of the first triangle should be equal respectively to two angles of the second; consequently, the third angle of the first triangle must equal the third angle of the second (B. I., T. V., C. III.). So that we shall have

$$A = A', B = B', C = C';$$

and consequently the triangles are similar, and we have this series of equal ratios (T. VI.):

$$BC : B'C' :: AC : A'C' :: AB : A'B'.$$

From the method of demonstration we see that the *homologous* sides BC and B′C′, AC and A′C′, AB and A′B′, are the *parallel* or the *perpendicular* sides.

THEOREM VIII.

*Two triangles, which have an angle of the one equal to an angle of the other, and the sides containing these angles proportional, are similar.*

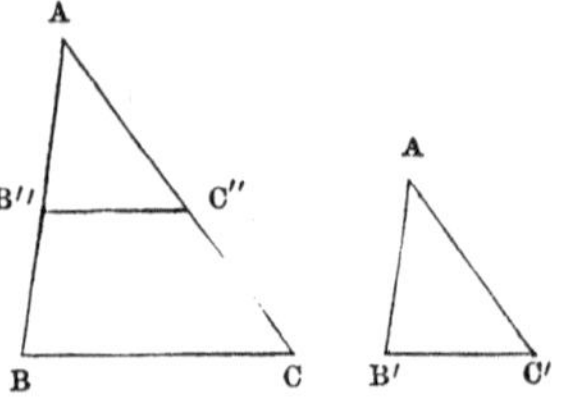


In the two triangles ABC, A'B'C', suppose we have A = A', and

AB : A'B' : : AC : A'C'.

Take on AB and AC, AB'' = A'B', AC'' = A'C', and draw B''C''. The two triangles A'B'C', AB''C'', are equal, having an equal angle included between equal sides each to each. Now, by hypothesis, we have

AB : A'B' : : AC : A'C',

consequently, AB : AB'' : : AC : AC'';

and B''C'' is parallel to BC (T. III., Cor. II.), and the triangle AB''C'' is therefore similar to ABC (T. IV.). Then, also, is A'B'C' similar to ABC.

---

## SIMILAR POLYGONS.

THEOREM IX.

*Two similar polygons have their homologous sides proportional, and the homologous angles respectively equal.*

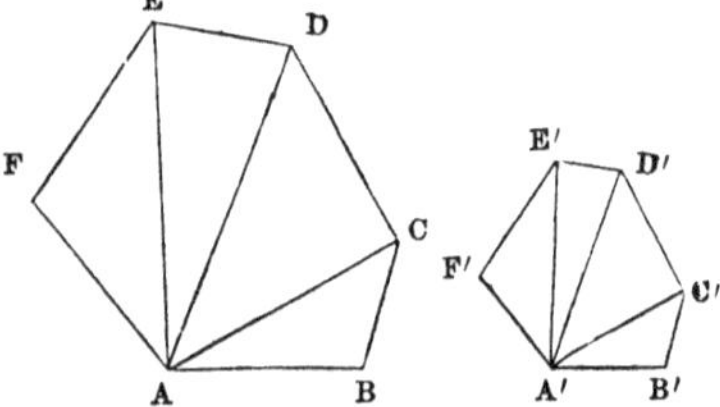


For, by the similarity of the triangles ABC and A'B'C', ACD and A'C'D', ADE and A'D'E', etc., we deduce the following series of equal ratios:

AB : A'B' : : AC : A'C' : : BC : B'C' : :, etc.
AC : A'C' : : AD : A'D' : : CD : C'D' : :, etc.
AD : A'D' : : AE : A'E' : : DE : D'E' : :, etc.
&c. &c. &c.

Omitting the ratios which are common in these series, we have

AB : A′B′ : : BC : B′C′ : : CD : C′D′ : : DE : D′E′ : :, etc.

Again, from the similarity of these same triangles, it follows that their homologous angles are equal (T. V.), and consequently the respective angles of the two polygons are equal, since they are composed of angles equal each to each. Thus, A = A′, B = B′, C = C′, D = D′, etc.

### THEOREM X.

*Two polygons are similar when they have their sides, taken in the same order, proportional, and their angles, also taken in the same order, equal each to each.*

That is to say, when we have

AB : A′B′ : : BC : B′C′ : : CD : C′D′ : :, etc.,

and A = A′, B = B′, C = C′, D = D′, etc.

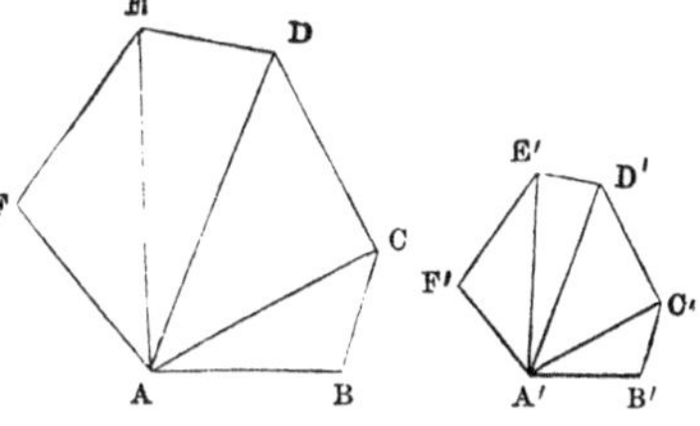


Let the two polygons be decomposed into triangles by drawing lines from A and A′.

The two triangles ABC, A′B′C′, are similar, having equal angles, B = B′, included between proportional sides AB, A′B′, BC, B′C′ (T. VIII.). Hence, angle ACB = angle A′C′B′, and

AC : A′C′ : : BC : B′C′.

But, by hypothesis, we have

BC : B′C′ : : CD : C′D′.

Consequently, AC : A′C′ : : CD : C′D′.

Now, comparing the two triangles ACD, A′C′D′, we see that the angles ACD, A′C′D′ are equal, being the differences between the equal angles BCD, B′C′D′, and ACB, A′C′B′, and we have just shown that the sides about these equal angles are proportional, hence these triangles are also similar. The same may be shown for all the couples of triangles, ADE and A′D′E′, AEF and A′E′F′. Hence the polygons are similar (D. III.).

6

*Scholium. Two parallelograms are similar, when they have an angle of the one equal to an angle of the other, and the sides containing the equal angles proportional.*

*Two rhombuses are similar, when they have an angle of the one equal to an angle of the other.*

*All squares are similar figures.*

*All regular polygons of the same number of sides are similar figures.*

---

## PROPORTIONAL LINES.—PROPERTIES OF THE SIDES OF TRIANGLES.

### THEOREM XI.

*The intercepted portions of two parallel lines, made by any number of lines proceeding from the same point, are proportional.*

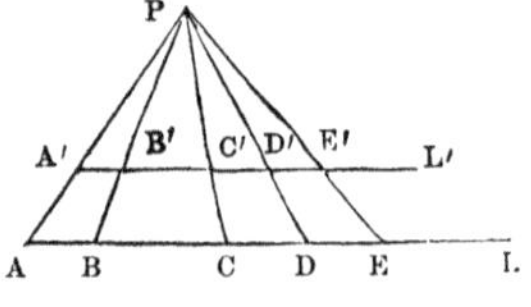


The comparison of the different couples of similar triangles PAB and PA′B′, PBC and PB′C′, etc., immediately gives the following series of equal ratios:

$$\begin{aligned} PA : PA' &:: AB : A'B' :: PB : PB', \\ PB : PB' &:: BC : B'C' :: PC : PC', \\ PC : PC' &:: CD : C'D' :: PD : PD', \\ \&c. \quad & \quad \&c. \quad \quad \&c. \end{aligned}$$

Now, all the ratios are equal, however far they may be extended, since the last in each line above is the first of the following line. Then, if we use only the intermediate ratios, we shall have

$$AB : A'B' :: BC : B'C' :: CD : C'D' :: DE : D'E' ::, \text{ etc.},$$

which establishes the proposition.

When the point P is within the parallels, the same demonstration will apply. In this case, the corresponding segments of the two lines AL and A′L′ will be situated in opposite directions.

THEOREM XII.

*If a line be drawn bisecting either angle of a triangle, it will divide the opposite side into two segments proportional to their adjacent sides.*

If the angle BAC is bisected by AD, we shall have

AB : AC : : BD : CD.

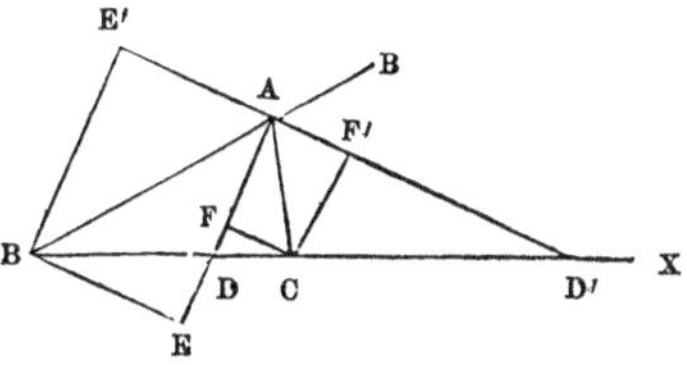


Draw BE and CF perpendicular to AD. The two triangles ABE, ACF are similar, since they are equiangular; hence

AB : AC : : BE : CF.

But the two triangles BDE, CFD are also similar, and give

BE : CF : : BD : CD.

Consequently, by equality of ratios, we have

AB : AC : : BD : CD.

*Cor.* The line AD′, bisecting the supplementary angle at A, that is to say, the line perpendicular to AD, will also determine, on the prolongation of the side BC, two segments BD′, CD′, such as to give the proportion,

AB : AC : : BD′ : CD′.

For, drawing BE′ and CF′ perpendicular to AD′, we have, by reason of similar triangles,

AB : AC : : BE′ : CF′; BE′ : CF′ : : BD′ : CD′;

consequently, AB : AC : : BD′ : CD′.

*Scholium.* From the two proportions,

AB : AC : : BD : CD,
AB : AC : : BD′ : CD′,

just demonstrated, we deduce the following:

BD : CD : : BD′ : CD′;

and the straight line BC is said to be divided *harmonically* at the points D and D′.

### THEOREM XIII.

*If from the right angle of a right-angled triangle a line is drawn perpendicular to the hypotenuse:*

I. *The two partial triangles thus formed will be similar to the whole triangle, and consequently similar to each other.*

II. *The perpendicular will be a mean proportional between the two segments of the hypotenuse.*

III. *Either side containing the right angle will be a mean proportional between its adjacent segment and the hypotenuse.*

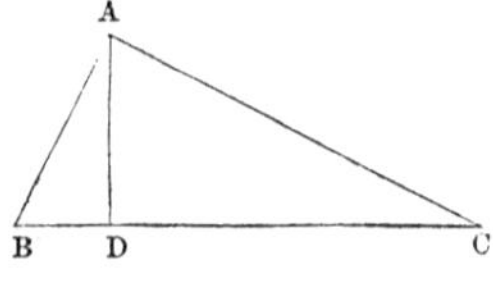


*First.* The two triangles ABC, ABD are similar, since they have the common angle at B, and the angle BAC of the one equal to the angle BDA of the other, each being a right angle. For the same reason, ACB is similar to ACD, consequently these triangles are similar to each other.

*Secondly.* Comparing the two partial triangles ABD, ACD, we have

$$BD : AD :: AD : DC.$$

*Thirdly.* Comparing the total triangle ABC with its partial triangle ABD, observing that in these triangles BC and AB are homologous, being the hypotenuses, that AB and BD are also homologous, being opposite the equal angles BCA, BAD, we shall obtain the proportion,

$$BC : AB :: AB : BD.$$

The comparison of the triangles ABC, ACD will give, in like manner,

$$BC : AC :: AC : DC.$$

### THEOREM XIV.

*In any right-angled triangle, the square, or second power, of the numerical value of the hypotenuse, is equal to the sum of the squares of the numerical values of the other two sides.*

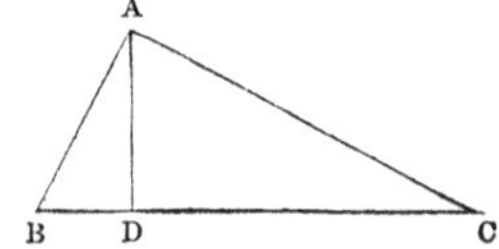


For, taking the product of the means and extremes of the last two proportions of the preceding Theorem, we have

$$AB^2 = BC \times BD\,;\ AC^2 = BC \times DC\,;$$

from which, by adding member to member,

$$AB^2 + AC^2 = BC \times (BD + DC) = BC \times BC\,;$$

that is, $$AB^2 + AC^2 = BC^2.$$

*Cor.* When $AB = AC$, we have $BC^2 = 2AB^2$; that is, in any right-angled isosceles triangle, the square of the hypotenuse is double the square of one of the sides. Consequently, the second power of the numerical value of the diagonal of a square is double that of the side. And the diagonal and side must be to each other in the ratio of $\sqrt{2}$ to 1; that is, they are incommensurable, as has already been shown (B. II., P. XIV.).

*Scholium* I. From the two equations,

$$AB^2 = BC \times BD,\ AC^2 = BC \times DC,$$

we deduce the proportion,

$$AB^2 : AC^2 :: BC \times BD : BC \times DC, \text{ or, } :: BD : DC.$$

In a similar manner, by combining the identical equation $BC^2 = BC^2$, with the same two equations above, we have

$$BC^2 : AB^2 :: BC \times BC : BC \times BD, \text{ or, } :: BC : BD;$$

and $$BC^2 : AC^2 :: BC \times BC : BC \times DC, \text{ or, } :: BC : DC.$$

From this we see that, in any right-angled triangle, the squares or second powers of the sides containing the right angle and of the hypotenuse, are proportional to the segments of this hypotenuse, and the hypotenuse itself. That is to say, we have

$$AB^2 : AC^2 : BC^2 :: BD : DC : BC.$$

*Scholium* II. The segments BD, DC, formed by the perpendicular AD on the hypotenuse BC, are called the *projections* of the two sides on the hypotenuse.

In general, *the projection* of any definite straight line MN, on another indefinite straight line AB, is the distance PQ of the intersection of the perpendiculars drawn through the extremities of the first line to the second.

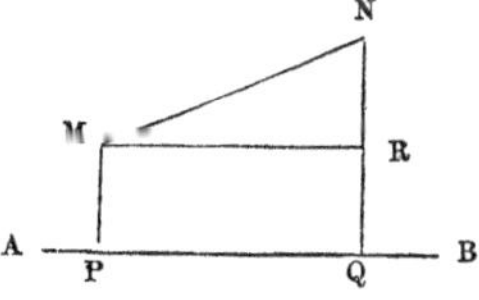


If through M we draw MR parallel to PQ, we shall have $MR = PQ$, and the right-angled triangle MNR will give

$$MN^2 = MR^2 + NR^2 = PQ^2 + (NQ - MP)^2.$$

That is—*The square of a straight line of determinate length is equal to the square of its projection on another line, plus the square of the difference of the perpendiculars which determine this projection.*

### THEOREM XV.

*In any obtuse-angled triangle, the square of the side opposite the obtuse angle is equal to the sum of the squares of the other two sides, increased by twice the product of either of the sides containing the obtuse angle into the projection of the other side on the prolongation of the first.*

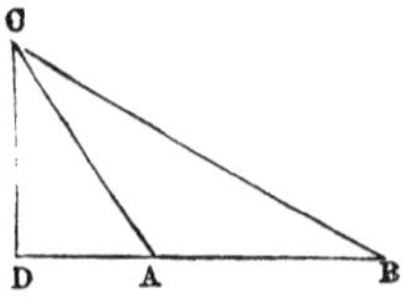

If the triangle ABC is obtuse-angled at A, we shall have

$$BC^2 = AB^2 + AC^2 + 2AB \times AD.$$

For, since $BD = AB + AD$, if we square both members of this equation, using the algebraic formula, we shall have

$$BD^2 = AB^2 + AD^2 + 2AB \times AD.$$

Adding now, to each member of this, $CD^2$, and observing that $BD^2 + CD^2 = BC^2$, and also that $AD^2 + CD^2 = AC^2$, we shall obtain

$$BC^2 = AB^2 + AC^2 + 2AB \times AD.$$

### THEOREM XVI.

*In any triangle, the square of the side opposite an acute angle is equal to the sum of the squares of the other two sides, diminished by twice the product of one of these sides, by the projection of the other on the preceding one, produced if necessary.*

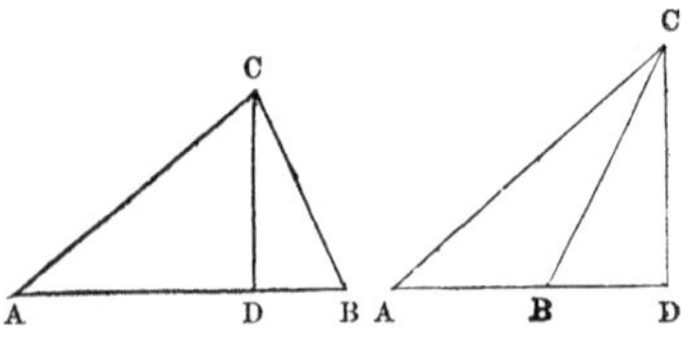

If the angle A is acute, we shall have

$$BC^2 = AB^2 + AC^2 - 2AB \times AD.$$

For, $BD = AB - AD$, or $AD - AB$, according to which diagram we use; but in both cases BD is the difference of AB and AD; consequently, if we take the squares of both members, by the well-known algebraic formula, we shall have

$$BD^2 = AB^2 + AD^2 - 2AB \times AD.$$

Adding to each member of this $DC^2$, and observing that $BD^2 + CD^2 = BC^2$, and also that $AD^2 + DC^2 = AC^2$, we shall obtain

$$BC^2 = AB^2 + AC^2 - 2AB \times AD.$$

This result will hold good even when the perpendicular CD coincides with the side CB, in which case $AD = AB$, and the above will become $BC^2 = AC^2 - AB^2$, or, $AC^2 = AB^2 + BC^2$, which we know to be true, since the triangle ABC is, in this case, right-angled at B (T. XIV.).

*Scholium.* A triangle is *right-angled*, *acute-angled*, or *obtuse-angled*, according as the square of the longest side is *equal*, *less*, or *greater* than the sum of the squares of the other two sides.

If a triangle has 3, 4, and 5 for the numerical values of its sides, it will be right-angled, since

$$5^2 = 3^2 + 4^2.$$

One having 4, 5, and 6 for its sides, is acute-angled, since

$$6^2 < 4^2 + 5^2.$$

One having 2, 3, and 5 for its sides, is obtuse-angled, since

$$5^2 > 2^2 + 3^2.$$

Thus we are furnished with a second very simple method for determining the kind of triangle, when its sides are given (B. I., T. XXXV., S.).

### THEOREM XVII.

*In any triangle, the sum of the squares of any two sides is equal to twice the square of half the third side, increased by twice the square of the line drawn from the middle of this third side to the opposite angle.*

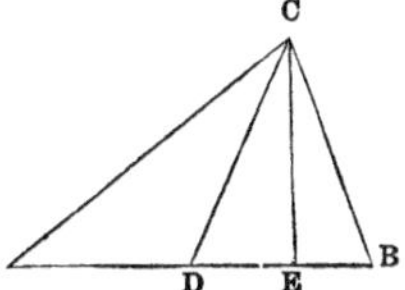


If CD is drawn bisecting AB, we shall have

$$AC^2 + BC^2 = 2AD^2 + 2CD^2.$$

For, the two triangles ADC, CDB, the

one obtuse-angled and the other acute-angled at D, give, by the two preceding Theorems,

$$AC^2 = AD^2 + CD^2 + 2AD \times DE,$$
$$BC^2 = BD^2 + CD^2 - 2DB + DE;$$

adding these, and observing that $AD = DB$, we have

$$AC^2 + BC^2 = 2AD^2 + 2CD^2.$$

*Cor. The sum of the squares of the four sides of any parallelogram is equal to the sum of the squares of its diagonals.*

This proposition is easily deduced from the above.

---

## DETERMINATION OF AREAS.

### THEOREM XVIII

*Two parallelograms having equal bases and the same altitude, are equivalent.*

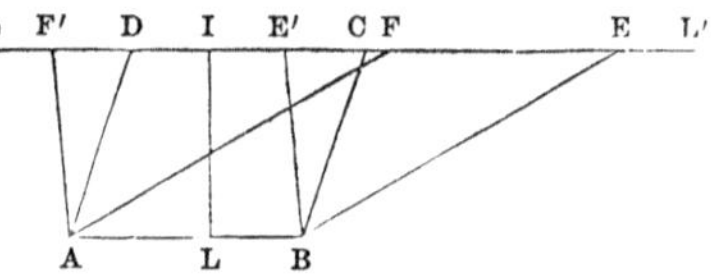


We will suppose the one figure placed upon the other, so that their inferior bases, which are equal, may coincide. Let ABCD be the first parallelogram, and ABEF or ABE′F′ the second. Since these parallelograms have the same altitude, their superior bases will be situated in the same indefinite line LL′, parallel to the lower base.

If now, we confine our attention to the two parallelograms ABCD, ABEF, we see that the triangles BCE, ADF are equal, having equal angles CBE and DAF (B. I., T. III.) included between equal sides, namely, $BE = AF$, $BC = AD$ (B. I., T. XXXI.). If, now, from the quadrilateral ABED we subtract the triangle BEC, we shall have left the parallelogram ABCD; but if from the same quadrilateral we subtract the equal triangle AFD, we shall have left the parallelogram ABEF. Hence the two parallelograms are equivalent.

We may in the same way prove that the parallelograms ABCD, ABE′F′ are equivalent.

*Cor.* A parallelogram is equivalent to a rectangle having the same base and the same altitude. Two rectangles of the same base and the same altitude are *equal*, and consequently *equivalent.*

THEOREM XIX.

*A triangle is equivalent to half of a parallelogram having the same base and the same altitude.*

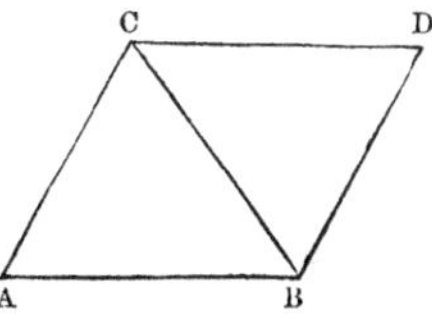


For, through B and C, draw BD parallel to AC, and CD parallel to AB; we thus form a parallelogram having the same base and the same altitude as the triangle.

The triangle is one half this parallelogram (B. I., T. XXXI., Cor I.).

*Cor.* Two triangles of the same base, or of equal bases, and the same altitude, are equivalent.

THEOREM XX.

*Two rectangles of the same altitude are to each other as their bases.*

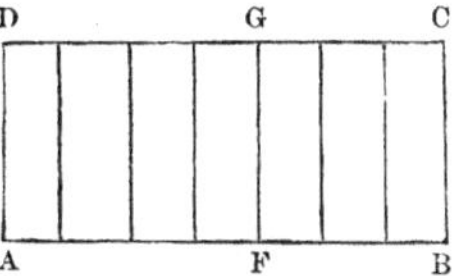


Let ABCD, AFGD be the two rectangles having the common altitude AD; then will they be to each other as their bases AB, AF.

We will first suppose the bases AB, AF to be commensurable; as, for example, suppose they are to each other as 7 to 4. If we divide AB into seven equal parts, AF will contain four of these parts; at each point of division draw lines perpendicular to the base, thus forming seven partial rectangles, which will be equal, since they have equal bases and the same altitude (T. XVIII.).

Now, as the rectangle ABCD contains seven of these equal rectangles, while AFGD contains only four, it follows, that

ABCD : AFGD : : 7 : 4;

but AB : AF : : 7 : 4;

therefore, ABCD : AFGD : : AB : AF.

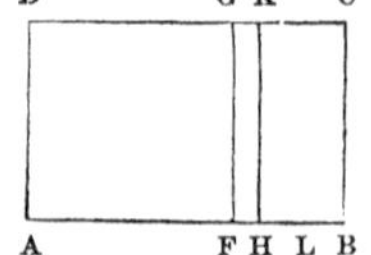


We will suppose, in the second place, that the bases AB, AF are incommensurable, still we shall have

ABCD : AFGD : : AB : AF.

For, if this proportion is not true, the first three terms remaining the same, the fourth term will be either greater or less than AF. Suppose it to be greater, and that we have

ABCD : AFGD : : AB : AL.

Divide the line AB into equal parts, each of which shall be less than FL. There will be at least one point of division, as at H, between F and L. Through this point H draw the perpendicular HK, then will the bases AB and AH be commensurable· and we shall have, by the first part of this proposition,

ABCD : AHKD : : AB : AH.

But, by supposition,

ABCD : AFGD : : AB : AL.

Since, in these two proportionals, the antecedents are equal, the consequents are proportional, and we have

AHKD : AFGD : : AH : AL.

But AL is greater than AH; hence, if this proportion is correct, we must have AFGD greater than AHKD; on the contrary, it is less, hence the above proportion is impossible. Therefore, ABCD cannot be to AFGD as AB is to a line greater than AF.

By similar reasoning, we can show that the fourth term of the proportion cannot be smaller than AF.

Hence, whatever may be the ratio of the bases, the two rectangles ABCD, AFGD, of the same altitude, will be to each other as their bases AB, AF.

### THEOREM XXI.

*Any two rectangles are to each other as the product of their bases multiplied by their altitudes.*

Let ABCD, AFGH be two rectangles, then will

ABCD : AFGH : : AB × AD : AH × AF.

Having placed the two rectangles so that the angles at A may be vertical and opposite, produce the sides CD, GF until they meet in K; the two rectangles ABCD, AFKD, having the same altitude AD, are to each other as their bases AB, AF. In like manner, the two rectangles AFKD, AFGH, having the same altitude AF, are to each other as their bases, AD, AF. Hence we have these two proportions:

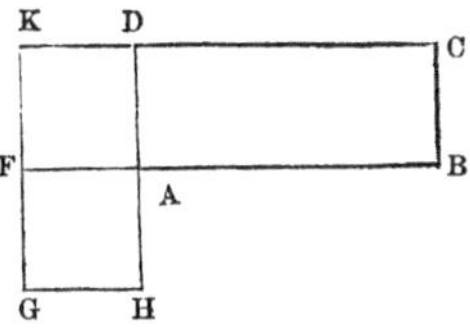


$$ABCD : AFKD :: AB : AF.$$
$$AFKD : AFGH :: AD : AH.$$

Multiplying the corresponding terms of these proportions together, and observing to omit AFKD, since it will occur in an antecedent and consequent, we shall have

$$ABCD : AFGH :: AB \times AD : AH \times AF.$$

*Scholium.* Hence we are at liberty to take for the measure of a rectangle, the product of its base by its altitude, provided we understand by this product the same as the product of two numbers, which numbers denote the linear units in the base and altitude respectively.

This measure, however, is not absolute, but only relative. It supposes that the area of any other rectangle is estimated in a similar manner, by measuring its sides by the same linear unit; we shall thus obtain a second product, and the ratio of these two products is the same as that of the two rectangles, in accordance with this proposition.

For example, if the base of the rectangle A is ten units, and its height three, the rectangle will be represented by the number $10 \times 3 = 30$, a number which, thus isolated, has no signification; but if we have a second rectangle B, whose base is twelve units and height seven, this second rectangle will be represented by the number $12 \times 7 = 84$. From which we conclude that the two rectangles A and B are to each other as 30 to 84. If we take the rectangle A as the unit of measure of surfaces, the rectangle B will have for its measure $\frac{84}{30}$; that is, it will be $\frac{84}{30}$ of our superficial units.

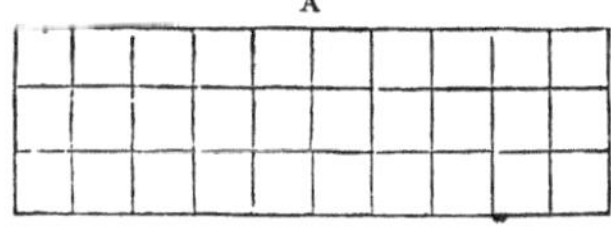

It is more common and more simple to take a square for the unit of surface, and we choose a square whose side is a unit of length. In this case, the measure which we have regarded as relative becomes absolute; for example, the number 30, which measured the rectangle A, represents 30 superficial units, or 30 squares, each side of which is a unit long.

In geometry, we frequently confound the product of two lines with that of their *rectangle*, and this expression is even employed in arithmetic to denote the product of two unequal numbers; but we use the term *square* to denote the product of a number multiplied by itself.

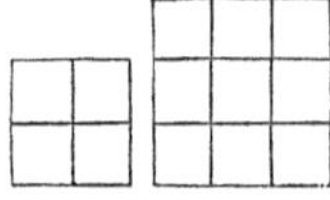

The squares of the numbers 1, 2, 3, etc., are 1, 4, 9, etc. From which we see that the square formed on a line of double length is quadruple; on a line of triple length it is nine times as great, and so of other squares.

### THEOREM XXII.

*The area of any parallelogram is equal the product of its base by its height.*

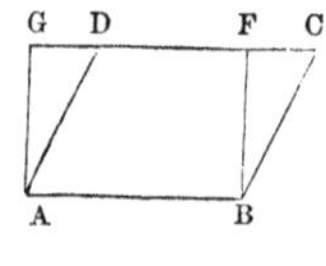


For the parallelogram ABCD is equivalent to the rectangle ABFG, which has the same base AB, and same altitude BF (T. XVIII., C.). But the rectangle ABFG is measured by AB × BF, consequently the parallelogram is measured by AB × BF.

*Cor.* Parallelograms of the same base are to each other as their altitudes, and parallelograms of the same altitude are to each other as their bases; and in general, parallelograms are to each other as the products of their bases multiplied by their altitudes.

### THEOREM XXIII.

*The area of a triangle is equal to the product of its base by half its altitude.*

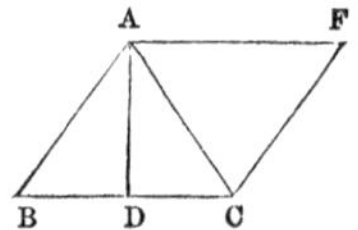


For, the triangle ABC is half the parallelogram ABCF (T. XIX.).

The parallelogram is measured by the base BC multiplied into the altitude AD; therefore, the triangle is measured by the base BC into half the altitude AD.

*Cor.* I. Triangles of the same altitude are to each other as their bases; and triangles of the same or equal bases are to each other as their altitudes.

*Cor.* II. The area of a trapezoid ABDC is equal to the product of half the sum of the bases, AB, CD, into the altitude IK; or, it is equal to the product IK into the straight line EF drawn at equal distance between the bases.

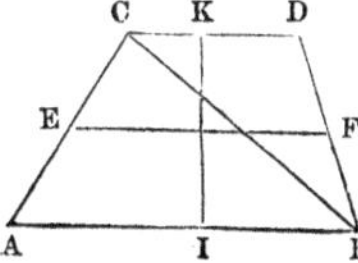


In effect this trapezoid is composed of two triangles CAB, BCD; and by the Theorem, we have

$$ABC=\frac{AB \times IK}{2},\ BCD=\frac{CD \times IK}{2};$$

then $$ABDC=\frac{AB \times IK}{2}+\frac{CD \times IK}{2}=\frac{AB+CD}{2} \times IK.$$

And as we have proved (B. I., T. XXXIII.) that $EF=\frac{AB+CD}{2}$, we have $ABDC=IK \times EF$.

THEOREM XXIV.

*The area of a triangle is equal to half the product of its perimeter by the radius of its inscribed circle.*

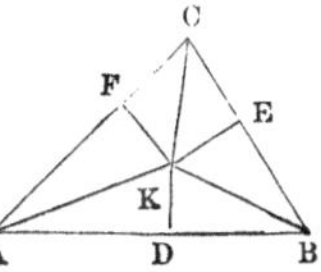


Let K be the centre of the inscribed circle. Draw KD, KE, KF respectively perpendicular to the sides AB, BC, AC. Also draw KA, KB, KC.

We obviously have

$$ABC=AKB+BKC+AKC.$$

But (T. XXIII.) we have

$$AKB=\frac{AB \times KD}{2},\ BKC=\frac{BC \times KE}{2},\ AKC=\frac{AC \times KF}{2}.$$

And since $KD=KE=KF$, we have

$$ABC=\frac{(AB+BC+AC) \times KD}{2}.$$

### THEOREM XXV.

*The area of a triangle is equal to the product of the three sides divided by twice the diameter of its circumscribed circle.*

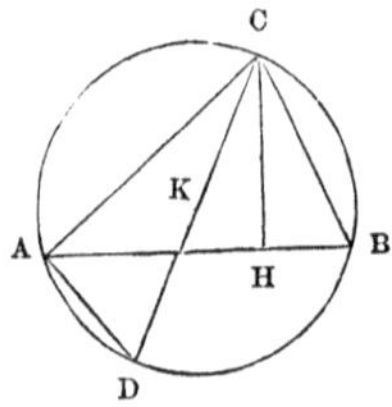


Let K be the centre of the circumscribed circle. Draw the diameter CKD, and the chord AD, and the perpendicular CH, which will be the altitude of the triangle.

The two triangles CAD, CHB are right-angled, the one at A (B. II., T. X., Cor. I.), and the other at H; moreover, the angles at D and B are equal, being inscribed in the same segment ADBC. Hence these triangles are similar, and their homologous sides being proportional, we have

$$CD : CB :: AC : CH.$$

Consequently, $$CH = \frac{CB \times AC}{CD}.$$

But we also have $$ABC = \frac{AB \times CH}{2},$$

in which, substituting the value of CH just found, we have

$$ABC = \frac{AB \times CB \times AC}{2CD}.$$

---

## RATIOS BETWEEN THE AREAS OF SIMILAR FIGURES

### THEOREM XXVI.

*The areas of two triangles which have an equal angle, are proportional to the rectangles of the sides containing the equal angle.*

Referring to either figure, let the two triangles ABC, ADE have the angle at A common; then we shall have

$$ABC : ADE :: AB \times AC : AD \times AE.$$

For, joining D and C, we have, since triangles of the same altitude are to each other as their bases,

$$ABC : ADC :: AB : AD,$$
$$ADC : ADE :: AC : AE.$$

Multiplying together the corresponding terms of these proportions, and omitting the common term ADC which enters into the antecedent and consequent of the first couplet, we have

$$ABC : ADE :: AB \times AC : AD \times AE.$$

*Cor.* Whenever the side DE of the second triangle is parallel to BC of the first, the triangle ADC is a mean proportional between the triangles ABC and ADE.

In effect we have, in this case (T. III.),

$$AB : AD :: AC : AE;$$

so that the two first proportions of the preceding Theorem have equal ratios, and give

$$ABC : ADC :: ADC : ADE.$$

THEOREM XXVII.

*The areas of similar triangles are proportional to the squares of their homologous sides.*

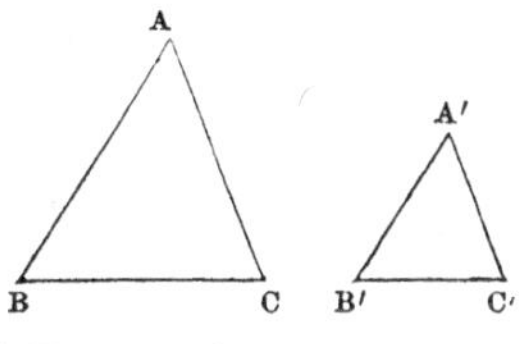

These triangles being similar, are equiangular (T. V.).

Hence, by the preceding Theorem, we have

A A′ B C B′ C′

$$ABC : A'B'C' :: AB \times AC : A'B' \times A'C';$$

but since the triangles are similar, we also have

$$AC : AC' :: AB : A'B'.$$

Multiplying the corresponding terms of these two proportions, omitting the common factor AC which enters into the antecedents, and the common factor A′C′ which enters into the consequents, we have

$$ABC : A'B'C' :: AB^2 : A'B'^2.$$

THEOREM XXVIII.

*The perimeters of two similar polygons are proportional to their homologous sides. And their areas are proportional to the squares of these sides.*

*First.* By reason of the similarity of these polygons, we have (T. IX.),

$$AB : A'B' :: BC : B'C' :: CD : C'D' ::, \text{etc.}$$

Now the sum of these antecedents, AB + BC + CD +, etc., which makes the perimeter of the first polygon, is to the sum of their consequents A′B′ + B′C′ + C′D′ +, etc., which makes the perimeter of the second polygon, as any one antecedent is to its corresponding consequent, and therefore as AB is to A′B′.

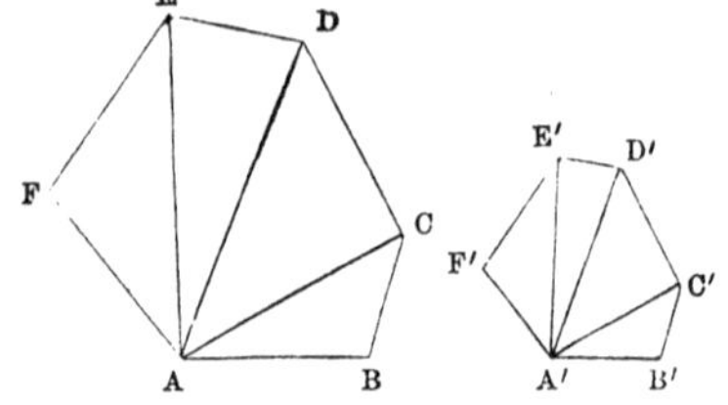


*Secondly.* Since these polygons are similar, they are each composed of the same number of similar triangles, which, compared together, are each to each, in the same ratio, that of the squares of the homologous sides. It follows, then, that the sum of the triangles which compose the first polygon, is to the sum of the triangles which compose the second polygon, as the squares of the homologous sides of the two polygons. That is, we shall have

$$ABCDEF : A'B'C'D'E'F' :: AB^2 : A'B'^2.$$

*Cor. The perimeters of similar figures are proportional to their homologous lines, and their areas are proportional to the squares of those lines.*

For, by the definition of similar figures, the ratio of any two homologous lines, which is called the ratio of similitude, is constantly the same.

## COMPARISON OF SQUARES CONSTRUCTED ON CERTAIN LINES.

### THEOREM XXIX.

*The square constructed on the hypotenuse of a right-angled triangle, is equivalent to the sum of the squares constructed respectively on the other two sides.*

This Theorem is not a fundamental one, like Theorem XIV., but its importance and its fecundity have given rise to several demonstrations founded solely upon the *equality* and *equivalence* of figures. We shall confine ourselves to the one generally given in works of geometry.

The three squares being constructed without the triangle ABC, we draw from the right angle at A the line AL perpendicular to the hypotenuse, and produce it to meet DE at M, thus dividing the square BCED into the two rectangles BDML, LCEM. We also draw AD and CF.

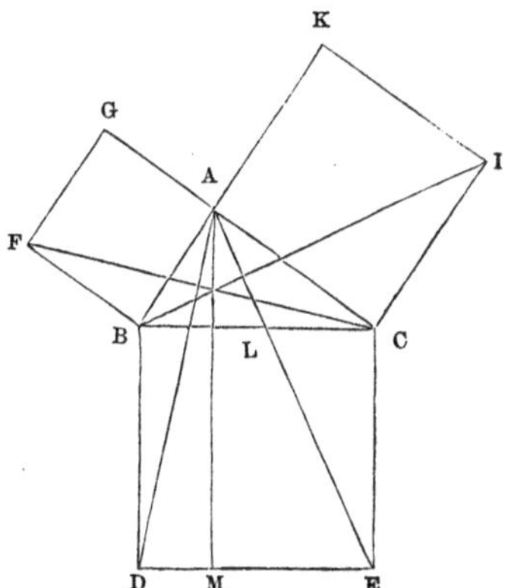


Now, the angles ABD, FBC are equal, being each composed of a right angle and of the common angle ABC; the lines AB and BF are equal, being sides of the same square, so also are BC and BD equal, for a like reason. Hence the two triangles BAD, BFC are equal (B. I., T. XX.). But the triangle ABD is half the rectangle BLMD, since they have the same base BD and the same altitude BL (T. XIX.). For the same reason the triangle BFC is half the square ABFG, for they have the same base BF and the same altitude AB; consequently,

$$BLMD = ABFG.$$

We can prove in the same manner, by drawing AE and BI, that

$$LMEC = ACIK;$$

and as the square BDEC is equivalent to BLMD + LMEC, there will result

$$\textit{square}\ \text{BDEC} = \textit{square}\ \text{ABFG} + \textit{square}\ \text{ACIK}.$$

*Cor.* I. The square MNPQ, constructed on the diagonal BD or AC of a square ABCD, is double the square itself.

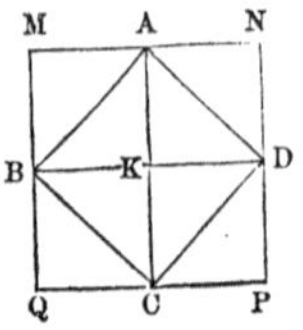

M A N B K D Q C P

This is obvious from the simple inspection of the figure. In effect, the four squares AKBM, AKDN, BKCQ, DKCP, are equal, and respectively the doubles of the triangles AKB, AKD, BKC, DKC, of which the sum is equal to the square ABCD.

*Cor.* II. We have already shown that the squares ABFG, ACIK (see first figure) are respectively equivalent to the rectangles BLMD, LMEC. And since these rectangles and the square BCED have the common altitude BD, they are to each other as their bases BL, LC, and BC (T. XX.). Hence we obtain this relation:

$$\text{AB}^2 : \text{AC}^2 : \text{BC}^2 :: \text{BL} : \text{LC} : \text{BC}.$$

*Cor.* III. Since the two squares ABFG, ACIK are respectively equivalent to the rectangles BLMD, LMEC, we have

$$\text{AB}^2 = \text{BC} \times \text{BL},\ \text{AC}^2 = \text{BC} \times \text{LC};$$

from which we deduce these two proportions:

$$\text{BC} : \text{AB} :: \text{AB} : \text{BL},\quad \text{BC} : \text{AC} :: \text{AC} : \text{LC}.$$

These two proportions, together with the relations of the preceding Corollary, correspond in all respects with the third portion of Theorem XIII., and Scholium I. of Theorem XIV. We have thus, by another method, established properties already demonstrated.

*Cor.* IV. In short, if on the three sides of a right-angled triangle ABC, we suppose three similar polygons to be constructed, as these polygons will be proportional to the squares of their homologous sides (T. XXVIII.), and we have this relation between these sides:

$$\text{BC}^2 = \text{AB}^2 + \text{AC}^2;$$

it follows, that, of the three similar polygons, the polygon constructed on the hypotenuse is equivalent to the sum of the polygons constructed on the sides containing the right angle.

*Scholium.* The algebraic formulas,

$$(p+q)^2=p^2+q^2+2pq,\ (p-q)^2=p^2+q^2-2pq,$$

which have been made the bases of Theorems XV. and XVI., and the other well-known formula,

$$(p+q)\times(p-q)=p^2-q^2,$$

being translated into geometrical language, will give rise to some new Theorems in regard to areas, with which we will close this Book.

### THEOREM XXX.

*The square constructed on the sum or on the difference of two lines, is equivalent to the sum of the squares constructed respectively on these lines, plus or minus twice their rectangle.*

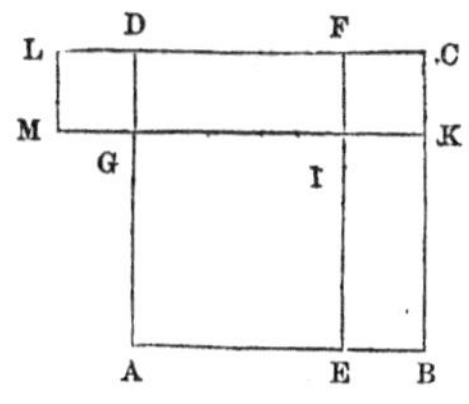


The simple inspection of the figure is nearly sufficient to satisfy one of the truth of this double proposition.

*First.* Let AE be the greater of the two lines, EB the less. Construct the squares AEIG, ABCD, and produce EI until it meets CD at F.

The square constructed on AB, the sum of the two lines, is evidently composed of the squares AEIG, IKCF, constructed on the line AE and on the line IK = EB, increased by the two rectangles BEIK, GIFD, which have for their respective bases GI = AE, EI = AE, and for altitudes GD = EB, IK = EB.

Hence, we have

*square* AB = *square* AE + *square* EB + *twice rectangle* AE × EB.

*Secondly.* Let AB be the greater line, BE the less, that which gives AE for the difference of these two lines. Construct the same figure as in the first case, and, in addition, construct the square GDLM equal to IKCF.

The square AEIG is equal to the square ABCD plus the square GDLM, minus the two rectangles EBCF, MIFL. Now, these two rectangles have respectively for bases FE = AB, IM = GK = AB, and for altitudes BE = IK = IF.

Hence we have

*square* AE = *square* AB + *square* EB − *twice rectangle* AB × BE.

### THEOREM XXXI.

*The rectangle constructed on the sum and the difference of two lines, is equivalent to the difference of the squares constructed on these two lines.*

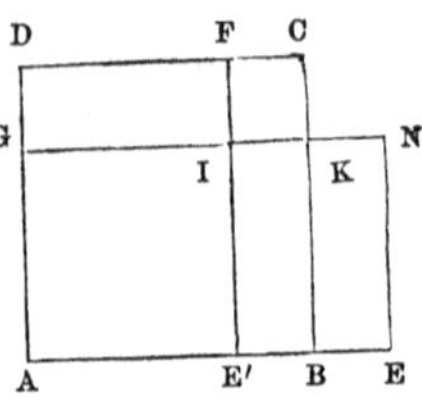


Let AB be the greater line, BE = BE′ the less, so that AE will represent the sum of these two lines, and AE′ their difference. Construct on AE as a base, and AG = AE′ as an altitude, the rectangle AENG, also the square ABCD, and draw E′IF perpendicular to AB, forming the square AE′IG.

The two rectangles BENK, GIFD are equal, having equal bases and equal altitudes, namely, BK = AG = GI = AE′, EB = E′B = IK = IF = DG; hence it follows that the rectangle AENG is equivalent to the figure DFIKBA. But this figure is the difference of the squares constructed on AB and on IK = BE′ = BE.

Hence, we have

*rectangle* AENG = *square* AB − *square* BE.

# FOURTH BOOK.

---

## THE PROPORTIONS OF LINES AND THE AREAS OF FIGURES IN CONNECTION WITH THE CIRCLE.

### DEFINITIONS.

I. *Similar arcs*, *Sectors*, or *Segments*, are those which, in different circles, correspond to equal angles at the centre.

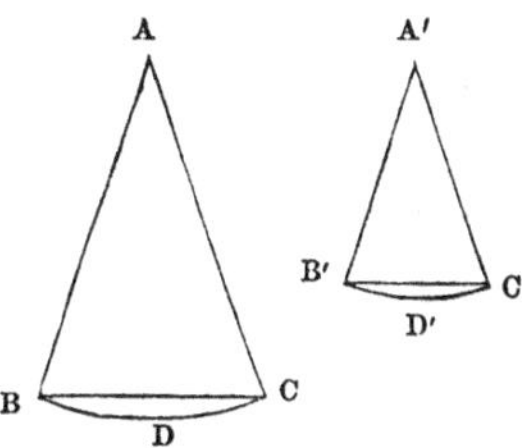


Thus, if the angles at A and A′ are equal, the arcs BC, B′C′ will be similar, the sectors BAC, B′A′C′ will be similar, so also will the segments BDC, B′D′C′.

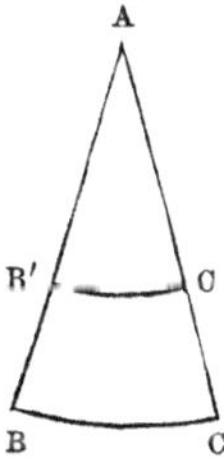


II. When two similar sectors are superposed so that their equal angles coincide, their difference is called a *circular trapezoid.* Thus the space BB′C′C is a circular trapezoid.

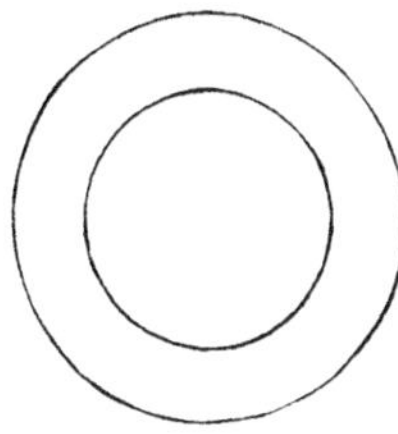

III. The space included between two concentric circumferences is called a *circular ring.*

IV. The arc of a circle is said to be *rectified*, when it is *developed;* that is, unfolded or drawn out into a straight line. It is only in this rectified form that we can conceive of its length, since it could not otherwise be compared with the linear unit.

---

## PROPORTIONAL LINES CONNECTED WITH THE CIRCLE.

### THEOREM I.

*When two chords intersect each other within a circle, the segments will be reciprocally or inversely proportional.*

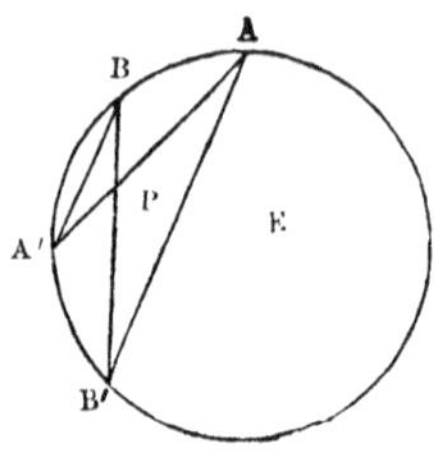


Let the two chords AA′, BB′ intersect at P. Drawing the auxiliary chords AB′, A′B, we thus obtain two triangles PAB′, PA′B, which are similar, since the angles at P are equal (B. I., T. I.); the angles at A and B are equal, so also are the angles at A′ and B′ (B. II., T. X.), hence these triangles are equiangular, and consequently similar (B. III., T. VI.), and their homologous sides give this proportion,

$$PA : PB :: PB' : PA'.$$

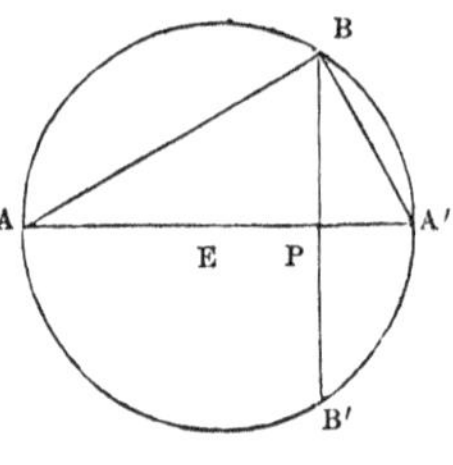


*Scholium* I. When one of the chords is a diameter, and the other is perpendicular to it, we have

$$PA : PB :: PB : PA'.$$

The line PB, drawn perpendicular to a diameter, is called an *ordinate* to this diameter. Hence, we have this condition:

*In a circle any ordinate to a diameter is a mean proportional between the two segments of this diameter.*

We can show that this property is one of the consequences of the properties of right-angled triangles; because if we join A, A′ with the point B, we shall form a right-angled triangle (B. II., T. X., C. I.), which gives (B. III., T. XIII.) this proportion,

$$PA : PB :: PB : PA'.$$

*Scholium* II. This same right-angled triangle ABA′ gives the proportion,

$$AA' : AB :: AB : AP.$$

That is,

*Any chord which is drawn through the extremity of a diameter is a mean proportional between its projection* (B. III., T. XIV., S. II.) *on the diameter and the diameter itself.*

THEOREM II.

*When two secants intersect each other, without a circle, they will be reciprocally proportional to their external segments.*

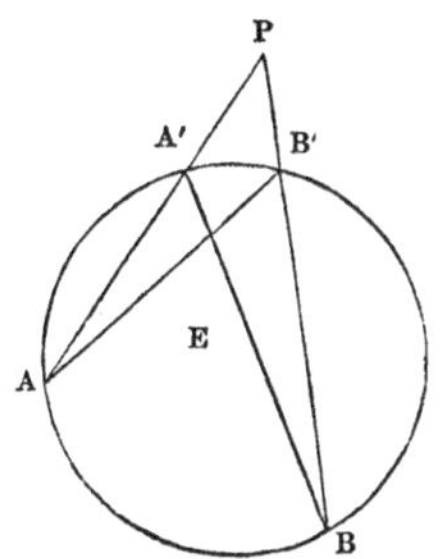


Let the two secants intersect each other at P. Draw, as in the preceding Theorem, the chords AB′, A′B, and the two triangles AB′P, BA′P, will be similar, since the angle at P is common, and A = B, each being measured by half of the arc A′B′ (B. II., T. X.), and the homologous sides give the proportion

$$PA : PB :: PB' : PA'.$$

THEOREM III.

*When a secant and tangent are drawn from the same point, the tangent is a mean proportional between the secant and its external segment.*

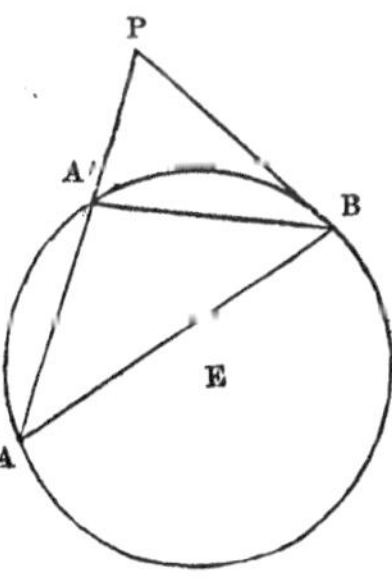


This Theorem is in reality only a particular case of the preceding, when the points B, B′ of the secant PB, are united in one point.

But we will give a direct demonstration, by drawing the chords AB, A′B. We have, in effect, two triangles, PAB, PA′B, similar, since the angle at P is common and the two angles at A and B are equal (B. II., T. X., C. II.).

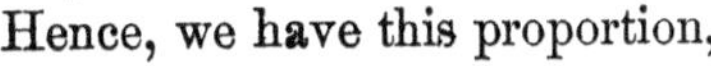

Hence, we have this proportion,

$$PA : PB :: PB : PA'.$$

*Scholium* I. As a particular case, let us suppose the secant to pass through the centre of the circle, and that the tangent is equal to the diameter; that is, PB = AA′ the proportion will then become

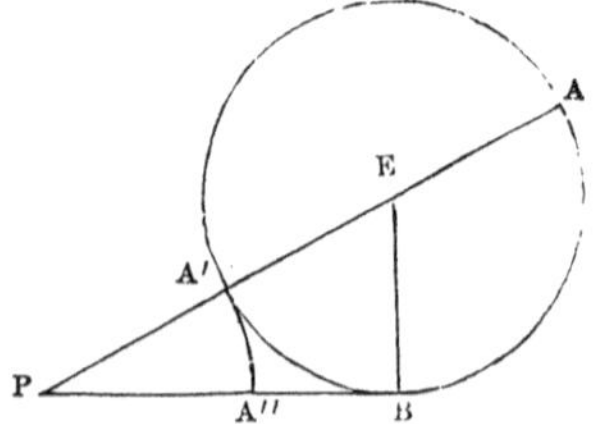


$$PA : AA' :: AA' : PA'.$$

When a straight line, as PA, is thus divided at a point A′, so that the greater segment AA′, is a mean proportional between the whole line and the other segment, it is said to be *divided into mean and extreme ratio.*

The above proportion will give

$$AA' : PA - AA' :: PA' : AA' - PA',$$

or,

$$AA' : PA' :: PA' : AA' - PA'.$$

Now if we take PA″ equal to PA′, and observe that AA′= PB, and consequently that AA′ − PA′ = PB − PA″ = A″B, we shall have

$$PB : PA'' :: PA'' : A''B.$$

From this we see that the straight line PB is also divided at the point A″ *into mean and extreme ratio.*

*Scholium* II. The three preceding Theorems give immediately these two equations,

$$PA \times PA' = PB \times PB', \quad PA \times PA' = PB^2,$$

which may be included in one single proposition, as follows:

*The product of the distances from the same point, either within or without a circle, to two points of its circumference, taken in the same straight line, is always the same.*

In the case of the tangent we must regard the two points of the circumference as united.

THEOREM IV.

*In a quadrilateral inscribed in a circle, the rectangle of the diagonals is equal to the sum of the rectangles of the opposite sides.*

Let ACBD be the inscribed quadrilateral, and we shall have

$$AB \times CD = AC \times BD + AD \times BC.$$

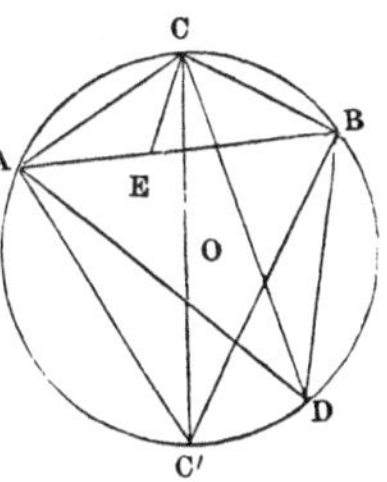


For, drawing CE, making the angle ACE equal to the angle BCD, and consequently the angle ECB equal to the angle ACD.

The two triangles CEA, CBD are similar, since the angle ACE equals the angle DCB, by construction, and CAE = CDB, since each is in the same segment (B. II., T. X.). Their homologous sides give this proportion,

$$AC : CD :: AE : BD;$$

consequently, $CD \times AE = AC \times BD.$

The comparison of the homologous sides of the two triangles BEC, DAC, which are similar, gives

$$AD : BE :: CD : BC,$$

and $CD \times BE = AD \times BC.$

Adding the corresponding members of these equations we obtain

$$CD \times (AE + BE) = AC \times BD + AD \times BC,$$

or $AB \times CD = AC \times BD + AD \times BC.$

*Scholium.* This Theorem has many important applications.

FIRST. *To find the chord* AB *of the sum of two arcs* AC, CB, *when their chords are known.*

For abridgment, let $a$, $b$, $c$ denote the three sides BC, AC, AB of the triangle ABC, and R the radius of its circumscribing circle. Draw the diameter COC′ = 2R, also the supplementary chords AC′, BC′.

Now the inscribed quadrilateral ACBC′ will give

$$AB \times CC' = AC \times BC' + AC' \times CB.$$

But the triangles CAC′, CBC′ being right-angled, since they are each in a semicircle, give

$$AC' = \sqrt{CC'^2 - AC^2}, \quad BC' = \sqrt{CC'^2 - BC^2};$$

and since CC′ = 2R, AC = $b$, BC = $a$, the foregoing expressions will become

$$AC' = \sqrt{4R^2 - b^2}, \quad BC' = \sqrt{4R^2 - a^2},$$

these values, substituted in the first condition, cause it to become

$$c \times 2R = b\sqrt{4R^2 - a^2} + a\sqrt{4R^2 - b^2};$$

hence, $$c = \frac{b}{2R}\sqrt{4R^2 - a^2} + \frac{a}{2R}\sqrt{4R^2 - b^2}. \qquad (1)$$

Secondly. *To find the chord* AB $= c$, *of double an arc, when we know the chord* BC $= a$ *of the arc itself.*

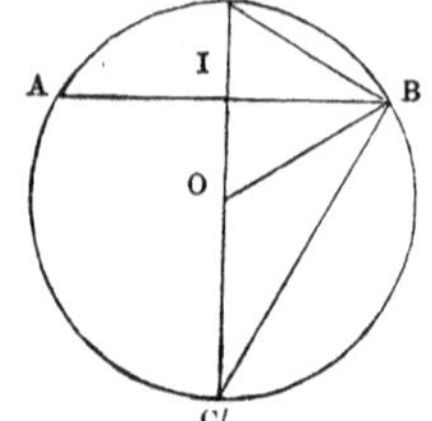


This will evidently be accomplished by making $b = a$ in the general expression just formed; so that, in this case, we have

$$c = \frac{a}{R}\sqrt{4R^2 - a^2}. \qquad (2)$$

Thirdly. *To find the chord of half an arc, when we know the chord of the arc itself.*

In the triangle CBC′, $CB^2 = CI \times CC'$ (B. III., T. XIII.).

Now, $CI = OC - OI = OC - \sqrt{OB^2 - BI^2}$ (B. III., T. XIV.);

or, $$CI = R - \sqrt{R^2 - \tfrac{1}{4}c^2}.$$

Hence, $$CB^2 = a^2 = 2R \times (R - \sqrt{R^2 - \tfrac{1}{4}c^2}) = 2R^2 - R\sqrt{4R^2 - c^2};$$

or, $$a = \sqrt{2R^2 - R\sqrt{4R^2 - c^2}}. \qquad (3)$$

It would not be difficult to deduce, algebraically, the value of $a$ in terms of $c$ and $b$, from formula (1); that is, to find the chord of the difference of two arcs, when we know their chords. When found, the expression is

$$a = \frac{c}{2R}\sqrt{4R^2 - b^2} - \frac{b}{2R}\sqrt{4R^2 - c^2}. \qquad (4)$$

We would also remark that formula (3) could have been obtained very readily from (2) by the ordinary rules of algebra.

## DETERMINATION OF THE SIDES AND OF THE AREAS OF REGULAR POLYGONS.

### THEOREM V.

*The area of any regular polygon is equal to half the product of its perimeter by its apothem.*

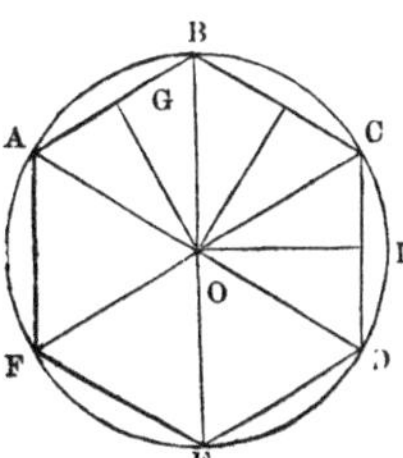


For, the equal isosceles triangles OAB, OBC, OCD, etc., give

$$\text{OAB} = \text{AB} \times \tfrac{1}{2} \text{ of OG},$$
$$\text{OBC} = \text{BC} \times \tfrac{1}{2} \text{ of OG},$$
$$\text{OCD} = \text{CD} \times \tfrac{1}{2} \text{ of OG},$$
&c., &c.

Consequently,

$$\textit{area } \text{ABCDEF} = (\text{AB} + \text{BC} + \text{CD} + \text{etc.}) \times \tfrac{1}{2}\text{OG};$$

or, more concisely, $\text{A} = \frac{1}{2}p \times r.$

A denoting the area of the polygon, $p$ its perimeter, and $r$ the apothem, or the radius of the inscribed circle.

*Scholium.* Calling R the radius of the polygon, $n$ the number of sides, and $a$ the length of one of the sides, we have, from the right-angled triangle OGA,

$$\text{OG} = \sqrt{\text{OA}^2 - \text{AG}^2}; \text{ that is, } r = \sqrt{\text{R}^2 - \tfrac{1}{4}a^2}.$$

We also have $p = na;$

consequently, $\text{A} = \dfrac{na\sqrt{4\text{R}^2 - a^2}}{4}.$

### THEOREM VI.

*The perimeters of two regular polygons of the same number of sides are proportional to the radii of their inscribed circles, or of their circumscribed circles; and their areas are proportional to the squares of those radii.*

Since these polygons are regular, and have the same number of sides, they are similar figures. Consequently, the isosceles triangles which have their vertices at the centre, and for bases the sides of the polygons, are each to each equiangular and similar. The radii of the inscribed circles and of the circumscribed circles are homologous lines. Therefore the proposition is true (B. III., T. XXVIII., C.).

Let R and R′, $r$ and $r'$, $a$ and $a'$, $p$ and $p'$, A and A′, denote, respectively, the radii, the apothems, the sides, the perimeters, and the areas of these two polygons, and we shall have as follows:

$$p : p' :: R : R' :: r : r' :: a : a',$$
$$A : A' :: R^2 : R'^2 :: r^2 : r'^2 :: a^2 : a'^2.$$

THEOREM VII.

*In a given circle, knowing the side of a regular inscribed polygon, we can obtain:*

I. *The value of the side of a regular inscribed polygon of half the number of sides.*

II. *The value of the side of a regular inscribed polygon of double the number of sides.*

III. *The radius and the side of a circumscribed polygon similar to the given polygon.*

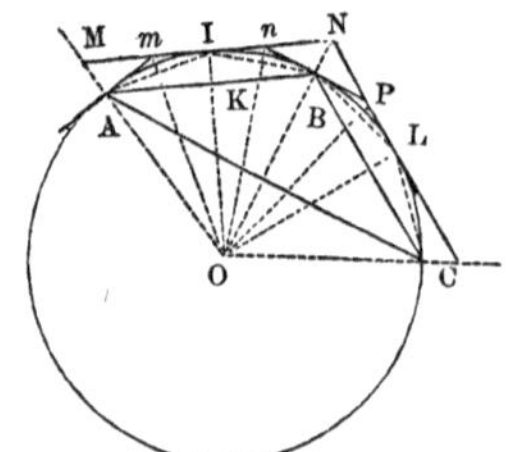


Using the same diagram as under Prob. XI., Book Second, let us represent the side AB, of the given polygon, by $a$, its radius OA by R.

*First.* Since AC subtends an arc double the arc subtended by $AB = a$, it is sufficient to substitute AC for $c$ in formula (2), scholium to T. IV. Which thus becomes

$$AC = \frac{a}{R}\sqrt{4R^2 - a^2}. \qquad (1)$$

*Secondly.* We obtain, in the same way, the value of AI, by means of formula (3), by substituting AI for $a$, and $a$ for $c$. Making these substitutions we have

$$AI = \sqrt{2R^2 - R\sqrt{4R^2 - a^2}}. \qquad (2)$$

*Thirdly.* As to the values of OM and MN, the two similar

triangles OMN, OAB, of which OI and OK are homologous lines, give

$$\text{OM} : \text{OA} :: \text{OI} : \text{OK}; \text{ hence, } \text{OM} = \frac{\text{OA} \times \text{OI}}{\text{OK}},$$

$$\text{MN} : \text{AB} :: \text{OI} : \text{OK}; \text{ hence, } \text{MN} = \frac{\text{AB} \times \text{OI}}{\text{OK}};$$

but we have $\text{OA} = \text{OI} = \text{R}$, $\text{AB} = a$, and

$$\text{OK} = \sqrt{\overline{\text{OA}}^2 - \overline{\text{AK}}^2} = \sqrt{\text{R}^2 - \tfrac{1}{4}a^2} = \tfrac{1}{2}\sqrt{4\text{R}^2 - a^2}.$$

Therefore,

$$\text{OM} = \frac{2\text{R}^2}{\sqrt{4\text{R}^2 - a^2}}, \quad \text{MN} = \frac{2a \times \text{R}}{\sqrt{4\text{R}^2 - a^2}}. \qquad (3)$$

*Scholium.* Knowing the radius and the side of each of the three new polygons, we are able to deduce the values of their areas by substituting, in the formula (T. V., S.),

$$\text{A} = \frac{na\sqrt{4\text{R}^2 - a^2}}{4}$$

for $n$, $a$, R, the respective values just obtained. $\frac{n}{2}$ must be used for $n$, for the first polygon, and $2n$ for $n$, for the second. It remains the same for the third.

### THEOREM VIII.

*The area of a regular inscribed polygon, and that of a regular circumscribed one of the same number of sides being known, we are always able to obtain the areas of two regular polygons, the one inscribed and the other circumscribed, of twice the number of sides.*

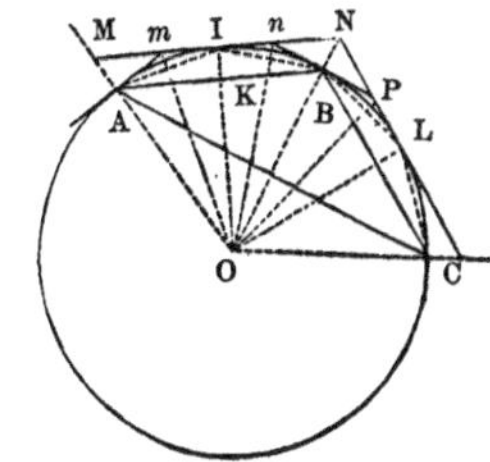


Let A and B denote respectively the areas of the inscribed and circumscribed polygons, each having a number of sides denoted by $n$, and A′ and B′, the areas respectively of the inscribed and circumscribed polygons of double the number of sides.

By inspecting the diagram we see that we have the following relations: $A = 2n \times OAK$, $B = 2n \times OMI$, $A' = 2n \times OAI$, $B' = 2n \times OAmI$; hence, the ratios between the areas A, B, A′, B′, taken two and two, are the same as those between the triangles OAK, OMI, OAI, and of the quadrilateral OA$m$I; thus the whole is reduced to the determination of the ratios which exists between these last four figures.

This being premised, we have immediately, in comparing the three triangles OMI, OAI, OAK, and observing that AK is parallel to MI,

$$OMI : OAI :: OAI : OAK \quad \text{(B. III., T. XXVI., C.)};$$

or, replacing these ratios of triangles by those of the polygons, we have

$$B : A' :: A' : A;$$

that is,

$$A' = \sqrt{A \times B}. \qquad (1)$$

In the second place, since in the triangle OMI, the straight line O$m$ divides the angle O into two equal parts, we have the proportion

$$Mm : mI :: OM : OI \quad \text{(B. III., T. XII.)}.$$

Now, the two triangles OM$m$, O$m$I, having the same altitude OI, give

$$OMm : OmI :: Mm : mI \quad \text{(B. III., T. XXIII., C. I.)}.$$

Also, since the triangles OAI, OAK have the same altitude AK, we have

$$OAI : OAK :: OI : OK :: OM : OA :: OM : OI\ ;$$

hence, we have

$$OMm : OmI :: OAI : OAK;$$

consequently,

$$OMm + OmI : OmI :: OAI + OAK : OAK;$$

or doubling the consequents, and observing that $OMm + OmI = OMI$, and $2 \times OmI = OAmI$, we have

$$OMI : OAmI :: OAI + OAK : 2 \times OAK.$$

Finally, substituting, in place of the three triangles OMI,

OAI, OAK, and of the quadrilateral OA$m$I, the polygons of which they are like parts, we shall obtain

$$B : B' : : A + A' : 2A;$$

consequently, $B' = \frac{2 \times A \times B}{A + A'}$, or $B' = \frac{2 \times A \times B}{A + \sqrt{A \times B}}$. (2)

*Scholium.* These two formulas show, at a glance, that $A' > A$ and $B' < B$.

*Thus the areas of the regular inscribed polygons in the same circle will continue to augment as the number of sides, arising from doubling each time, increases. But the areas of the circumscribed polygons will decrease as the sides increase.*

*However great may be the number of sides of the regular inscribed and circumscribed polygons, it is evident that the area of an inscribed polygon cannot exceed the area of the circle, and the area of a circumscribed polygon cannot be less than the circle.*

The same is true in regard to their perimeters as compared with the circumference of the circle.

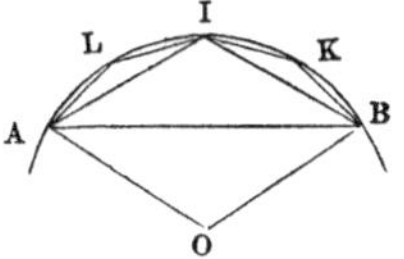


To prove this, it will be necessary only to consider the portion of the figure corresponding to the arc AB, since the same course of reasoning may be repeated for each of the corresponding portions BC, CD, etc.

Let the arc AB be divided into two, four, eight, etc., equal parts. Draw the chords of these new arcs.

From the definition of a straight line, we evidently have

*chord* AB < AI + IB < AL + LI + IK + KB <, etc., < *arc* AB.

*Hence, the perimeters of the regular inscribed polygons, in the same circle, will continue to augment as the number of sides increase, without being able to surpass the circumference of the circle.*

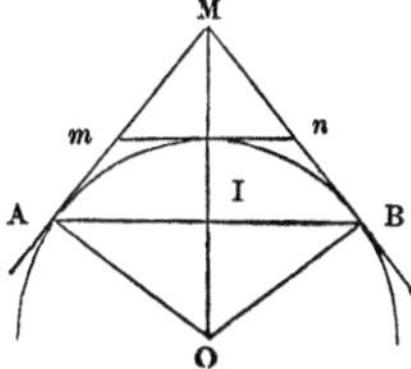


Now, let AM and BM be two half sides of the first regular circumscribed polygon corresponding to the arc AB; $mn$, and A$m$, B$n$ a side and half sides of the regular circumscribed polygon of double the number of sides.

We shall have $mM + Mn > mn$; adding $Am + nB$ to each member, we have

$$Am + mM + Mn + nB, \text{ or } AM + MB > Am + mn + nB.$$

*Hence, the perimeters of the regular circumscribed polygons will continue to decrease as the number of sides increase, without being able to become less than the circumference.*

This being premised, we give, in mathematics, the name of LIMIT to a constant and determinate magnitude, towards which a variable magnitude is constantly converging, either by increasing or diminishing. Hence, we say that:

*The area of a circle is the* SUPERIOR *limit of the areas of the regular inscribed polygons, and the* INFERIOR *limit of the areas of the regular circumscribed polygons.*

*The circumference of a circle is the superior limit of the perimeters of the first polygons, and the inferior limit of the second.*

We feel authorized to regard the circle as a regular polygon of an infinite number of sides, infinitely small; these sides are called the *elements* of the curve.

And this new definition of a circle, if it does not at first appear very rigorous, has at least the advantage of giving greater simplicity and precision to our demonstrations.

---

## THE DETERMINATION OF THE SIDES AND OF THE AREAS OF REGULAR POLYGONS OF A PARTICULAR KIND.

### THEOREM IX.

*The side of a regular inscribed hexagon is equal to the radius.*

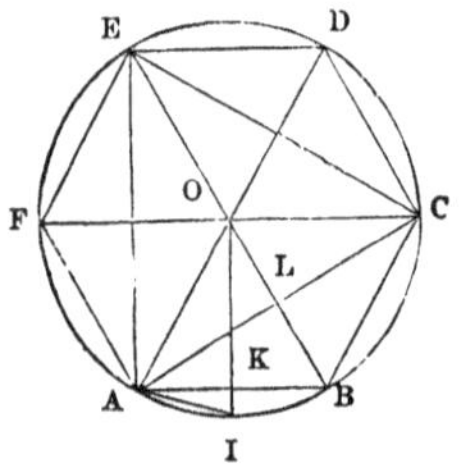

E D O F C L K A B I

Draw OA and OB; the value of the angle O, of the triangle OAB, is $\frac{4}{6} = \frac{2}{3}$ of a right angle, hence the sum of the other two angles is $2 - \frac{2}{3} = \frac{4}{3}$ of a right angle, and since $OA = OB$, it follows that the angle $OAB = $ angle $OBA = \frac{1}{2}$ of $\frac{4}{3} = \frac{2}{3}$ of a right angle; hence the triangle OAB is equilateral, and gives $AB = OA = OB = R$.

*Cor.* I. *The side* AC *of an inscribed equilateral triangle is to the radius as* $\sqrt{3} : 1$.

If in formula (2), of Scholium to T. IV., we make $a = R$, we shall find

$$AC = \frac{R}{R}\sqrt{3R^2} = R\sqrt{3};$$

hence, $$AC : R :: \sqrt{3} : 1.$$

*Cor.* II. The side of a circumscribed equilateral triangle is double the side of the inscribed one.

To show this, it is sufficient to substitute $R\sqrt{3}$ for $a$ in formula (3) of T. VII., which thus becomes

$$MN = \frac{2R^2\sqrt{3}}{\sqrt{4R^2 - 3R^2}} = 2R\sqrt{3}.$$

We may remark, that the altitude EL of the inscribed triangle is equal to $\frac{3}{2}R$. For, we have $EL = EO + OL = R + \frac{1}{2}R$, since the figure OABC is a lozenge, consequently $EL = \frac{3}{2}R$.

It follows, then, that the altitude of the circumscribed triangle is equal to 3R.

*Cor.* III. If, in formula (2) of T. VII., we substitute R for $a$, we shall find

$$AI = (2R^2 - R\sqrt{4R^2 - R^2})^{\frac{1}{2}} = R\sqrt{2 - \sqrt{3}},$$

*for the side of a regular inscribed dodecagon.*

*Cor.* IV. Finally, in formula (3), T. VII., substituting R for $a$, we find the *radius* and a *side* of a *regular circumscribed hexagon*, as follows:

$$OM = \tfrac{2}{3}R\sqrt{3};\ MN = \tfrac{2}{3}R\sqrt{3}.$$

*Scholium.* We are now prepared, by the aid of the general formula (T. V.),

$$A = \frac{na\sqrt{4R^2 - a^2}}{4},$$

to calculate the areas of these different polygons.

Thus, for example, if we make $n = 6$, $a = R$, we shall obtain for the expression of the *area of the regular inscribed hexagon*,

$$A = \tfrac{3}{2}R^2\sqrt{3}.$$

8

By making $n = 3$, $a = \mathrm{R}\sqrt{3}$, we find

$$\mathrm{A} = \tfrac{3}{4}\mathrm{R}^2\sqrt{3},$$

for the area of the inscribed equilateral triangle.

In a similar manner we proceed for the other cases.

### THEOREM X.

*The side of the inscribed square is to the radius in the ratio of $\sqrt{2}$ to 1.*

Draw the two diameters AB, CD perpendicular to each other, and draw the chords AC, CB, BD, DA. The figure ADBC is evidently a square; and we have

$$\mathrm{AC}^2 = \mathrm{AO}^2 + \mathrm{OC}^2 = 2\mathrm{R}^2;$$

consequently $\mathrm{AC} = \mathrm{R}\sqrt{2}$, and

$$\mathrm{AC} : \mathrm{R} :: \sqrt{2} : 1.$$

*Scholium.* In drawing through the points A, D, B, C, tangents, we form the circumscribed square; and we have

$$\mathrm{MN} = \mathrm{AB} = 2\mathrm{R}.$$

That is—*The side of the circumscribed square is equal to the diameter of the circle.*

The areas of these two polygons are expressed respectively by $2\mathrm{R}^2$ and $4\mathrm{R}^2$.

### THEOREM XI.

*The side of a regular inscribed decagon is equal to the greater segment of the radius, when divided into mean and extreme ratio.*

Draw the radii OA, OB to the extremities of a side AB. The angle O of the triangle OAB is equal to $\frac{4}{10} = \frac{2}{5}$ of a right angle; there remains, then, $2 - \frac{2}{5} = \frac{8}{5}$ of a right angle, for the sum of the other two angles; and as $\mathrm{OA} = \mathrm{OB}$, it follows that the angle $\mathrm{OAB} = \mathrm{OBA} = \frac{4}{5}$ of a right angle.

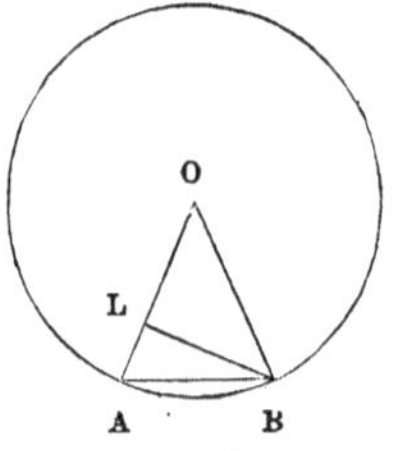

Thus, each of the angles at the base of the triangle OAB is double the angle at the vertex.

This being premised, draw BL bisecting the angle OBA, and we shall have this proportion,

$$AL : LO :: AB : OB \quad \text{(B. III., T. XII.)}.$$

But since the angle LBO = LBA = LOB = $\frac{2}{5}$, and consequently ALB = 2LOB = $\frac{4}{5}$, the two triangles OLB, ALB are also isosceles, and give OL = LB = AB, and also having OB = OA, the above proportion becomes

$$AL : OL :: OL : OA.$$

From which we see that the point L divides the radius OA into mean and extreme ratio (T. III., S. I.), and that the greater segment OL is equal to AB a side of the regular decagon.

*Scholium* I. We will deduce the numerical value of this side algebraically. Denoting it by $x$, we have (T. III., S. I.),

$$R : x :: x : R - x,$$

$$\text{or, } x^2 + Rx = R^2 \text{ and } x = \frac{R}{2}(\pm \sqrt{5} - 1).$$

That is, the numerical value of the side of a regular inscribed decagon is $\frac{R}{2}(\sqrt{5} - 1)$.

*Scholium* II. If for $a$ in formula (2) of Scholium to T. IV. we substitute $\frac{R}{2}(\sqrt{5} - 1)$, which we have just found for the value of a side of a regular inscribed decagon, we shall find $\frac{R}{2}(10 - 2\sqrt{5})^{\frac{1}{2}}$ for the side of a regular inscribed pentagon.

Now, since $\left[\frac{R}{2}(10 - 2\sqrt{5})^{\frac{1}{2}}\right]^2 = \left[\frac{R}{2}(\sqrt{5} - 1)\right]^2 + R^2$, it follows, that *the square of the side of a regular inscribed pentagon is equal to the square of the side of the regular inscribed decagon increased by the square of the radius.*

### THEOREM XII.

*The side of a regular inscribed pentadecagon is the chord of the difference between the arcs subtended respectively by the sides of a hexagon and a decagon.*

For, we have $\frac{1}{6}-\frac{1}{10}=\frac{5}{30}-\frac{3}{30}=\frac{2}{30}=\frac{1}{15}$; which shows that —*The difference between the arcs subtended by the sides of a hexagon and of the decagon is equal to one fifteenth of the entire circumference.*

If we wish the numerical value of the side of the pentadecagon, we may employ formula (4) of T. IV., in which, for $c$, we must substitute the value of the side of a hexagon, and for $b$ we must use the side of a decagon.

*General Scholium.* It results from what has been said in the last four Theorems, and of the principles before established, that we have geometrical methods for constructing regular inscribed and circumscribed polygons of 3, 6, 12, 24, etc., of 4, 8, 16, 32, etc., of 5, 10, 20, 40, etc., in short, of 15, 30, 60, etc., number of sides.

That we also have arithmetical methods for calculating the sides, and afterwards the areas, of all these polygons, if not exactly—since nearly all the numbers which enter into our calculations are incommensurable—yet to as close a degree of approximation as may be desired.

---

## MEASURE OF THE CIRCLE AND OF ITS CIRCUMFERENCE.

### THEOREM XIII.

*The area of a circle is equal to half the product of the circumference into its radius.*

In effect, let us consider a series of regular circumscribed polygons, each having double the number of sides of the preceding one. The superficies of these polygons have for their respective measures half the product of each perimeter into its apothem—that is, into the radius of the circle. Hence, the area of the circle, which is the *limit* (T. VIII., S.) of these polygons, has for its measure half the product of the circumference, which is the *limit* of the perimeters, into its radius.

Or, considering the circle as a regular polygon of an infinite number of sides, it immediately follows that its area is equal to half the product of its perimeter, that is, of its circumference, into its radius.

*Scholium* I. Let R, C, and S denote respectively the radius, circumference, and surface of any circle. We shall have

$$S = \tfrac{1}{2}C \times R = C \times \tfrac{1}{2}R.$$

We must keep in mind that S is an *abstract number*, expressing the ratio of the surface of the circle to the unit of surface; C and R are ratios of the circumference and of the radius to the linear unit.

*Scholium* II. We may also say that the area of a circle is equal to that of a triangle having for its base the *rectified* circumference (D. IV.), and for its altitude the radius of the circle (B. III., T. XXIII.).

### THEOREM XIV.

*In two circles, the circumferences are proportional to their radii or to their diameters, and their areas are proportional to the squares of their radii.*

*First.* Let us conceive a series of regular polygons, each having double the number of sides of the preceding one, to circumscribe the circumference C; also another series of similar polygons to circumscribe the circumference C′; and let us designate by R, R′ the constant apothems of these two series of polygons; by $p$, $p'$ the perimeters of two similar polygons, taken, the one in the first series, the other in the second.

This being supposed, we shall have this proportion,

$$p : p' :: R : R',$$

which is applicable to any two similar polygons whatever; it must therefore hold good at the *limit* of the perimeters, that is, when $p$, $p'$ become C, C′; hence,

$$C : C' :: R : R', \text{ or, } :: 2R : 2R'.$$

*Secondly.* Multiplying this last proportion by the obvious proportion,

$$\tfrac{1}{2}R : \tfrac{1}{2}R' :: R : R', \text{ or}, :: 2R : 2R',$$

it becomes, $C \times \tfrac{1}{2}R : C' \times \tfrac{1}{2}R' :: R^2 : R'^2$, or $4R^2 : 4R'^2$.

Or, since (T. XIII., S. I.) $C \times \tfrac{1}{2}R = S$, $C' \times \tfrac{1}{2}R' = S'$, this becomes

$$S : S' :: R^2 : R'^2, \text{ or } 4R^2 : 4R'^2.$$

Or, considering the circle as a regular polygon of an infinite number of sides, the above follows immediately from T. XXVIII., B. III.

*Cor.* The proportion $C : C' :: 2R : 2R'$, may be written

$$C : 2R :: C' : 2R',$$

and being applicable to any number whatever of circumferences, we have

$$C : 2R :: C' : 2R' :: C'' : 2R'' :: C''' : 2R''' ::, \text{ etc.},$$

from which we infer that

*The ratio of the circumference to the diameter is a constant number.*

We usually denote by $\pi$ this constant ratio.

*Scholium.* If R, C, and S denote respectively the radius, the circumference, and the surface of any circle whatever, we shall have this proportion

$$\pi : 1 :: C : 2R;$$

consequently, $C = 2\pi R$, from which, by multiplying each term by $\tfrac{1}{2}R$, we find $C \times \tfrac{1}{2}R = S = \pi R^2$.

From this, we see that the circumference of any circle may be found by multiplying its diameter (2R) by $\pi$; that its surface may be obtained by multiplying the square of its radius by $\pi$.

### THEOREM XV.

*The area of a circular sector is equal to half the product of its arc into its radius.*

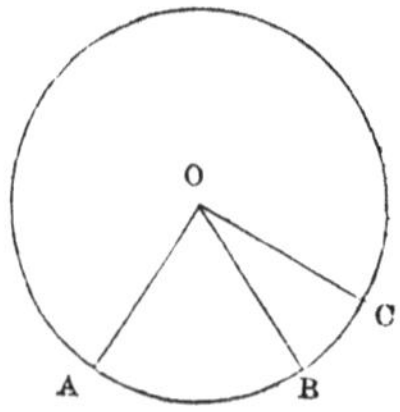

In effect, from the definition of a circular sector, it follows that in the same circle two sectors are proportional to their angles, and consequently to their corresponding arcs.

O
C
A
B

Hence, comparing the given sector AOB

with the sector AOC, which has its angle AOC right, we have this proportion,

$$sector\ AOB : sector\ AOC :: arc\ AB : AC;$$

or, multiplying the two consequents by four, and denoting by C, S, the circumference and surface, or area of the circle, we have

$$sector\ AOB : S :: arc\ AB : C.$$

Again, multiplying the terms of the second couplet of this last proportion by $\frac{1}{2}$R, we obtain,

$$sector\ AOB : S :: arc\ AB \times \tfrac{1}{2}R : C \times \tfrac{1}{2}R.$$

But, $C \times \frac{1}{2}R = S$ (T. XIII., S. I.); consequently,

$$sector\ AOB = arc\ AB \times \tfrac{1}{2}R = \tfrac{1}{2}\ arc\ AB \times R.$$

*Scholium.* The area of a sector is equal to that of a triangle having for its base the *rectified arc* of the sector, and for its altitude the radius of the circle.

THEOREM XVI.

*The areas of two similar sectors are proportional to the squares of their radii, or of their arcs.*

The sectors being similar, have their angles equal. We will, therefore, suppose the one placed upon the other so that their angles may coincide.

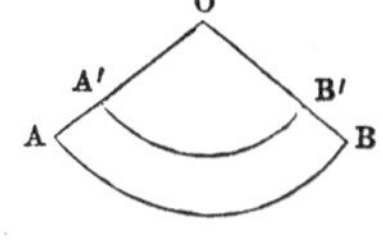


Since (T. XV.) $sector\ AOB = arc\ AB \times \frac{1}{2}OA$, $sector\ A'OB' = arc\ A'B' \times \frac{1}{2}OA'$, we have

$$sector\ AOB : sector\ A'OB' :: arc\ AB \times \tfrac{1}{2}OA : arc\ A'B' \times \tfrac{1}{2}OA';$$

but the arcs AB, A'B' being evidently in the same ratio as the entire circumferences of which they are parts, which circumferences are in the ratio of their radii OA, OA' (T. XIV.), there will result

$$arc\ AB : arc\ A'B' :: OA : OA',$$

hence, $arc\ AB \times \frac{1}{2}OA : arc\ A'B' \times \frac{1}{2}OA' :: OA^2 : OA'^2;$

consequently, $sector\ AOB : sector\ A'OB' :: OA^2 : OA'^2,$

also, $sector\ AOB : sector\ A'OB' :: (arc\ AB)^2 : (arc\ A'B')^2.$

*Scholium* I. The difference AA′B′B between the similar sectors OAB, OA′B′, which is called a *circular trapezoid* (D. II.), *has for its measure the product of half the sum of its bases into the difference of their radii.*

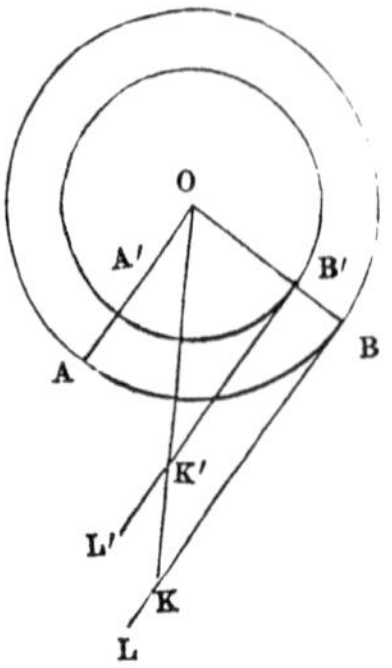

To prove this, draw the indefinite tangents BL, B′L′, and take on BL the portion BK equal to the rectified arc BA, and draw KO meeting B′L′ in K′. Then, by reason of the similarity of figures, we shall have

$$\text{OB} : \text{OB}' :: \text{BK} : \text{B}'\text{K}';$$

$$\text{OB} : \text{OB}' :: \textit{arc}\ \text{BA} : \textit{arc}\ \text{B}'\text{A}';$$

consequently,

$$\text{BK} : \text{B}'\text{K}' :: \textit{arc}\ \text{BA} : \textit{arc}\ \text{B}'\text{A}'.$$

But BK = *arc* BA by construction, therefore B′K′ = *arc* B′A′. Hence the two sectors OAB, OA′B′ are respectively equivalent to the two triangles OBK, OB′K′ (T. XV., S.); it therefore follows that the circular trapezoid is equivalent to the right-angled trapezoid KBB′K′, which has for its measure $\frac{\text{BK} + \text{B}'\text{K}'}{2} \times \text{BB}'$ (B. III., T. XXIII., C. II.). Consequently we have

$$\textit{circular trapezoid}\ \text{AA}'\text{B}'\text{B} = \frac{\textit{arc}\ \text{AB} + \textit{arc}\ \text{A}'\text{B}'}{2} \times \text{BB}'.$$

It would be, moreover, easy to prove that the half sum of the bases, or the mean difference between the two bases, is a concentric arc with AB, A′B′, at equidistance from the same.

*Scholium* II. The *circular ring* (D. III.), that is to say, the space between the two concentric circumferences (see last diagram), is only a particular case of the circular trapezoid, consequently its area is expressed by *the product of the mean circumference into its width* BB′, *the difference of the radii.*

## RATIO OF THE CIRCUMFERENCE TO THE DIAMETER.

### FIRST METHOD.

We know that the side of a regular inscribed hexagon is equal to the radius (T. IX.). If we suppose, then, the radius $R=1$, we can find the side of a regular inscribed polygon of twelve sides, by making $a=R=1$, in formula (2) of Theorem VII., that which gives $\sqrt{2-\sqrt{3}}$ for a side of a regular inscribed dodecagon. Its perimeter is $12\sqrt{2-\sqrt{3}}$; consequently the ratio of this perimeter to the diameter is $6\sqrt{2-\sqrt{3}}$.

Were we to substitute anew, in the same formula, 1 for R, and $\sqrt{2-\sqrt{3}}$ for $a$, we should find a side of an inscribed regular polygon of twenty-four sides. Thus we might continue to find the sides of the successive polygons, each having double as many sides as the preceding one, to any extent we please.

As the number of sides of the polygons increase, their perimeters will approximate more nearly to the circumference of the circle; we are thus enabled, by this method, to obtain the value of $\pi$ to as close a degree of approximation as we please.

### SECOND METHOD.

We have demonstrated (T. XIV., S.) that the area of a circle which has R for its radius, is equal to $\pi R^2$.

Now, if we make $R=1$ in this expression, it will become $\pi$, which shows that

*The ratio of the circumference to the diameter is equal to the area of a circle whose radius is unity.*

This being supposed, let us start with an inscribed and circumscribed square, whose areas are respectively 2 and 4 (T. X., S.). And the formulas

$$A' = \sqrt{A \times B}, \quad B' = \frac{2 \times A \times B}{A + A'}$$

(T. VIII.), if we suppose $A=2$, $B=4$, will become

$$A' = \sqrt{8}, \quad B' = \frac{16}{2+\sqrt{8}} = 4(\sqrt{8}-2),$$

which, converted into decimals, will give 2.8284271, etc.,

3.3137085, etc., for the areas, respectively, of the inscribed and circumscribed octagons.

Using these values in the same formulas, we shall obtain the areas of the inscribed and circumscribed polygons of sixteen sides, which in turn will make known the areas of polygons of thirty-two sides, and so on, till we arrive at an inscribed and circumscribed polygon, differing from each other, and consequently from the circle, so little that either may be considered as equivalent to it. The subjoined table exhibits the *area*, or numerical expression for the surface of each succeeding polygon, carried to seven places of decimals.

| No. of sides. | Area of the inscribed polygon. | Area of the circumscribed polygon. |
|---|---|---|
| 4 | 2.0000000 | 4.0000000 |
| 8 | 2.8284271 | 3.3137085 |
| 16 | 3.0614674 | 3.1825979 |
| 32 | 3.1214451 | 3.1517249 |
| 64 | 3.1365485 | 3.1441184 |
| 128 | 3.1403311 | 3.1422236 |
| 256 | 3.1412772 | 3.1417504 |
| 512 | 3.1415138 | 3.1416321 |
| 1024 | 3.1415729 | 3.1416025 |
| 2048 | 3.1415877 | 3.1415951 |
| 4096 | 3.1415914 | 3.1415933 |
| 8192 | 3.1415923 | 3.1415928 |

It appears, then, that the inscribed and circumscribed polygons of 8192 sides differ so little from each other, that the numerical value of each, as far as six places of decimals, is absolutely the same; and as the circle is between the two, it cannot, strictly speaking, differ from either so much as they do from each other; so that the number 3.141592 expresses the area of a circle whose radius is 1: that is, the value of $\pi$ to six places of decimals is 3.141592.

### THIRD METHOD.

Instead of seeking the approximate value of the circumference, or of the area of a circle, of which the radius is 1, *we shall seek the value of the radius corresponding to a given circumference.*

For this purpose, we will premise the following

### LEMMA.

*Having given the radius* R *and the apothem* $r$ *of a regular polygon, it is required to find the radius* R′ *and the apothem* $r'$ *of a second regular polygon having the same perimeter as the first and double the number of sides.*

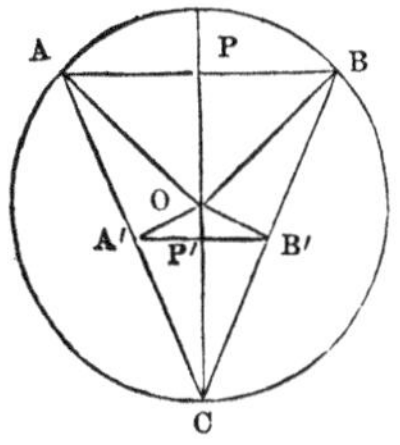

Suppose the given polygon to be circumscribed, and that AB is one of the sides, AOB the angle at the centre, OA = R = the radius, and OP = $r$ = the apothem.

This being supposed, the angle at the centre of the second polygon is half the angle AOB. If we produce PO until it meets the circumference at C, and draw the chords CA, CB, the angle ACB, half of AOB is the angle at the centre of the second polygon.

Also, if we draw OA′ perpendicular to CA, and draw A′B′ parallel to AB, we shall have A′B′ = ½AB (B. I., T. XXXIII., Cor.) for the side of the second polygon, and CA′ will be its radius, and CP′ its apothem.

The similar triangles CAP, CA′P′, give

$$\text{CP}' = \tfrac{1}{2}\text{CP} = \tfrac{1}{2}(\text{CO} + \text{OP}) = \tfrac{1}{2}(\text{OA} + \text{OP}),$$

consequently, $r' = \frac{1}{2}(\text{R} + r)$.

Again, the right-angled triangle OA′C gives

$$\text{CA}'^2 = \text{CO} \times \text{CP}' = \text{OA} \times \text{CP}';$$

that is, $\text{R}' = \sqrt{\text{R} \times r'}$; and as $r'$ has already been found, the problem is resolved.

These formulas are much more simple than those already used in the first and second methods.

Now, to make an application of these formulas, we will take for our given polygon a square whose side is 1, and consequently its perimeter is 4. Its radius is $\frac{1}{2}\sqrt{2}$, being obviously half its diagonal, and its apothem is $\frac{1}{2}$.

If, then, in the formula $r' = \frac{1}{2}(\text{R} + r)$, we make $\text{R} = \frac{1}{2}\sqrt{2} = 0.7071068$, etc., $r = \frac{1}{2} = 0.5$, we shall find $r' = 0.6035534$, etc. Formula $\text{R}' = \sqrt{\text{R} \times r'}$ will now give $\text{R}' = 0.6532815$, etc. We thus find the radius and apothem of a regular polygon of eight sides, and whose perimeter is 4.

Using these values in the same formulas, they will in turn make known the radius and apothem of a regular polygon of sixteen sides, having the same perimeter, 4. We give the results of these successive operations, to seven places of decimals, in the following table:

| No. of sides. | Apothems. | Radii. |
|---|---|---|
| 4 | 0.5000000 | 0.7071068 |
| 8 | 0.6035534 | 0.6532815 |
| 16 | 0.6284174 | 0.6407289 |
| 32 | 0.6345731 | 0.6376435 |
| 64 | 0.6361083 | 0.6368754 |
| 128 | 0.6364919 | 0.6366836 |
| 256 | 0.6365878 | 0.6366357 |
| 512 | 0.6366117 | 0.6366237 |
| 1024 | 0.6366177 | 0.6366207 |
| 2048 | 0.6366192 | 0.6366199 |
| 4096 | 0.6366195 | 0.6366197 |
| 8192 | 0.6366196 | 0.6366196 |

From this table, we see that a circle whose circumference is 4 has for its radius 0.6366196, etc., and consequently for its diameter 1.2732392, etc. Hence, the ratio of this circumference to its diameter, or the value of $\pi$, is

$$\frac{4}{1.2732392,\text{ etc.,}} = 3.1415931,\text{ etc.}$$

This value of $\pi$ differs less than a unit in the sixth decimal place from the true value.

There are other methods, depending upon a knowledge of the higher branches of mathematics, by which this value of $\pi$ has been extended to more than two hundred decimal places. The value to thirty-one decimals is

$$\pi = 3.1415926535897932384626433832795.$$

The following *original geometrical construction* is very simple, and gives this ratio sufficiently accurate for all practical purposes:

Let ACB be the diameter of the given circle: produce it towards N; take BD and DE each equal to AB; through E draw

EG perpendicular to AE, and take EF and FG each equal to AB; join AG, AF, DG, and DF. Set off on the line EN, from E, the distances EH and HK, each equal to AG; then set off, in the opposite direction, the distance KL equal to AF, and from L set off LM equal to DG; also set off MN equal to DF Then bisect EN at the point P; bisect EP at the point R, and, finally, trisect ER at the point T; then will CT be the circumference of the circle, nearly.

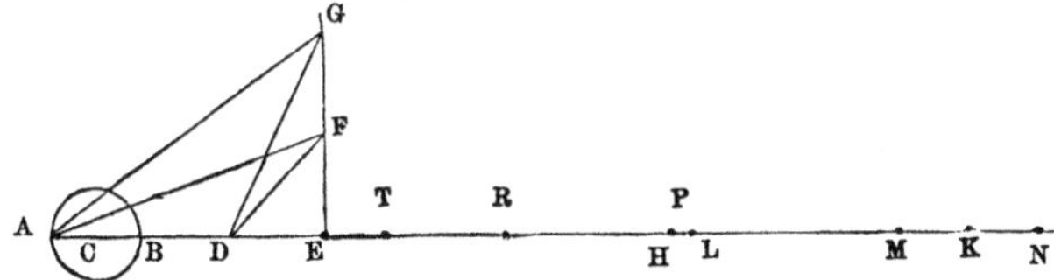


For, by construction, we have, if we call the diameter a unit, $CE = 2\frac{1}{2}$; $EL = 2EH - KL = 2\sqrt{13} - \sqrt{10}$; $LM = \sqrt{5}$; $MN = \sqrt{2}$. Therefore,

$$EN = 2\sqrt{13} - \sqrt{10} + \sqrt{5} + \sqrt{2};$$

and $$ET = \tfrac{1}{12}(2\sqrt{13} - \sqrt{10} + \sqrt{5} + \sqrt{2});$$

and, therefore,

$$CT = 2\tfrac{1}{2} + \tfrac{1}{12}(2\sqrt{13} - \sqrt{10} + \sqrt{5} + \sqrt{2}) = 3.1415922, \text{ etc.},$$

which is the ratio true to six decimals. For simplicity and accuracy, a better graphic method of finding this ratio can hardly be expected, or even desired.

# PROBLEMS,

## WHICH REFER TO THE THIRD AND FOURTH BOOKS.

---

## CONSTRUCTION OF PROPORTIONAL LINES.

### PROBLEM I.

*To divide a given line into any number of equal parts.*

To make a definite case, suppose we wish to divide AB into five equal parts.

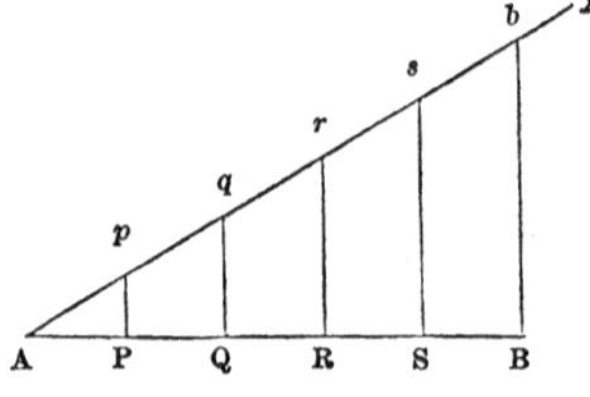


Through A draw any indefinite line AX, making an angle with AB. Lay off on this line any convenient length five times, as A*p*, *pq*, *qr*, *rs*, *sb*. Join the last extremity *b* with B, and through the other points of division draw parallel to B*b*, lines cutting AB in the points P, Q, R, and S (B. II., P. VI.).

The straight line will thus be divided into five equal parts (B. III., T. I.).

### PROBLEM II.

*To divide a given line into parts proportional to given lines.*

As a definite case, suppose we wish to divide AB into three parts, which shall be to each other as the three lines P, Q, R.

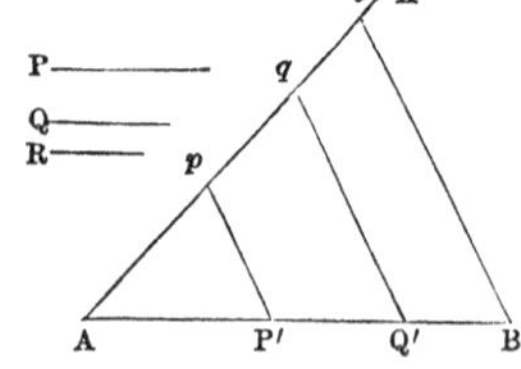


As in the last Problem, draw any indefinite line AX, making an angle with AB. Make A*p*, *pq*, *qb* respectively equal to P, Q, and R. Draw B*b*, and parallel to it draw *p*P′, *q*Q′, and the line AB will be divided at P′ and Q′ as required (B. III., T. III.)

*Scholium.* As a particular case, suppose we wish to divide a straight line AB into two parts proportional to the lines M and N.

Through the points A and B draw any two parallel lines AX, BY, so that the alternate angles BAX, ABY may be equal. Take on AX and BY portions AC, BC′, respectively equal to M and N; then draw CC′ meeting AB in D.

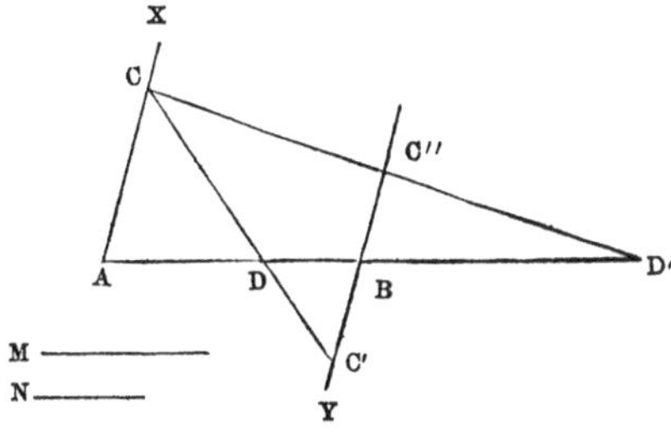


The straight line AB will thus be divided at D in the ratio required.

For, the two triangles DAC, DBC′, are evidently similar, and give

$$AD : DB :: AC : BC' :: M : N.$$

If, instead of taking BC′, equal to N, in the direction of BY, opposite to AX, we had taken BC″, equal to N, in the same direction with AX, then the line CC″ would have met AB produced at D′, which is called the *conjugate* point of the point D (B. III., T. XII., S.).

PROBLEM III.

*To find a fourth proportional to three given lines* M, N, P.

Form any angle, as XAY, and take on AX, AB = M, BC = N, and on AY, AD = P; then draw BD, and through the point C draw CE parallel to BD. The line DE will be the fourth proportional required.

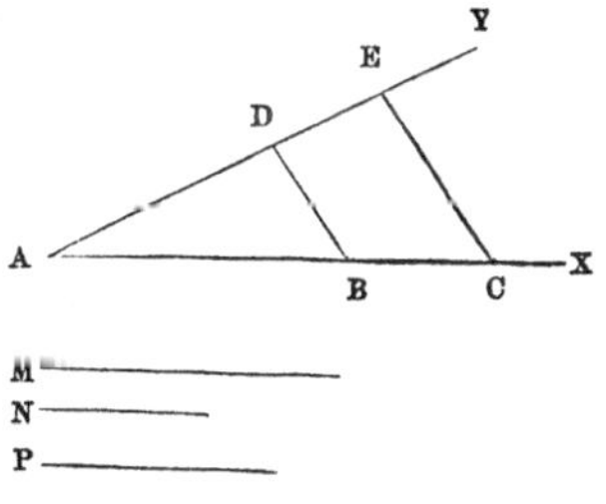


For, we have (B. III., T. III.),

$$AB : BC :: AD : DE, \quad \text{or } M : N :: P : DE.$$

*Cor.* I. If P were equal to N, the above proportion would become

$$M : N :: N : DE.$$

That is, DE would in this case be a third proportional to the two lines M and N.

*Cor.* II. We deduce immediately the following problem:

*A point* O *being given within an angle* YAX, *to draw through* O *a straight line* DOE, *such that the segments* DO, OE *may be to each other in the ratio of* M *to* N.

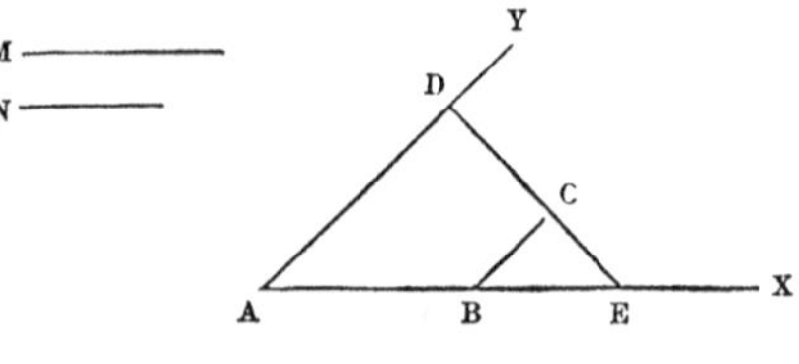


Through the point O draw OB parallel to AY; find a fourth proportional to the three lines M, N, AB; make BE equal to this fourth proportional, and draw EOD, which will be the line required.

For, since OB is parallel to AD, we have

$$OD : OE :: AB : BE :: M : N.$$

When $M = N$, it is sufficient to take on AX, $BE = AB$, and then to draw EOD.

### PROBLEM IV.

*To find a mean proportional between two given lines* M *and* N.

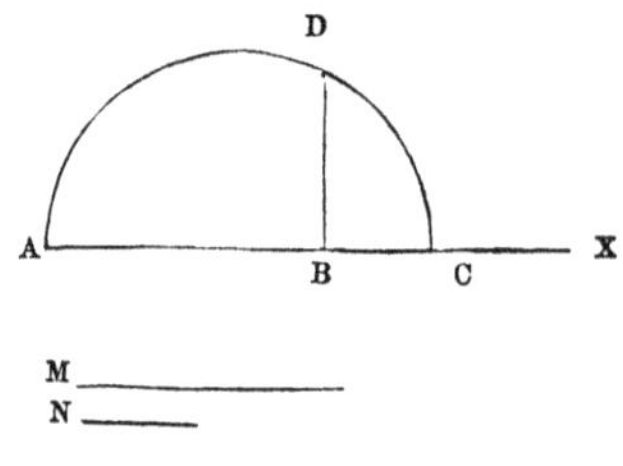


On the indefinite line AX, take $AB = M$, $BC = N$, and on AC, as a diameter, describe the semi-circumference ADC. Through B draw BD perpendicular to AC, and BD will be the mean proportional sought. For, we have (T. I., S. I.),

$$AB : BD :: BD : BC, \quad \text{or } M : BD :: BD : N.$$

### PROBLEM V.

*To divide a given line into mean and extreme ratio.*

We have already noticed this kind of division (T. III., S. I.). And we have also actually solved this problem algebraically (T. XI., S. I.).

We will now proceed with its geometrical solution.

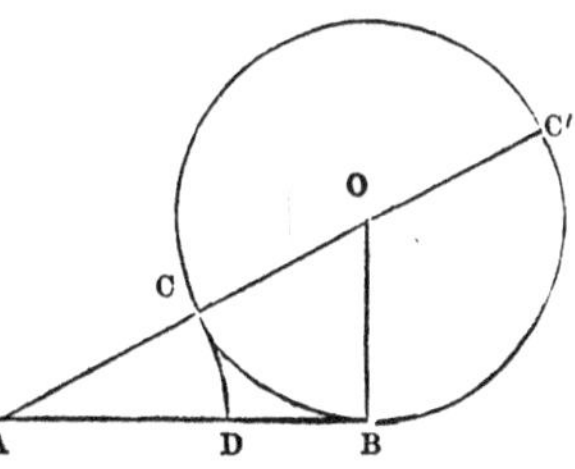


Let AB be the line which we are required to divide.

Draw BO perpendicular to AB and equal to one half of AB; then draw AO, with O as a centre; with OB for a radius, describe the circumference meeting AO in C. Finally, make AD equal to AC, and the line AB will be divided at the point D as required.

In effect, from the construction, AB is a tangent to the circle OB; and if AO is produced until it meets the circumference in C′, we shall have (T. III.),

$$AC' : AB :: AB : AC.$$

Consequently,

$$AC' - AB : AB :: AB - AC : AC.$$

Now, $AC' = AC + CC' = AC + AB$; consequently, $AC' - AB = AC = AD$, and $AB - AC = AB - AD = BD$· hence the proportion evidently becomes

$$AD : AB :: BD : AD,$$

or interchanging means and extremes,

$$AB : AD :: AD : BD.$$

### PROBLEM VI.

*To draw a tangent common to two circumferences.*

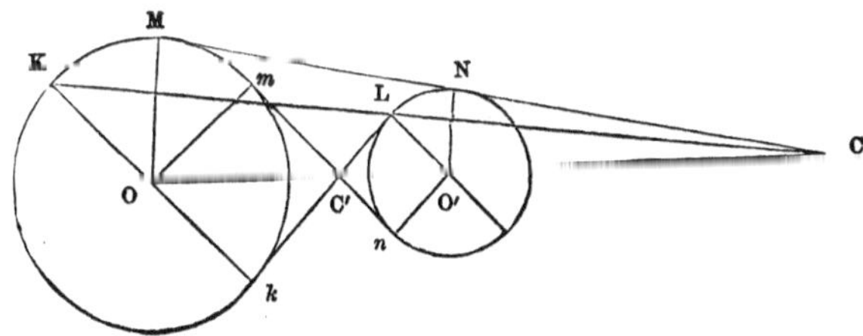


We will suppose the problem solved, and that MN, *mn* are the two tangents, the one *exterior* and the other *interior*, meeting the line of the centres OO′ in two points C, C′.

It is evident that, if these points were known, it would be sufficient to draw through each of them a tangent to one of the circles (B. II., P. XII.), and it would of necessity be tangent to the other; the problem would thus become resolved.

Now, in drawing the radii OM and O′N, O$m$ and O′$n$, we shall evidently obtain two couple of similar triangles OMC and O NC, O$m$C and O′$n$C′, which give the proportions

$$OC : O'C :: OM : O'N,$$
$$OC' : O'C' :: OM : O'N.$$

But the radii OM, O′N, are given lines. We see then that the points C, C′ are the *conjugate* points (B. III., T. XII., S.; also P. II., S.), which divide the distance OO′ in the ratio of OM to O′N. From this results the following construction:

Draw any diameter KO$k$ of the circle O, and through the point O′ draw the radius O′L parallel to OK; join the points K and $k$ with the point L.

The points C, C′, where the straight lines KL, $k$L meet the line of the centres, are the points sought; since we have (B. III., T. IV.),

$$OC : O'C :: OM : O'N,$$
$$OC' : O'C' :: OM : O'N.$$

Now draw through each of these points a tangent to either of these circles, and it will be tangent to the other.

*Scholium.* This problem is evidently susceptible of *four*, *three*, *two*, or *only one* solution, or it may not admit of *any*, according to the *five* relative positions of the two circumferences (B. II., T. XVIII., S.).

---

## PROBLEMS OF AREAS.

### PROBLEM VII.

*To transform a polygon into another having one side less than the first, and finally into a triangle.*

Let ABCDE, etc., be any polygon whatever, which we here represent by a broken line, in order that the generality of the construction may be the better shown.

Through the point A draw the diagonal AC cutting off the triangle ABC. Draw afterwards, through the point B, the line BI parallel to AC, meeting DC, produced, in I, then draw AI.

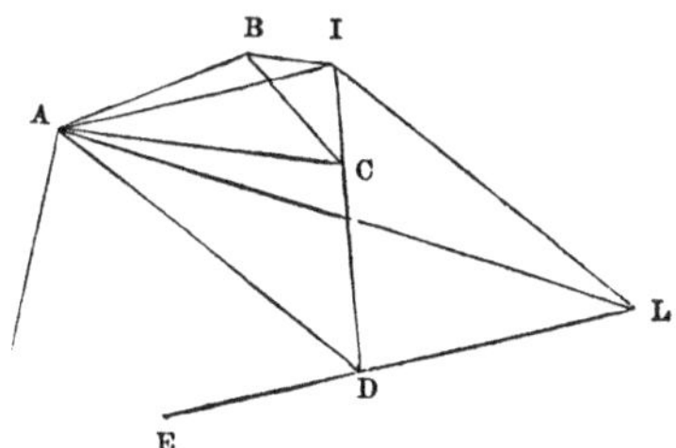


The polygon AIDE, etc., is equivalent to the given polygon ABCDE, etc., and it has one side less.

For, since BI is parallel to AC, the two triangles AIC, ABC are equivalent (B. III., T. XIX., C.), and if to these two triangles we add the portion of the surface ACDE, etc., we shall have, in the first case, the polygon AIDE, etc., and in the second case, the polygon ABCDE, etc.; consequently these two polygons are equivalent.

It is evident, moreover, since the side CI of the triangle AIC is the prolongation of the side DC, that the two sides AB, BC of the first polygon have been replaced by the single side AI; hence the second figure has one side less than the first.

Actually operate upon the polygon AIDE, etc., as upon the first; that is to say, draw the diagonal AD cutting off the triangle AID; afterwards draw IL parallel to AD meeting ED produced in L; we shall thus obtain a new polygon ALE, etc., equivalent to the second and having one side less.

In continuing, in this way, we shall necessarily reach a polygon of three sides, and the problem will be resolved.

### PROBLEM VIII.

*To transform a polygon into a square.*

If the given polygon is a *triangle*, let $b$ represent its base, and $h$ its height, and $x$ the side of the square sought.

We have this condition,

$$x^2 = b \times \tfrac{1}{2}h,$$

from which we obtain the following proportion:

$$b : x :: x : \tfrac{1}{2}h.$$

Thus, the *side* of the square sought is *a mean proportional between the base and half the height of the triangle.*

This line being constructed (P. IV.), we easily obtain the construction of the square.

Whatever may be the polygon, we commence by transforming it into a triangle (P. VII.); afterwards we transform, as above, the triangle into a square.

When the given polygon is a *parallelogram*, or a *rectangle*, or in short, any figure whose area depends immediately upon the *product of two lines*, all that is necessary in order to obtain the side of an equivalent square is *to construct a mean proportional between these two lines.*

Thus, in the case of a regular polygon, it is sufficient, after having *developed* on an indefinite straight line the perimeter of the polygon, to seek a mean proportional between the *half perimeter and the radius of its inscribed circle.*

To transform a circle into a square, it is necessary first to *rectify* the circumference, and afterwards to determine a mean proportional between the *radius* and the *rectified semi-circumference.*

*Scholium.* The *quadrature* of the circle is wholly dependent upon the *rectification* of a circumference; and thus, up to the present, we have not been able, by the assistance of the Ruler and the Compass, to construct rigorously a *square equivalent to a circle*, as we can do for rectilinear figures, since all known methods give only approximate values for the *ratio of the circumference to the diameter.*

---

## CONSTRUCTION OF SIMILAR POLYGONS UNDER CERTAIN CONDITIONS.

### PROBLEM IX.

*Upon a given line, to construct a polygon similar to a given polygon.*

Let *ab* be the given line, and ABCDEF the given polygon.

After decomposing the given polygon into triangles by drawing diagonals from the corner A to the other corners, form at the points $a$ and $b$ two angles $cab$, $abc$, respectively equal to the angles CAB, ABC. We shall thus obtain a triangle $abc$ similar

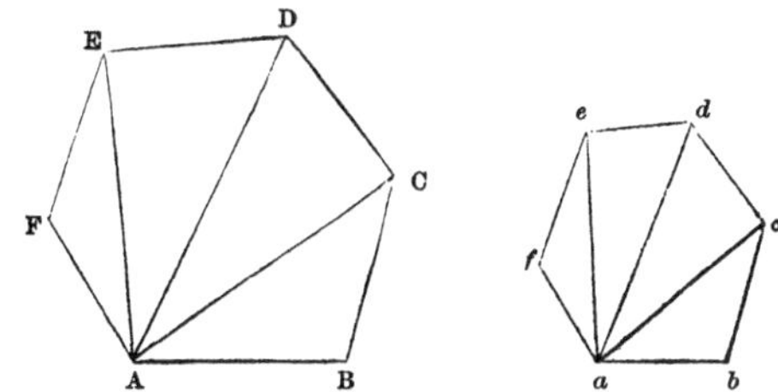


to the triangle ABC. Construct, in the same manner, on $ac$ a triangle $acd$ similar to the triangle ACD; and so on for the other triangles.

The polygon $abcdef$ thus obtained will be similar to the given polygon ABCDEF.

### PROBLEM X.

*Two similar polygons being given, it is required to construct a third polygon similar to the two first, and equivalent to their sum or to their difference.*

The solution of this problem is an immediate consequence of C. IV., T. XXIX., B. III.

Let $a$, $a'$ be two homologous sides of the given polygons A, A′. Upon these sides, regarded as the sides about the right angle, or as the hypotenuse and one of the sides of the right angle, construct a right-angled triangle; afterwards, on the third side $a''$ of the triangle thus obtained, construct (P. IX.) a polygon A″ similar to one of the given polygons.

The polygon thus constructed will be the polygon required.

For, by construction, $a''^2 = a^2 + a'^2$, or $a''^2 = a^2 - a'^2$; hence, (B. III., T. XXIX., C. IV.),

$$A'' = A + A', \text{ or } A'' = A - A'.$$

### PROBLEM XI.

*To find a square which shall be to a given square in the ratio of two given lines.*

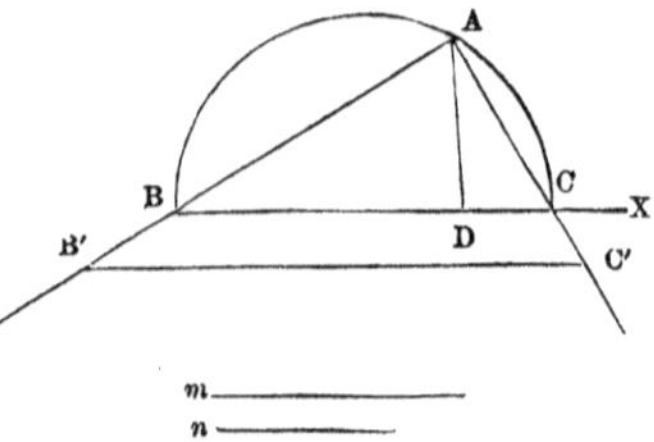


On an indefinite straight line BX take $BD = m$, one of the given lines, $DC = n$, the other given line, and on $BC = m + n$, as a diameter, describe a semi-circumference; having drawn DA perpendicular to BC, draw the chords AB, AC, and prolong them if necessary beyond B and C. Upon AB take $AB' = a$, the side of the given square, which we here suppose to be greater than AB, and draw B′C′ parallel to BC. The line AC′ will be the side of the square sought.

In effect, the two similar triangles ABC, AB′C′, give the proportion

$$AB : AC :: AB' : AC'.$$

Consequently, $AB^2 : AC^2 :: AB'^2 : AC'^2$;

but we have (B. III., T. XIV., S. I.),

$$AB^2 : AC^2 :: BD : DC, \text{ or } :: m : n;$$

hence, $AB'^2 : AC'^2 :: m : n$;

or, putting $a$ for AB′, and changing the order of the terms of the proportion, we have

$$m : n :: a^2 : AC'^2.$$

PROBLEM XII.

*A polygon* P *being given, to construct a second polygon* X *similar to the first, and such that the first may be to the second as m to n.*

Let $a$ denote the side of the given polygon, and $x$ the homologous side of the sought polygon. We shall have

$$P : X :: a^2 : x^2,$$
$$P : X :: m : n;$$

consequently, $m : n :: a^2 : x^2$.

The question is thus made to depend upon the preceding problem; $x$ being known, it is sufficient then to construct on this line a polygon similar to the given polygon.

### PROBLEM XIII.

*Having given two polygons,* P *and* Q, *to construct a third polygon* X *similar to the first, and such that the second may be to the third in the ratio of m to n.*

Let $a$ denote a side of the polygon P, and $x$ the homologous side of the polygon X.

Then we shall have, by the conditions of the problem,

$$P : X :: a^2 : x^2,$$
$$Q : X :: m : n.$$

This being supposed, conceive the polygons P and Q transformed into equivalent squares (P. VIII.), having respectively for their sides $p$ and $q$. In the same manner let $y$ represent the side of a square equivalent to X; the preceding proportions will be changed into the following:

$$p^2 : y^2 :: a^2 : x^2;$$

hence,
$$p : y :: a : x,$$

and
$$p^2 : y^2 :: m : n.$$

Now, the last of these proportions will make known $y$ by the construction of Problem XI., and the second will then give $x$ as a fourth proportional to the lines $p$, $y$, and $a$, which will be the side of the polygon sought, homologous to $a$.

### PROBLEM XIV.

*To construct a rectangle equivalent to a given square* $m^2$, *and such that the sum or the difference of two adjacent sides may be equal to a given line a.*

*First case.* The given line $a$ being the sum of the adjacent sides.

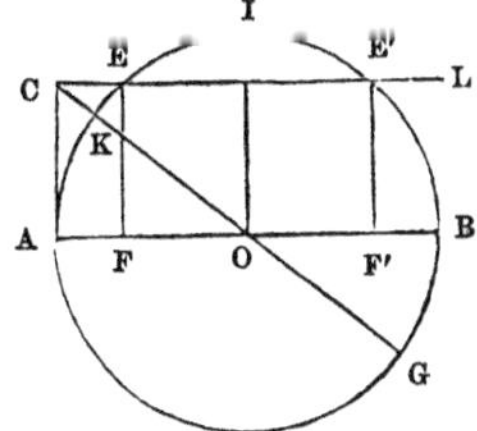


Upon a straight line AB $= a$, as a diameter, describe a circumference; through A the extremity of the diameter AB, draw the perpendicular AC$=m$. Through C draw CL parallel to AB, meeting the circumference in the points E, E'. From the points E and E' draw EF, E'F' perpendicular to AB.

The distances AF and FB, or AF′ and F′B, will be the two adjacent sides of the rectangle required to be constructed.

In effect, we have (T. I., S. I.),

$$AF \times FB = EF^2 = AC^2 = m^2,$$

and
$$AF + FB = AB = a.$$

The problem is *impossible* if AC > OI or OA; that is, when $m > \frac{1}{2}a$.

If $m = \frac{1}{2}a$, the points E, E′ will unite at I, and the rectangle will become a square.

This demonstrates that *the greatest rectangle which we can form by dividing a given line into two parts for the adjacent sides, is the square constructed on half the given line.*

*Second case.* The given line being the difference of two adjacent sides.

After having drawn $AC = m$, as in the first case, we join C and O, and produce it so as to meet the circumference at the points K and G. The two lines CG, CK will be the adjacent sides of the rectangle sought.

We have, by the construction (T. III.),

$$CA^2 = CG \times CK, \text{ or } CG \times CK = m^2,$$

and
$$CG - CK = KG = AB = a.$$

In this case, the problem is always possible for all values of $a$ and $m$.

# FIFTH BOOK.

---

## OF THE PLANE, AND ITS COMBINATION WITH THE STRAIGHT LINE.

### DEFINITIONS.

I. The *intersection* of two planes is the line in which they meet to cut each other.

It is obvious, from our definition of a plane (B. I., D. XVII.), that this intersection is a straight line.

II. A line is *perpendicular to a plane*, when it is perpendicular to all the lines in that plane which meet it.

III. One *plane is perpendicular* to another, when every line in the one which is perpendicular to their intersection is perpendicular to the other plane.

IV. A line is *parallel to a plane*, when, if both are produced to any distance, they do not meet; and, conversely, the plane is then also parallel to the line.

V. Two *planes are parallel to each other*, when, both being produced to any distance, they do not meet.

VI. The inclination of two planes which intersect, is called a *diedral angle*.

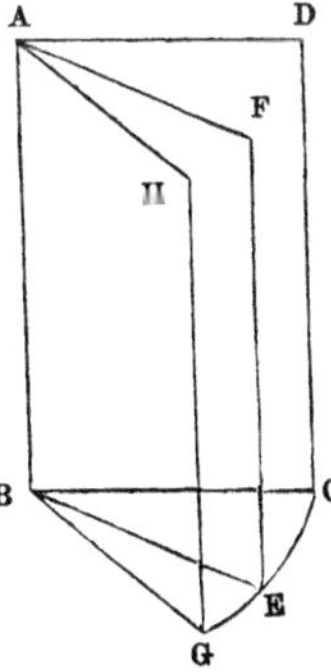


The planes forming a diedral angle are called the *faces* of the angle, and their intersection is called its *edge*.

If a rectangle ABCD is revolved about its side AB as an axis, so as to assume in succession the positions ABEF, ABGH, etc., the point C will describe the circumference of a circle. In fact, any point of the plane of this rectangle, not situated in the axis AB, will, in a similar manner, describe the circumference of a circle having its centre in the axis. During this revolution, the arc CE will increase in the same ratio as the opening or inclination of the planes ABCD, ABEF, increases; that is, as the *diedral angle* DABE increases. Now, the arc CE is the measure of the angle CBE (B. II., T. IX., S.), hence a diedral angle has for its measure the angle contained by two lines drawn from any point of its edge, perpendicular to the same, one in each face.

VII. When several planes pass through a common point, the angular space included between these planes is called a *polyedral angle.* Each plane is called a *face;* the line in which any two faces intersect is called an *edge;* and the common point through which the planes pass is called the *vertex.*

Three planes, at least, are necessary for forming a polyedral angle. In the case of three planes, the angle is called a *triedral angle.*

THEOREM I.

*One part of a straight line cannot be in a plane, and another part out of it.*

For (B. I., D. XVII.), when a straight line has two points common with a plane, it must coincide with the plane.

THEOREM II.

*Two straight lines which intersect each other, are situated in the same plane, and determine its position.*

Let AB, AC be two straight lines which intersect each other in A; and conceive a plane passing through one of the lines, as AB, and if also AC be in this plane, then it is clear that the two lines, according to the terms of the proposition, are in the same plane; but if not, let the plane passing through AB be supposed to be turned round AB till it passes through the point C, then the line AC, which has two of its points, A and C, in this plane, coincides with it; and

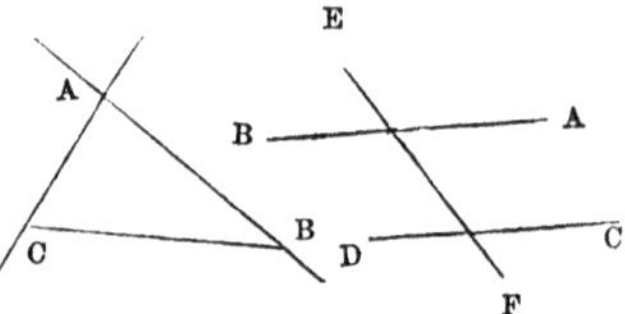

hence the position of the plane is determined by the single condition of containing the two straight lines AB, AC.

*Cor.* I. A triangle ABC, or any three points not in a straight line, determines the position of a plane.

*Cor.* II. Hence, also, two parallels AB, CD determine the position of a plane; for, drawing the secant EF, the plane of the two straight lines AB, EF is that of the parallels AB, CD.

### THEOREM III.

*The intersection of two planes is a straight line.*

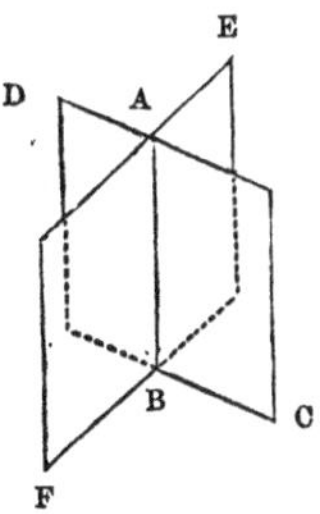


Let DC and EF be two planes cutting each other, and A, B two points in which the planes meet. Draw the straight line AB; this line is the intersection of the two planes.

For, since the straight line touches the two planes in the points A and B, it lies wholly in both these planes, or is common to both of them; that is, the intersection of the two planes is in a straight line.

### THEOREM IV.

*If a straight line is perpendicular to each of two straight lines at their point of intersection, it is perpendicular to the plane of these lines.*

Let AP be perpendicular to the two lines PB, PC, at their point of intersection P; then will it be perpendicular to MN the plane of the lines.

Through P draw in the plane MN any line, as PQ; and through any point of this line, as Q, draw BQC, so that BQ = QC (B. IV., P. III., C. II.); join AB, AQ, AC.

The base BC being divided into two equal parts at the point Q, the triangle BPC (B. III., T. XVII.) will give

$$PC^2 + PB^2 = 2PQ^2 + 2QC^2.$$

The triangle BAC will, in like manner, give

$$AC^2 + AB^2 = 2AQ^2 + 2QC^2.$$

Taking the first equation from the second, and observing that the triangles APC, APB, which are both right-angled at P, give $AC^2 - PC^2 = AP^2$, and $AB^2 - PB^2 = AP^2$, we shall have

$$2AP^2 = 2AQ^2 - 2PQ^2,$$

or,

$$AP^2 = AQ^2 - PQ^2;$$

that is,

$$AQ^2 = AP^2 + PQ^2.$$

Hence the triangle APQ is right-angled at P, and therefore AP is perpendicular to PQ.

*Scholium.* Thus it is evident, not only that a straight line may be perpendicular to all the straight lines which pass through its foot in a plane, but that it always must be so, whenever it is perpendicular to two straight lines drawn in the plane.

*Cor.* At a given point P on a plane, it is impossible to draw more than one perpendicular to that plane. For, if there could be two perpendiculars at the same point P, draw along these two perpendiculars a plane, whose intersection with the plane MN is PQ; then those two perpendiculars would be perpendicular to the line PQ at the same point, and in the same plane, which is impossible.

It is also impossible to draw, from a given point out of a plane, two perpendiculars to that plane. For, let AP, AQ be these two perpendiculars, then the triangle APQ would have two right angles APQ, AQP, which is impossible.

### THEOREM V.

*If from a point without a plane, a perpendicular and several oblique lines be drawn to this plane:*

I. *The perpendicular will be shorter than any oblique line.*

II. *Any two oblique lines which terminate at equal distances from the foot of the perpendicular, will be equal.*

III. *Of two oblique lines, terminating at unequal distances from the foot of the perpendicular, the one at the greater distance will be the longer.*

Let O be the given point, MN the given plane, OP perpendicular, and OA, OA′, OA″, and OB oblique lines.

*First.* Since OA is an oblique line, the right-angled triangle OPA gives OP < OA (B. I., T. XXVII.).

*Secondly.* If PA = PA′, the two right-angled triangles OPA, OPA′ will be equal (B. I., T. XX., C.), and give OA′ = OA.

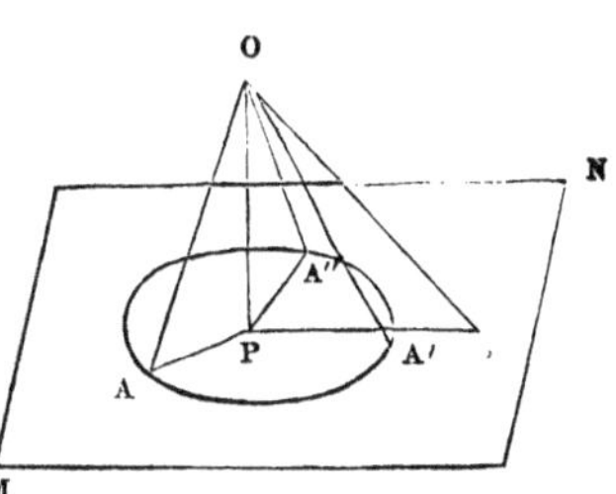


*Thirdly.* If PB > PA, take on PB a distance PA′ = PA, and draw OA′, and we shall have OB > OA′ (B. I., T. XII.); but we have OA′ = OA, consequently OB > OA.

*Cor.* I. The perpendicular drawn from any point to a plane, measures the true distance of the point from the plane.

*Cor.* II. Any point whatever in a line perpendicular to a plane, is equally distant from all the points of the circumference described about the foot of the perpendicular as a centre, with any radius whatever.

*Scholium* I. This perpendicular is called the *axis* of the circle. All the points of the axis may be regarded as centres of the circle, and the oblique lines as corresponding radii, any one of which may be used for describing the circle.

*Scholium* II. By means of an extended thread, one extremity being fixed at the point O, and the other extremity attached to a pencil, we may determine any three points, as A, A′, A″, equally distant from the point O. If we then describe the circumference of a circle passing through these three points, its centre will be the foot of the perpendicular drawn from O to the plane.

*Scholium* III. When a plane passes perpendicularly through the middle of a straight line:

1. All points of this plane are equally distant from the extremities of this line.

2. All points situated without this plane are unequally distant from these extremities.

THEOREM VI.

*If, from a point without a plane, a perpendicular be drawn to that plane, and from the foot of the perpendicular a line be drawn perpendicular to a line in the plane, and from the point of intersection a line be drawn to the first point, this last line will be perpendicular to the line in the plane.*

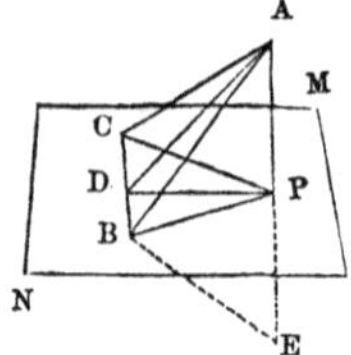


Let AP be a line perpendicular to the plane MN, and PD perpendicular to BC; then will AD be perpendicular to BC.

Take DB = DC, and join PB, PC, AB, AC. Since PD is perpendicular at D, the middle point of BC, PB is equal to PC (B. I., T. XII.); consequently the oblique lines AB, AC are equal, and the two triangles ADB, ADC have the three sides of the one equal to the three sides of the other; therefore they are equal (B. I., T. XXV.), and the angle ADB = ADC; hence each is a right angle, and AD is perpendicular to BC (B. I., D. XII.).

*Cor.* It is evident, likewise, that BC is perpendicular to the plane APD, since BC is at once perpendicular to the two straight lines AD, PD (T. IV.).

*Scholium.* The two straight lines AE, BC afford an instance of two lines not parallel which do not meet, because they are not situated in the same plane. The shortest distance between these lines is the straight line PD, which is perpendicular both to the line AP and to the line BC. For if we join any other two points, such as A and B, we shall have AB > AD, AD > PD, therefore AB > PD.

The two lines AE, CB, though not situated in the same plane, are conceived as forming a right angle with each other, because AE and the line drawn through one of its points parallel to BC would make with each other a right angle. In the same manner, the line AB and the line PD, which represent any two straight lines not situated in the same plane, are supposed to form with each other the same angle which would be formed by AB and a straight line parallel to PD drawn through any point of AB.

### THEOREM VII.

*If one of two parallel lines is perpendicular to a plane, the other will also be perpendicular to this plane.*

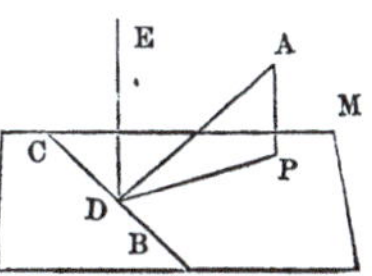


Let AP and ED be parallel lines, of which AP is perpendicular to the plane MN; then will ED be also perpendicular to this plane.

Along the parallels AP, DE extend a plane; its intersection with the plane MN will be PD. In the plane MN draw BC perpendicular to PD, and join AD.

Then BC is perpendicular to the plane APDE (T. VI., Cor.), and therefore the angle BDE is right. But the angle EDP is right also, since AP is perpendicular to PD, and DE parallel to AP (B. I., T. XV., Cor.); therefore the line DE is perpendicular to the two straight lines DP, DB, hence it is perpendicular to their plane MN (T. IV.).

*Cor.* I. Conversely, if the straight lines AP, DE are perpendicular to the same plane MN, they will be parallel. For, if they be not so, draw through the point D a line parallel to AP; this parallel will be perpendicular to the plane MN; therefore, through the same point D more than one perpendicular might be drawn to the same plane, which (T. IV., C.) is impossible.

*Cor.* II. Two lines, A and B, parallel to a third line C, are parallel to each other. For, conceive a plane perpendicular to the line C; the lines A and B, being parallel to C, will be perpendicular to the same plane; therefore, by the preceding Corollary, they will be parallel to each other.

### THEOREM VIII.

*If a straight line without a plane is parallel to a line in the plane, it is parallel to the plane itself.*

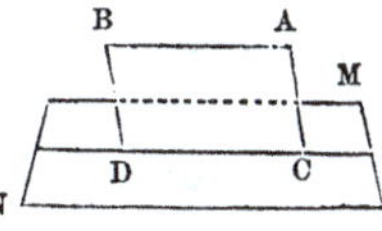


Let the straight line AB, without the plane MN, be parallel to the line CD of this plane; then will AB be parallel to the plane MN.

For, if the line AB, which lies in the plane ABDC, could

meet the plane MN, this could only be in some point of the line CD, the intersection of the two planes; but AB cannot meet CD, since they are parallel; hence it will not meet the plane MN, therefore (D. IV.) it is parallel to that plane.

### THEOREM IX.

*If two planes are perpendicular to the same line, they are parallel.*

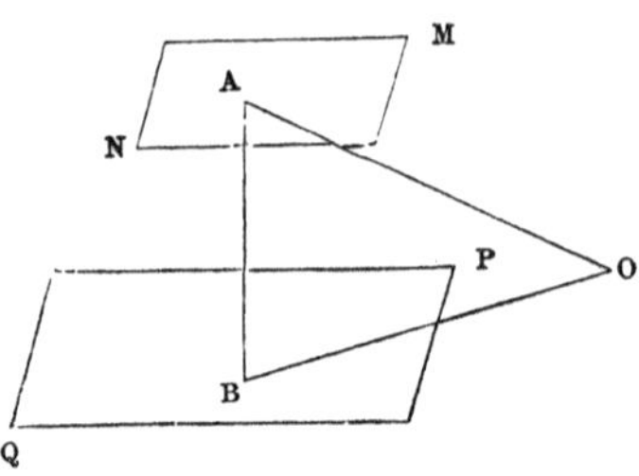


Let the planes MN and PQ be each perpendicular to AB; then will they be parallel.

For, if they can meet anywhere, let O be one of their common points, and join OA, OB. The line AB, which is perpendicular to the plane MN, is perpendicular to the straight line AO drawn through its foot in that plane. For the same reason, AB is perpendicular to BO. Therefore OA and OB are two perpendiculars drawn from the same point O, upon the same straight line, which is impossible; hence the planes MN, PQ cannot meet each other, and consequently they are parallel.

### THEOREM X.

*The intersections of two parallel planes with a third plane are parallel.*

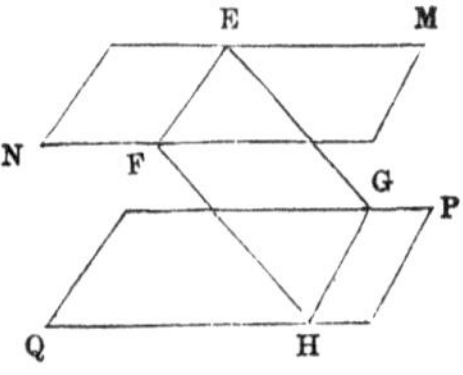


Let the two parallel planes MN and PQ intersect the plane EH; then will EF be parallel to HG.

For, if the lines EF, GH, lying in the same plane, were not parallel, they would meet each other when produced; therefore the planes MN, PQ, in which those lines are situated, would also meet, and the planes would not be parallel.

### THEOREM XI.

*A line which is perpendicular to one of two parallel planes, is perpendicular to the other also.*

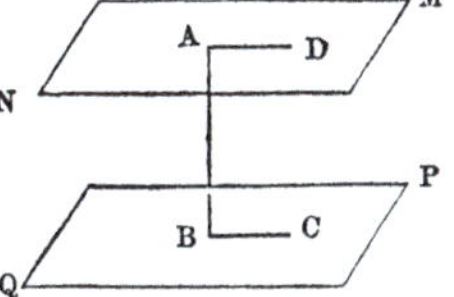


Let the two planes MN and PQ be parallel; then if the line AB is perpendicular to the plane MN, it will also be perpendicular to PQ.

Having drawn through B any line BC in the plane PQ, along the lines AB and BC, extend a plane ABC, intersecting the plane MN in AD; the intersection AD will be parallel to BC (T. X.). But the line AB, being perpendicular to the plane MN, is perpendicular to the straight line AD; therefore, also, it is perpendicular to its parallel BC (B. I., T. XV., Cor.). Hence the line AB, being perpendicular to any line BC drawn through its foot in the plane PQ, is consequently perpendicular to that plane.

### THEOREM XII.

*Two parallel lines, included between two parallel planes, are equal.*

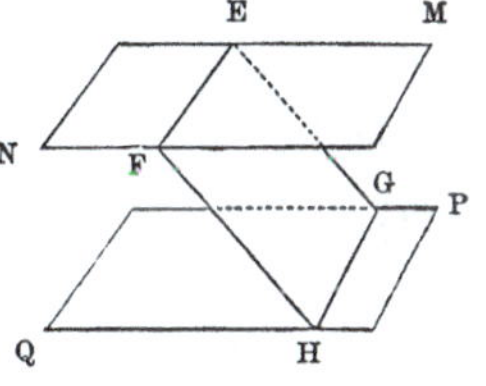


Let EG, FH be two parallel lines included between the two parallel planes MN, PQ; then will these lines be equal.

Through the parallels EG, FH, draw the plane EGHF to meet the parallel planes in EF and GH. The intersections EF, GH (T. X.) are parallel to each other, so likewise are EG, FH; therefore the figure EGHF is a parallelogram, and EG = FH.

*Cor.* Hence it follows, that *two parallel planes are everywhere equidistant;* for, if EG and FH are perpendicular to the two planes MN, PQ, they will be parallel to each other (T. VII., C.), and therefore equal.

### THEOREM XIII.

*If two angles, not situated in the same plane, have their sides parallel and lying in the same direction, they will be equal, and the planes in which they are situated will be parallel.*

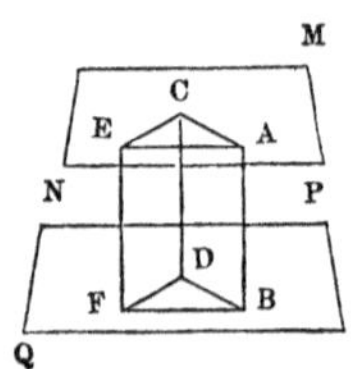


Let CAE, DBF be two angles not situated in the same plane, having AC parallel to BD and lying in the same direction, and AE parallel to BF, and also lying in the same direction; then will these angles be equal, and their planes will be parallel.

Make AC = BD, AE = BF; and join CE, DF, AB, CD, EF. Since AC is equal and parallel to BD, the figure ABDC is a parallelogram (B. I., T. XXX.), therefore CD is equal and parallel to AB. For a similar reason, EF is equal and parallel to AB; hence, also, CD is equal and parallel to EF. The figure CEFD is therefore a parallelogram, and the side CE is equal and parallel to DF; therefore the triangles CAE, DBF have their corresponding sides equal, and consequently the angle CAE = DBF.

Again, the plane ACE is parallel to the plane BDF. For, suppose the plane parallel to BDF, drawn through the point A, were to meet the lines CD, EF in points different from C and E, for instance in G and H, then (T. XII.) the three lines AB, GD, HF would be equal. But the lines AB, CD, EF are already known to be equal, hence CD = GD, and HF = EF, which is absurd; hence the plane ACE is parallel to BDF.

*Cor.* If two parallel planes MN, PQ are met by two other planes not parallel CABD, EABF, the angles CAE, DBF formed by the intersections of the parallel planes will be equal; for (T. X.) the intersection AC is parallel to BD, and AE to BF, and therefore the angle CAE = DBF.

### THEOREM XIV.

*If three straight lines, not situated in the same plane, are equal and parallel, the triangles formed by joining their corresponding opposite extremities will be equal, and their planes will be parallel.*

Let ACE, BDF be two triangles formed by joining the opposite extremities of the three equal and parallel lines AB, CD, EF; then will these triangles be equal, and their planes will be parallel.

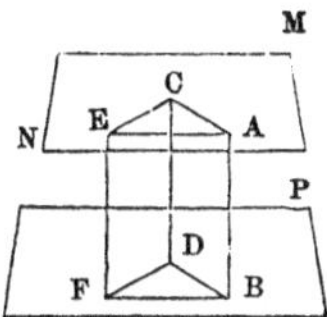


For, since AB is equal and parallel to CD, the figure ABDC is a parallelogram (B. I., T. XXX.); hence the side AC is equal and parallel to BD. For a like reason, the sides AE, BF are equal and parallel, as also CE, DF. Therefore the two triangles ACE, BDF are equal; and, consequently, as in the last proposition, their planes are parallel.

### THEOREM XV.

*If two straight lines are cut by three parallel planes, they will be divided proportionally.*

Suppose the line AB to be cut by the parallel planes MN, PQ, RS, at the points A, E, B; and the line CD to be cut by the same planes at the points C, F, D; then

AE : EB : : CF : FD.

Draw AD meeting the plane PQ in G, and join AC, EG, GF, BD. The intersections EG, BD, of the parallel planes PQ, RS in the plane ABD, are parallel (T. X.); therefore,

AE : EB : : AG : GD (B. III., T. III.).

In like manner, the intersections AC, GF being parallel,

AG : GD : : CF : FD.

The ratio AG : GD is the same in both; hence

AE : EB : : CF : FD.

### THEOREM XVI.

*If a straight line is perpendicular to a plane, then every plane passing through this line will also be perpendicular to the first plane.*

Let the line AP be perpendicular to the plane MN; then any plane, as AB, passing through this line, will also be perpendicular to the plane MN.

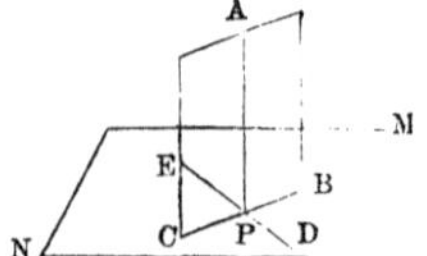


For, let BC be the intersection of the planes AB, MN. In the plane MN draw DE perpendicular to BP; then the line AP, being perpendicular to the plane MN, will be perpendicular to each of the two straight lines BC, DE. But the angle APD, formed by the two perpendiculars PA, PD at their common intersection BP, is the measure of the angle of the two planes (D. VI.); and since, in the present case, the angle is a right angle, the two planes are perpendicular to each other.

*Scholium.* When the three lines, such as AP, BP, DP, are perpendicular to each other, each of these lines is perpendicular to the plane of the other two, and the planes themselves are perpendicular to each other.

THEOREM XVII.

*If two planes are perpendicular to each other, every line drawn in one of them perpendicular to their common intersection, will be perpendicular to the other plane.*

Let the planes AB, MN be perpendicular to each other; and in the plane AB, let PA be drawn perpendicular to the common intersection PB; then will it be perpendicular to the plane MN.

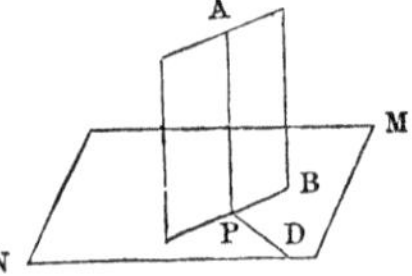


For, in the plane MN draw PD perpendicular to PB; then because the planes are perpendicular, the angle APD is a right angle; therefore the line AP is perpendicular to the two straight lines PB, PD, and consequently perpendicular to their plane MN.

*Cor.* If the plane AB be perpendicular to the plane MN, and if at a point P of the common intersection a perpendicular be drawn to the plane MN, that perpendicular will be in the plane AB. For, if not, then in the plane AB we might draw AP perpendicular to PB their common intersection, and this AP at the same time would be perpendicular to the plane MN; therefore, at the same point P there would be two perpendiculars to the plane MN, which is impossible.

### THEOREM XVIII.

*If two planes be perpendicular to a third plane, their common intersection will be also perpendicular to the third plane.*

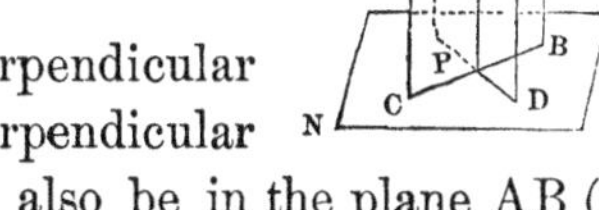


Let AB, AD be perpendicular to MN; then will their common intersection AP be perpendicular to the same plane MN.

For, at the point P draw the perpendicular to the plane MN; then that perpendicular must be in the plane AD; it must also be in the plane AB (T. XVII.), therefore it is their common intersection AP.

### THEOREM XIX.

*If a polyedral angle is formed by three plane angles, the sum of any two of these angles will be greater than the third.*

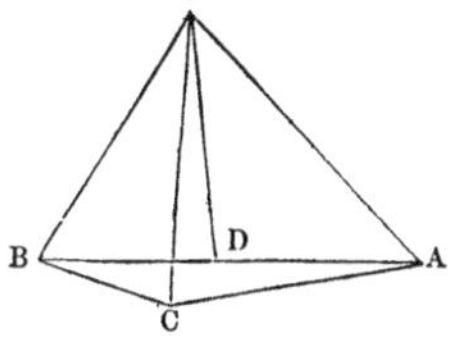


The proposition requires demonstration only when the plane angle, which is compared to the sum of the other two, is greater than either of them. Therefore, suppose the polyedral angle S to be formed by three plane angles ASB, ASC, BSC, whereof the angle ASB is the greatest: we are to show that $ASB < ASC + BSC$.

In the plane ASB make the angle $BSD = BSC$; draw the straight line ADB at pleasure; and, having taken $SC = SD$, join AC, BC.

The two sides BS, SD are equal to the two sides BS, SC; the angle $BSD = BSC$; therefore the triangles BSD, BSC are equal, and $BD = BC$. But $AB < AC + BC$; taking BD from the one side, and from the other its equal BC, there remains $AD < AC$. The two sides AS, SD are equal respectively to the two AS, SC; the third side AD is less than the third side AC; therefore the angle $ASD < ASC$. Adding $BSD = BSC$, we shall have $ASD + BSD$, or $ASB < ASC + BSC$.

### THEOREM XX.

*The sum of the plane angles which form any polyedral angle, is always less than four right angles.*

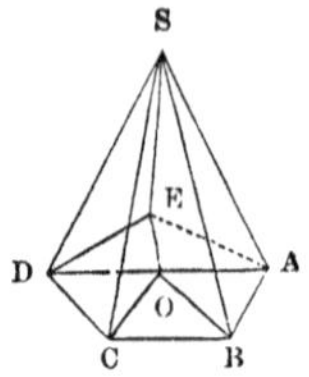

Conceive the polyedral angle S to be cut by any plane ABCDE; from O a point in that plane, draw to the several angles straight lines AO, OB, OC, OD, OE.

The sum of the angles of the triangles ASB, BSC, etc., formed about the vertex S, is equivalent to the sum of the angles of an equal number of triangles AOB, BOC, etc., formed about the point O. But at the point B the angles ABO, OBC, taken together, make the angle ABC (T. XIX.) less than the sum of the angles ABS, SBC. In the same manner, at the point C we have $BCO + OCD < BCS + SCD$, and so with all the angles of the polygon ABCDE. Whence it follows that the sum of all the angles at the bases of the triangles whose common vertex is in O, is less than the sum of all the angles at the bases of the triangles whose common vertex is in S; hence, to make up the deficiency, the sum of the angles formed about the point O, is greater than the sum of the angles about the point S. But the sum of the angles about the point O is equal to four right angles (B. I., T. I., C. III.); therefore the sum of the plane angles, which form the polyedral angle S, is less than four right angles.

*Scholium.* This demonstration is founded on the supposition that the polyedral angle is convex, or that the plane of no one surface produced can ever meet the polyedral angle. If it were otherwise, the sum of the plane angles would no longer be limited, and might be of any magnitude.

### THEOREM XXI.

*If two polyedral angles are composed of three plane angles respectively equal to each other, the planes which contain the equal angles will be equally inclined to each other.*

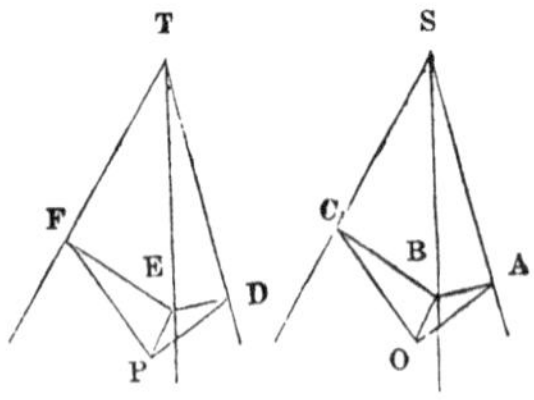

Let the angle $ASC = DTF$, the angle $ASB = DTE$, and the angle $BSC = ETF$; then will the inclination of the planes ASC, ASB be equal to that of the planes DTF, DTE.

Having taken SB at pleasure, draw BO perpendicular to the plane ASC; from the point O at which that perpendicular meets the plane, draw OA, OC perpendicular to SA, SC; join AB, BC; next take TE = SB; draw EP perpendicular to the plane DTF; from the point P draw PD, PF perpendicular to TD, TF; lastly, join DE, EF.

The triangle SAB is right-angled at A, and the triangle TDE is right-angled at D (T. VI.), and since the angle ASB = DTE, we have SBA = TED. Likewise SB = TE, therefore the triangle SAB is equal to the triangle DTE; therefore SA = TD, and AB = DE. In like manner it may be shown that SC = TF, and BC = EF. That granted, the quadrilateral SAOC is equal to the quadrilateral TDPF; for, place the angle ASC upon its equal DTF, because SA = TD, and SC = TF, the point A will coincide with D, and the point C with F; and at the same time AO, which is perpendicular to SA, will coincide with PD, which is perpendicular to TD, and in like manner CO with FP; wherefore the point O will coincide with the point P, and AO will be equal to DP. But the triangles AOB, DPE are right-angled at O and P; the hypotenuse AB = DE, and the side AO = DP; hence those triangles are equal—therefore the angle OAB = PDE. The angle OAB is the inclination of the two planes ASB, ASC, and the angle PDE is that of the two planes DTE, DTF; hence these two inclinations are equal to each other.

It must, however, be observed, that the angle A of the right-angled triangle AOB is properly the inclination of the two planes ASB, ASC, only when the perpendicular BO falls on the same side of SA as SC falls; for if it fell on the other side, the angle of the two planes would be obtuse, and, added to the angle A of the triangle OAB, it would make two right angles. But, in the same case, the angle of the two planes TDE, TDF would also be obtuse, and, added to the angle D of the triangle PDE, it would make two right angles; and the angle A being thus always equal to the angle at D, it would follow, in the same manner, that the inclination of the two planes ASB, ASC must be equal to that of the two planes DTE, DTF.

# SIXTH BOOK.

## BODIES BOUNDED BY PLANES.

### DEFINITIONS.

I. A solid, bounded by polygons, is called a *Polyedron*. The bounding polygons are called *faces*, the straight lines formed by the intersection of any two faces are called *edges*, and the points where three or more faces meet are called *corners*.

A straight line joining any two corners not situated in the same face is called a *diagonal*.

II. A polyedron, having two of its faces equal and parallel, and all the other faces parallelograms, is called a *Prism*. The two equal and parallel faces are called the *bases* of the prism, usually referred to as the *lower base* and the *upper base;* and the parallelograms which make up the other faces, taken together, constitute the *lateral surface* of the prism.

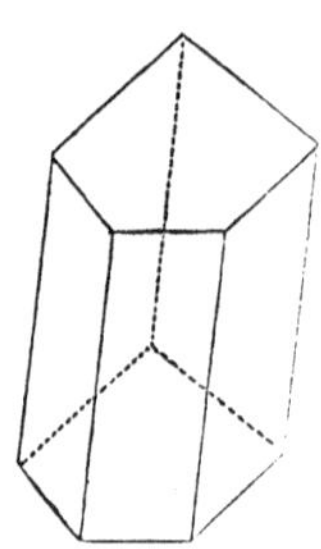

III. A prism is *triangular*, *quadrangular*, *pentagonal*, etc., according as the bases are triangles, quadrilaterals, pentagons, etc.

IV. When the edges formed by the intersection of the lateral faces of a prism are perpendicular to its bases, it is a *right prism:* each edge is then equal to its altitude. In all other cases the prism is said to be *oblique*, and the edges are greater than the altitude.

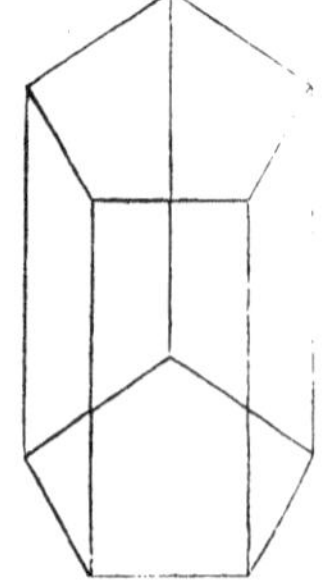

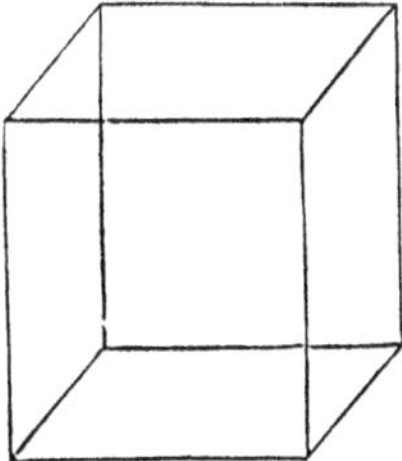

V. A prism, whose bases are parallelograms, is called a *Parallelopipedon*. When all the faces are rectangles it is called a *Rectangular Parallelopipedon*. When all the faces are squares it is called a *Cube*, or *regular hexaedron*.

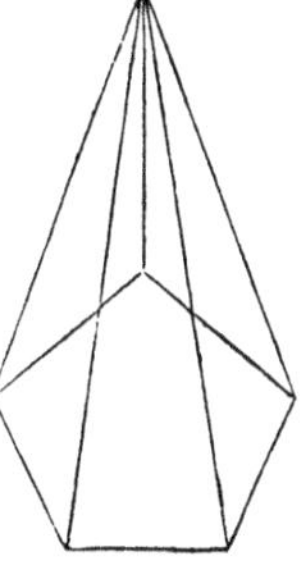

VI. A *Pyramid* is a solid formed by several triangular planes meeting in a point, called the *vertex*, and terminating in the sides of a polygon, forming its *base*. The triangles meeting at the vertex, taken together, constitute the *lateral surface* of the pyramid. The perpendicular distance from the vertex to the base is called its *altitude*.

VII. A pyramid is *triangular*, *quadrilateral*, *pentagonal*, etc., according as the base is a triangle, quadrilateral, pentagon, etc.

VIII. When the base of a pyramid is a regular polygon, and the triangles forming the lateral surface are isosceles, it is a *right pyramid*. The perpendicular drawn from the vertex to the base will, in this case, pass through its centre, and it is then called the *axis* of the pyramid. A line drawn from the vertex of a right pyramid to the middle point of one of the sides of the polygon constituting its base, is called its *slant height*.

IX. A triangular pyramid, having in all four triangular faces, is called a *Tetraedron*.

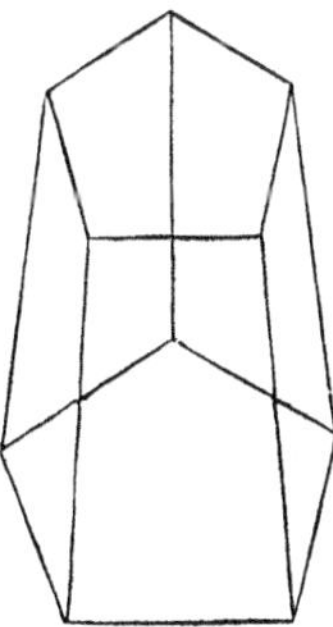

X. If a pyramid is cut by a plane parallel to its base it will be divided into two portions, one of which will be a second pyramid having the same vertex as the first. The remaining portion, having two parallel *bases*, is called a *truncated pyramid*, or *frustum of a pyramid*. The *altitude* of a frustum of a pyramid is the perpendicular distance between its bases. The *slant height* of a frustum of a right pyramid is the portion of its slant height included between its bases.

XI. *Two tetraedrons are similar* when they have their edges proportional and arranged in the same order. Consequently the triangular faces of two similar tetraedrons are similar triangles each to each (B. III., D. II.).

Any two polyedrons whatever, are similar when they are capable of being decomposed into the same number of similar tetraedrons each to each, having the same order of arrangement.

XII. A *regular Polyedron* is one having all its faces equal and regular polygons; and all its diedral angles equal.

THEOREM I.

*The lateral surface of a right prism is equal to the perimeter of its base multiplied by its altitude.*

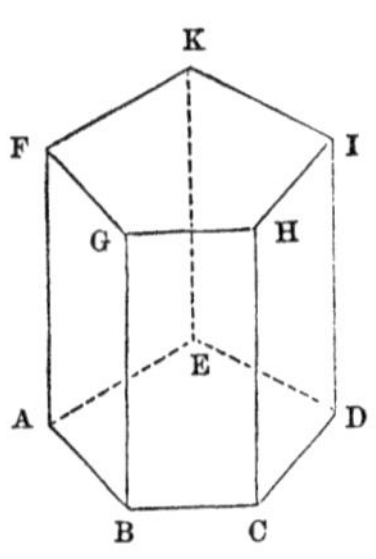


For this surface is equal to the sum of the rectangles AFGB, BGHC, CHID, etc., which compose it. Now the altitudes AF, BG, CH, etc., of those rectangles, are equal to the altitude of the prism; their bases AB, BC, CD, etc., taken together, make up the perimeter of the prism's base: hence the sum of these rectangles, or the lateral surface of the prism, is equal to the perimeter of its base multiplied by its altitude.

*Cor.* If two right prisms have the same altitude, their lateral surfaces will be to each other as the perimeters of their bases.

THEOREM II.

*Two prisms are equal when a polyedral angle in each is contained by three planes, which are respectively equal in both, and similarly situated.*

Let the base ABCDE be equal to the base *abcde;* the parallelogram ABGF equal to the parallelogram *abgf*, and the parallelogram BCHG equal to the parallelogram *bchg;* then will the prism ABCI be equal to the prism *abci.*

For, apply the base ABCDE upon its equal *abcde*, so that

they may coincide. But the three plane angles which form the polyedral angle B are respectively equal to the three plane angles which form the polyedral angle *b*; that is, ABC $=abc$, ABG $=abg$, and GBC $=gbc$, and they are also similarly situated; therefore the polyedral angles at B and *b* are equal (B. V., T. XXI.); hence BG will coincide with its equal *bg*. And it is likewise evident, since the parallelograms ABGF and *abgf* are equal, that the side GF will coincide with its equal *gf*: and in the same manner GH will coincide with *gh*; therefore the upper base FGHIK will coincide with its equal *fghik*, and the two solids will be identical, since their vertices are the same.

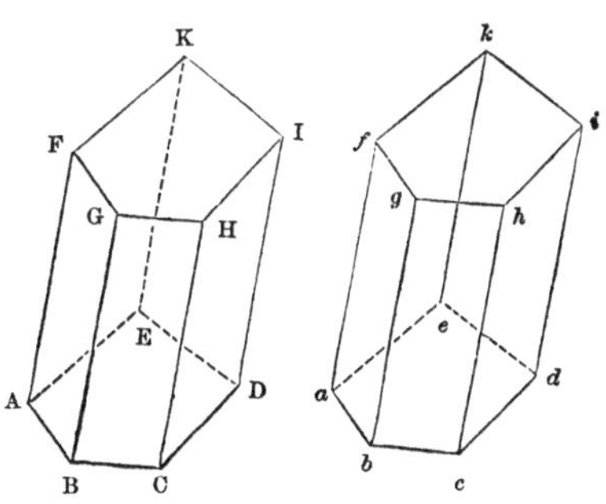


*Cor. Two right prisms, which have equal bases and equal altitudes, are equal.* For, since the side AB is equal to *ab*, and the altitude BG to *bg*, the rectangle ABGF will be equal to *abgf*, and in the same way, the rectangle BGHC will be equal to *bghc*; and thus the three planes which form the solid angle B will be equal to the three planes which form the solid angle *b*: hence the two prisms are equal.

### THEOREM III.

*In every parallelopipedon the opposite faces are equal and parallel.*

By the definition of this solid, the bases ABCD, EFGH are equal parallelograms, and their sides are parallel: it remains only to show that the same is true of any two opposite lateral faces, such as AEHD, BFGC. Now AD is equal and parallel to BC, because the figure ABCD is a parallelogram; for a like reason, AE is parallel to BF: hence the angle DAE is equal to the angle CBF, and the planes DAE, CBF are parallel; hence also the parallelogram DAEH is equal to the parallelogram CBFG.

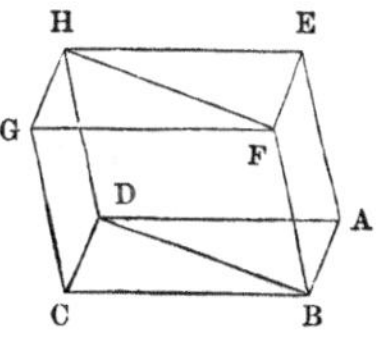

In the same way, it may be shown that the opposite parallelograms ABFE, DCGH are equal and parallel.

*Cor.* Since the parallelopipedon is a solid bounded by six planes, whereof those lying opposite to each other are equal and parallel, it follows that any face and the one opposite to it may be assumed as the bases of the parallelopipedon.

*Scholium.* If three straight lines AB, AE, AD, passing through the same point A, and making given angles with each other, are known, a parallelopipedon may be formed on those lines. For this purpose a plane must be extended through the extremity of each line, and parallel to the plane of the other two; that is, through the point B a plane parallel to DAE, through D a plane parallel to BAE, and through E a plane parallel to BAD. The mutual intersections of these planes will form the parallelopipedon required.

### THEOREM IV.

*In every prism the sections formed by parallel planes are equal polygons.*

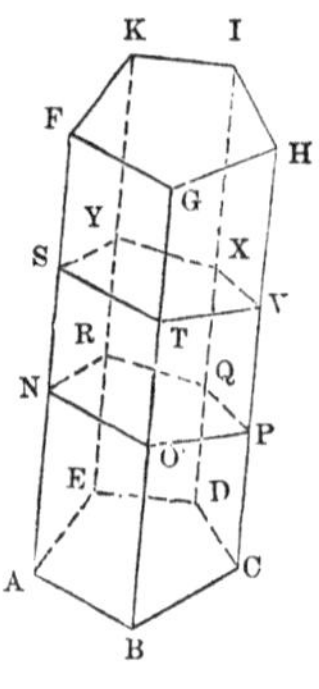


In the prism ABCI, let the sections NOPQR, STVXY be formed by parallel planes; then will these sections be equal polygons.

For the sides ST, NO are parallel, being the intersections of two parallel planes with a third plane ABGF; moreover the sides ST, NO are included between the parallels NS, OT, which are sides of the prism: hence NO is equal to ST. For like reasons, the sides OP, PQ, QR, etc., of the section NOPQR, are respectively equal to the sides TV, VX, XY, etc., of the section STVXY; and since the equal sides are at the same time parallel, it follows that the angles NOP, OPQ, etc., of the first section are respectively equal to the angles STV, TVX, etc., of the second: hence the two sections NOPQR, STVXY are equal polygons.

*Cor.* Every section in a prism, if made parallel to the base, is also equal to that base.

### THEOREM V.

*If a plane be made to pass through the diagonal and opposite edges of a parallelopipedon, so as to divide it into two triangular prisms, those prisms will be equal.*

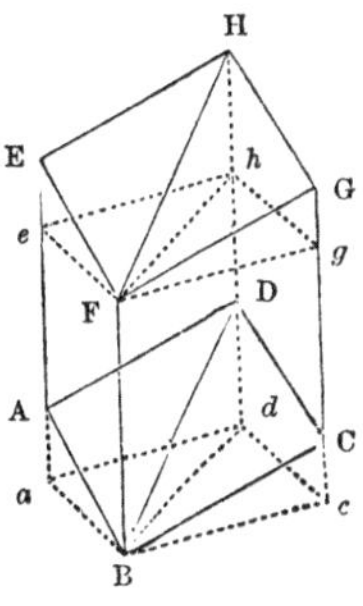


Let the parallelopipedon ABCG be divided by the plane BDHF into the two triangular prisms ABDHEF, BCDFGH; then will those prisms be equal.

Through the vertices B and F, draw the planes B*adc*, F*ehg* at right angles to the side BF, and meeting AE, DH, CG, the three other sides of the parallelopipedon, in the points *a*, *d*, *c* towards one direction, and in *e*, *h*, *g* towards the other: then the sections B*adc*, F*ehg* will be equal parallelograms; being equal, because they are formed by planes perpendicular to the same straight line, and consequently parallel; and being parallelograms, because *a*B, *dc*, two opposite sides of the same section, are formed by the meeting of one plane with two parallel planes ABFE, DCGH.

For a like reason, the figure B*ae*F is a parallelogram; so also are BF*gc*, *cdhg*, and *adhe*, the other lateral faces of the solid B*adc*F*ehg*: hence that solid is a prism (D. IV.), and that prism is a right one, because the side BF is perpendicular to its base.

This being proved, if the right prism B*h* be divided by the plane BFHD into two right-triangular prisms, *a*B*de*F*h*, B*dc*F*hg*, it will remain to be shown that the oblique-triangular prism ABDEFH will be equal to the right-triangular prism *a*B*de*F*h*; and since those two prisms have a part ABD*he*F in common, it will only be requisite to prove that the remaining parts, namely, the solids B*a*AD*d*, F*e*EH*h*, are equal.

Now, by reason of the parallelograms ABFE, *a*BF*e*, the sides AE, *ae* being equal to their parallel BF, are equal to each other; and taking away the common part A*e*, there remains A*a* = E*e*. In the same manner we could prove D*d* = H*h*.

Let us now place the base F*eh* on its equal B*ad*; the point *e* coinciding with *a*, and the point *h* with *d*, the sides *e*E, *h*H will coincide with their equals *a*A, *d*D, because they are perpendicular to the same plane B*ad*. Hence the two solids in question will coincide exactly with each other, and the oblique prism BADFEH is therefore equal to the right one B*ad*F*eh*.

In the same manner might the oblique prism BDCFHG be proved equal to the right prism B*dc*F*hg*. But (T. I.) the two right prisms B*ad*F*eh*, B*dc*F*hg* are equal, since they have the same altitude BF, and since their bases B*ad*, B*dc* are halves of the same parallelogram. Hence the two triangular prisms BADFEH, BDCFHG, being equal to the equal right prisms, are equal to each other.

*Cor.* Every triangular prism ABDHEF is half of the parallelopipedon AG described with the same polyedral angle A, and with the same edges AB, AD, AE.

### THEOREM VI.

*Two parallelopipedons having a common base, and their upper bases in the same plane and between the same parallels, are equivalent to each other.*

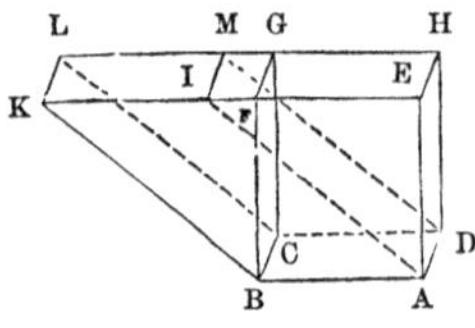


Let the two parallelopipedons AG, AL have the common base ABCD; and let their upper bases EFGH, IKLM be in the same plane, and between the same parallels EK, HL; then will these parallelopipedons be equivalent.

There may be three cases to this proposition, according as EI is greater, less than, or equal to EF; but the demonstration is the same for all. In the first place, then, we shall show that the triangular prism AEIDHM is equal to the triangular prism BFKCGL.

Since AE is parallel to BF, and HE to GF, the angle AEI = BFK, HEI = GFK, and HEA = GFB. Of these six angles, the first three form the polyedral angle E, and the last three the polyedral angle F; therefore, the plane angles being respectively equal and similarly arranged, the polyedral angles F and E must be equal. Now if the prism AEM be laid on the prism BFL, the base AEI being placed on the base BFK, will coincide with it, because they are equal; and since the polyedral angle E is equal to the polyedral angle F, the side EH will coincide with its equal FG; and nothing more is required to prove the coincidence of the two prisms throughout their whole extent, for (T. II.) the base AEI and the edge EH determine the prism AEM, as the base BFK and the edge FG determine the prism BFL; hence these prisms are equal.

But if the prism AEM is taken away from the solid AL, there will remain the parallelopipedon AIL; and if the prism BFL is taken away from the same solid, there will remain the parallelopipedon AEG; hence these two parallelopipedons AIL, AEG are equivalent.

### THEOREM VII.

*Two parallelopipedons having the same base and same altitude, are equivalent to each other.*

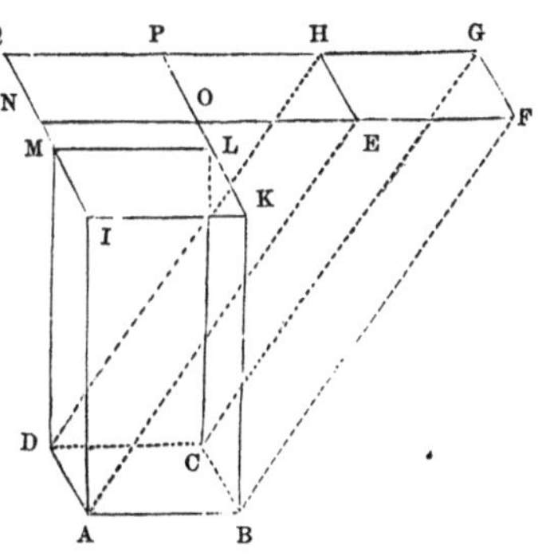


Let ABCD be the common base of the two parallelopipedons AG, AL: since they have the same altitude, their upper bases EFGH, IKLM will be in the same plane. Also the sides EF and AB will be equal and parallel, as well as IK and AB; hence EF is equal and parallel to IK: for a like reason, GF is equal and parallel to LK. Let the sides EF, HG be produced, and likewise LK, IM, till, by their intersections, they form the parallelogram NOPQ: this parallelogram will evidently be equal to either of the bases EFGH, IKLM. Now, if a third parallelopipedon be conceived, having ABCD for its lower base and NOPQ for its upper, this third parallelopipedon will (T. VI.) be equivalent to the parallelopipedon AG; since, with the same lower base, their upper bases lie in the same plane and between the same parallels GQ, FN. For the same reason, this third parallelopipedon will also be equivalent to the parallelopipedon AL; hence the two parallelopipedons AG, AL, which have the same base and the same altitude, are equivalent.

THEOREM VIII.

*Any parallelopipedon may be changed into an equal rectangular parallelopipedon having the same altitude and an equivalent base.*

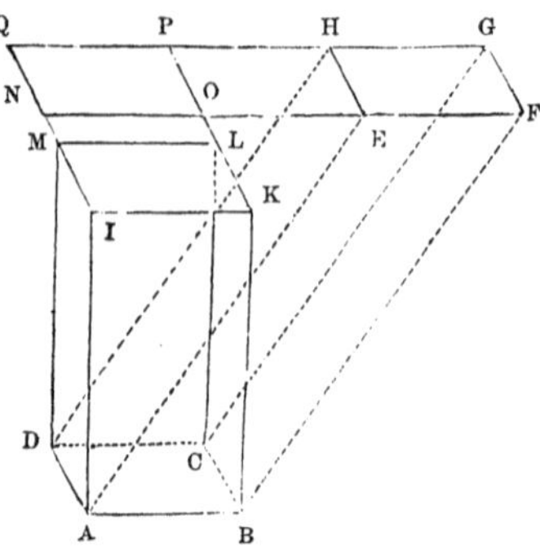


Let AG be the parallelopipedon proposed. From the points A, B, C, D, draw AI, BK, CL, DM perpendicular to the plane of the base; and we shall thus form the parallelopipedon AL equivalent to AG, and having its lateral faces AK, BL, etc., rectangular. Hence, if the base ABCD be a rectangle, AL will be the rectangular parallelopipedon equivalent to AG, the parallelopipedon proposed.

But if ABCD is not a rectangle, draw AO and BN perpendicular to CD, and OQ and NP perpendicular to the base; then the solid ABNOIKPQ will be a rectangular parallelopipedon; for, by construction, the base ABNO and its opposite IKPQ are rectangles; so also are the lateral faces, the edges AI, OQ, etc., being perpendicular to the plane of the base; hence the solid AP is a rectangular parallelopipedon. But the two parallelopipedons AP, AL may be conceived as having the same base ABKI, and the same altitude AO; hence the parallelopipedon AG, which was at first changed into an equivalent parallelopipedon AL, is again changed into an equivalent rectangular parallelopipedon AP, having the same altitude AI, and a base ABNO equivalent to the base ABCD.

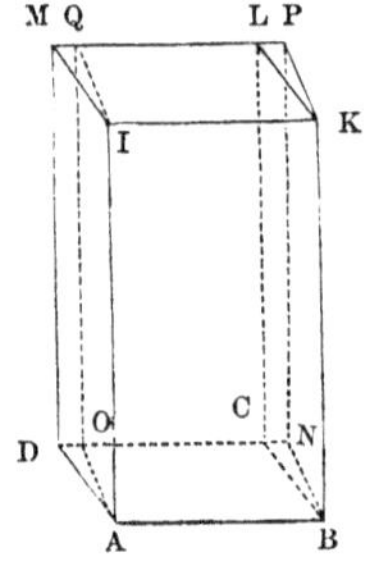

### THEOREM IX.

*Two rectangular parallelopipedons having the same base, are to each other as their altitudes.*

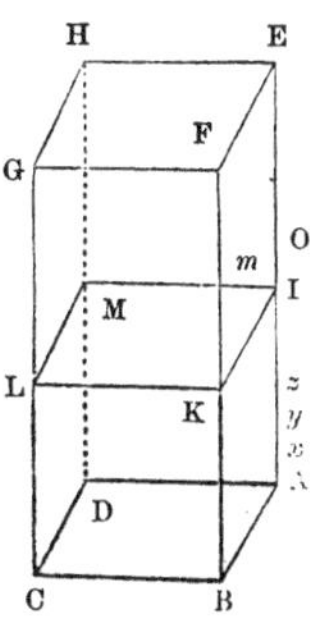


Let the parallelopipedons AG, AL have the common base ABCD; then will they be to each other as their altitudes AE, AI.

First, suppose the altitudes AE, AI to be to each other as two whole numbers; for example, as 15 is to 8. Divide AE into 15 equal parts, whereof AI will contain 8; and through $x$, $y$, $z$, etc., the points of division, draw planes parallel to the base. These planes will cut the solid AG into 15 partial parallelopipedons, all equal to each other, having equal bases and equal altitudes: equal bases, because every section MIKL made parallel to the base ABCD of a prism, is equal to that base (T. IV., C.); and equal altitudes, because these altitudes are the equal divisions A$x$, $xy$, $yz$, etc. But of those 15 equal parallelopipedons, 8 are contained in AL; hence the solid AG is to the solid AL as 15 is to 8; or, generally, as the altitude AE is to the altitude AI.

Again, if the ratio of AE to AI cannot be expressed in whole numbers, it is to be shown that, notwithstanding, we shall have

solid AG : solid AL : : AE : AI.

For, if this proportion is not correct, suppose we have

solid AG : solid AL : : AE : AO,

greater than AI. Divide AE into equal parts, such that each shall be less than OI; there will be at least one point of division $m$ between O and I. Let P be the parallelopipedon, whose base is ABCD and altitude A$m$. Since the altitudes AE, A$m$ are to each other as two whole numbers, we shall have

solid AG : P : : AE : A$m$.

But, by hypothesis, we have

solid AG : solid AL : : AE : AO;

therefore, solid AL : P : : AO : A$m$.

But AO is greater than A$m$: hence, if the proportion is correct, the solid AL must be greater than P. On the contrary, however, it is less; hence the fourth term of this proportion, solid AG : solid AL : : AE : $x$, cannot possibly be a line greater than AI.

By the same mode of reasoning, it might be shown that the fourth term cannot be less than AI; therefore it is equal to AI. Hence rectangular parallelopipedons having the same base, are to each other as their altitudes.

### THEOREM X.

*Two rectangular parallelopipedons having the same altitude, are to each other as their bases.*

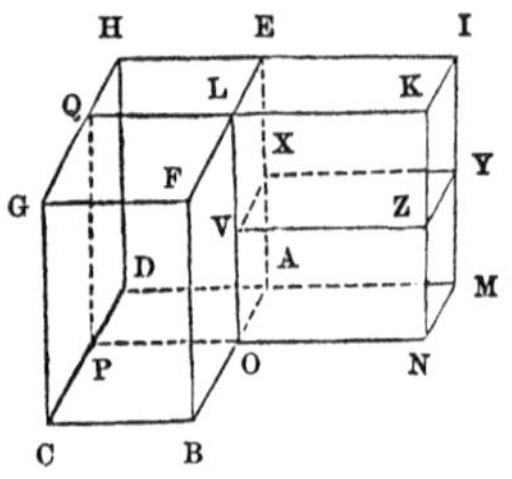


Let the two rectangular parallelopipedons AG, AK have the same altitude; then will they be to each other as their bases ABCD, AMNO.

Produce the plane ONKL till it meets the plane DCGH in PQ; we shall thus have a third parallelopipedon AQ, which may be compared with each of the parallelopipedons AG, AK. The two solids AG, AQ having the same base AEHD, are to each other as their altitudes AB, AO. In like manner, the two solids AQ, AK having the same base AOLE, are to each other as their altitudes AD, AM. Hence we have the two proportions

sol. AG : sol. AQ : : AB : AO;
sol. AQ : sol. AK : : AD : AM.

Multiply together the corresponding terms of these proportions, omitting in the result the common multiplier sol. AQ: we shall have

sol. AG : sol. AK : : AB × AD : AO × AM.

But AB × AD represents the base ABCD, and AO × AM represents the base AMNO; hence two rectangular parallelopipedons of the same altitude are to each other as their bases.

THEOREM XI.

*Any two rectangular parallelopipedons are to each other as the products of their bases by their altitudes; that is to say, as the products of their three dimensions.*

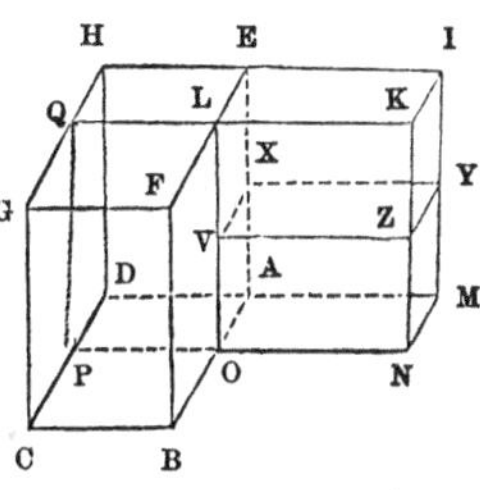


For, having placed the two solids AG, AZ, so that their surfaces have the common angle BAE, produce the interior planes necessary for completing the third parallelopipedon AK, having the same altitude with the parallelopipedon AG. By the last proposition, we shall have

sol. AG : sol. AK : : ABCD : AMNO.

But the two parallelopipedons AK, AZ having the same base AMNO, are to each other as their altitudes AE, AX; hence we have

sol. AK : sol. AZ : : AE : AX.

Multiply together the corresponding terms of these proportions, omitting in the result the common multiplier sol. AK: we shall have

sol. AG : sol. AZ : : ABCD × AE : AMNO × AX.

Instead of the bases ABCD and AMNO, put AB × AD and AO × AM; it will give

sol. AG : sol. AZ : : AB × AD × AE : AO × AM × AX.

Hence any two rectangular parallelopipedons are to each other, etc.

*Scholium.* We are consequently authorized to assume, as the measure of a rectangular parallelopipedon, the product of its base by its altitude; in other words, the product of its three dimensions.

In order to comprehend the nature of this measurement, it is necessary to reflect, that by the product of two or more lines is always meant the product of the numbers which represent them; those numbers themselves being determined by their linear unit, which may be assumed at pleasure. Upon this principle, the product of the three dimensions of a parallelopipedon is a number, which signifies nothing of itself, and would be different if a different linear unit had been assumed; but if the three dimensions of another parallelopipedon are valued according to the same linear unit, and multiplied together in the same manner, the two products will be to each other as the solids, and will serve to express their relative magnitude.

The measure of a solid, or its magnitude, is called its volume. Hence we say the volume of a rectangular parallelopipedon is equal to the product of its base by its altitude, or to the product of its three dimensions.

As the cube has all its three dimensions equal, if the side is 1, the volume will be $1 \times 1 \times 1 = 1$; if the side is 2, the volume will be $2 \times 2 \times 2 = 8$; if the side is 3, the volume will be $3 \times 3 \times 3 = 27$; and so on. Hence, if the sides of a series of cubes are to each other as the numbers 1, 2, 3, etc., the cubes themselves (or their volumes) will be as the numbers 1, 8, 27, etc. Hence it is, that in arithmetic the *cube* of a number is the name given to the product which results from three factors, each equal to this number.

If it were proposed to find a cube double of a given cube, the side of the required cube would have to be to that of the given one, as the cube root of two is to unity. Now, by a geometrical construction, it is easy to find the square root of 2; but the cube root of it cannot be so found, at least not by the simple operations of elementary geometry, which consist in employing nothing but straight lines, two points of which are known, and circles whose centres and radii are determined.

Owing to this difficulty, the problem of *the duplication of the cube* became celebrated among the ancient geometers, as well as that of *the trisection of an angle*, which is nearly of the same species. The solutions of which such problems are susceptible, have, however, long since been discovered; and though less simple than the constructions of elementary geometry, they are not, on that account, less rigorous or less satisfactory.

### THEOREM XII.

*The volume of a parallelopipedon, and generally of any prism, is equal to the product of its base by its altitude.*

For, in the first place, any parallelopipedon (T. VII.) is equivalent to a rectangular parallelopipedon having the same altitude and an equivalent base. Now the volume of the latter is equal to its base multiplied by its height; hence the volume of the former is, in like manner, equal to the product of its base by its altitude.

In the second place, and for a like reason, any triangular prism is half of the parallelopipedon so constructed as to have the same altitude and a double base; but the volume of the latter is equal to its base multiplied by its altitude; hence that of a triangular prism is also equal to the product of its base multiplied into its altitude.

In the third place, any prism may be divided into as many triangular prisms of the same altitude, as there are triangles capable of being formed in the polygon which constitutes its base; but the volume of each triangular prism is equal to its base multiplied by its altitude; and since the altitude is the same for all, it follows that the sum of all the partial prisms must be equal to the sum of all the partial triangles which constitute their bases, multiplied by the common altitude.

Hence the volume of any polygonal prism is equal to the product of its base by its altitude.

*Cor.* Comparing two prisms which have the same altitude, the products of their bases by their altitudes will be as the bases simply. Hence *two prisms of the same altitude are to each other as their bases.* For a like reason, *two prisms of the same base are to each other as their altitudes.*

## THEOREM XIII.

*Similar prisms are to one another as the cubes of their homologous sides.*

Let P and $p$ be two prisms, of which BC, $bc$ are homologous sides: the prism P is to the prism $p$ as the cube of BC to the cube of $bc$.

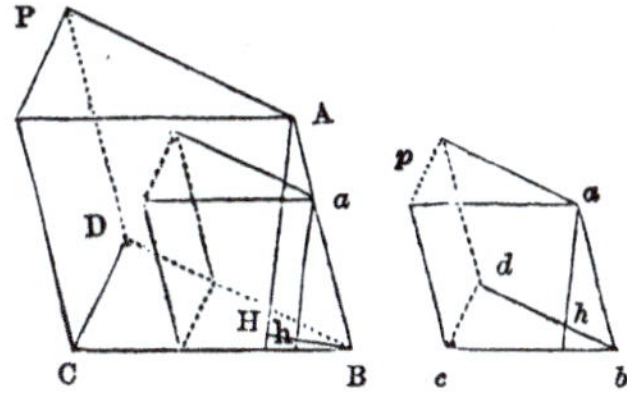


From A and $a$, homologous angles of the two prisms, draw

AH, *ah* perpendicular to the bases BCD, *bcd*: join BH; take B*a* = *ba*, and in the plane BHA draw *ah* perpendicular to BH: then *ah* will be perpendicular to the plane CBD, and equal to *ah* the altitude of the other prism; for if the solid angles B and *b* were applied the one to the other, the planes which contain them, and consequently the perpendiculars *ah*, *ah* would coincide.

Now because of the similar triangles ABH, *abh*, and the similar figures P, *p*, we have

$$\text{AH} : ah :: \text{AB} : ab :: \text{BC} : bc;$$

and because of these similar bases, the

$$\text{base BCD} : \text{base } bcd :: \text{BC}^2 : bc^2. \quad \text{(B. III., T. XXVII.)}$$

Taking the product of the corresponding terms of these proportions, we have

$$\text{AH} \times \text{base BCD} : ah \times \text{base } bcd :: \text{BC}^3 : bc^3.$$

But AH × base BCD expresses the volume of the prism P, and *ah* × base *bcd* expresses the volume of the other prism *p*; therefore

$$\text{prism P} : \text{prism } p :: \text{BC}^3 : bc^3.$$

## THEOREM XIV.

*If a pyramid be cut by a plane parallel to its base, the section thus formed will be a polygon similar to the base, and the lateral edges and the altitude will be cut into proportional parts.*

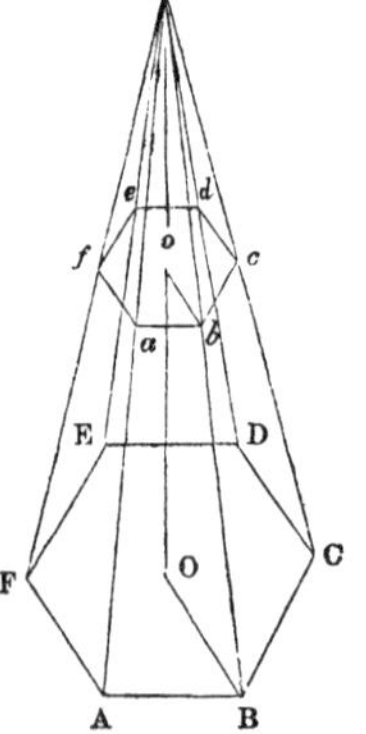


First, since the planes ABCDEF, *abcdef* are parallel, it follows that AB and *ab*, BC and *bc*, CD and *cd*, etc., are parallel (B. V., T. X.); consequently the angles ABC and *abc*, BCD and *bcd* are respectively equal (B. V., T. XIII.). Moreover the couples of similar triangles SAB and S*ab*, SBC and S*bc*, etc., give the following equal ratios:

$$\text{SA} : \text{S}a :: \text{AB} : ab :: \text{SB} : \text{S}b,$$
$$\text{SB} : \text{S}b :: \text{BC} : bc :: \text{SC} : \text{S}c,$$
$$\text{SC} : \text{S}c :: \text{CD} : cd :: \text{SD} : \text{S}d,$$

&c. &c. &c.

Hence, by equality of ratios, we have

$$AB : ab :: BC : bc :: CD : cd :: \text{ etc.}$$

Consequently the two polygons ABCDEF, *abcdef* are similar (B. III., T. X.).

Now, considering only the ratios between the lateral edges, we have

$$SA : Sa :: SB : Sb :: SC : Sc :: \text{etc.},$$

from which we see that the edges are cut into proportional parts.

Finally, if we pass a plane through the edge SB and the altitude SO, its intersection *bo* with the plane *abcde* will be parallel to BO (B. V., T. X.), and the two similar triangles SBO, S*bo* will give

$$SO : So :: SB : Sb :: SA : Sa :: \text{etc.},$$

which establishes the theorem.

*Cor.* When two pyramids S – ABCDE, T – MNP have equivalent bases situated in the same plane, and equal altitudes, the sections made by a plane parallel to their bases, are also equivalent.

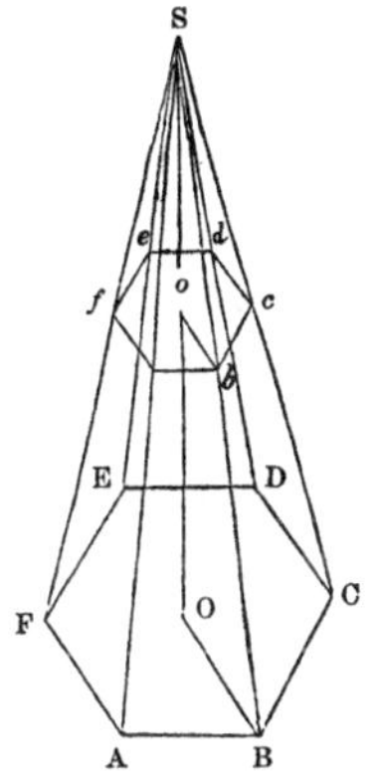


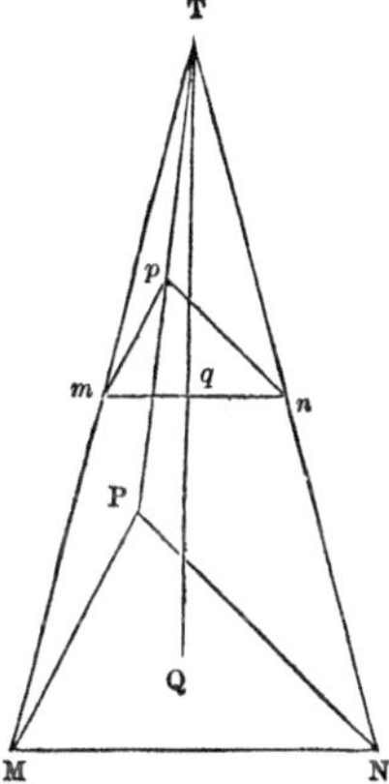


For, since the polygons ABCDE, *abcde* are similar, we have the proportion

$$ABCDE : abcde :: AB^2 : ab^2 ;$$

we also have, by reason of the foregoing relations,

$$ABCDE : abcde :: SO^2 : So^2.$$

In the same manner the two polygons MNP, *mnp* give

$$MNP : mnp :: TQ^2 : Tq^2,$$

we have, moreover, by supposition, $SO = TQ$, $So = Tq$; hence,

$$ABCDE : abcde :: MNP : mnp.$$

Now, by hypothesis, $ABCDE = MNP$; consequently, $abcde = mnp$.

THEOREM XV.

*The lateral surface of a right pyramid is equal to the perimeter of its base multiplied by half the slant height.*

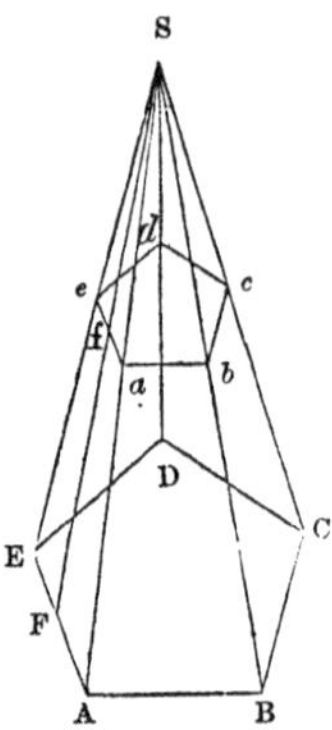


Since the lateral surface is composed of equal isosceles triangles SAB, SBC, SCD, etc., each one of which is measured by its base into one half its altitude, which altitude is the same as SF the slant height of the pyramid, it follows that the lateral surface is equal to the sum of the bases of all these triangles, or the perimeter of the pyramid's base, into one half the slant height.

*Cor.* I. If two right pyramids have the same slant heights, their lateral surfaces will be to each other as the perimeters of their bases.

*Cor.* II. *The lateral surface of a frustum of a regular pyramid is measured by half the sum of the perimeters of its two bases multiplied by its slant height.*

For, since the frustum is formed from a regular pyramid, the two bases are similar and regular polygons (T. XIV.). Consequently the lateral surface of a frustum of a right pyramid is composed of equal trapezoids, the altitude of each being the same as the slant height of the frustum.

The area of each trapezoid is measured by half the sum of its parallel bases multiplied into its altitude (B. III., T. XXIII., C. II.). Hence the sum of all these trapezoids, or the lateral surface of the frustum, is measured by half the sum of the perimeters of its bases into its slant height.

### THEOREM XVI.

*Two triangular pyramids having equivalent bases and equal altitudes, are equivalent, or equal in volume.*

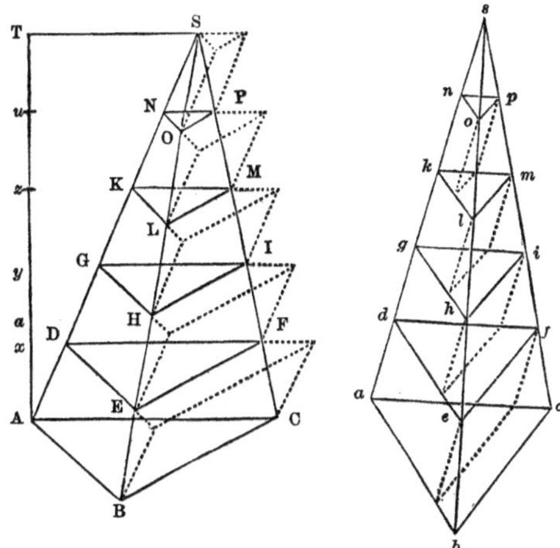


Let S – ABC, S – *abc* be those two pyramids; let their equivalent bases ABC, *abc* be situated in the same plane, and let AT be their common altitude. If they are not equivalent, let S – *abc* be the smaller; and suppose A*a* to be the altitude of a prism, which, having ABC for its base, is equal to their difference.

Divide the altitude AT into equal parts A*x*, *xy*, *yz*, etc., each less than A*a*, and let *k* be one of those parts; through the points of division, pass planes parallel to the plane of the bases: the corresponding sections formed by these planes in the two pyramids will be respectively equivalent (T. XIV., C.), namely, DEF to *def*, GHI to *ghi*, etc.

This being granted, upon the triangles ABC, DEF, GHI, etc., taken as bases, construct exterior prisms, having for edges the parts AD, DG, GK, etc., of the edge SA. In like manner, on the bases *def*, *ghi*, *klm*, etc., in the second pyramid, construct interior prisms, having for edges the corresponding parts of *sa*. It is plain that the sum of all the exterior prisms of the pyramid S – ABC will be greater than this pyramid; and, also, that the sum of all the interior prisms of the pyramid *s* – *abc* will be less than this. Hence the difference between the sum of all the exterior prisms and the sum of all the interior ones, must be greater than the difference between the two pyramids themselves.

Now, beginning with the bases ABC, *abc*, the second exterior prism DEFG is equivalent to the first interior prism *defa*, because they have the same altitude *k*, and their bases DEF, *def* are equivalent: for like reasons, the third exterior prism GHIK, and the second interior prism *ghid*, are equivalent; the fourth exterior, and the third interior; and so on, to the last in each series. Hence all the exterior prisms of the pyramid S – ABC, excepting the first prism DABC, have equivalent corresponding ones in the interior prisms of the pyramid *s* – *abc*: hence the prism DABC is the difference between the sum of all the exterior prisms of the pyramid S – ABC, and the sum of all the interior prisms of the pyramid *s* – *abc*. But the difference between these two sets of prisms has already been proved to be greater than that of the two pyramids, which latter difference we supposed to be equal to the prism *a*ABC: hence the prism DABC must be greater than the prism *a*ABC; but in reality it is less, for they have the same base ABC, and the altitude A*x* of the first is less than A*a* the altitude of the second. Hence the supposed inequality between the two pyramids cannot exist; hence the two pyramids S – ABC, *s* – *abc*, having equal altitudes and equivalent bases, are themselves equivalent.

### THEOREM XVII.

*Every triangular pyramid is the third of the triangular prism having the same base and altitude.*

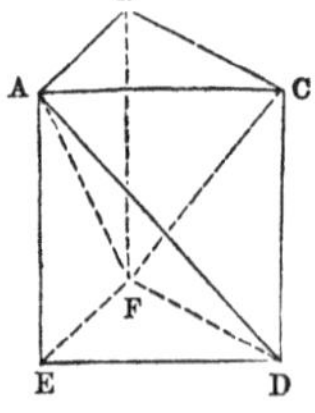


Let F – ABC be a triangular pyramid, ABCDEF a triangular prism of the same base and altitude; the pyramid will be equal to one third of the prism.

Conceive the pyramid F—ABC to be cut off from the prism by a section made along the plane FAC, and there will remain the solid FACDE, which may be considered as a quadrangular pyramid whose vertex is F, and base the parallelogram ACDE. Draw the diagonal AD, and extend the plane FAD, which will cut the quadrangular pyramid into two triangular ones F — ACD, F — ADE. These two triangular pyramids have for their common altitude the perpendicular drawn from F to the plane ACDE; they have equal bases, the triangles ACD, ADE being halves of the same parallelogram; hence the two pyramids F—ACD, F—ADE are equivalent. But the pyramid F—ADE and the pyramid F — ABC have equal bases, ABC, DEF; they have also the same altitude, namely, the distance of the parallel planes ABC, DEF: hence the two pyramids are equivalent. Now the pyramid F — ADE has already been proved equivalent to F — ACD; hence the three pyramids F — ABC, F — ADE, F — ACD, which compose the prism ABCD, are all equivalent. Hence the pyramid F — ABC is the third part of the prism ABCD, which has the same base and the same altitude.

*Cor.* The volume of a triangular pyramid is equal to a third part of the product of its base by its altitude.

### THEOREM XVIII.

*The volume of any pyramid has for its measure the area of its base multiplied into one third of its altitude.*

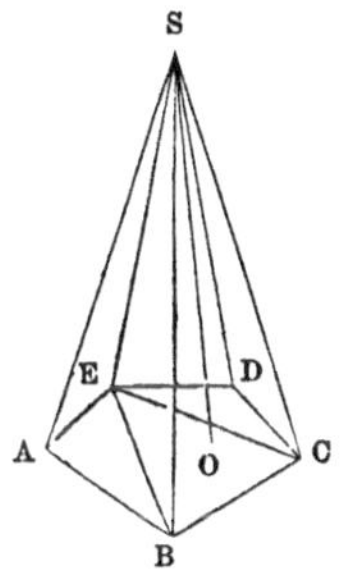


Let S — ABCDE be a pyramid, having the altitude SO; then will it be measured by the base ABCDE into one third of the altitude SO.

For, extending the planes SEB, SEC through the diagonals EB, EC, the polygonal pyramid S—ABCDE will be divided into several triangular pyramids, all having the same altitude SO. But (T. XVII., C.) each of these pyramids is measured by multiplying its base ABE, BCE, or CDE by the third part of its altitude SO; hence the sum of these triangular pyramids, or the polygonal pyramid S — ABCDE will be measured by the sum of the triangles ABE, BCE, CDE, or the polygon ABCDE, multiplied by $\frac{1}{3}$SO. Hence every pyramid is measured by a third part of the product of its base by its altitude.

*Cor.* I. Every pyramid is the third part of the prism which has the same base and the same altitude.

*Cor.* II. Two pyramids having the same altitude, are to each other as their bases.

*Scholium.* The volume of any polyedral body may be computed, by dividing the body into pyramids; and this division may be accomplished in various ways. One of the simplest is to make all the planes of division pass through the vertex of one solid angle; in that case, there will be formed as many partial pyramids as the polyedron has faces, *minus* those faces which form the polyedral angle whence the planes of division proceed.

### THEOREM XIX.

*The volume of a frustum of a pyramid is equivalent to three pyramids having the common altitude of the frustum, and for bases, the lower base of the frustum, the upper base, and a mean proportional between them.*

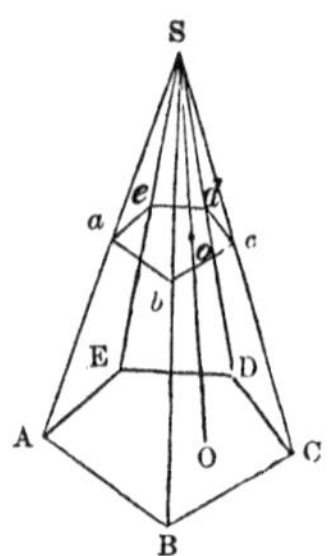


We will represent the lower base ABCDE of the frustum by A, and the upper base *abcde* by $a$; also, we will denote the altitude $o$O of the frustum by $h$. If we denote the altitude SO of the pyramid, whose base is ABCDE by $x$, we shall have $x-h$ for the altitude S$o$ of the pyramid whose base is *abcde*.

Since the two bases of the frustum are similar polygons (T. XIV.), their areas are to each other as the squares of their homologous sides AB and $ab$ (B. III., T. XXVIII.), which sides are to each other as SO to S$o$; consequently we have

$$A : a :: x^2 : (x-h)^2.$$

Extracting the square root of each term, we have

$$A^{\frac{1}{2}} : a^{\frac{1}{2}} :: x : x-h.$$

This proportion immediately gives $x = \dfrac{A^{\frac{1}{2}}}{A^{\frac{1}{2}} - a^{\frac{1}{2}}} \times h$, and consequently, $x - h = \dfrac{a^{\frac{1}{2}}}{A^{\frac{1}{2}} - a^{\frac{1}{2}}} \times h$.

Hence the volume of the larger pyramid is $A \times \frac{1}{3}x = \frac{A^{\frac{3}{2}}}{A^{\frac{1}{2}} - a^{\frac{1}{2}}} \times \frac{1}{3}h$. The volume of the smaller pyramid is

$$a \times \frac{1}{3}(x - h) = \frac{a^{\frac{3}{2}}}{A^{\frac{1}{2}} - a^{\frac{1}{2}}} \times \frac{1}{3}h.$$

If we denote the volume of the frustum, which is the difference of these pyramids by V, we shall have $V = \frac{A^{\frac{3}{2}} - a^{\frac{3}{2}}}{A^{\frac{1}{2}} - a^{\frac{1}{2}}} \times \frac{1}{3}h$, which becomes $V = (A + a + A^{\frac{1}{2}} a^{\frac{1}{2}}) \times \frac{1}{3}h = A \times \frac{1}{3}h + a \times \frac{1}{3}h + \sqrt{A \times a} \times \frac{1}{3}h$, which establishes the Theorem.

THEOREM XX.

*Two similar pyramids are to each other as the cubes of their homologous sides.*

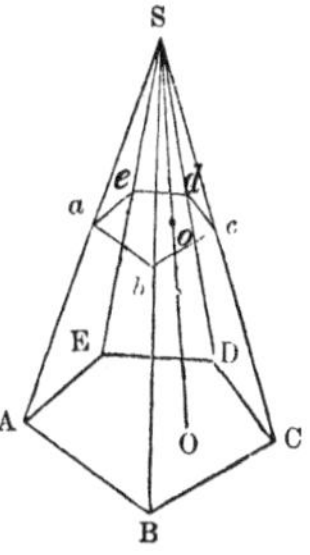


For, two pyramids being similar, the smaller may be placed within the greater, so that the angle S shall be common to both. In that position the bases ABCDE, *abcde* will be parallel; because, since the homologous faces are similar, the angle S*ab* is equal to SAB, and S*bc* to SBC; hence the plane ABC is parallel to the plane *abc*. This granted, let SO be the perpendicular drawn from the vertex S to the plane ABC, and *o* the point where this perpendicular meets the plane *abc*: from what has already been shown (T. XIV.), we shall have

$$SO : So :: SA : Sa :: AB : ab;$$

and, consequently, $\frac{1}{3}SO : \frac{1}{3}So :: AB : ab.$

But the bases ABCDE, *abcde* being similar figures, we have

$$ABCDE : abcde :: AB^2 : ab^2.$$

Multiply the corresponding terms of these two proportions; there results the proportion,

$$ABCDE \times \frac{1}{3}SO : abcde \times \frac{1}{3}So :: AB^3 : ab^3.$$

Now $ABCDE \times \frac{1}{3}SO$ is the volume of the pyramid S–ABCDE, and $abcde \times \frac{1}{3}So$ is that of the pyramid S–*abcde* (T. XVII., C.); hence two similar pyramids are to each other as the cubes of their homologous sides.

# SEVENTH BOOK.

## THE THREE ROUND BODIES.

### DEFINITIONS.

I. A *cylinder* is a solid, which may be produced or generated by the revolution of a rectangle ABCD, conceived to revolve about the side AB.

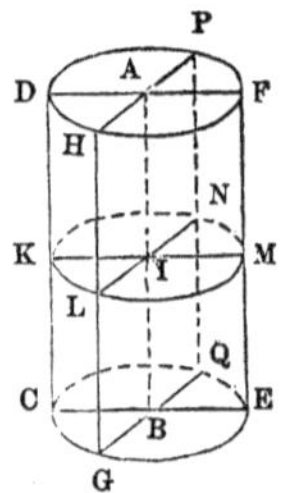


In this rotation, the sides AD, BC, continuing always perpendicular to AB, describe equal circular planes DHP, CGQ, which are called *the bases of the cylinder;* the side CD at the same time describing *the convex surface.*

The immovable line AB is called *the axis of the cylinder.*

Every section KLM made in the cylinder, at right angles to the axis, is a circle equal to either of the bases; for, while the rectangle ABCD revolves about AB, the line KI, perpendicular to AB, describes a circular plane, equal to the base, which is a section made perpendicular to the axis at the point I.

Every section PQGH passing through the axis is a rectangle, and is double of the generating rectangle ABCD.

II. A *cone* is a solid, which may be produced or generated by the revolution of a right-angled triangle SAB, conceived to revolve about the side SA.

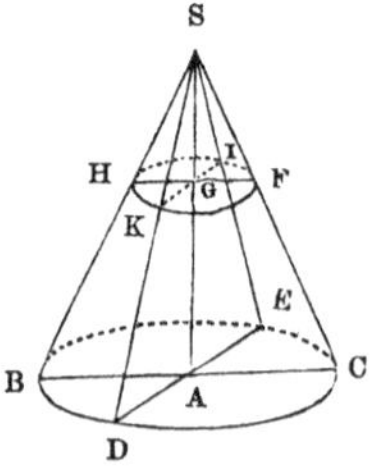


In this rotation, the side AB describes a circular plane BDCE, named *the base of the cone;* and the hypotenuse SB describes its *convex surface.*

The point S is named *the vertex of the cone;* SA its *axis*, or *altitude.*

Every section HKFI formed at right angles to the axis, is a circle. Every section SDE passing through the axis, is an isosceles triangle double of the generating triangle SAB.

III. If, from the cone SCDB, the cone SFKH be cut off by a section parallel to the base, the remaining solid CBHF is called *a truncated cone*, or *the frustum of a cone.* We may conceive it to be described by the revolution of a trapezium ABHG, whose angles A and C are right, about the side AG. The immovable line AG is called *the axis* or *altitude of the frustum;* the circles BDC, HFK are its *bases*, and BH is its *slant height.*

IV. Two cylinders or two cones, are *similar*, when their axes are to each other as the diameters of their bases.

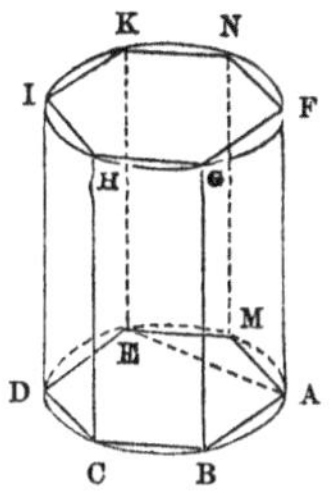


V. If, in the circle ACD which forms the base of a cylinder, a polygon ABCDEM be inscribed, a right prism, constructed on this base ABCDEM, and equal in altitude to the cylinder, is said to be *inscribed in the cylinder*, or the cylinder to be *circumscribed about the prism.*

The edges AF, BG, CH, etc., of the prism, being perpendicular to the plane of the base, are evidently included in the convex surface of the cylinder. Hence the prism and the cylinder touch one another along these edges.

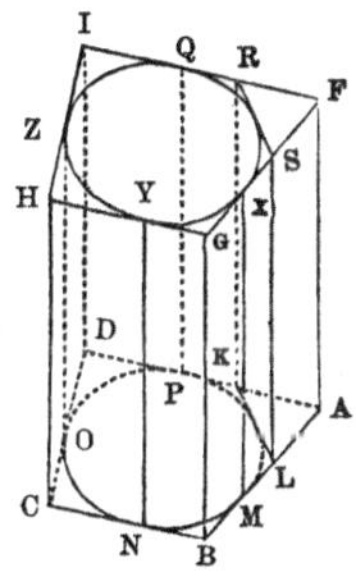


VI. In like manner, if ABCD is a polygon circumscribed about the base of a cylinder, a right prism, constructed on this base ABCD, and equal in altitude to the cylinder, is said to be *circumscribed about the cylinder*, or the cylinder to be *inscribed in the prism.*

Let M, N, etc., be the points of contact in the sides AB, BC, etc.; and through the points M, N, etc., let MX, NY, etc., be drawn perpendicular to the plane of the base: those perpendiculars will evidently lie both in the surface of the cylinder, and in that of the circumscribed prism; hence they will be their lines of contact.

VII. A *sphere* is a solid terminated by a curved surface, all the points of which are equally distant from a point within, called the *centre.*

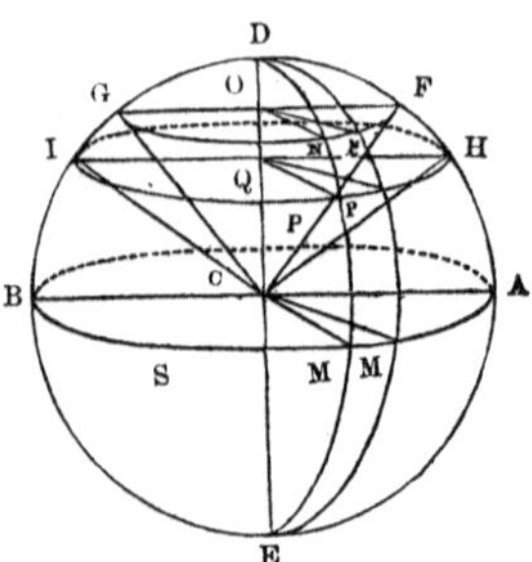


The sphere may be conceived to be generated by the revolution of a semicircle DAE about its diameter DE; for the surface described in this movement by the curve DAE will have all its points equally distant from the centre C.

*The radius of a sphere* is a straight line drawn from the centre to any point in the surface; *the diameter*, or *axis*, is a line passing through this centre, and terminated on both sides by the surface.

All the radii of a sphere are equal. All the diameters are equal, and double of the radius.

VIII. *A great circle of the sphere*, is a section which passes through the centre; *a small circle*, one which does not pass through it.

IX. *A plane* is *a tangent* to a sphere, when their surfaces have but one point in common.

X. *A zone* is the portion of the surface of the sphere included between two parallel planes, which form its *bases*. One of these planes may be a tangent to the sphere, in which case the zone has only a single base.

XI. *A spherical segment* is the portion of the solid sphere included between two parallel planes which form its bases. One of those planes may be a tangent to the sphere, in which case the segment has only a single base.

XII. *The altitude of a zone*, or of *a segment*, is the distance between the two parallel planes, which form the bases of the zone or segment.

XIII. While the semicircle DAE, revolving round its diameter DE, describes the sphere, any circular sector, as DCF or FCH, describes a solid, which is named *a spherical sector.*

NOTE.—The cylinder, the cone, and the sphere, are *the three round bodies* treated of in the elements of geometry.

### THEOREM I.

*The lateral or convex surface of a cylinder has for its measure the product of its circumference into its altitude.*

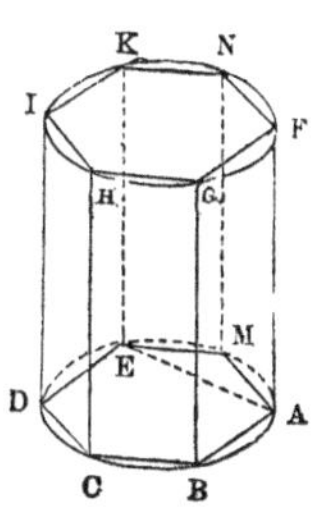


In the cylinder, suppose a right prism to be inscribed, having a regular polygon for its base. The lateral surface of this prism has for its measure the perimeter of its base multiplied by its altitude (B. VI., T. I.). When the number of sides of the polygon forming the base of the inscribed prism is indefinitely increased, its limit will become the circle constituting the base of the cylinder (B. IV., T. VIII., S.). We may, therefore, regard a cylinder as a right prism having a regular polygon of an infinite number of infinitely small sides for its base; and since the lateral surface of a right prism will always have for its measure the perimeter of its base into its altitude, it follows that the lateral surface of a cylinder has for its measure the circumference of its base into its altitude.

### THEOREM II.

*The volume of a cylinder has for its measure the product of its base into its altitude.*

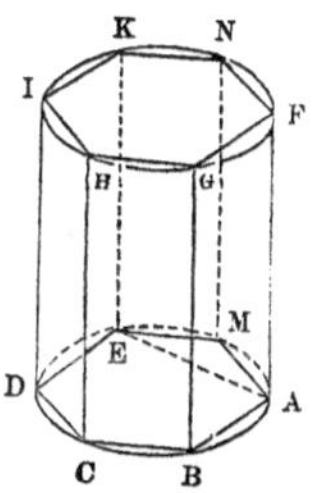


As in the last Theorem, if we regard a cylinder as a right prism, having a regular polygon of an infinite number of sides for its base, and recall to mind that the volume of a right prism is the product of its base into its altitude (B. VI., T. XII.), we shall at once see that the volume of a cylinder is equal to the product of its base into its altitude.

*Cor.* I. Cylinders of the same altitude are to each other as their bases; and cylinders of the same base are to each other as their altitudes.

*Cor.* II. Similar cylinders are to each other as the cubes of their altitudes, or as the cubes of the diameters of their bases. For the bases are as the squares of their diameters; and the cylinders being similar, the diameters of their bases (D. IV.) are to each other as the altitudes: hence the bases are as the squares of the altitudes; consequently, the bases multiplied by the altitudes, or the cylinders themselves, are as the cubes of the altitudes.

*Scholium.* Let R be the radius of a cylinder's base, and H the altitude. The area of the base (B. IV., T. XIV., S.) will be $\pi R^2$; and the volume of the cylinder will be $\pi R^2 \times H$, or $\pi R^2 H$, where $\pi = 3.141592$, etc.

### THEOREM III.

*The convex, or lateral surface of a cone, is equal to the product of the circumference of its base into half its slant height.*

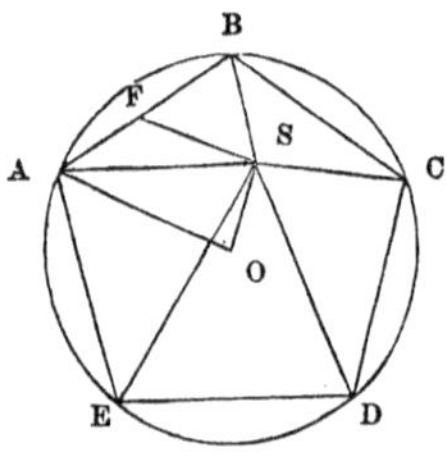


Let the circle whose radius is OA be the base of the cone, S its vertex, and SA its slant height. Then will the convex surface of the cone have for its measure *circ.* OA $\times \frac{1}{2}$SA.

For, conceive a regular polygon inscribed in the circle OA; and on this polygon, as a base, a pyramid having S for its vertex, to be constructed. The lateral surface of this pyramid will have for its measure the perimeter of the polygon, constituting its base, into one half of SF its slant height (B. VI., T. XV.). When the number of sides of the inscribed polygon is indefinitely increased, its perimeter will be *limited* by the circumference of the circle, its slant height will be *limited* by the slant height of the cone, and the *limit* of the lateral surface will be the convex surface of the cone. A cone may thus be regarded as a right pyramid, having a regular polygon of an infinite number of infinitely short sides for its base. And since the lateral surface of a pyramid will always have for its measure the perimeter of its base into half its slant height, however great may be the number of sides in the polygon forming its base, it follows, that the convex surface of a cone has for its measure the circumference of its base into half its slant height.

*Scholium.* Let L be the slant height of a cone, R the radius of its base. The circumference of this base will be $2\pi R$; and the convex surface of the cone will be $2\pi R \times \frac{1}{2}L$, or $\pi RL$.

THEOREM IV.

*The convex surface of a truncated cone is equal to its side multiplied by half the sum of the circumferences of its two bases.*

In the plane SAB which passes through the axis SO, draw the line AF perpendicular to SA, and equal to the circumference having AO for its radius; join SF, and draw DH parallel to AF.

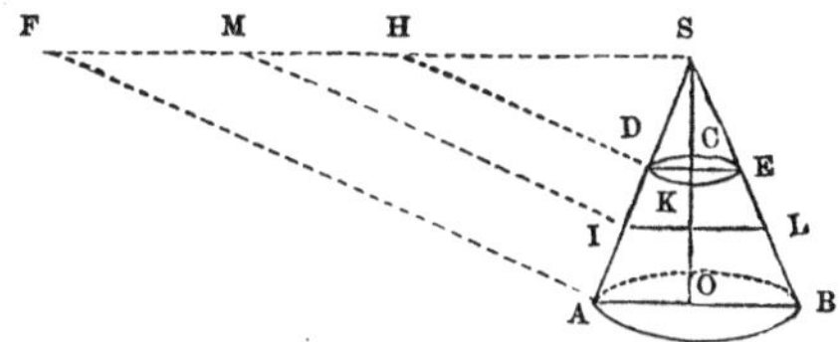


From the similar triangles SAO, SDC, we have

$$AO : DC :: SA : SD;$$

and by the similar triangles SAF, SDH,

$$AF : DH :: SA : SD;$$

hence, $$AF : DH :: AO : DC;$$

or (B. IV., T. XIV.), as circ. AO is to circ. DC. But, by construction, AF = circ. AO; hence DH = circ. DC. Hence the triangle SAF, measured by $AF \times \frac{1}{2}SA$, is equal to the surface of the cone SAB, which is measured by circ. $AO \times \frac{1}{2}SA$. For a like reason, the triangle SDH is equal to the surface of the cone SDE. Therefore the surface of the truncated cone ADEB is equal to that of the trapezium ADHF; but the latter is measured by $AD \times \left(\frac{AF + DH}{2}\right)$ (B. III., T. XXIII., C. II.). Hence the surface of the truncated cone ADEB is equal to its side AD multiplied by half the sum of the circumferences of its two bases.

*Scholium.* If a line AD, lying wholly on one side of the line OC, and in the same plane, make a revolution around OC, the surface described by AD will have for its measure

$$AD \times \left(\frac{\text{circ. } AO + \text{circ. } DC}{2}\right),$$

the lines AO, DC being perpendiculars, drawn from the extremities of the axis OC.

For, if AD and OC are produced till they meet in S, the surface described by AD is evidently that of a truncated cone having AO and DC for the radii of its bases, the vertex of the whole cone being S. Hence this surface will be measured as above.

This measure will always hold good, even when the point D falls on S, and thus forms a whole cone; and also when the line AD is parallel to the axis, and thus forms a cylinder. In the first case, DC would be nothing; in the second, DC would be equal to AO and to IK.

*Cor.* I. Through I, the middle point of AD, draw IKL parallel to AB, and IM parallel to AF: it may be shown, as above, that IM = circ. IK; but the

$$\text{trapezium ADHF} = \text{AD} \times \text{IM} = \text{AD} \times \text{circ. IK}.$$

Hence it may also be asserted, that *the surface of a truncated cone is equal to its side multiplied by the circumference of a section at equal distances from the two bases.*

*Cor.* II. The point I being the middle of AB, and IK a perpendicular drawn from the point I to the axis, the surface described by AB, by the last Corollary, will have for its measure AB × circ. IK. Draw AX parallel to the axis; the triangles ABX, OIK will have their sides perpendicular each to each, namely, OI to AB, IK to AX, and OK to BX; hence these triangles are similar, and give the proportion

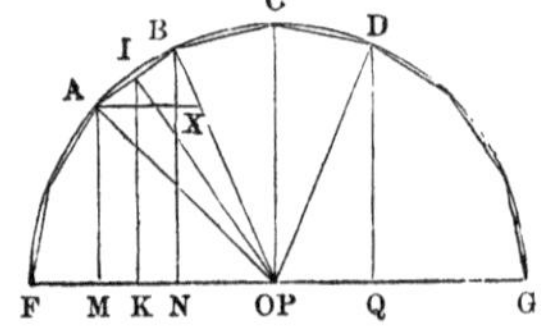


$$\text{AB} : \text{AX or MN} :: \text{OI} : \text{IK},$$

or as circ. OI to circ. IK; hence AB × circ. IK = MN × circ. OI. Whence it is plain that the surface described by the partial polygon ABCD is measured by (MN + NP + PQ) × circ. OI, or by MQ × circ. OI; hence it is equal to the altitude multiplied by the circumference of the inscribed circle.

*Cor.* III. If the whole polygon has an even number of sides, and if the axis FG passes through two opposite vertices F and G, the whole surface described by the revolution of the half polygon FACG will be equal to its axis FG multiplied by the circumference of the inscribed circle. This axis FG will, at the same time, be the diameter of the circumscribed circle.

THEOREM V.

*The volume of a cone has for its measure the product of its base into one third of its altitude.*

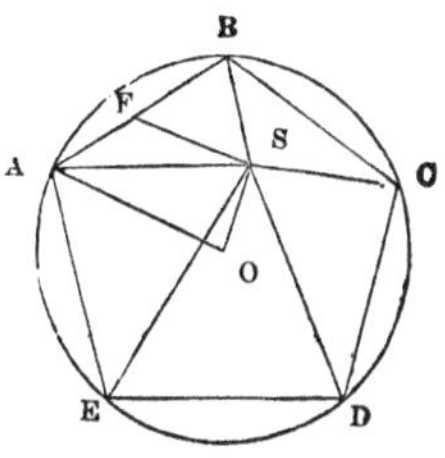


Let the cone have OA for the radius of its base, and SO for its altitude; then will its volume be represented by *area* OA $\times \frac{1}{3}$SO.

As in Theorem III., conceive a regular pyramid to be inscribed in the cone. The volume of this pyramid will have for its measure the area of the polygon constituting its base multiplied by one third of its altitude (B. VI., T. XVIII.). When the number of sides of the polygon forming the base of the inscribed pyramid is indefinitely increased, the area of the polygon will have for its *limit* the area of the circle forming the base of the cone, and the *limit* of the inscribed pyramid will be the cone. Hence, the volume of a cone has for its measure the product of its base into one third of its altitude.

*Cor.* I. A cone is the third of a cylinder having the same base and the same altitude. Whence it follows:

1. That cones of equal altitudes are to each other as their bases;

2. That cones of equal bases are to each other as their altitudes;

3. That similar cones are as the cubes of the diameters of their bases, or as the cubes of their altitudes.

*Scholium.* If R be the radius of a cone's base, and H its altitude, the volume of the cone will be $\pi R^2 \times \frac{1}{3}H$, or $\frac{1}{3}\pi R^2 H$.

*Cor.* II. If a plane be drawn parallel to the base of a cone, cutting it so as to form a frustum of a cone, it will at the same time cut the inscribed pyramid, forming also a frustum of a regular pyramid inscribed in the frustum of the cone. The *limit* of the inscribed frustum will be the frustum of the cone. And since the volume of the frustum of a pyramid is equivalent to the volumes of three pyramids having the common altitude of the frustum, and for bases the lower base of the frustum, the upper base of the frustum, and a mean proportional between them (B. VI., T. XIX.), it follows, that *the volume of the frustum of a cone is equivalent to three cones having the common altitude of the frustum, and for bases the lower base of the frustum, the upper base, and a mean proportional between them.*

If the radii of the two bases of a frustum of a cone be represented by R and R′, its altitude by $h$, and its volume by V, we shall have $V = \frac{1}{3}\pi h(R^2 + R'^2 + RR')$.

### THEOREM VI.

*Every section of a sphere, made by a plane, is a circle.*

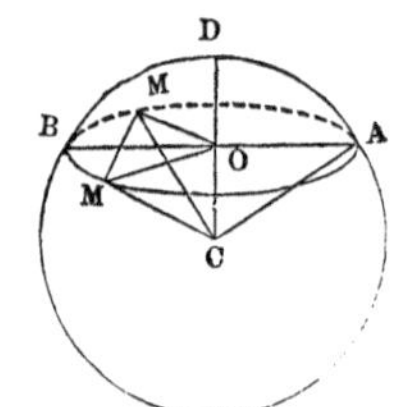


Let AMB be the section, made by a plane in the sphere whose centre is C. From the point C draw CO perpendicular to the plane AMB; and draw lines CM, CM to different points of the curve AMB, which terminates the section.

The oblique lines CM, CM, CA being equal, being radii of the sphere, they are equally distant from the perpendicular CO (B. V., T. V.); hence all the lines OM, OM, OB are equal; hence the section AMB is a circle, whose centre is O.

*Cor.* I. If the section passes through the centre of the sphere, its radius will be the radius of the sphere; hence all great circles are equal.

*Cor.* II. Two great circles always bisect each other; for their common intersection, passing through the centre, is a diameter.

*Cor.* III. Every great circle divides the sphere and its surface into two equal parts; for, if the two hemispheres were separated, and afterwards placed on the common base, with their convexities turned the same way, the two surfaces would exactly coincide, no point of the one being nearer the centre than any point of the other.

*Cor.* IV. The centre of a small circle, and that of the sphere, are in the same straight line perpendicular to the plane of the little circle.

*Cor.* V. Small circles are the less the further they lie from the centre of the sphere; for, the greater CO is, the less is the chord AB, the diameter of the small circle AMB.

*Cor.* VI. An arc of a great circle may always be made to pass through any two given points in the surface of the sphere; for the two given points and the centre of the sphere make three points, which determine the position of a plane. But if the two given points were at the extremities of a diameter, these two points and the centre would then lie in one straight line, and an infinite number of great circles might be made to pass through the two given points.

### THEOREM VII.

*A plane perpendicular to the radius at its extremity, is tangent to the sphere.*

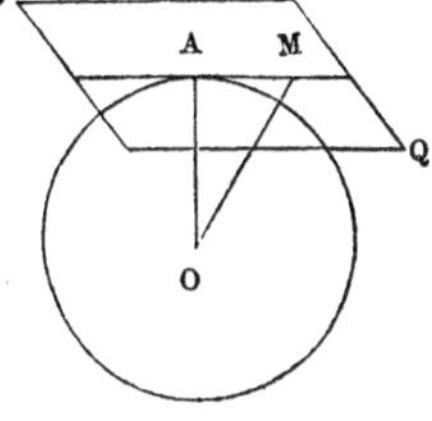


Let PQ be a plane perpendicular to the radius OA; at its extremity, A, it will be tangent to the sphere.

For, taking any other point, as M, in this plane, if we draw OM, it will be longer than the perpendicular OA (B. V., T. V.); consequently the point M is situated without the surface of the sphere. Hence the plane PQ can have only the point A common with this surface; it is therefore tangent to the sphere (D. IX.).

### THEOREM VIII.

*When two spherical surfaces cut each other, the line of intersection is a circumference of a circle, of which the plane is perpendicular to the line of their centres, and whose centre is situated on this line.*

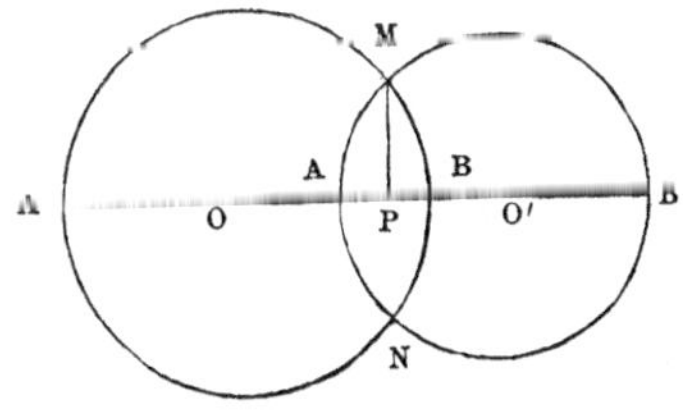


Let O, O′ be the centres of two spheres, M a point common to their surfaces, and MP a perpendicular drawn to the line of their centres. If we draw a plane through the three points O, O′, M, it will intersect the surface of the two spheres in arcs of great circles. Now, suppose the two semicircles AMB, A′MB′ to make a complete revolution about the common line AB′ as an axis, it is evident that the perpendicular MP will describe a circle common to the two spheres thus engendered. From which we see that the surface of the spheres cut each other in the circumference of a circle, whose radius is the perpendicular drawn from the point M to this axis.

*Scholium.* Whatever may be the relative positions of two spheres, since all planes passing through the line of their centres give two circumferences whose centres and whose radii are those of the spheres themselves, it follows that the condition of *contact* and of *intersection* of their surfaces are, in all respects, identical with those in reference to the two circumferences. So that to enumerate and demonstrate these different conditions, it is sufficient to refer to Book II., Theorems XVIII., XIX., XX., and XXI. We will confine ourselves in this place to noticing the condition relative to *contact: When two spheres touch each other, the distance between their centres is equal to the sum or the difference of their radii. They have at their point of contact a common tangent plane.*

THEOREM IX.

*The surface of a sphere is equal to its diameter multiplied by the circumference of a great circle.*

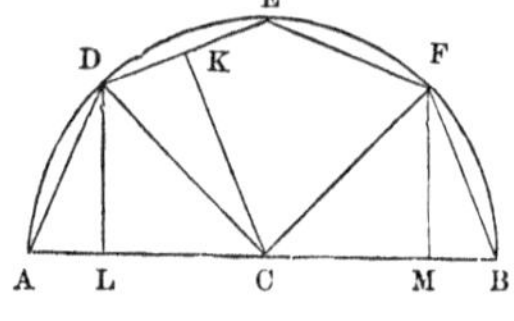


Let the surface of the sphere be engendered by the revolution of the semi-circumference ADEFB about its diameter AB; it will have for its measure its diameter multiplied by its entire circumference.

Suppose a regular semi-polygon to be inscribed in the semi-circle, and to revolve simultaneously with the semicircle; then will the surface engendered by the semi-polygon have for its measure the diameter AB multiplied by the circ. of its apothem CK (T. IV., C. III.). Now, when the number of sides of this inscribed semi-polygon is indefinitely increased, the *limit* of its perimeter will become the semi-circumference, its apothem will become the radius of the semicircle, and the surface engendered by the revolution of this semi-polygon will become the surface of the sphere. Hence, the surface of a sphere has for its measure the product of its diameter into its circumference.

*Cor.* I. The surface of the great circle is measured by multiplying its circumference by half the radius, or by a fourth of the diameter; hence, *the surface of a sphere is four times that of its great circle.*

*Cor.* II. *The surface of a zone has for its measure its altitude multiplied by the circumference of a great circle.*

For, the surface described by any portion of the inscribed polygon, as DE + EF, has for its measure LM into the circ. of its apothem CK (T. IV., C. II.); which, at the *limit*, gives the surface of the zone equal to its altitude LM multiplied into its circumference CA.

*Cor.* III. Two zones, taken in the same sphere, or in equal spheres, are to each other as their altitudes; and any zone is to the surface of the sphere, as the altitude of that zone is to the diameter of the sphere.

### THEOREM X.

*If a triangle and a rectangle, having the same base and the same altitude, turn simultaneously about the common base, the solid described by the revolution of the triangle will be a third of the cylinder described by the revolution of the rectangle.*

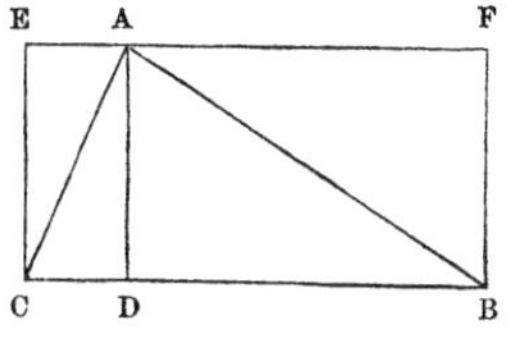


Let ABC be the triangle, and EB the rectangle.

To the axis draw the perpendicular AD; the cone described by the triangle ABD is the third part of the cylinder described by the rectangle AFBD (T. V., C. I.); also the cone described by the triangle ADC is the third part of the cylinder described by the rectangle ADCE: hence the sum of the two cones, or the solid described by ABC, is the third part of the two cylinders taken together, or of the cylinder described by the rectangle BCEF.

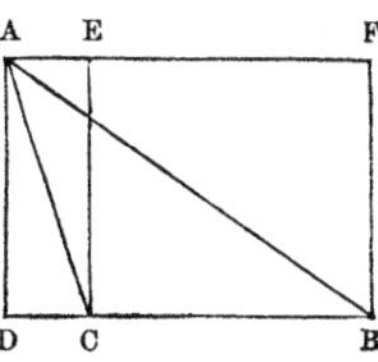


If the perpendicular AD falls without the triangle, the solid described by ABC will be the difference of the two cones described by ABD and ACD; but, at the same time, the cylinder described by BCEF will be the difference of the two cylinders described by

AFBD and AECD. Hence the solid described by the revolution of the triangle will still be a third part of the cylinder described by the revolution of the rectangle having the same base and altitude.

*Scholium.* The circle of which AD is radius has for its measure $\pi \times AD^2$; hence $\pi \times AD^2 \times BC$ measures the cylinder described by BCEF, and $\frac{1}{3}\pi \times AD^2 \times BC$ measures the solid described by the triangle ABC.

### THEOREM XI.

*If a triangle be revolved about a line drawn at pleasure through its vertex, the solid described by the triangle will have for its measure the area of the triangle multiplied by two thirds of the circumference traced by the middle point of the base.*

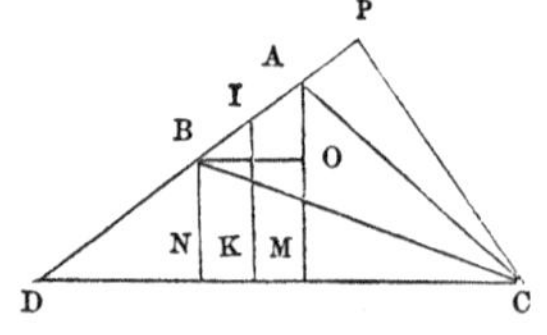


Let CAB be the triangle, and CD the line about which it revolves.

Produce the side AB till it meets the axis CD in D; from the points A and B, draw AM, BN perpendicular to the axis.

The solid described by the triangle CAD is measured (T. X., S.) by $\frac{1}{3}\pi \times AM^2 \times CD$; the solid described by the triangle CBD is measured by $\frac{1}{3}\pi \times BN^2 \times CD$: hence the difference of those solids, or the solid described by ABC, will have for its measure $\frac{1}{3}\pi (AM^2 - BN^2) \times CD$.

To this expression another form may be given. From I, the middle point of AB, draw IK perpendicular to CD; and through B draw BO parallel to CD: we shall have $AM + BN = 2IK$, and $AM - BN = AO$; hence $(AM + BN) \times (AM - BN)$, or $AM^2 - BN^2 = 2IK \times AO$. Hence the measure of the solid in question is expressed by $\frac{2}{3}\pi \times IK \times AO \times CD$. But if CP is drawn perpendicular to AB, the triangles ABO, DCP will be similar, and give the proportion,

$$AO : CP :: AB : CD;$$

hence $AO \times CD = CP \times AB$, but $CP \times AB$ is double the area of the triangle ABC; hence we have $AO \times CD = 2ABC$; hence the solid described by the triangle ABC is also measured by

$\frac{4}{3}\pi \times ABC \times IK$, or, which is the same thing, by $ABC \times \frac{2}{3}$ circ. IK, circ. IK being equal to $2\pi \times IK$. Hence, *the solid described by the revolution of the triangle* ABC *has for its measure the area of this triangle multiplied by two thirds of the circumference traced by* I, *the middle point of the base.*

*Cor.* I. If the side $AC = CB$, the line CI will be perpendicular to AB, the area ABC will be equal to $AB \times \frac{1}{2}CI$, and the volume $\frac{4}{3}\pi \times ABC \times IK$ will become $\frac{2}{3}\pi \times AB \times IK \times CI$. But the triangles ABO, CIK are similar, and give the proportion,

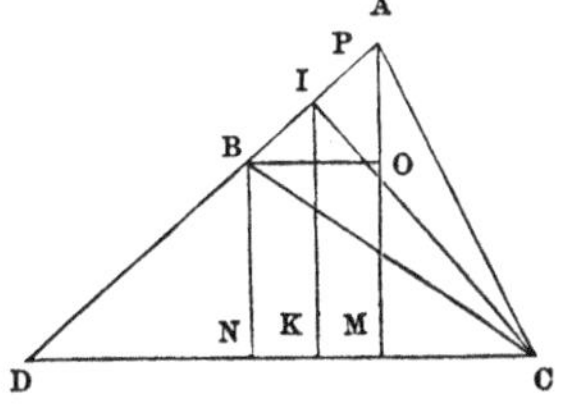


$$AB : BO \text{ or } MN :: CI : IK;$$

hence $AB \times IK = MN \times CI$: hence the solid described by the isosceles triangle ABC will have for its measure $\frac{2}{3}\pi \times MN \times CI^2$.

*Cor.* II. The general solution appears to include the supposition that AB produced will meet the axis; but the results would be equally true though AB were parallel to the axis.

Thus the cylinder described by AMNB is equal to $\pi . AM^2 . MN$; the cone described by ACM is equal to $\frac{1}{3}\pi AM^2 . CM$, and the cone described by BCN to $\frac{1}{3}\pi . AM^2 . CN$. Add the first two solids, and take away the third: we shall have the measure of the solid described by ABC equal to $\pi . AM^2 . (MN + \frac{1}{3}CM - \frac{1}{3}CN)$; and since $CN - CM = MN$, this expression is reducible to $\pi . AM^2 . \frac{2}{3}MN$, or $\frac{2}{3}\pi CP^2 . MN$, which agrees with the conclusion above drawn.

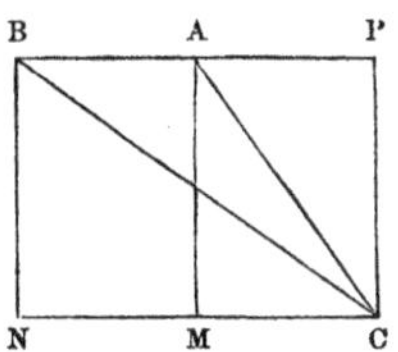


*Cor.* III. Let AB, BC, CD be the several successive sides of a regular polygon, O its centre, and OI the radius of the inscribed circle; if the polygonal sector AOD lying all on one side of the diameter FG be supposed to perform a revolution about this diameter, the solid so described will have for its measure $\frac{2}{3}\pi . OI^2 . MQ$, MQ being that portion of the axis which is included by the extreme perpendiculars AM, DQ.

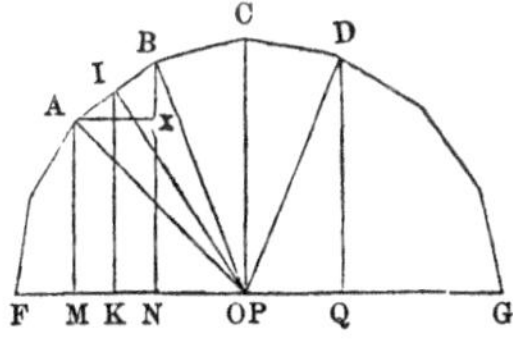


For, since the polygon is regular, all the triangles, AOB, BOC, etc., are equal and isosceles. Now, by Corollary I., the solid produced by the isosceles triangle AOB has for its measure $\frac{2}{3}\pi \,.\, OI^2 \,.\, MN$; the solid described by the triangle BOC has for its measure $\frac{2}{3}\pi \,.\, OI^2 \,.\, NP$, and the solid described by the triangle COD has for its measure $\frac{2}{3}\pi \,.\, OI^2 \,.\, PQ$: hence the sum of those solids, or the whole solid described by the polygonal sector AOD will have for its measure $\frac{2}{3}\pi \,.\, OI^2 \,.\, (MN + NP + IQ)$, or $\frac{2}{3}\pi \,.\, OI^2 \,.\, MQ$.

*Cor.* IV. The solid generated by the revolution of the entire polygon will have for its measure $\frac{2}{3}\pi \times OI^2 \times FG$. But the surface generated by the entire polygon has for its measure $2\pi \times OI \times FG$ (T. IV., C. III.), since $2\pi \times OI$ is the circumference of the inscribed circle. Hence the solid is measured by $2\pi \times OI \times FG$ into $\frac{1}{3}OI$; that is, the solid generated by the revolution of a semi-polygon inscribed in a semicircle, has for its measure its surface multiplied by one third the apothem of the generating polygon.

### THEOREM XII.

*The volume of a sphere is measured by its surface multiplied by one third its radius.*

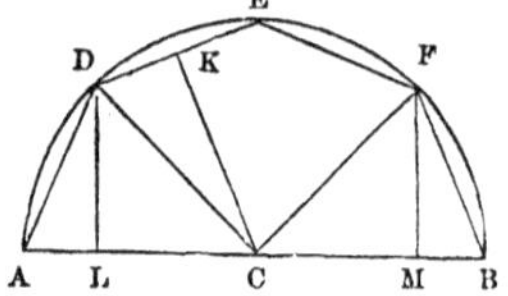


As in Theorem IX., let the sphere be generated by the revolution of the semicircle whose radius is CA, about its diameter AB. The regular inscribed polygon will generate a solid having for its measure the surface of the solid multiplied by one third the apothem of the polygon (T. XI., C. IV.). When we pass to the *limit*, the surface of the solid generated by the polygon will be the surface of the sphere generated by the semicircle (T. IX.), and the apothem of the polygon will become the radius. Consequently, the volume of a sphere is measured by its surface multiplied into one third of its radius.

*Cor. The volume of a spherical sector has for its measure the zone which forms its base multiplied into one third of its radius.*

For, the solid generated by any sectorial portion of the inscribed polygon, as CDEF, has for its measure $\frac{2}{3}\pi \times CK^2 \times LM$

(T. XI., C. III.). Hence at the limit we have, for the measure of the spherical sector generated by the plane sector CDF, $\frac{2}{3}\pi \times CA^2 \times LM' = 2\pi \times CA \times LM \times \frac{1}{3}CA$. But $2\pi \times CA \times LM$ is the measure of the surface of the zone forming the base of the spherical sector; hence the volume of a spherical sector is measured by its surface into one third of its radius.

*Scholium.* Let R be the radius of a sphere: its surface will be $4\pi R^2$; its solidity, $4\pi R^2 \times \frac{1}{3}R$, or $\frac{4}{3}\pi . R^3$. If the diameter is named D, we shall have $R = \frac{1}{2}D$, and $R^3 = \frac{1}{8}D^3$: hence the solidity may likewise be expressed by $\frac{4}{3}\pi . \frac{1}{8}D^3$, or $\frac{1}{6}\pi D^3$.

### THEOREM XIII.

*The surface of a sphere is to the whole surface of the circumscribed cylinder (including its bases), as 2 is to 3; and the volumes of these two bodies are to each other in the same ratio.*

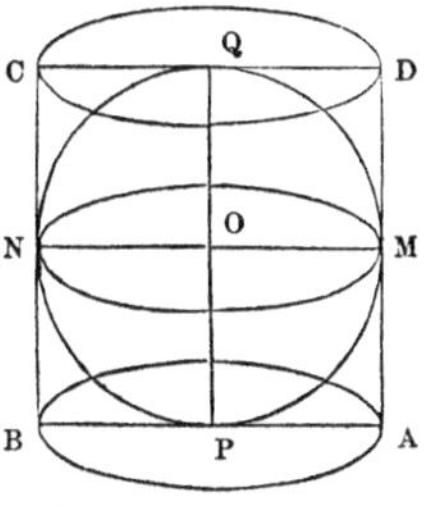


Let MPNQ be a great circle of the sphere, and ABCD the circumscribed square. If the semicircle PMQ and the half square PADQ are at the same time made to revolve about the diameter PQ, the semicircle will generate the sphere, while the half square will generate the cylinder circumscribed about that sphere.

The altitude AD of that cylinder is equal to the diameter PQ, the base of the cylinder is equal to the great circle, its diameter AB being equal to MN: hence (T. I.), the convex surface of the cylinder is equal to the circumference of the great circle multiplied by its diameter. This measure (T. IX.) is the same as that of the surface of the sphere: hence, *the surface of the sphere is equal to the convex surface of the circumscribed cylinder.*

But the surface of the sphere is equal to four great circles, hence the convex surface of the cylinder is also equal to four great circles; and adding the two bases, each equal to a great circle, the total surface of the circumscribed cylinder will be equal to six great circles: hence the surface of the sphere is to the total surface of the circumscribed cylinder as 4 is to 6, or as 2 is to 3, which is the first branch of the proposition.

In the next place, since the base of the circumscribed cylinder is equal to a great circle, and its altitude to the diameter, the volume of the cylinder (T. II.) will be equal to a great circle multiplied by its diameter. But (T. XII.) the volume of the sphere is equal to four great circles multiplied by a third of the radius; in other terms, to one great circle multiplied by $\frac{4}{3}$ of the radius, or by $\frac{2}{3}$ of the diameter. Hence the sphere is to the circumscribed cylinder as 2 to 3, and consequently the volumes of these two bodies are as their surfaces.

# EIGHTH BOOK.

## SPHERICAL GEOMETRY.

### DEFINITIONS.

I. *A spherical triangle* is a portion of the surface of a sphere, bounded by three arcs of great circles.

Those arcs, named *the sides of the triangle*, are always supposed to be each less than a semi-circumference; the angles, which their planes form with each other, are *the angles of the triangle.*

II. A spherical triangle takes the name of *right-angled*, *isosceles*, *equilateral*, in the same cases as a rectilineal triangle.

III. *A spherical polygon* is a portion of the surface of a sphere, terminated by several arcs of great circles.

IV. *A lune* is that portion of the surface of a sphere, which is included between two great semicircles meeting in a common diameter.

V. *A spherical wedge*, or *ungula*, is that portion of the solid sphere which is included between the same great semicircles, and has the lune for its base.

VI. *A spherical pyramid* is a portion of the solid sphere, included between the planes of a polyedral angle whose vertex is the centre; *the base* of the pyramid is the spherical polygon intercepted by the same planes.

VII. *The pole of a circle* is a point on the surface of the sphere equally distant from all the points of the circumference.

### THEOREM I.

*In every spherical triangle, any side is less than the sum of the other two.*

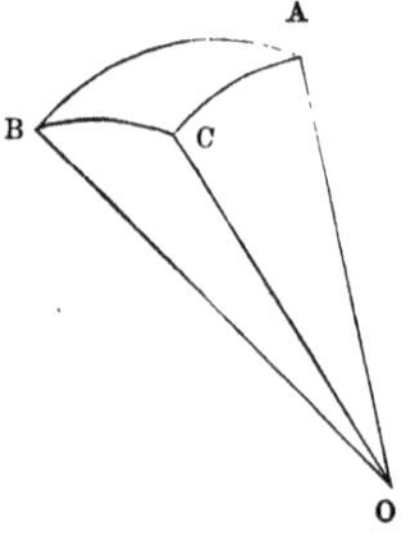


Let O be the centre of the sphere; and draw the radii OA, OB, OC. Imagine the planes AOB, AOC, COB; those planes will form a polyedral angle at the point O; and the angles AOB, AOC, COB will be measured by AB, AC, BC, the sides of the spherical triangle. But (B. V., T. XIX.) each of the three plane triangles composing a polyedral angle is less than the sum of the other two; hence, any side of the triangle ABC is less than the sum of the other two.

### THEOREM II.

*The sum of all the three sides of a spherical triangle is less than the circumference of a great circle.*

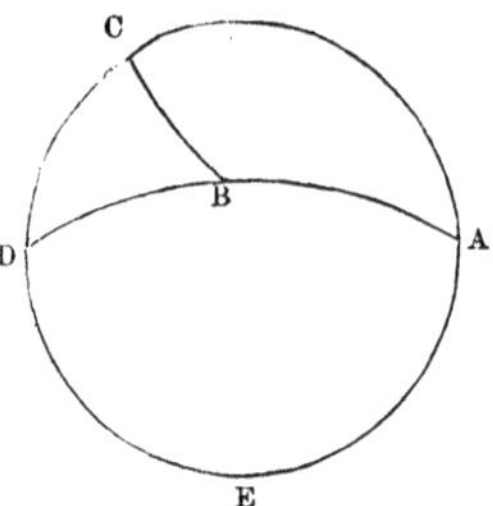


Let ABC be any spherical triangle; produce the sides AB, AC till they meet again in D. The arcs ABD, ACD will be semi-circumferences— since (B. VII., T. VI., C. II.) two great circles always bisect each other. But in the triangle BCD we have (T. I.) the side $BC < BD + CD$: add $AB + AC$ to each; we shall have $AB + AC + BC < ABD + ACD$, that is to say, less than a circumference.

### THEOREM III.

*The sum of all the sides of any spherical polygon is less than the circumference of a great circle.*

Let us take, for example, the pentagon ABCDE. Produce the sides AB, DC till they meet in F; then, since BC is less than BF + CF, the perimeter of the pentagon ABCDE will be less than that of the quadrilateral AEDF. Again, produce the sides AE, FD till they meet in G; we shall have ED < EG + DG: hence the perimeter of the quadrilateral AEDF is less than that of the triangle AFG, which last is itself less than the circumference of a great circle (T. II.): hence the perimeter of the polygon ABCDE is less than this same circumference.

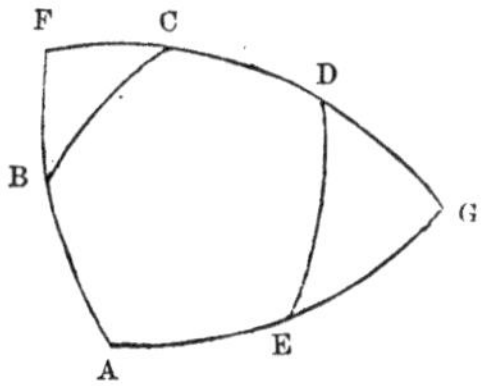


THEOREM IV.

*If a diameter be drawn perpendicular to the plane of a great circle, its extremities will be the poles of that circle, and also of all small circles parallel to it.*

For, DC being perpendicular to the plane AMB, is perpendicular to all the straight lines CA, CM, CB, etc., drawn through its foot in this plane; hence all the arcs DA, DM, DB, etc., are quarters of the circumference. So likewise are all the arcs EA, EM, EB, etc.; hence the points D and E are each equally distant from all the points of the circumference AMB; therefore (D. VII.) they are the poles of that circumference.

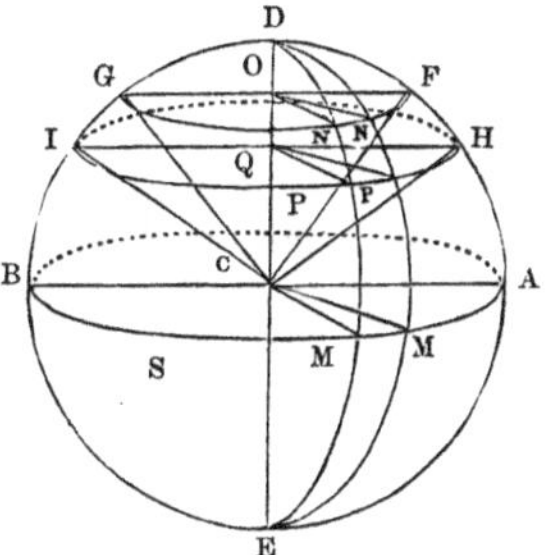


Again, the radius DC, perpendicular to the plane AMB, is perpendicular to its parallel FNG; hence (B. VII., T. VI., C. IV.) it passes through O, the centre of the circle FNG; therefore, if the oblique lines DF, DN, DG be drawn, they will be equal (B. V., T. V.); but, the chords being equal, the arcs are equal: hence the point D is the pole of the small circle FNG; and, for like reasons, the point E is the other pole.

*Cor.* I. Every arc DM drawn from a point in the arc of a great circle AMB to its pole, is a quarter of the circumference, which, for the sake of brevity, is usually named *a quadrant;* and this quadrant, at the same time, makes a right angle with the arc AM. For (B. V., T. XVI.), the line DC being perpendicular to the plane AMC, every plane DMC passing through the line DC is perpendicular to the plane AMC; hence the an gles of these planes, or the angle AMD, is a right angle.

*Cor.* II. To find the pole of a given arc AM, draw the indefinite arc MD perpendicular to AM; take MD equal to a quadrant: the point D will be one of the poles of the arc AMD. Or thus: at the two points A and M, draw the arcs AD and MD perpendicular to AM; their point of intersection, D, will be the pole required.

*Cor.* III. Conversely, if the distance of the point D from each of the points A and M be equal to a quadrant, the point D will be the pole of the arc AM; and also the angles DAM, DMA will be right angles.

For, let C be the centre of the sphere, and draw the radii CA, CD, CM. Since the angles ACD, MCD are right, the line CD is perpendicular to the two straight lines CA, CM; it is, therefore, perpendicular to their plane: hence the point D is the pole of the arc AM, and consequently the angles DAM, DMA are right.

*Scholium.* The properties of these poles enable us to describe arcs of a circle on the surface of a sphere with the same facility as on a plane surface. It is evident, for instance, that by turning the arc DF, or any other line extending to the same distance, round the point D, the extremity F will describe the small circle FNG; and by turning the quadrant DFA round the point D, its extremity A will describe the arc of the great circle AM.

If the arc AM were required to be produced, and nothing were given but the points A and M through which it was to pass, we should first have to determine the pole D, by the intersection of two arcs described from the points A and M as centres, with a distance equal to a quadrant; the pole D being found, we might describe the arc AM and its prolongation from D as a centre, and with the same distance as before.

Lastly, if it be required from a given point P to draw a perpendicular on the given arc AM, produce this arc to S till the distance PS be equal to a quadrant; then from the pole S, and with the same distance, describe the arc PM, which will be the perpendicular required.

### THEOREM V.

*The angle formed by two arcs of great circles is equal to the angle formed by the tangents of these arcs at the point of intersection, and is therefore measured by the arc described from the point of intersection as a pole, between the sides, produced if necessary.*

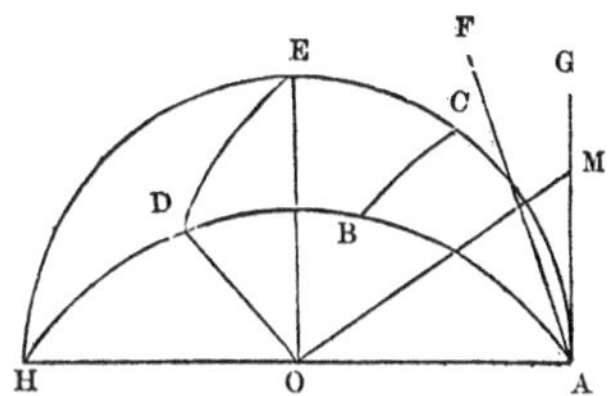


For the tangent AF drawn in the plane of the arc AB is perpendicular to the radius AO; and the tangent AG drawn in the plane of the arc AC is perpendicular to the same radius AO: hence (B. V., D. VI.) the angle FAG is equal to the angle contained by the planes OAB, OAC, which is that of the arcs AB, AC, and is named BAC.

In like manner, if the arcs AD and AE are both quadrants, the lines OD, OE will be perpendicular to AO, and the angle DOE will still be equal to the angle of the planes AOD, AOE; hence the arc DE is the measure of the angle contained by these planes, or of the angle CAB.

*Cor.* The angles of spherical triangles may be compared together, by means of the arcs of great circles described from their vertices as poles, and included between their sides; hence it is easy to make an angle of this kind equal to a given angle.

*Scholium.* Vertical angles, such as ACO and BCN, are equal; for either of them is still the angle formed by the two planes ACB, OCN.

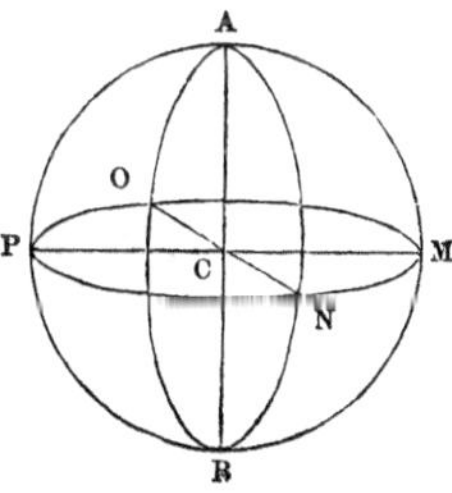


It is further evident, that, in the intersection of two arcs ACB, OCN, the two adjacent angles ACO, OCB taken together are equal to two right angles.

### THEOREM VI.

*If, with the vertices of a given triangle as poles, arcs be described, forming a new triangle, then will the vertices of this new triangle be the poles respectively of the sides of the given triangle.*

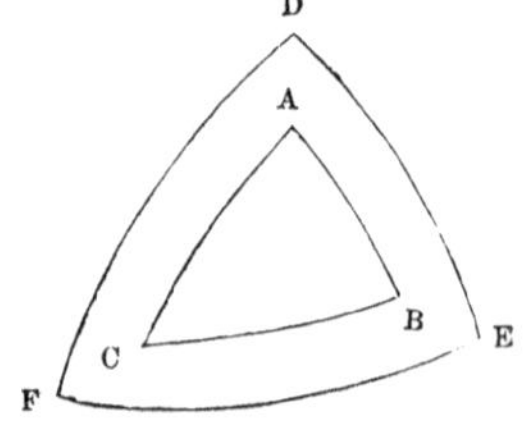


For, the point A being the pole of the arc EF, the distance AE is a quadrant; the point C being the pole of the arc DE, the distance CE is likewise a quadrant: hence the point E is removed the distance of a quadrant from each of the points A and C; it is, therefore, the pole of the arc AC. It might be shown by the same method, that D is the pole of the arc BC, and F that of the arc AB.

*Cor.* Hence the triangle ABC may be described by means of DEF, as DEF may by means of ABC.

### THEOREM VII.

*The same supposition being made as in the last Theorem, each angle in either one of the triangles will be measured by a semi-circumference* MINUS *the side lying opposite to it in the other triangle.*

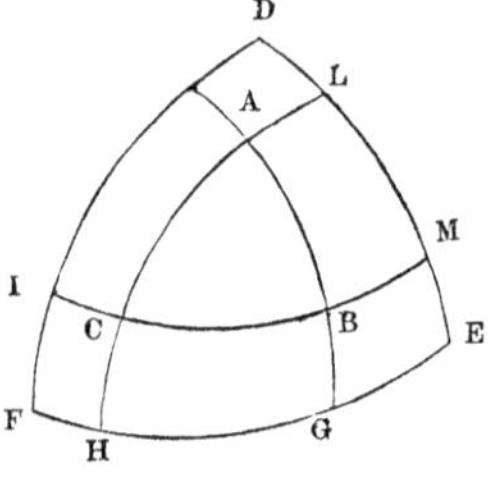


Produce the sides AB, AC, if necessary, till they meet EF in G and H. The point A being the pole of the arc GH, the angle A will be measured by that arc; but the arc EH is a quadrant, and likewise GF, E being the pole of AH, and F of AG: hence EH + GF is equal to a semi-circumference. Now, EH + GF is the same as EF + GH; hence the arc GH, which measures the angle A, is equal to a semi-circumference *minus* the side EF. In like manner, the angle B will be measured by $\frac{1}{2}$ circ. — DF; the angle C, by $\frac{1}{2}$ circ. — DE.

And this property must be reciprocal in the two triangles, since each of them is described in a similar manner by means of the other. Thus we shall find the angles D, E, F of the triangle DEF to be measured respectively by $\frac{1}{2}$ circ. − BC, $\frac{1}{2}$ circ. − AC, $\frac{1}{2}$ circ. − AB. The angle D, for example, is measured by the arc MI; but MI + BC = MC + BI = $\frac{1}{2}$ circ.; hence the arc MI, the measure of D, is equal to $\frac{1}{2}$ circ. − BC; and so of the other angles.

*Scholium.* It must further be observed, that besides the triangle DEF, three others might be formed by the intersection of the three arcs DE, EF, DF. But the proposition immediately before us is applicable only to the central triangle, which is distinguished from the other three by the circumstance that the two angles A and D lie on the same side of BC, the two B and E on the same side of AC, and the two C and F on the same side of AB.

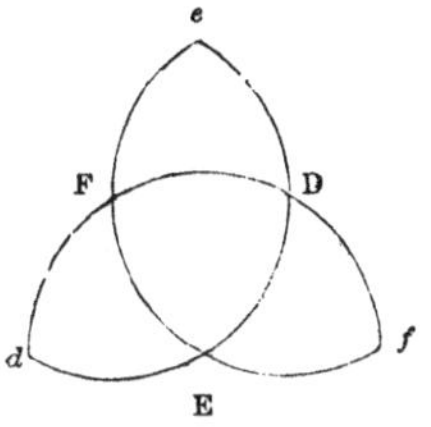


Various names have been given to the triangles ABC, DEF, but they are now more generally denominated *polar triangles.*

THEOREM VIII.

*Any triangle on a sphere being given, another triangle may be constructed which shall have all its parts equal respectively to the corresponding parts of the given triangle.*

Let ABC be the given triangle. With A as a pole, describe the arc CED passing through C; and with B as a pole, describe another arc CFD passing through C, and intersecting the former arc at D. Then will the triangle ADB have all its parts equal to the corresponding parts of the triangle ABC.

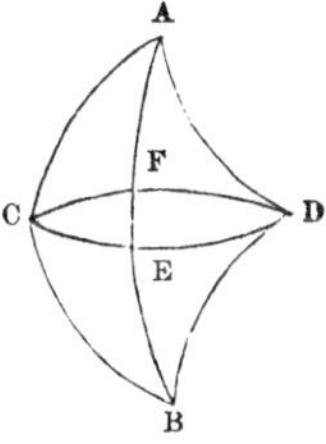


For, by construction, the side AD = AC, BD = BC, and AB is common; hence those two triangles have their sides equal each to each, and it is to be shown that the angles opposite these equal sides are also equal.

If the centre of the sphere is supposed to be at O, a polyedral angle may be conceived as formed at O by the three plane angles AOB, AOC, BOC; likewise another polyedral angle may be conceived as formed by the three plane angles AOB,

AOD, BOD. And because the sides of the triangle ABC are equal to those of the triangle ADB, the plane angles forming the one of these polyedral angles must be equal to the plane angles forming the other, each to each; but in this case the planes, in which the equal angles lie, are equally inclined to each other (B. V., T. XXI.); hence all the angles of the spherical triangle DAB are respectively equal to those of the triangle CAB, namely, DAB = CAB, DBA = CBA, and ADB = ACB; therefore the sides and the angles of the triangle ADB are equal to the sides and the angles of the triangle ACB. These triangles are called *symmetrical.*

### THEOREM IX.

*Two triangles on the same sphere, or on equal spheres, are equal in all their parts, when they have each an equal angle included between equal sides.*

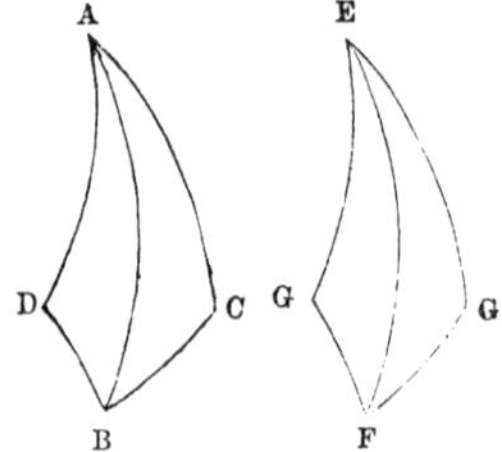


Suppose the side AB = EF, the side AC = EG, and the angle BAC = FEG; the triangle EFG may be placed on the triangle ABC, or on ABD symmetrical with ABC, just as two rectilineal triangles are placed upon each other when they have an equal angle included between equal sides. Hence all the parts of the triangle EFG will be equal to all the parts of the triangle ABC; that is, besides the three parts equal by hypothesis, we shall have the side BC = FG, the angle ABC = EFG, and the angle ACB = EGF.

### THEOREM X.

*Two triangles on the same sphere, or on equal spheres, are equal in all their parts, when two angles and the included side of the one are equal to two angles and the included side of the other.*

For one of those triangles, or the triangle symmetrical with it, may be placed on the other, and be made to coincide with it, as is obvious.

### THEOREM XI.

*If two triangles on the same sphere, or on equal spheres, have all their sides respectively equal, their angles will likewise be all respectively equal, the equal angles lying opposite the equal sides.*

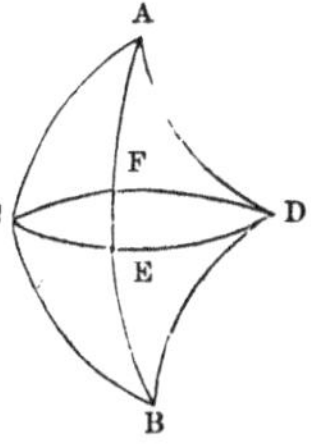


The truth is evident by T. VIII., where it was shown that, with three given sides, AB, AC, BC, there can only be two triangles ACB, ABD different as to the position of their parts, and equal as to the magnitude of those parts. Hence those two triangles, having all their sides respectively equal in both, must either be absolutely equal, or at least *symmetrically* so; in both of which cases, their corresponding angles must be equal, and lie opposite to equal sides.

### THEOREM XII.

*If two triangles on the same sphere, or on equal spheres, are mutually equiangular, they will also be mutually equilateral.*

Let A and B be the two given triangles; P and Q their polar triangles. Since the angles are equal in the triangles A and B, the sides will be equal in the polar triangles P and Q (T. VIII.); but since the triangles P and Q are mutually equilateral, they must also (T. XI.) be mutually equiangular; and, lastly, the angles being equal in the triangles P and Q, it follows (T. VIII.) that the sides are equal in their polar triangles A and B. Hence the mutually equiangular triangles A and B are at the same time mutually equilateral.

*Scholium.* This proposition is not applicable to rectilineal triangles, in which equality among the angles indicates only proportionality among the sides. Nor is it difficult to account for the difference observable, in this respect, between spherical and rectilineal triangles. In the proposition now before us, as well as in the three last, which treat of the comparison of triangles, it is expressly required that the arcs be traced on the same sphere, or on equal spheres. Now, similar arcs are to each other as their radii: hence, on equal spheres, two triangles cannot be similar without being equal; therefore it is not strange that equality among the angles should produce equality among the sides.

The case would be different if the triangles were drawn upon unequal spheres; there, the angles being equal, the triangles would be similar, and the homologous sides would be to each other as the radii of their spheres.

THEOREM XIII.

*In every isosceles spherical triangle, the angles opposite the equal sides are equal; and, conversely, if two angles of a spherical triangle are equal, the triangle will be isosceles.*

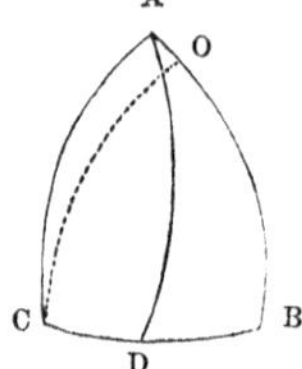


*First.* Suppose the side AB = AC, we shall have the angle C = B. For, if the arc AD be drawn from the vertex A to the middle point D of the base, the two triangles ABD, ACD will have all the sides of the one respectively equal to the corresponding sides of the other, namely, AD common, BD = CD, and AB = AC; hence, by the last proposition, their angles will be equal; therefore B = C.

*Secondly.* Suppose the angle B = C; we shall have the side AC = AB. For, if not, let AB be the greater of the two; take BO = AC, and join OC. The two sides BO, BC are equal respectively to the two AC, BC; the angle OBC contained by the first two is equal to ACB contained by the second two. Hence (T. IX.) the two triangles BOC, ACB have all their other parts equal; hence the angle OCB = ABC; but, by hypothesis, the angle ABC = ACB; hence we have OCB = ACB, which is absurd; therefore AB is not different from AC; that is, the sides AB, AC, opposite to the equal angles B and C, are equal.

*Scholium.* The same demonstration proves that the angle BAD = DAC, and the angle BDA = ADC. Hence the two last are right angles; consequently, *the arc drawn from the vertex of an isosceles spherical triangle to the middle of the base, is at right angles to that base, and bisects the opposite angle.*

### THEOREM XIV.

*In any spherical triangle, the greater side is opposite the greater angle; and, conversely, the greater angle is opposite the greater side.*

*First.* Suppose the angle A>B: make the angle BAD = B; then we shall have AD = DB (T. XII.); but AD + DC is greater than AC: hence, putting DB in place of AD, we shall have DB + DC, or BC > AC.

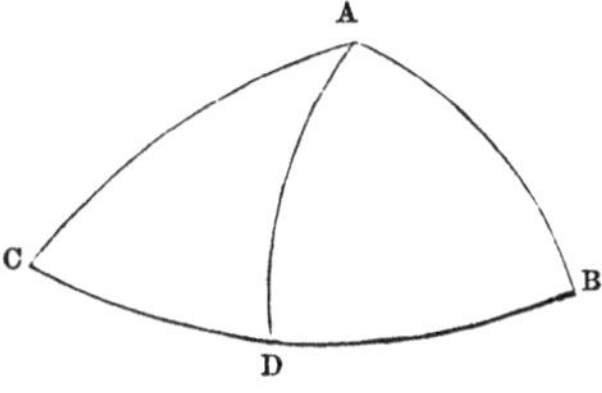


*Secondly.* If we suppose BC > AC, the angle BAC will be greater than ABC. For, if BAC were equal to ABC, we should have BC = AC; if BAC were less than ABC, we should then, as has just been shown, find BC < AC. Both these conclusions are false; hence the angle BAC is greater than ABC.

### THEOREM XV.

*The sum of all the angles in any spherical triangle is less than six right angles, and greater than two.*

For, in the first place, every angle of a spherical triangle is less than two right angles (see the following Scholium); hence the sum of all the three is less than six right angles.

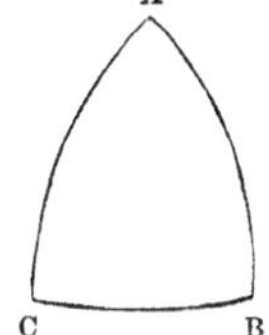


Secondly, the measure of each angle in the spherical triangle (T. VII.) is equal to the semi-circumference *minus* the corresponding side of the polar triangle; hence the sum of all the three is measured by three semi-circumferences *minus* the sum of all the sides of the polar triangle. Now (T. II.), this latter sum is less than a circumference; therefore, taking it away from three semi-circumferences, the remainder will be greater than one semi-circumference, which is the measure of two right angles. Hence, in the second place, the sum of all the angles in a spherical triangle is greater than two right angles.

*Cor.* I. The sum of all the angles in a spherical triangle is not constant, like that of all the angles in a rectilineal triangle; it varies between two right angles and six, without ever arriving at either of these limits. Two given angles, therefore, do not serve to determine the third.

*Cor.* II. A spherical triangle may have two or even three angles right, two or three obtuse.

If the triangle ABC have two right angles B and C, the vertex A will (T. IV.) be the pole of the base BC, and the sides AB, AC will be quadrants.

If the angle A is also right, the triangle ABC will have all its angles right, and its sides quadrants. The tri-rectangular triangle is contained eight times in the surface of the sphere, as is evident from the figure in the next proposition, supposing the arc MN to be a quadrant.

*Scholium.* In all the preceding observations, we have supposed, in conformity with D. I., that our spherical triangles have always each of their sides less than a semi-circumference; from which it follows that any one of their angles is always less than two right angles. For (see the figure of T. II.), if the side AB is less than a semi-circumference, and AC is so likewise, both those arcs will require to be produced before they can meet in D. Now, the two angles ABC, CBD taken together, are equal to two right angles; hence the angle ABC itself is less than two right angles.

We may observe, however, that some spherical triangles do exist, in which certain of the sides are greater than a semi-circumference, and certain of the angles greater than two right angles. Thus, if the side AC is produced so as to form a whole circumference ACDE, the part which remains after subtracting the triangle ABC from the hemisphere, is a new triangle also designated by ABC, and having AB, BC, AEDC for its sides. Here, it is plain, the side AEDC is greater than the semi-circumference AED; and, at the same time, the angle B opposite to it exceeds two right angles by the quantity CBD.

The triangles, whose sides and angles are so large, have been excluded from our definition; but the only reason was, that the solution of them, or the determination of their parts, is always reducible to the solution of such triangles as are comprehended by the definition. Indeed, it is evident enough, that if the sides and angles of the triangle ABC are known, it will be easy to discover the angles and sides of the triangle which bears the same name, and is the difference between a hemisphere and the former triangle.

### THEOREM XVI.

*The surface of a lune is to the entire surface of the sphere, as the angle of the lune is to four right angles, or as the arc which measures the angle of the lune is to the circumference.*

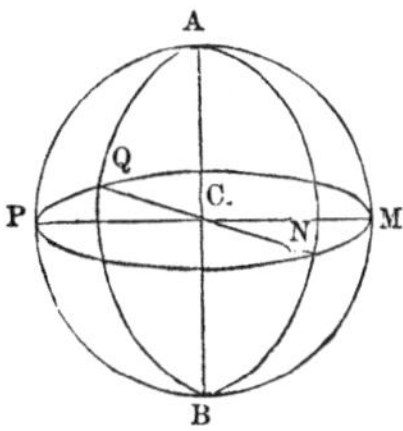


Suppose, in the first place, the arc MN to be to the circumference MNPQ as some one rational number is to another, as 5 to 48 for example. The circumference MNPQ being divided into 48 equal parts, MN will contain 5 of them; and if the pole A were joined with the several points of division, by as many quadrants, we should in the hemisphere AMNPQ have 48 triangles, all equal, because having all their parts equal. Hence the whole sphere must contain 96 of those partial triangles, and the lune AMBNA will contain 10 of them; hence the lune is to the sphere as 10 is to 96, or as 5 to 48; in other words, as the arc MN is to the circumference.

If the arc MN is not commensurable with the circumference, we may still show, by the mode of reasoning employed in the case of incommensurable magnitudes, that in this instance, also, the lune is to the sphere as MN is to the circumference.

*Cor.* I. Two lunes are to each other as their respective angles.

*Cor.* II. It was shown (T. XV., C. II.) that the whole surface of the sphere is equal to eight tri-rectangular triangles; hence, if the area of one such triangle is taken for unity, the surface of the sphere will be represented by 8. This granted, the surface of the lune whose angle is A will be expressed by 2 A (the angle A being always estimated from the right angle assumed as unity), since 2 A : A : : 8 : 4. Thus we have here two different unities: one for angles, being the right angle; the other for surfaces, being the tri-rectangular spherical triangle, or the triangle whose angles are all right, and whose sides are quadrants.

*Scholium.* The spherical ungula bounded by the planes AMB,

ANB, is to the whole sphere, as the angle A is to four right angles; for, the lunes being equal, the spherical ungulas will also be equal; hence two spherical ungulas are to each other as the angles formed by the planes which bound them.

### THEOREM XVII.

*Two symmetrical spherical triangles are equal in surface.*

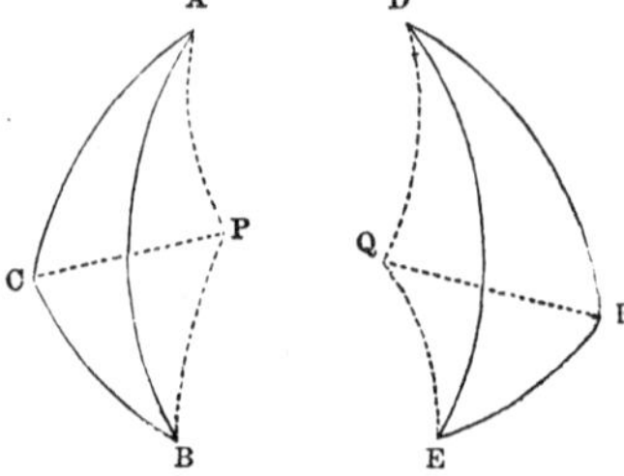


Let ABC, DEF be two symmetrical triangles; that is to say, two triangles having their sides

$$AB = DE,$$
$$AC = DF,$$
$$BC = EF,$$

and yet incapable of coinciding with each other: we are to show that the surface ABC is equal to the surface DEF.

Let P be the pole of the little circle passing through the three points A, B, C; from this point, draw (T. IV., S.) the equal arcs PA, PB, PC; at the point F, make the angle DFQ = ACP, the arc FQ = CP; and join DQ, EQ.

The sides DF, FQ are equal to the sides AC, CP; the angle DFQ = ACP; hence (T. IX.) the two triangles DFQ, ACP are equal in all their parts; hence the side DQ = AP, and the angle DQF = APC.

In the proposed triangles DEF, ABC, the angles DFE, ACB opposite to the equal sides DE, AB being equal (T. VIII.), if the angles DFQ, ACP, which are equal by construction, be taken away from them, there will remain the angle QFE equal to PCB. Also the sides QF, FE are equal to the sides PC, CB; hence the two triangles FQE, CPB are equal in all their parts: hence the side QE = PB, and the angle FQE = CPB.

Now, observing that the triangles DFQ, ACP, which have their sides respectively equal, are at the same time isosceles, we shall see them to be capable of mutual adaptation, when applied to each other; for, having placed PA on its equal QF, the side PC will fall on its equal QD, and thus the two triangles will exactly coincide; hence they are equal, and the surface DQF=

APC. For a like reason, the surface FQE = CPB, and the surface DQE = APB; hence we have

$$DQF + FQE - DQE = APC + CPB - APB,$$

that is, $$DEF = ABC;$$

therefore, the two symmetrical triangles ABC, DEF are equal in surface.

*Scholium.* The poles P and Q might lie within the triangles ABC, DEF; in which case, it would be requisite to add the three triangles DQF, FQE, DQE together, in order to make up the triangle DEF; and, in like manner, to add the three triangles APC, CPB, APB together, in order to make up the triangle ABC. In all other respects, the demonstration and the result would still be the same.

### THEOREM XVIII.

*If two great circles intersect each other on the surface of a hemisphere, the sum of the opposite triangles thus formed will be equivalent to the lune whose angle is equal to the angle formed by the circles.*

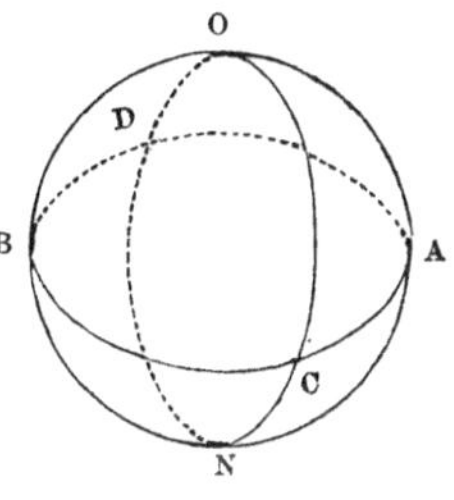


Let the circumference AOB, COD intersect on the hemisphere OACBD; then will the opposite triangles AOC, BOD be equal to the lune whose angle is BOD.

For, producing the arcs OB, OD in the other hemisphere, till they meet in N, the arc OBN will be a semi-circumference, and AOB one also; and taking OB from each, we shall have BN = AO. For a like reason, we have DN = CO, and BD = AC. Hence the two triangles AOC, BDN have their three sides respectively equal; besides, they are so placed as to be symmetrical; hence (T. XVII.) they are equal in surface, and the sum of the triangles AOC, BOD is equal to the lune OBNDO. whose angle is BOD.

THEOREM XIX.

*The surface of any spherical triangle is measured by the excess of the sum of its three angles above two right angles.*

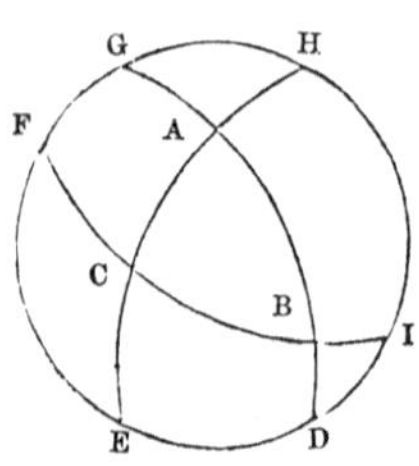


Let ABC be the proposed triangle: produce its sides till they meet the great circle DEFGHI, drawn anywhere without the triangle. By the last proposition, the two triangles ADE, AGH are together equal to the lune whose angle is A, and which is measured (T. XVI., C. II.) by 2 A; hence we have ADE + AGH = 2 A; and, for a like reason, BGF + BID = 2 B, and CIH + CFE = 2 C. But the sum of those six triangles exceeds the hemisphere by twice the triangle ABC, and the hemisphere is represented by 4; therefore twice the triangle ABC is equal to 2 A + 2 B + 2 C − 4, and consequently once ABC = A + B + C − 2; hence, every spherical triangle is measured by the sum of all its angles *minus* two right angles.

*Cor.* I. However many right angles there be contained in this measure, just so many tri-rectangular triangles, or eighths of the sphere, which (T. XVI., C. II.) are the unit of surface, will the proposed triangle contain. If the angles, for example, are each equal to $\frac{4}{3}$ of a right angle, the three angles will amount to 4 right angles, and the proposed triangle will be represented by 4 − 2, or 2; therefore it will be equal to two tri-rectangular triangles, or to the fourth part of the whole surface of the sphere.

*Cor.* II. The spherical triangle ABC is equal to the lune whose angle is $\frac{A+B+C}{2}-1$. Likewise the spherical pyramid which has ABC for its base, is equal to the spherical ungula whose angle is $\frac{A+B+C}{2}-1$.

*Scholium.* While the spherical triangle ABC is compared with the tri-rectangular triangle, the spherical pyramid, which has ABC for its base, is compared with the tri-rectangular pyramid, and the same ratio is found to subsist between them. The polyedral angle at the vertex of the pyramid is, in like manner, compared with the polyedral angle at the vertex of the tri-rectangular pyramid. These comparisons are founded on the coincidence of the corresponding parts. If the bases of the pyramids coincide, the pyramids themselves will evidently coincide, and likewise the polyedral angles at their vertices. From this, the following consequences are deduced:

*First.* Two triangular spherical pyramids are to each other as their bases; and since a polygonal pyramid may always be divided into a certain number of triangular ones, it follows that any two spherical pyramids are to each other as the polygons which form their bases.

*Secondly.* The polyedral angles at the vertices of those pyramids are also as their bases; hence, for comparing any two polyedral angles, we have merely to place their vertices at the centres of two equal spheres, and the polyedral angles will be to each other as the spherical polygons intercepted between their planes or faces.

The vertical angle of the tri-rectangular pyramid is formed by three planes at right angles to each other: this angle, which may be called *a right polyedral angle*, will serve as a very natural unit of measure for all other polyedral angles; and if so, the same number that exhibits the area of a spherical polygon, will exhibit the measure of the corresponding polyedral angle. If the area of the polygon is $\frac{3}{4}$, for example; in other words, if the polygon is $\frac{3}{4}$ of the tri-rectangular polygon, then the corresponding polyedral angle will also be $\frac{3}{4}$ of the right polyedral angle.

### THEOREM XX.

*The surface of a spherical polygon is measured by the sum of all its angles,* MINUS *the product of two right angles by the number of sides in the polygon* MINUS *two.*

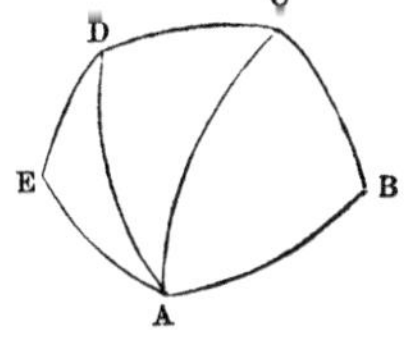


From one of the vertices A, let diagonals AC, AD be drawn to all the other vertices; the polygon ABCDE will be divided into as many triangles, *minus* two, as it has sides. But the surface of each triangle is measured by the sum of all its angles *minus* two right angles, and the sum of the angles in all the triangles is evidently the same as that of all the angles in the polygon; hence the surface of the polygon is equal to the sum of all its angles, diminished by twice as many right angles as it has sides *minus* two.

*Scholium.* Let $s$ be the sum of all the angles in a spherical polygon, and $n$ the number of its sides; the right angle being taken for unity, the surface of the polygon will be measured by $s - 2(n-2)$, or $s - 2n + 4$.

---

## REGULAR POLYEDRONS.

Since the faces of any regular polyedron are all equal and regular polygons (B. VI., D. XII.), it follows that all its polyedral angles must be equal.

*There can be only five kinds of regular polyedrons.*

For, we already know that *three* faces at least are necessary to form a polyedral angle; and moreover that the sum of all the plane angles which form the polyedral angle is less than 4 right angles. It is sufficient then to consider all the cases in which the angles of equal and regular polygons, taken 3 at a time, 4 at a time, 5 at a time, etc., do not equal or exceed 4 right angles. Now,

1. The angle of an equilateral triangle being $\frac{2}{3}$ of a right angle, we may construct a polyedral angle by using *three*, *four*, or *five* equilateral triangles for faces. But we cannot use six or a greater number, since $\frac{2}{3}$ of a right angle $\times 6 = 4$ right angles.

2. The angle of a square being 1 right angle, we may form a polyedral angle by using *three* squares for the faces. But we could not use four or a greater number, since 1 right angle $\times 4 =$ 4 right angles.

3. The angle of a regular pentagon being $\frac{6}{5}$, a polyedral angle may be formed with *three* regular pentagons, which gives $\frac{6}{5} \times 3 = 3\frac{3}{5}$.

Since the angle of a regular hexagon is $\frac{4}{3}$, and $\frac{4}{3} \times 3 = 4$, it follows, that a polyedral angle cannot be formed with regular hexagons for faces. And of course, regular polygons of a greater number of sides cannot be used in forming a polyedral angle.

Thus, the only possible regular polyedrons are those of which each polyedral angle is formed by *three*, *four*, or *five* equilateral triangles, by *three* squares, and by *three* regular pentagons, making in all *five* polyedrons.

When *three* equilateral triangles are used in forming each polyedral angle, the whole number of faces of the regular polyedron will be *four*, and it is called a *Tetraedron.*

When *four* equilateral triangles unite in forming each polyedral angle, the polyedron will have *eight* faces, and it is called an *Octaedron.*

When *five* equilateral triangles are used in forming a polyedral angle, the polyedron will have *twenty* faces, and it is then called an *Icosaedron.*

When *three* squares are used in forming each polyedral angle of a regular polyedron, there will be *six* faces, and the polyedron is called a *Hexagon*, or *Cube* (B. VI., D. V.).

When *three* regular pentagons are used for each polyedral angle, the number of faces will be *twelve*, and the solid is called a *Dodecaedron.*

These five regular solids may obviously be inscribed in a sphere, as well as circumscribed about a sphere; or a sphere may be inscribed in each, and circumscribed about each. The inscribed and circumscribed spheres have a common centre, which may be regarded as the centre of the polyedron.

If we suppose planes to pass through the centre of a regular polyedron and each of its edges, they will divide the polyedron into as many equal pyramids as the polyedron has faces, since each face of the polyedron will thus become a base of a pyramid, whose altitude will be the radius of the inscribed sphere; and since each pyramid is measured by the area of its base multiplied by one third of its altitude, it follows that the volume of any regular polyedron has for its measure its surface multiplied by one third the radius of its inscribed sphere.

Were we to regard the sphere as a regular polyedron of an infinite number of infinitely small and equal faces, we should, from the above, at once infer that the volume of a sphere has for its measure its surface multiplied by one third its radius.

Whether we regard the sphere as a regular polygon or not, it is essentially a *regular body.*

Hence, there are then six *regular bodies:* the *Tetraedron*, *Hexaedron*, *Octaedron*, *Dodecaedron*, *Icosaedron*, and the *Sphere.*

## APPLICATION OF ALGEBRA TO THE SOLUTION OF GEOMETRICAL PROBLEMS.

### PROBLEM I.

*In an equilateral triangle, having given the lengths of the three perpendiculars drawn from a certain point within it to the three sides, to determine its side.*

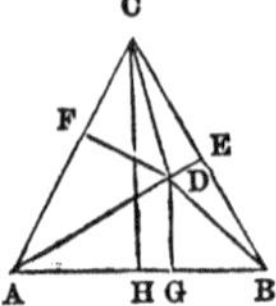


Let ABC be the equilateral triangle, and DE, DF, DG the perpendiculars from the point D upon the sides respectively. Denote these perpendiculars by $a$, $b$, $c$, in order, and the side of the triangle ABC by $2x$. Then, if the perpendicular CH be drawn,

$$\text{CH} = \sqrt{\text{AC}^2 - \text{AH}^2} = \sqrt{4x^2 - x^2} = x\sqrt{3}.$$

The area of the triangle $\text{ADB} = \frac{1}{2}\text{AB.GD} = cx$. Similarly the triangle $\text{BDC} = ax$, the triangle $\text{CDA} = bx$, and the triangle $\text{ACB} = \frac{1}{2}\text{AB.CH} = x^2\sqrt{3}$. Also, $\text{BDC} + \text{CDA} + \text{ADB} = \text{ABC}$; that is, in symbols,

$$x^2\sqrt{3} = (a + b + c)\,x, \text{ and } x = \frac{a + b + c}{\sqrt{3}},$$

which is half the side of the triangle sought.

*Cor.* From the resulting equation, we have

$$x\sqrt{3} = a + b + c:$$

we also had $\text{CH} = x\sqrt{3}$. Hence $\text{CH} = a + b + c$; or the whole perpendicular CH is equal to the sum of the three smaller perpendiculars from D upon the sides, whenever the point D is taken *within* the triangle. Had the point D been taken *without* the triangle, the perpendicular upon the side which subtends the angle within which the point lies would become negative. Thus, had the point been without the triangle, but between the sides AB, AC *produced*, then $\text{CH} = \text{DF} + \text{DG} - \text{DE}$.

PROBLEM II.

*A May-pole was broken off by the wind, and its top struck the ground twenty feet from the base; and, being repaired, was broken a second time five feet lower, and its top struck the ground ten feet farther from the base. What was the height of the May-pole?*

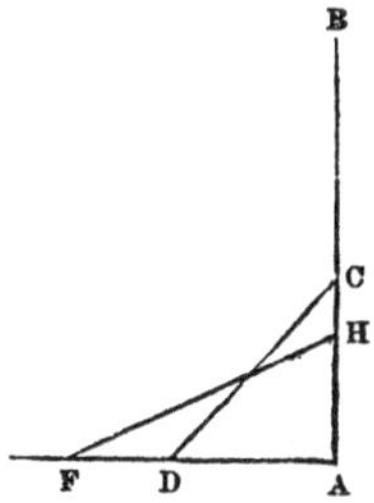

Let AB be the unbroken May-pole, C and H the points in which it was successively broken, and D and F the corresponding points at which the top B struck the ground. Then will CAD and HAF be right-angled triangles.

Put BC = CD = $x$, CA = $y$, AD = $a$, AF = $b$, and CH = $c$. Then AB = $x+y$, BH = HF = $x+c$, and HA = $y-c$; therefore (B. III., T. XIV.), we have

$$y^2+a^2=x^2, \qquad (1)$$

$$(y-c)^2+b^2=(x+c)^2. \qquad (2)$$

Expanding (2) and subtracting (1) from it, we have, after a slight reduction,

$$x+y=\frac{b^2-a^2}{2c}=50 \text{ feet, the required height.}$$

PROBLEM III.

*A statue eighty feet high stands on a pedestal fifty feet high, and to a spectator on the horizontal plane, they subtend equal angles; required the distance of the observer from the base, the height of the eye being five feet.*

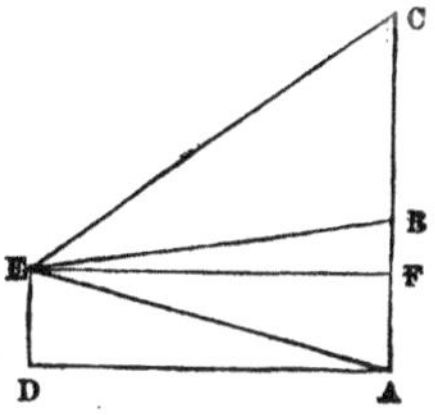

Let AB = $a$, the height of the pedestal;
BC = $b$, the height of the statue;
DE = $c$, the height of the eye from ground,
and DA = EF = $x$, the distance sought.
Then

$$EC^2=EF^2+CF^2=x^2+(a+b-c)^2,$$

and

$$EA^2=EF^2+ED^2=x^2+c^2.$$

But, since the angle CEB = BEA, we have (B. III., T. XII.)

$$EC : EA :: CB : BA, \text{ or,}$$
$$EC^2 : EA^2 :: CB^2 : BA^2;$$

or, in symbols, $x^2 + (a + b - c)^2 : x^2 + c^2 :: b^2 : a^2$.

From this proportion, we readily deduce

$$x = \pm \sqrt{\left(\frac{a(a-c)^2 + b(a^2 - c^2)}{b - a}\right)};$$

the double sign merely indicating that this value of $x$ may be measured either way, from A towards D, or from D towards A.

### PROBLEM IV.

*To determine the area of a triangle when the three sides are given.*

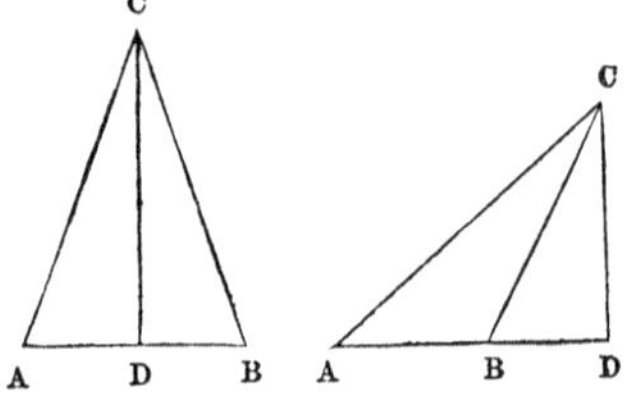


Let ABC be the triangle. From the angle C draw the perpendicular CD, which in the second figure falls without the triangle, and meets the base AB produced.

Denote the sides respectively opposite the angles A, B, C, by $a$, $b$, $c$, and the perpendicular CD by $p$. The right-angled triangles CDA, CDB give

$$b^2 = DA^2 + p^2; \quad a^2 = DB^2 + p^2; \text{ consequently,}$$
$$b^2 - a^2 = DA^2 - DB^2 = (DA + DB)(DA - DB).$$

Hence, in the first figure,

$$DA - DB = \frac{b^2 - a^2}{c},$$

and in the second figure,

$$DA + DB = \frac{b^2 - a^2}{c}.$$

Now, since half the difference of two quantities added to half their sum gives the greater, we have, in both cases,

$$\text{the greater segment DA} = \frac{b^2 - a^2}{2c} + \frac{c}{2} = \frac{b^2 + c^2 - a^2}{2c}.$$

Hence, $\quad p = \sqrt{b^2 - DA^2} = \dfrac{\sqrt{4b^2c^2 - (b^2 + c^2 - a^2)^2}}{2c}.$

Since the difference of two squares is equal to the product of the sum of the two roots into their difference, we have

$$p = \frac{\sqrt{(b^2+2bc+c^2-a^2)(a^2-b^2+2bc-c^2)}}{2c}$$

$$= \frac{\sqrt{[(b+c)^2-a^2]\times[a^2-(b-c)^2]}}{2c}$$

$$= \frac{\sqrt{(a+b+c)(-a+b+c)(a-b+c)(a+b-c)}}{2c}. \quad (1)$$

Multiplying this perpendicular by half the base, we have for the area,

$$\tfrac{1}{4}\sqrt{(a+b+c)(-a+b+c)(a-b+c)(a+b-c)}, \text{ or}$$

$$\left[\left(\frac{a+b+c}{2}\right)\left(\frac{-a+b+c}{2}\right)\left(\frac{a-b+c}{2}\right)\left(\frac{a+b-c}{2}\right)\right]^{\frac{1}{2}}. \quad (2)$$

Hence we may find the area of a triangle, when the three sides are known, by this

RULE.—*Take half the sum of the three sides, and from this half sum subtract each side separately; then take the square root of the continued product of the half sum and the three remainders, and it will be the area.*

## PROBLEM V.

*Given the three sides $a$, $b$, $c$ of a triangle, to find:*

1. *The three perpendiculars from the angles upon the opposite sides;*
2. *The area of the triangle;*
3. *The radius of the circumscribed circle;*
4. *The radius of the inscribed circle;*
5. *The radii of the escribed circles.*

Let ABC be the triangle; and let $a$, $b$, $c$ denote the sides opposite the angles A, B, C respectively, and $P_1$, $P_2$, $P_3$ the perpendiculars drawn from the angles A, B, C; $\Delta$ the area of the triangle; R the radius of the circumscribing circle; $r$ that of the inscribed circle; and $r_1$, $r_2$, $r_3$ the radii of the three escribed circles, which touch the sides $a$, $b$, $c$ externally.

An *escribed* circle has already been defined as a circle which touches one of the sides of a triangle exteriorly, and the other two sides produced. (See B. II., T. XII., S. II.)

We have already found (Prob. IV.) the perpendiculars to be

$$P_1=\frac{\sqrt{(a+b+c)(-a+b+c)(a-b+c)(a+b-c)}}{2a}, \quad (1)$$

$$P_2=\frac{\sqrt{(a+b+c)(-a+b+c)(a-b+c)(a+b-c)}}{2b}, \quad (2)$$

$$P_3=\frac{\sqrt{(a+b+c)(-a+b+c)(a-b+c)(a+b-c)}}{2c}. \quad (3)$$

We have, also, under the same Problem, found the area to be

$$\Delta=\left\{\left(\frac{a+b+c}{2}\right)\left(\frac{-a+b+c}{2}\right)\left(\frac{a-b+c}{2}\right)\left(\frac{a+b-c}{2}\right)\right\}^{\frac{1}{2}}. \quad (4)$$

We have (B. III., T. XXV.) $\Delta=\frac{abc}{4R}$, we also have $\Delta=\frac{P_1a}{2}$;

therefore, $$\frac{abc}{4R}=\frac{P_1a}{2}.$$

Hence, $$R=\frac{bc}{2P_1}.$$

Substituting for $P_1$ its value already found, we have

$$R=\frac{abc}{\sqrt{(a+b+c)(-a+b+c)(a-b+c)(a+b-c)}}. \quad (5)$$

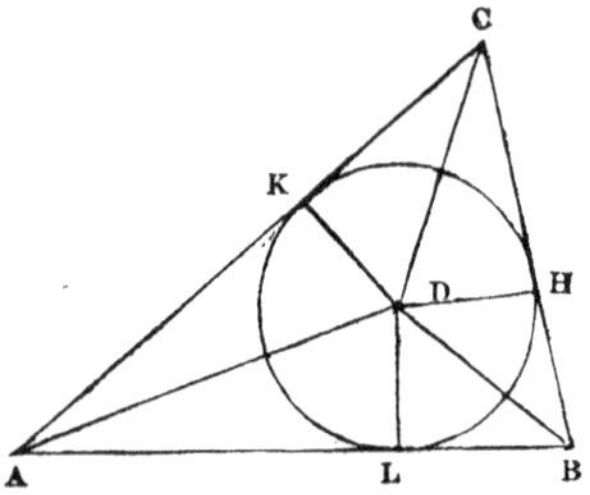

The sum of the areas of the three triangles ADB, BDC, CDA equals the area ABC. But

$$ADB=\tfrac{1}{2}rc,$$
$$BDC=\tfrac{1}{2}ra,$$
$$CDA=\tfrac{1}{2}rb;$$

hence, $$\frac{a+b+c}{2}\times r=\Delta, \text{ and} \quad (6)$$

$$r=\frac{2\Delta}{a+b+c}=\left\{\frac{(-a+b+c)(a-b+c)(a+b-c)}{4(a+b+c)}\right\}^{\frac{1}{2}}. \quad (7)$$

We will now seek the radius $r_1$ of the escribed circle, whose centre is at E.

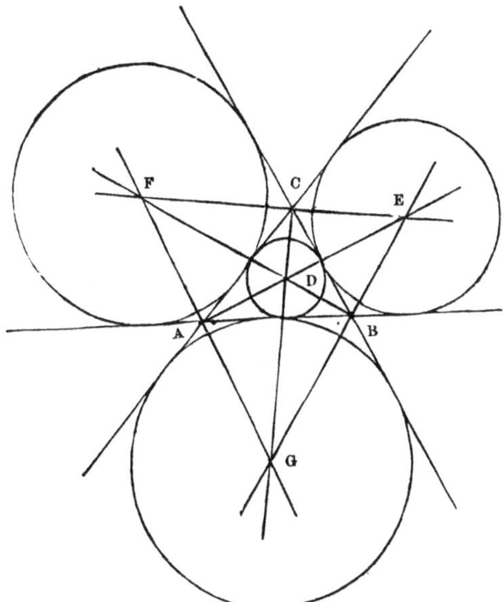


Area EBA + ECA − EBC is evidently equal to the area ABC. Area EBA is equal to the base AB multiplied by half the perpendicular drawn from E upon AB produced; hence area EBA $= c \times \frac{1}{2} r_1 = \frac{1}{2} c r_1$. In a similar way, we find area ECA $= \frac{1}{2} b r_1$, and area EBC $= \frac{1}{2} a r_1$.

Therefore we have $\frac{b+c-a}{2} \times r_1 = \Delta$, and (8)

$$r_1 = \frac{2\,\Delta}{b+c-a} = \left\{ \frac{(a+b+c)\,(a-b+c)\,(a+b-c)}{4\,(-a+b+c)} \right\}^{\frac{1}{2}}. \quad (9)$$

By simply permuting, we obtain

$$r_2 = \left\{ \frac{a+b+c)\,(-a+b+e)\,(a+b-c)}{4\,(a-b+c)} \right\}^{\frac{1}{2}}, \quad (10)$$

$$r_3 = \left\{ \frac{(a+b+c)\,(-a+b+c)\,(a-b+c)}{4\,(a+b-c)} \right\}^{\frac{1}{2}}. \quad (11)$$

Equation (8) readily gives

$$\frac{1}{r_1} = \frac{-a+b+c}{2\,\Delta};$$

and, in a similar manner,

$$\frac{1}{r_2} = \frac{a-b+c}{2\Delta}, \text{ and } \frac{1}{r_3} = \frac{a+b-c}{2\Delta};$$

therefore, $$\frac{1}{r_1}+\frac{1}{r_2}+\frac{1}{r_3}=\frac{a+b+c}{2\,\Delta}.$$

But we have already from equation (7),

$$r=\frac{2\,\Delta}{a+b+c}, \text{ or } \frac{1}{r}=\frac{a+b+c}{2\,\Delta};$$

therefore, we have $$\frac{1}{r}=\frac{1}{r_1}+\frac{1}{r_2}+\frac{1}{r_3}. \qquad (12)$$

Since any side of a triangle, multiplied by the perpendicular which meets it from the opposite angle, gives double the area of the triangle, we have $2\,\Delta=a\,\mathrm{P}_1$, or

$$\frac{2\,\Delta}{\mathrm{P}_1}=a. \qquad (13)$$

In a similar manner, we have

$$\frac{2\,\Delta}{\mathrm{P}_2}=b, \qquad (14)$$

$$\frac{2\,\Delta}{\mathrm{P}_3}=c. \qquad (15)$$

Taking the product of (13), (14), and (15), we have

$$\frac{8\,\Delta^3}{\mathrm{P}_1\mathrm{P}_2\mathrm{P}_3}=abc. \qquad (16)$$

Again, we have $$2\,\mathrm{RP}_1=bc, \qquad (17)$$

$$2\,\mathrm{RP}_2=ac, \qquad (18)$$

$$2\,\mathrm{RP}_3=ab. \qquad (19)$$

Taking the product of (17), (18), and (19), we have

$$8\,\mathrm{R}^3\mathrm{P}_1\mathrm{P}_2\mathrm{P}_3=a^2b^2c^2. \qquad (20)$$

Extracting the cube root of the product of (16) and (20), we find

$$4\,\mathrm{R}\,\Delta=abc. \qquad (21)$$

Dividing (20) by the square of (16), we find

$$\frac{\mathrm{R}^3\mathrm{P}_1^3\mathrm{P}_2^3\mathrm{P}_3^3}{8\,\Delta^6}=1, \text{ or } \mathrm{R}\,\mathrm{P}_1\mathrm{P}_2\mathrm{P}_3=2\,\Delta^2. \qquad (22)$$

By taking the continued product of (7), (9), (10), and (11), we have

$$rr_1r_2r_3=\frac{(a+b+c)\,(-a+b+c)\,(a-b+c)\,(a+b-c)}{16}=\Delta^2. \qquad (23)$$

By multiplying (5) and (7) together, or (7) and (21), we have

$$2\,\mathrm{R}r=\frac{abc}{a+b+c}. \qquad (24)$$

By a similar multiplication, we find

$$\left.\begin{aligned} 2\,\mathrm{R}r_1 &= \frac{abc}{-a+b+c}, \\ 2\,\mathrm{R}r_2 &= \frac{abc}{a-b+c}, \\ 2\,\mathrm{R}r_3 &= \frac{abc}{a+b-c}. \end{aligned}\right\} \qquad (25)$$

By combining the values of (9), (10), and (11), by two and two, we find

$$r_1r_2 + r_1r_3 + r_2r_3 = \left(\frac{a+b+c}{2}\right)^2. \qquad (26)$$

We may also deduce

$$\frac{1}{\mathrm{P}_1} + \frac{1}{\mathrm{P}_2} + \frac{1}{\mathrm{P}_3} = \frac{1}{r_1} + \frac{1}{r_2} + \frac{1}{r_3} = \frac{1}{r}. \qquad (27)$$

$$\tfrac{1}{2}\,\mathrm{R}r = \frac{\Delta^2}{\mathrm{P}_1\mathrm{P}_2 + \mathrm{P}_2\mathrm{P}_3 + \mathrm{P}_3\mathrm{P}_1}. \qquad (28)$$

$$\left.\begin{aligned} \mathrm{P}_1 &= \frac{2\,r_2r_3}{r_2+r_3}, \\ \mathrm{P}_2 &= \frac{2\,r_1r_3}{r_1+r_3}, \\ \mathrm{P}_3 &= \frac{2\,r_1r_2}{r_1+r_2}. \end{aligned}\right\} \qquad (29)$$

$$\left.\begin{aligned} \frac{1}{\mathrm{P}_1} + \frac{1}{\mathrm{P}_2} &= \frac{1}{\mathrm{P}_3} + \frac{1}{r_3}, \\ \frac{1}{\mathrm{P}_1} + \frac{1}{\mathrm{P}_3} &= \frac{1}{\mathrm{P}_2} + \frac{1}{r_2}, \\ \frac{1}{\mathrm{P}_2} + \frac{1}{\mathrm{P}_3} &= \frac{1}{\mathrm{P}_1} + \frac{1}{r_1}. \end{aligned}\right\} \qquad (30)$$

$$r_1 + r_2 + r_3 = 4\,\mathrm{R} + r. \qquad (31)$$

Any of the foregoing expressions, when properly translated into common language, leads to a theorem. We will translate some of the most interesting ones.

Equation (12) gives the following

THEOREM. *The reciprocal of the radius of the inscribed circle is equal to the sum of the reciprocals of the radii of the three escribed circles.*

Equation (21) yields the following

THEOREM. *Four times the radius of the circumscribed circle, into the area of the triangle, is equal to the continued product of the three sides.*

Equation (22) gives this

THEOREM. *The radius of the circumscribed circle, into the continued product of the three perpendiculars, is equal to twice the square of the area.*

Equation (23) gives this

THEOREM. *The radius of the inscribed circle, into the continued product of the radii of the three escribed circles, is equal to the square of the area.*

Equation (27) gives this

THEOREM. *The sum of the reciprocals of the three perpendiculars is equal to the sum of the reciprocals of the three radii of the escribed circles.*

By combining the equations already formed, new ones would arise, which might still afford interest. Thus, by a comparison of equations (22) and (23), we find

$$R\,P_1P_2P_3 = 2\,rr_1r_2r_3; \qquad (32)$$

which gives this

THEOREM. *The radius of the circumscribed circle, into the continued product of the three perpendiculars, is equal to the diameter of the inscribed circle, into the continued product of the three radii of the escribed circles.*

## PROBLEM VI.

*Given the three sides of a triangle, to find the three lines drawn from the angles to the middle points of the opposite sides.*

Let ABC be the triangle. Denote the sides opposite the angles A, B, C, by $a$, $b$, $c$ respectively; also denote the lines drawn from the angles A, B, C to the middle points of the opposite sides, by $m_1$, $m_2$, $m_3$ respectively. Then (B. III., T. XVII.), we shall have

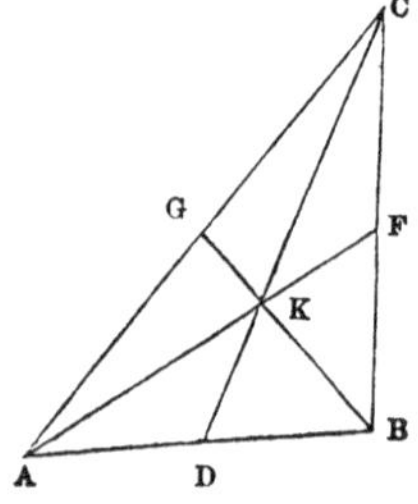

$$2\,AF^2 + 2\,BF^2 = AB^2 + AC^2, \text{ or}$$

$$2\,m_1^2 + \tfrac{1}{2}a^2 = c^2 + b^2;$$

which gives $$4\,m_1^2 = -a^2 + 2\,b^2 + 2\,c^2. \qquad (1)$$

In a similar manner, we find

$$4\,m_2^2 = -b^2 + 2\,c^2 + 2\,a^2, \qquad (2)$$

$$4\,m_3^2 = -c^2 + 2\,a^2 + 2\,b^2. \qquad (3)$$

Equations (1), (2), and (3), readily give

$$\left.\begin{aligned} m_1 &= \tfrac{1}{2}\sqrt{-a^2 + 2\,b^2 + 2\,c^2}, \\ m_2 &= \tfrac{1}{2}\sqrt{-b^2 + 2\,c^2 + 2\,a^2}, \\ m_3 &= \tfrac{1}{2}\sqrt{-c^2 + 2\,a^2 + 2\,b^2}, \end{aligned}\right\} \qquad (4)$$

We will now deduce a few remarkable relations, which, when properly translated, will give some beautiful theorems.

Taking the sum of (1), (2), and (3), we obtain

$$4\,(m_1^2 + m_2^2 + m_3^2) = 3\,(a^2 + b^2 + c^2). \qquad (5)$$

If we take the product of (1) and (2), we shall have

$$16\,m_1^2 m_2^2 = -2\,a^4 + 5\,a^2b^2 + 2\,a^2c^2 - 2\,b^4 + 2\,b^2c^2 + 4\,c^4. \qquad (6)$$

By simply permuting, we find

$$16\,m_2^2 m_3^2 = -2\,b^4 + 5\,b^2c^2 + 2\,b^2a^2 - 2\,c^4 + 2\,c^2a^2 + 4\,a^4, \qquad (7)$$

$$16\,m_3^2 m_1^2 = -2\,c^4 + 5\,c^2a^2 + 2\,c^2b^2 - 2\,a^4 + 2\,a^2b^2 + 4\,b^4. \qquad (8)$$

Taking the sum of (6), (7), and (8), we obtain

$$16\,(m_1^2m_2^2 + m_2^2m_3^2 + m_3^2m_1^2) = 9\,(a^2b^2 + b^2c^2 + c^2a^2). \qquad (9)$$

If, from the square of (5), we subtract twice (9), we shall have

$$16\,(m_1^4 + m_2^4 + m_3^4) = 9\,(a^4 + b^4 + c^4). \qquad (10)$$

The point where these lines trisect each other, is the centre of gravity of the triangle. If we denote the distances AK, BK, CK, by $d_1$, $d_2$, $d_3$ respectively, we shall have $m_1 = \frac{3}{2}\,d_1$, $m_2 = \frac{3}{2}\,d_2$, $m_3 = \frac{3}{2}\,d_3$ (B. I., T. XXXVIII.). These values substituted in (5), (9), and (10), cause them to become

$$3\,(d_1^2 + d_2^2 + d_3^2) = a^2 + b^2 + c^2, \qquad (11)$$

$$9\,(d_1^2\,d_2^2 + d_2^2\,d_3^2 + d_3^2\,d_1^2) = a^2\,b^2 + b^2\,c^2 + c^2\,a^2, \qquad (12)$$

$$9\,(d_1^4 + d_2^4 + d_3^4) = a^4 + b^4 + c^4. \qquad (13)$$

Equation (5) is equivalent to the following

THEOREM. *Four times the sum of the squares of the lines drawn from the angles of a triangle to the middle points of the opposite sides, is equal to three times the sum of the squares of the sides.*

Equation (9) gives this

THEOREM. *Sixteen times the sum of the products, taken two at a time, of the squares of the lines drawn from the angles of a triangle to the middle points of the opposite sides, is equal to nine times the sum of the products, taken two at a time, of the squares of the sides.*

Equation (10) gives this

THEOREM. *Sixteen times the sum of the fourth powers of the lines drawn from the angles of a triangle to the middle points of the opposite sides, is equal to nine times the sum of the fourth powers of the sides.*

Equations (11), (12), and (13) would lead to beautiful theorems in reference to the distances of the centre of gravity of three equal bodies, and the mutual distances of the bodies themselves. For it is evident that, instead of considering the triangle as material, we may suppose equal weights placed at its vertices; and, under this point of view, equation (11) will give the following

THEOREM. *Three times the sum of the squares of the distances of three equal bodies from their common centre of gravity, is equal to the sum of the squares of their mutual distances.*

## PROBLEM VII.

*Given the three sides of a triangle, to find the lines bisecting the angles, and terminating in the opposite sides.*

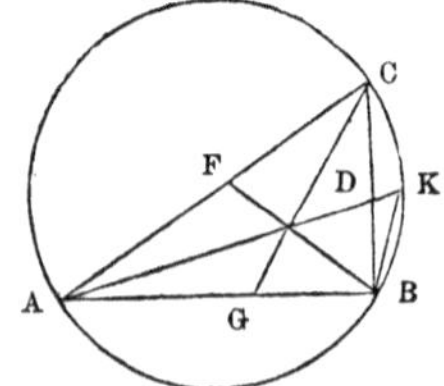


Let the sides opposite the angles A, B, C be denoted by $a$, $b$, $c$ respectively; also denote the lines which bisect the angles A, B, C respectively by $l_1$, $l_2$, $l_3$.

Pass a circumference through the three points A, B, C, and produce $l_1$ to meet it at K; join BK.

The two triangles ABK, ADC are similar, and give

$$AB : AD :: AK : AC;$$

consequently we have

$$AB \times AC = AD \times AK = AD \times (AD + DK) = AD^2 + AD \times DK;$$

which finally becomes

$$AB \times AC = AD^2 + BD \times DC,$$

since BD × DC = AD × DK (B. IV., T. III., S. II.). In symbols this becomes

$$bc = l_1{}^2 + BD \times DC. \qquad (1)$$

Again, we have (B. III., T. XII.)

$$AB : AC :: BD : DC.$$

Consequently,

$$AB + AC : AB :: BD + DC : BD,$$
$$AB + AC : AC :: BD + DC : DC;$$

or, symbols,

$$c + b : c :: a : BD,$$
$$c + b : b :: a : DC.$$

Hence

$$BD = \frac{ac}{c+b}; \ DC = \frac{ab}{c+b}.$$

These values of BD, DC, cause (1) to become

$$cb = l_1^2 + \frac{a^2bc}{(c+b)^2}.$$

This readily gives $l_1^2 = cb - \frac{a^2bc}{(c+b)^2}$, or

$$l_1^2 = \frac{cb\,(a+b+c)\,(-a+b+c)}{(c+b)^2}. \qquad (2)$$

In a similar manner, we find

$$l_2^2 = \frac{ac\,(a+b+c)\,(a-b+c)}{(a+c)^2}, \qquad (3)$$

$$l_3^2 = \frac{ba\,(a+b+c)\,(a+b-c)}{(b+a)^2}. \qquad (4)$$

Equations (2), (3), (4), give

$$\left.\begin{aligned} l_1 &= \frac{\sqrt{cb\,(a+b+c)\,(-a+b+c)}}{c+b}, \\ l_2 &= \frac{\sqrt{ac\,(a+b+c)\,(a-b+c)}}{a+c}, \\ l_3 &= \frac{\sqrt{ba\,(a+b+c)\,(a+b-c)}}{b+a}. \end{aligned}\right\} \qquad (5)$$

Taking the continued product of these values given by (5), we find

$$l_1 l_2 l_3 = \frac{abc\,(a+b+c)\,\sqrt{(a+b+c)\,(-a+b+c)\,(a-b+c)\,(a+b-c)}}{(a+b)\,(b+c)\,(c+a)}.$$

The expression

$$\sqrt{(a+b+c)(-a+b+c)(a-b+c)(a+b-c)},$$

is equal to four times the area of the triangle (see P. IV.), which we will denote by $4\Delta$; so that we shall have

$$l_1 l_2 l_3 = \frac{4\,abc\,(a+b+c)\,\Delta}{(a+b)(b+c)(c+a)}. \qquad (6)$$

PROBLEM VIII.

*To determine a right-angled triangle, having given the hypotenuse and difference of two lines drawn from the two acute angles to the centre of the inscribed circle.*

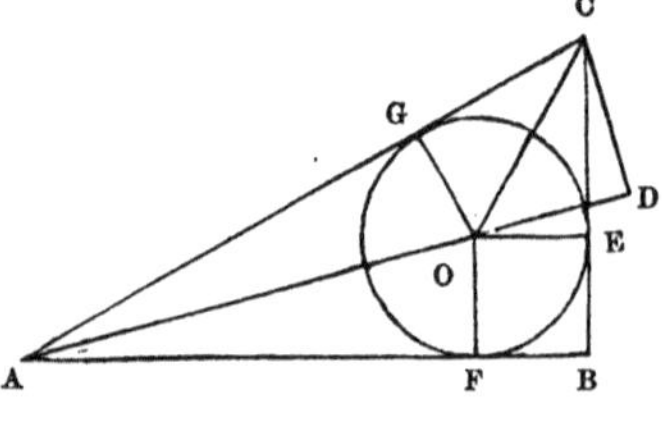


Let $AC = h$; $AO = x + d$; $CO = x - d$; $d$ being half the difference of the lines AO and CO. Produce AO, and draw CD perpendicular to AO thus produced. Then, since the angle COD is equal to the sum of CAO and ACO, and since CAO is half the angle CAB, and ACO the half of ACB, it follows that COD is half a right angle; consequently OCD is also half a right angle: therefore OD is to OC as the side of a square is to its diagonal; that is, as 1 to $\sqrt{2}$. Hence

$$OD = CD = \frac{x-d}{\sqrt{2}}, \text{ and } AD = x + d + \frac{x-d}{\sqrt{2}}.$$

Again, $AC^2 = AD^2 + CD^2$; or which is in symbols

$$h^2 = \left(x + d + \frac{x-d}{\sqrt{2}}\right)^2 + \left(\frac{x-d}{\sqrt{2}}\right)^2. \qquad (1)$$

This readily gives $x = \left(\frac{h^2 - (2-\sqrt{2})\,d^2}{2+\sqrt{2}}\right)^{\frac{1}{2}}$. (2)

Having found $x$, we of course know AO and CO. Let AO be denoted by $a$, and CO by $b$; also let the radius of the inscribed circle be denoted by $r$.

Then $$\mathrm{AF} = \sqrt{\mathrm{AO}^2 - \mathrm{OF}^2} = \sqrt{a^2 - r^2},$$

$$\mathrm{CE} = \sqrt{\mathrm{CO}^2 - \mathrm{OE}^2} = \sqrt{b^2 - r^2}.$$

But AF = AG, and CE = CG; therefore,

$$\sqrt{a^2 - r^2} + \sqrt{b^2 - r^2} = h. \qquad (3)$$

This readily gives

$$r = \left\{ b^2 - \left(\frac{h^2 + b^2 - a^2}{2h}\right)^2 \right\}^{\frac{1}{2}}, \text{ or}$$

$$r = \frac{\sqrt{(a+b+h)(-a+b+h)(a-b+h)(a+b-h)}}{2h}. \qquad (4)$$

This value of $r$ might have been found as follows: In the triangle AOC, all the sides are known; and it is required to find the perpendicular OG, which is $r$, the radius of the circle. Under P. IV. we have found the perpendiculars when the three sides are known. The perpendicular OG found in this way will be precisely the same as the above value of $r$.

Now, having found $r$, we will denote AB by $x$, and CB by $y$; then AF $= x - r$, and CE $= y - r$. Hence,

$$\mathrm{AF} + \mathrm{CE} = h = x + y - 2r. \qquad (5)$$

Again, $xy =$ double the area of the triangle ABC; also, $(x + y + h)r =$ double the area. Therefore,

$$xy = (x + y + h)r. \qquad (6)$$

Equations (5) and (6) readily make known $x$ and $y$, or the sides of the triangle; these values are

$$\left.\begin{aligned} x &= \frac{h + 2r + \sqrt{h^2 - 4hr - 4r^2}}{2}, \\ y &= \frac{h + 2r - \sqrt{h^2 - 4hr - 4r^2}}{2}. \end{aligned}\right\} \qquad (7)$$

### PROBLEM IX.

*In a right-angled triangle, having given the perimeter, and the perpendicular drawn from the right angle upon the hypotenuse; to find the sides of the triangle.*

C

A D B

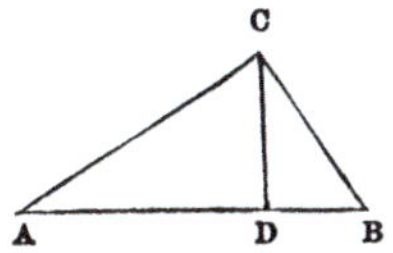

Let S = sum of sides, or perimeter; P = perpendicular CD; AC $= x$, and BC $= y$. Then will $\sqrt{x^2 + y^2} = \mathrm{AB}$. $xy =$ double the area of triangle ABC; also, $P\sqrt{x^2+y^2}=$ double the area of triangle ABC; therefore,

$$xy = P\sqrt{x^2+y^2}. \tag{1}$$

We also have
$$x+y+\sqrt{x^2+y^2}=S. \tag{2}$$

Equation (2) becomes, by transposing and squaring,

$$(x+y)^2=(S-\sqrt{x^2+y^2})^2, \text{ or}$$

$$x^2+2xy+y^2=S^2-2S\sqrt{x^2+y^2}+x^2+y^2, \text{ or}$$

$$2xy=S^2-2S\sqrt{x^2+y^2}. \tag{3}$$

Taking the double of (1), we have

$$2xy=2P\sqrt{x^2+y^2}. \tag{4}$$

Equating right-hand members of (3) and (4), we have

$$2P\sqrt{x^2+y^2}=S^2-2S\sqrt{x^2+y^2}, \text{ or}$$

$$\sqrt{x^2+y^2}=\frac{S^2}{2(P+S)}. \tag{5}$$

Using this value in (1), we find

$$xy=\frac{PS^2}{2(P+S)}. \tag{6}$$

Squaring (5), we have

$$x^2+y^2=\frac{S^4}{4(S+P)^2}. \tag{7}$$

Adding twice (6) to (7), and also subtracting twice (6) from (7), we have

$$x^2+2xy+y^2=\frac{S^2(S^2+4PS+4P^2)}{4(P+S)^2}, \tag{8}$$

$$x^2-2xy+y^2=\frac{S^2(S^2-4PS-4P^2)}{4(P+S)^2}. \tag{9}$$

Extracting the square root of (8) and (9), we have

$$x+y=\frac{S(S+2P)}{2(P+S)}, \tag{10}$$

$$x-y=\frac{S\sqrt{S^2-4PS-4P^2}}{2(P+S)}. \tag{11}$$

Hence,

$$x=\frac{S(S+2P)+S\sqrt{S^2-4PS-4P^2}}{4(P+S)}, \tag{12}$$

$$y=\frac{S(S-2P)-S\sqrt{S^2-4PS-4P^2}}{4(P+S)}. \tag{13}$$

PROBLEM X.

*To determine a right-angled triangle; having given the hypotenuse, and side of the inscribed square.*

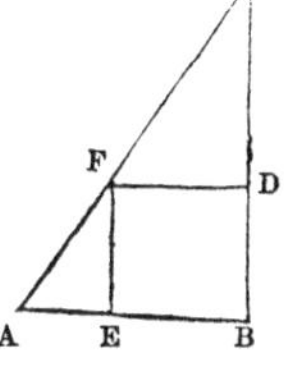


Let $h =$ hypotenuse; $s =$ side of inscribed square; $AB = x$, and $BC = y$.

Then $AE : AB :: EF : BC$;

that is, $x - s : x :: s : y$;

or, $$xy = (x + y)s. \qquad (1)$$

Again, $$x^2 + y^2 = h^2. \qquad (2)$$

Add twice (1) to (2), and we obtain

$$x^2 + 2xy + y^2 = 2s(x + y) + h^2; \text{ or}$$

$$(x + y)^2 - 2s(x + y) = h^2, \qquad (3)$$

which is a quadratic in terms of $x + y$: hence we find

$$x + y = s + \sqrt{s^2 + h^2}. \qquad (4)$$

Substituting this value of $x + y$ in (1), we shall obtain

$$xy = s^2 + s\sqrt{s^2 + h^2}. \qquad (5)$$

From the square of (4) subtracting four times (5), we obtain

$$x^2 - 2xy + y^2 = h^2 - 2s^2 - 2s\sqrt{s^2 + h^2}. \qquad (6)$$

Extracting the square root of (6), we have

$$x - y = \sqrt{h^2 - 2s^2 - 2s\sqrt{s^2 + h^2}}. \qquad (7)$$

Taking half the sum of (4) and (7), and also half their difference, we obtain

$$x = \frac{s + \sqrt{s^2 + h^2} + (h^2 - 2s^2 - 2s\sqrt{s^2 + h^2})^{\frac{1}{2}}}{2}, \qquad (8)$$

$$y = \frac{s + \sqrt{s^2 + h^2} - (h^2 - 2s^2 - 2s\sqrt{s^2 + h^2})^{\frac{1}{2}}}{2}. \qquad (9)$$

### PROBLEM XI.

*To determine a right-angled triangle; having given the hypotenuse, and the radius of the inscribed circle.*

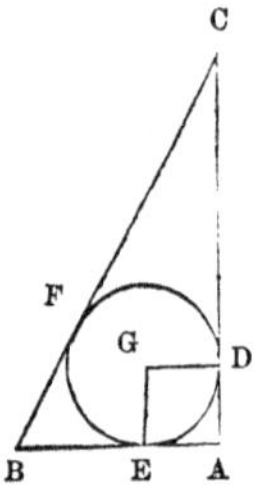

Let $h$ = hypotenuse; $r$ = radius of the inscribed circle; $AB = x$, and $AC = y$.

Then $\qquad BF = BE = x - r,$

and $\qquad CF = CD = y - r;$

therefore $\quad BC = BF + CF = x + y - 2r = h,$

or, $\qquad x + y = 2r + h. \qquad (1)$



Again, $\qquad xy$ = double the area;

$(x + y + h)r = (x + y)r + hr$ = double the area:

therefore $\qquad xy = (x + y)r + hr. \qquad (2)$

In (2), for $x + y$ substitute its value given by (1), and it will become

$$xy = 2r(r + h). \qquad (3)$$

Equations (1) and (3) readily give

$$x = \frac{2r + h + \sqrt{h^2 - 4hr - 4r^2}}{2}, \qquad (4)$$

$$y = \frac{2r + h - \sqrt{h^2 - 4hr - 4r^2}}{2}. \qquad (5)$$

### PROBLEM XII.

*To determine a right-angled triangle; having given the lengths of the lines drawn from the acute angles to the middle points of the opposite sides.*

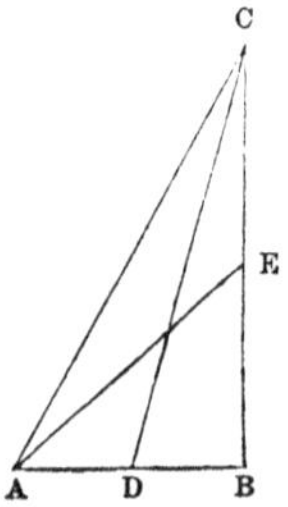

Let $AE = a$, and $CD = b$; also $AB = x$, and $BC = y$. Then will

$$AB^2 + BE^2 = a^2, \text{ or}$$

$$x^2 + \tfrac{1}{4}y^2 = a^2. \qquad (1)$$

In a similar way we find

$$y^2 + \tfrac{1}{4}x^2 = b^2. \qquad (2)$$

Equations (1) and (2), when cleared of fractions, become

$$4x^2 + y^2 = 4a^2, \qquad (3)$$

$$x^2 + 4y^2 = 4b^2. \qquad (4)$$

These equations give

$$x = 2\sqrt{\frac{4a^2 - b^2}{15}}, \qquad (5)$$

$$y = 2\sqrt{\frac{4b^2 - a^2}{15}}. \qquad (6)$$

### PROBLEM XIII.

*To determine a triangle; having given the base, the perpendicular, and the difference of the two other sides.*

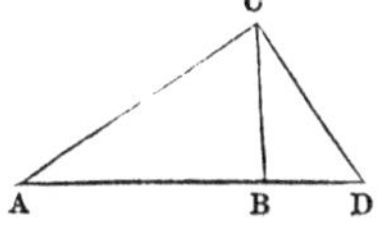


Let $p=$ the perpendicular; $d=$ half the difference of sides; $x+d=\text{AC}$; $x-d=\text{BC}$. $b=$ half the base; $b+y=\text{AD}$; $b-y=\text{DB}$.

Then, from the right-angled triangles ADC, BDC, we have

$$(b+y)^2 + p^2 = (x+d)^2, \qquad (1)$$

$$(b-y)^2 + p^2 = (x-d)^2. \qquad (2)$$

Expanding (1) and (2), and then taking half their sum and one fourth of their difference, we obtain

$$b^2 + y^2 + p^2 = x^2 + d^2, \qquad (3)$$

$$by = dx. \qquad (4)$$

These equations readily give

$$x = b\left(\frac{b^2 - d^2 + p^2}{b^2 - d^2}\right)^{\frac{1}{2}}, \qquad (5)$$

$$y = d\left(\frac{b^2 - d^2 + p^2}{b^2 - d^2}\right)^{\frac{1}{2}}. \qquad (6)$$

Hence,

$$\text{AC} = b\left(\frac{b^2 - d^2 + p^2}{b^2 - d^2}\right)^{\frac{1}{2}} + d, \qquad (7)$$

$$\text{BC} = b\left(\frac{b^2 - d^2 + p^2}{b^2 - d^2}\right)^{\frac{1}{2}} - d. \qquad (8)$$

### PROBLEM XIV.

*To determine the radii of three equal circles, described in a given circle, to touch each other, and also to touch the circumference of the given circle.*

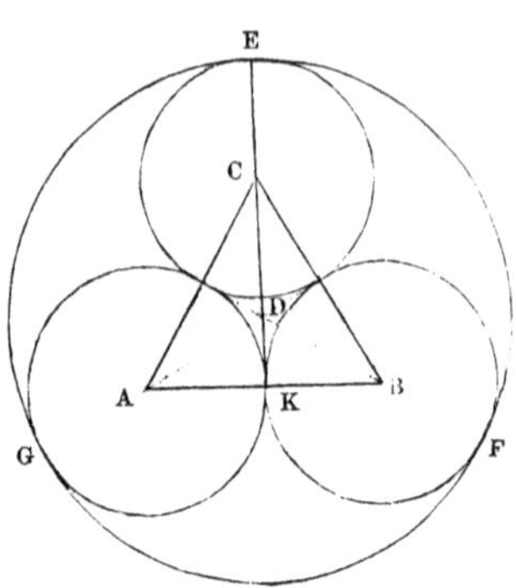


Let D be the centre of the given circle, whose radius we will denote by R; also, let A, B, C be the centres of the three equal circles, whose common radius we will denote by $r$. Then joining A, B, and C, we have the triangle ABC equilateral, each of whose sides is $2r$. Drawing CK perpendicular to AB, the right-angled triangle AKC gives

$$CK^2 = AC^2 - AK^2,$$

or, in symbols, $$CK^2 = 4r^2 - r^2 = 3r^2;$$

consequently, $$CK = r\sqrt{3}. \qquad (1)$$

Now (B. I., T. XXXVIII.),

$$CD = \tfrac{2}{3}CK = \tfrac{2}{3}r\sqrt{3}. \qquad (2)$$

But $$ED = EC + CD;$$

that is, $$R = r + \tfrac{2}{3}r\sqrt{3}: \qquad (3)$$

from which we readily find

$$r = \frac{3R}{3 + 2\sqrt{3}} = \frac{R}{1 + 2\sqrt{\tfrac{1}{3}}}. \qquad (4)$$

### PROBLEM XV.

*Given the radius, r, of a circle, to find the sides of the inscribed and circumscribed pentagons and decagons.*

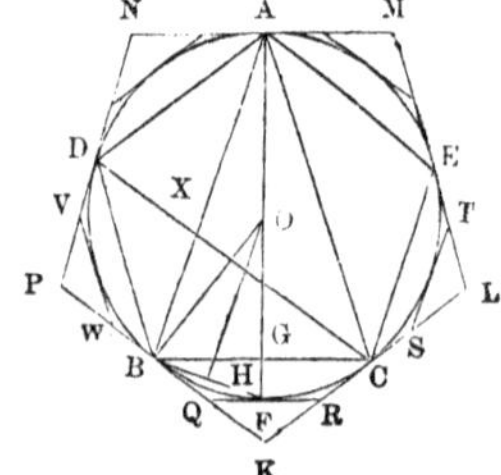


1. *The inscribed pentagon.* Let ADBCE be the pentagon inscribed in the circle, and let O be the centre of the circumscribing circle. Join AB, AC, and draw AF perpendicular to BC; then, by known properties, BC is bisected in G, and the line AF passes through the centre O of the circle.

Since the angle ADX is measured by half the arc AEC (B. II., T. X.), and the angle AXD is measured by half the sum of the arcs AD and BC (B. II., T. XI.), it follows that these angles are equal, and the triangle DAX is isosceles.

Again, the angles DAX and BAC are equal, being measured by halves of the equal arcs DB, BC; hence the triangles ABC and DAX are similar. But the triangle DAX is obviously equal to BCX; therefore,

$$AB : BC :: BC : BX;$$

but $$BX = AB - AX = AB - BC;$$

therefore, $$AB : BC :: BC : AB - BC. \qquad (1)$$

Put $BC = 2x$, or $BG = x$; $BA = y$, and $BF = z$; and the above proportion will become

$$y : 2x :: 2x : y - 2x;$$

which gives $$y = (1 + \sqrt{5})\,x. \qquad (2)$$

Now, we have $OG = \sqrt{r^2 - x^2}$, $AG = \sqrt{y^2 - x^2}$; and hence $AG = AO + OG$ gives

$$\sqrt{y^2 - x^2} = r + \sqrt{r^2 - x^2},$$

or, squaring, $$y^2 - 2r^2 = 2r\sqrt{r^2 - x^2}.$$

Again, squaring, we have

$$y^4 - 4r^2y^2 + 4r^2x^2 = 0. \qquad (3)$$

Substituting the value of $y$ as given by (2), we find the side of an inscribed pentagon (see B. IV., T. XI., S. II.),

$$2x = \tfrac{1}{2}r\sqrt{10 - 2\sqrt{5}}. \qquad (4)$$

We have $$y = x(1 + \sqrt{5}) = \tfrac{1}{2}r\sqrt{10 + 2\sqrt{5}}, \qquad (5)$$

$$OG = \sqrt{r^2 - x^2} = \sqrt{r^2 - \tfrac{1}{8}(5 - \sqrt{5})r^2} = \tfrac{1}{4}r(1 + \sqrt{5}.) \quad (6)$$

2. *The inscribed decagon.* Join BF; then, since AF bisects the line BC at right angles, it bisects the arc BFC in F; and hence BF is the side of the inscribed decagon. But ABF being a right angle, since it is in a semicircle, we have

$$BF^2 = FA^2 - AB^2; \text{ or, in symbols,}$$

$$z^2 = 4r^2 - y^2 = 4r^2 - \tfrac{1}{4}(10 + 2\sqrt{5})r^2 = \tfrac{1}{2}(3 - \sqrt{5})r^2;$$

or, extracting the square root, we have, for the side of the inscribed decagon (see B. IV., T. XI., S. I.),

$$z = \tfrac{1}{2}r(\sqrt{5} - 1). \qquad (7)$$

3. *The circumscribing pentagon.* The inscribed and circumscribed pentagons being regular, are similar figures, and their sides are as the perpendiculars from the centre upon the sides. That is, if PK be a side of the circumscribing pentagon, we shall have

$$\text{OG} : \text{OB} :: \text{BC} : \text{PK}; \text{ or, in symbols,}$$

$$\text{PK} = \frac{\text{OB} \times \text{BC}}{\text{OG}} = \frac{r \cdot \frac{1}{2}r\sqrt{10 - 2\sqrt{5}}}{\frac{1}{4}r(1 + \sqrt{5})} = 2r\sqrt{5 - 2\sqrt{5}}. \quad (8)$$

4. *The circumscribing decagon.* Let QR be one of the sides; and draw OH perpendicular to BF, which it bisects in H. Also by similar triangles ABF, OHF, we have

$$\text{OH} = \tfrac{1}{2}\text{AB} = \tfrac{1}{2}y = \tfrac{1}{4}r\sqrt{10 + 2\sqrt{5}}.$$

Also, as in the last case,

$$\text{OH} : \text{OF} :: \text{BF} : \text{QR}; \text{ which gives}$$

$$\text{QR} = \frac{\text{OF} \times \text{BF}}{\text{OH}} = \frac{r \cdot \frac{1}{2}r(\sqrt{5} - 1)}{\frac{1}{4}r\sqrt{10 + 2\sqrt{5}}} = 2r\sqrt{\frac{5 - 2\sqrt{5}}{5}}.$$

### PROBLEM XVI.

*Given the lengths of three lines drawn from a point to the three angles of an equilateral triangle, to find its side.*

Let ABC be the triangle, and D the point.

Put $\text{AD} = a$,
$\text{BD} = b$,
$\text{CD} = c$;
$\text{AB} = 2x$,
$\text{EF} = y$,
$\text{FD} = z$.

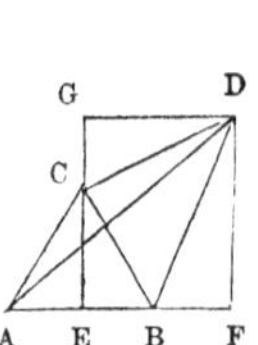


Then will $\text{AF} = x + y$,
and $\text{BF} = x - y$ or $y - x$.

$$\text{CE} = \sqrt{\text{AC}^2 - \text{AE}^2} = \sqrt{4x^2 - x^2} = x\sqrt{3};$$

$$\text{CG} = x\sqrt{3} - z, \text{ or } z - x\sqrt{3}.$$

Hence we have the following relations true, whether the point D is *within* the equilateral triangle, or *without* it.

$$(x + y)^2 + z^2 = a^2, \quad (1)$$

$$(x - y)^2 + z^2 = b^2, \quad (2)$$

$$(x\sqrt{3} - z)^2 + y^2 = c^2. \quad (3)$$

Subtracting (2) from (1), we find

$$4xy = a^2 - b^2, \text{ or } y = \frac{a^2 - b^2}{4x}; \tag{4}$$

consequently,

$$y^2 = \frac{(a^2 - b^2)^2}{16x^2}. \tag{5}$$

Equation (1) gives immediately

$$z = \sqrt{a^2 - (x + y)^2};$$

in which, substituting for $y$ its value given by (4), we have

$$z = \sqrt{a^2 - \left(x + \frac{a^2 - b^2}{4x}\right)^2}. \tag{6}$$

This value of $z$, and the value of $y^2$ given by (5), being substituted in (3), will cause it to become

$$\left\{x\sqrt{3} - \sqrt{a^2 - \left(x + \frac{a^2 - b^2}{4x}\right)^2}\right\}^2 + \frac{(a^2 - b^2)^2}{16x^2} = c^2. \tag{7}$$

This equation contains only the unknown $x$, which value may therefore be found: the reduction leads to

$$16x^4 - 4(a^2 + b^2 + c^2)x^2 = -a^4 - b^4 - c^4 + a^2b^2 + b^2c^2 + c^2a^2. \tag{8}$$

This equation, when solved by the rule for quadratics, gives

$$2x = \left\{\frac{a^2 + b^2 + c^2 \pm \sqrt{6(a^2b^2 + b^2c^2 + c^2a^2) - 3(a^4 + b^4 + c^4)}}{2}\right\}^{\frac{1}{2}}. \tag{9}$$

We must use the + sign when the point is within the triangle, as in the first figure; and the − sign when the point is without, as in the second figure.

If we suppose a triangle to be formed with the three lines $a$, $b$, $c$, and denote its area by $\triangle$, expression (9) will become

$$2x = \left(\frac{a^2 + b^2 + c^2 \pm 4\triangle\sqrt{3}}{2}\right)^{\frac{1}{2}}. \tag{10}$$

### PROBLEM XVII.

*Suppose* AB *to be a tree, standing on the horizontal plane* AC; *it is required to find at what point it must break, so that, by falling, the top may strike the ground at* C.

If we join BC, and bisect it with the perpendicular DF, we shall have F for the point sought.

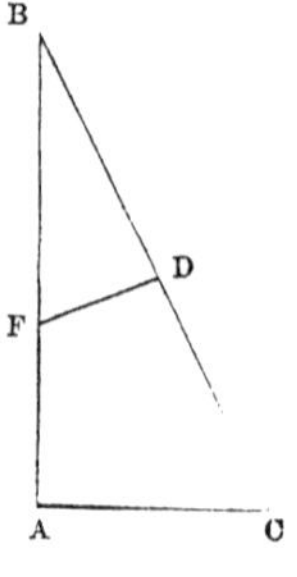


Let the height of the tree be denoted by $h$, and the distance AC by $b$; then BC will equal

$$\sqrt{b^2 + h^2};$$

and, consequently,

$$CD = \tfrac{1}{2} BC = \tfrac{1}{2}\sqrt{b^2 + h^2}.$$

Again, the two triangles BDF and BAC are similar, having the angle B common, and the angle BDF equal to the angle BAC, each being a right angle; therefore,

$$BA : BC :: BD : BF,$$

$$h : \sqrt{b^2 + h^2} :: \tfrac{1}{2}\sqrt{b^2 + h^2} : BF.$$

Consequently, $$BF = \frac{b^2 + h^2}{2h},$$

and $$AF = AB - BF = h - \frac{b^2 + h^2}{2h} = \frac{h^2 - b^2}{2h}.$$

### PROBLEM XVIII.

*Suppose* AC *and* BD *to represent two trees standing on the horizontal plane* AB; *it is required to find a point in this plane, situated on the line* AB, *equally distant from the tops* C *and* D.

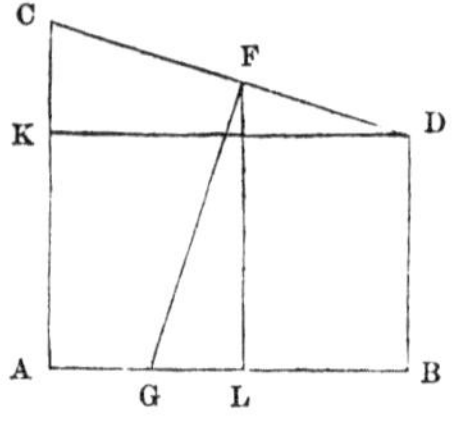


Join CD, and bisect it with the perpendicular FG; then G is the point sought.

This point may be found arithmetically as follows:

Draw DK parallel to AB, and FL perpendicular to AB. Let the height of the tree AC be denoted by $h_1$, the height of the tree BD by $h_2$, and the distance AB by $d$; then will

$$KC = AC - BD = h_1 - h_2, \quad FL = \tfrac{1}{2}(h_1 + h_2).$$

Since the sides of the two triangles DKC, FLG are respectively perpendicular to each other, they are similar, and we have

$$DK : KC :: FL : LG;$$

or, in symbols, $$d : h_1 - h_2 :: \tfrac{1}{2}(h_1 + h_2) : LG.$$

Consequently, $$LG = \frac{h_1^2 - h_2^2}{2d};$$

and $$AG = AL - LG = \tfrac{1}{2}d - \frac{h_1^2 - h_2^2}{2d} = \frac{d^2 - h_1^2 + h_2^2}{2d},$$

$$BG = BL + LG = \tfrac{1}{2}d + \frac{h_1^2 - h_2^2}{2d} = \frac{d^2 + h_1^2 - h_2^2}{2d}.$$

## PROBLEM XIX.

*Suppose three trees to stand vertically upon a horizontal plane at the points* A, B, *and* C. *It is required to find a point in this plane equally distant from their tops.*

Pass a plane through the two trees at A and B, and suppose this plane to revolve about the line AB until it coincides with the horizontal plane, the trees being then represented by AD and BE. Join DE and bisect it with the perpendicular HL;

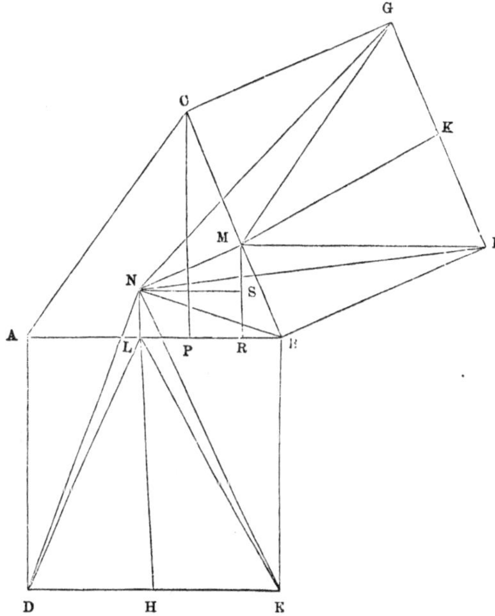


then will the point L be equally distant from the points D and E. Also, pass a plane through the two trees at B and C, and suppose this plane to be revolved about the line BC until it coincides with the horizontal plane, the trees then taking the position of BF and CG. Join FG and bisect it with the perpendicular KM; then will the point M be equally distant from the points F and G.

Draw LN and MN respectively perpendicular to AB and BC, meeting at the point N. Then will N be the point sought.

For, suppose the planes ADEB and BFGC to be brought back to the first position, so as to be perpendicular to the horizontal plane ABC. Then join ND, NE, NF, and NG; also, join LD, LE, MF, and MG. Since BE is equal to BF, each representing the height of the tree standing at B, it follows that NE is equal to NF. NL will be perpendicular to the plane ADEB, consequently the triangles DLN, ELN are right-angled, and DN and LN are respectively equal to EL and LN; hence their hypotenuses DN and EN are equal, which proves N to be equally distant from the tops of the trees standing at A and B.

In a similar manner it may be shown that the right-angled triangles FMN, GMN are equal, and consequently the side FN is equal to GN. But FN has already been shown to be equal to EN, consequently DN, EN, FN, and GN are each of the same length; that is, N is a point in the horizontal plane equally distant from the tops of the three trees standing at the points A, B, and C.

As an illustration, let us suppose AD = 114 feet, BE = 110 feet = BF, CG = 98 feet. Also, suppose AB = 112, BC = 104, AC = 120. Then we find BL = 60, BM = 40 (see P. XVIII.).

The perpendicular CP we find (P. IV.) to be 96, and BP = 40. If we draw MR perpendicular to AB, and NS perpendicular to MR, we shall have the triangles CPB, MRB, and NSM right-angled and similar. Hence, by the similar triangles CPB, MRB, we have

$$BC : BM :: BP : BR, \text{ or}$$

$$104 : 40 :: 40 : BR = 15\tfrac{5}{13}.$$

And $$LR = BL - BR = 60 - 15\tfrac{5}{13} = 44\tfrac{8}{13}.$$

Again, by the similar triangles CPB, NSM, we have

$$CP : CB :: NS\,(= LR) : NM, \text{ or}$$

$$96 : 104 :: 44\tfrac{8}{13} : NM = 48\tfrac{1}{3}.$$

Since the triangle NMB is right-angled, we have

$$BN^2 = NM^2 + BM^2 = 2336\tfrac{1}{9} + 1600 = 3936\tfrac{1}{9}.$$

Again, since the triangle NBE, or its equal NBF, is right-angled at B, we have

$$NE^2 = NF^2 = BN^2 + BE^2 = 3936\tfrac{1}{9} + 12100 = 16036\tfrac{1}{9}.$$

And $NE = NF = NG = ND = \sqrt{16036\tfrac{1}{9}} = 126.63$, nearly, for the number of feet from the point N to the top of each tree.

END OF PART I.

PART SECOND.

---

# PLANE AND SPHERICAL TRIGONOMETRY,

ETC.

# TRIGONOMETRY.

## CHAPTER I.

### FIRST PRINCIPLES OF PLANE TRIGONOMETRY.

§ **1.** TRIGONOMETRY is that science which investigates the relations of the sides and angles of triangles.

When confined to plane triangles, it is called *Plane Trigonometry.*

*Definition of an Angle.*

§ **2.** When two straight lines, AB, AC, meet each other, the opening between these lines is called an angle. This angle or opening, which is read BAC or CAB, or simply the angle at A, does not depend upon the lengths of these lines, but only upon the difference of their directions.

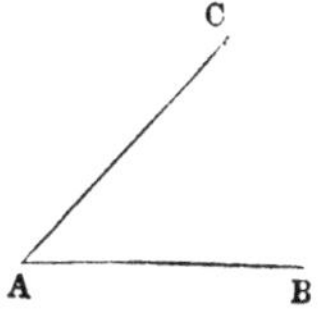


§ **3.** If any straight line, as AB, revolve about A as a centre, in such a manner that the part $AC_0$ shall take the successive positions $AC_1$, $AC_2$, $AC_3$, &c., then will the arc described by the point $C_0$ increase at the same rate as the angle increases; so that the arc $C_0C_1$ is to the arc $C_0C_2$, as the angle $C_0AC_1$ is to the angle $C_0AC_2$; and the same for other arcs and their corresponding angles. Hence, the arcs thus described may be taken as measures of their corresponding angles.

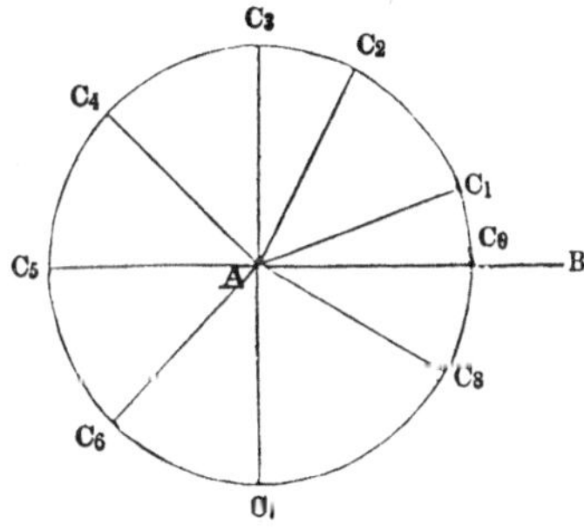


§ **4.** When the line $AC_0$ has reached the position $AC_3$, so that the arc $C_0C_3$ is a *quadrant*, or one-fourth of the entire circumference, the angle $C_0AC_3$ measured by this quadrant, is called a *right angle.* The right angle is usually taken as the unit angle; and all angles less than a right angle are called *acute angles*, and are each measured by an arc less than one-fourth of an entire circumference.

When the line $AC_0$ has reached the position $AC_5$, the point $C_0$ has passed over an arc equal to one-half of the entire circumference, or over an arc which is the measure of two right angles. Hence, when two lines meet from opposite directions, they are said to form an angle equal to two right angles.

An angle such as $C_0AC_4$, which is greater than a right angle and less than two right angles, is called an *obtuse angle.*

§ **5.** All the foregoing relations hold good, whatever be the length of the radius $AC_0$. For simplicity, we shall hereafter use such arcs, for the measure of their corresponding angles, as are described with a *unit* radius.

The Greek letter $\pi$ is usually employed to denote the semi-circumference of a circle whose radius is a unit, or, which is the same thing, it represents the ratio of any circumference to its diameter. Hence $\frac{\pi}{2}$ is the length of an arc called a quadrant, which is the measure of a right angle. The numerical value of $\pi$ is as follows:

$\pi$=3·14159265358979323 84, &c.=semi-circumference, when the radius=1. (See Geom. B. IV., Lemma.)

§ **6.** If the whole circumference be divided into 360 equal parts, one of those parts is called a *degree.* Degrees are also divided into 60 equal parts, called *minutes;* minutes are similarly divided into 60 equal parts, called *seconds*, and so on for other subdivisions. Thus 43° 14′ 12″ denotes an angle whose measuring arc consists of 43 degrees, 14 minutes, and 12 seconds. In the same way 90° denotes an angle whose measuring arc consists of ¼ of 360 degrees, and whose length is therefore $\frac{\pi}{2}$ = 1·57079632679489661 92, &c. From the above value of $\pi$, which is the arc of 180°, it is easy to find the length of an arc corresponding to any angle expressed in degrees, minutes, and seconds. Thus, we find,

$$\tfrac{1}{180} \text{ of } \pi = 1^\circ = 0{\cdot}0174532925199432957\text{7, \&c.}$$
$$\tfrac{1}{60} \text{ of } \tfrac{1}{180} \text{ of } \pi = 1' = 0{\cdot}00029088820866572160\text{, \&c.}$$
$$\tfrac{1}{60} \text{ of } \tfrac{1}{60} \text{ of } \tfrac{1}{180} \text{ of } \pi = 1'' = 0{\cdot}000004848136811095\text{36, \&c.}$$

Modern French writers, instead of using the *sexagesimal* division, use the *centesimal;* and it is to be regretted that the latter is not universally used, on account of the great ease with which all calculations are made in that division.

If E denote the number of English degrees in a given angle, and F denote the number of French grades in the same angle, we shall obviously have

$$\frac{E}{90}=\frac{F}{100}, \text{ or } \frac{E}{9}=\frac{F}{10}. \quad \text{Consequently,}$$

$$E = \tfrac{9}{10} \text{ of } F = F - \tfrac{1}{10} \text{ of } F; \text{ and } F = \tfrac{10}{9} \text{ of } E = E + \tfrac{1}{9} \text{ of } E.$$

§ **7.** When the sum of two angles is equal to a right angle, those angles are mutually *complements* of each other. In such case, the sum of their measuring arcs is $\frac{\pi}{2}$, and the sum of the degrees which these arcs contain is 90°. When the sum of two angles is equal to two right angles, these angles are mutually *supplement* of each other: the sum of the arcs which measure these angles is equal to $\pi$, and the sum of the degrees which they contain is 180°. Thus the angles $C_0AC_1$, $C_1AC_3$, are mutually complements of each other. Also if we denote any arc by $a$, its complement will be $\frac{\pi}{2}-a$, and its supplement will be $\pi-a$. If an angle is expressed in degrees, as A°, where A° denotes the number of degrees in the arc which measures the given angle, then will 90° — A° denote the number of degrees in the arc which measures its complement. For brevity, we say that $a$ and $\frac{\pi}{2}-a$ are complementary arcs, and that A° and 90°—A° are complementary degrees. For a like reason, $a$ and $\pi-a$ are supplementary arcs, and A° and 180°—A° are supplementary degrees.

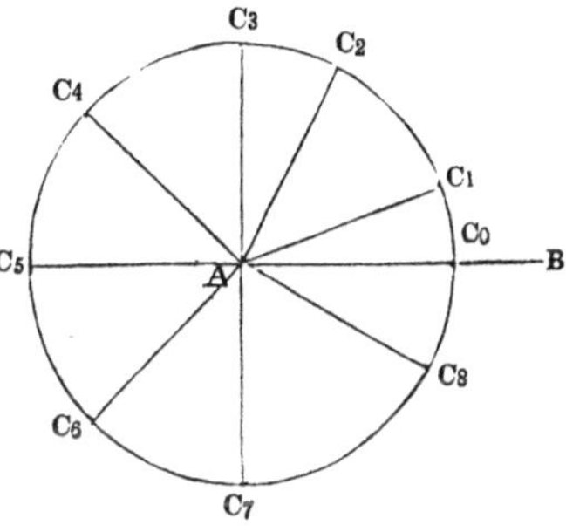

*Definition of Sines, Tangents, Secants, &c.*

§ **8.** In a right triangle the *sine* of either of the acute angles is the quotient obtained by dividing the side opposite the given angle by the hypotenuse.

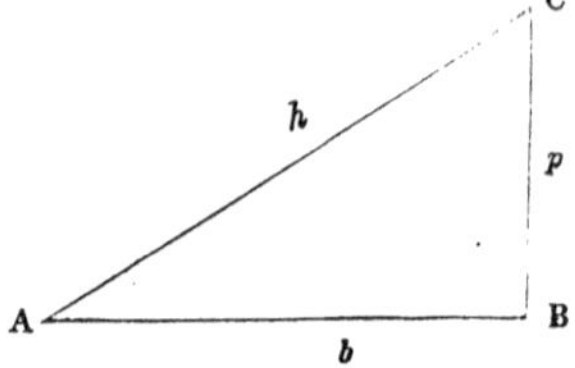


Thus, in the right triangle ABC, if we represent the base, perpendicular, and hypotenuse, by $b$, $p$, and $h$ respectively, we shall have

$$\text{sine A} = \frac{p}{h}; \text{ sine C} = \frac{b}{h}.$$

The *tangent* of either of the acute angles is the quotient obtained by dividing the side opposite the angle by the adjacent side. Thus,

$$\text{tangent A} = \frac{p}{b}; \text{ tangent C} = \frac{b}{p}.$$

The *secant* of either of the acute angles is the quotient obtained by dividing the hypotenuse by the side adjacent the given angle. Thus,

$$\text{secant A} = \frac{h}{b}; \text{ secant C} = \frac{h}{p}.$$

The *cosine*, *cotangent*, and *cosecant* of any angle are respectively the sine, tangent, and secant of its complement. The two acute angles of a right triangle being complements of each other, it follows that we shall have these relations :

$$\left.\begin{array}{ll}
\text{sin. A} = \text{cos. C} = \frac{p}{h}. & (1) \\
\text{cos. A} = \text{sin. C} = \frac{b}{h}. & (2) \\
\text{tang. A} = \text{cotan. C} = \frac{p}{b}. & (3) \\
\text{cotan. A} = \text{tang. C} = \frac{b}{p}. & (4) \\
\text{sec. A} = \text{cosec. C} = \frac{h}{b}. & (5) \\
\text{cosec. A} = \text{sec. C} = \frac{h}{p}. & (6)
\end{array}\right\} \quad . \; . \; . \; \textbf{(A)}.$$

From the above relations, we immediately find

$$\left.\begin{array}{ll}\dfrac{\sin. A}{\cos. A}=\dfrac{p}{b}=\text{tang. } A. & (1)\\ \dfrac{\cos. A}{\sin. A}=\dfrac{b}{p}=\cot. \ A. & (2)\\ \dfrac{1}{\cos. A}=\dfrac{h}{b}=\sec. \ A. & (3)\\ \dfrac{1}{\sin. A}=\dfrac{h}{p}=\text{cosec. } A. & (4)\\ \sin.^2 A+\cos.^2 A=\dfrac{p^2+b^2}{h^2}=1 & (5)\end{array}\right\} \quad . \ . \ . \ (B).$$

The exponent 2, equation (5) of (B), which is used to denote the square of the *sine* and *cosine* of the angle A, is placed immediately after *sin.* and *cos.*, so that $\sin.^2 A$ is the same as $(\sin. A)^2$, $\cos.^2 A$ is the same as $(\cos. A)^2$.

From the above relations, we see that the *sine* and *cosecant* are reciprocals of each other; *cosine* and *secant* are also reciprocals; so also are the *tangent* and *cotangent.*

§ **9.** If from one extremity of the arc which measures an angle, a perpendicular be drawn to the diameter which passes through the other extremity, such perpendicular will be the sine of that angle. Thus, $C_1D_1$ is the sine of the angle $C_0AC_1$. For we have (§ 8) defined the sine of this angle to be the quotient obtained by dividing $C_1D_1$ by $AC_1$, which quotient becomes $C_1D_1$, since the divisor is the radius of the circle, and therefore (§ 5) equal to unity. In the same way it may be shown that $C_2D_2$ is the sine of the angle $C_0AC_2$, that $C_4D_4$ is the sine of the angle $C_0AC_4$. As the arc is the measure of angles, we say that $C_1D_1$, $C_2D_2$, $C_4D_4$, are respectively the sines of the arcs $C_0C_1$, $C_0C_2$, $C_0C_4$.

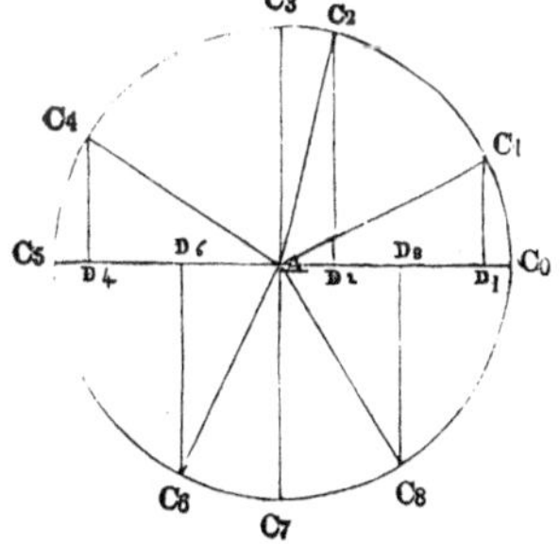


Passing from the angle to the arc, we also say, when the arc exceeds the semi-circumference, that $C_6D_6$, $C_8D_8$, are respectively the sines of the arcs $C_0C_5C_6$, $C_0C_5C_7C_8$.

In all these cases we suppose the arc to be estimated from the point $C_0$, at the extreme right of the circle, upwards and around towards the left. When angles are estimated from $C_0$ downwards around towards the left, they are considered as negative arcs. Thus, $C_0AC_8$ is considered as a negative angle, measured by the negative arc $C_0C_8$. By reference to the diagram, it will be readily seen that so long as a positive arc does not exceed $\pi$ or 180°, the sines are all situated above the horizontal diameter, and are considered as positive; for arcs greater than 180° and less than 360°, their sines are situated below the horizontal diameter, and are considered as negative.

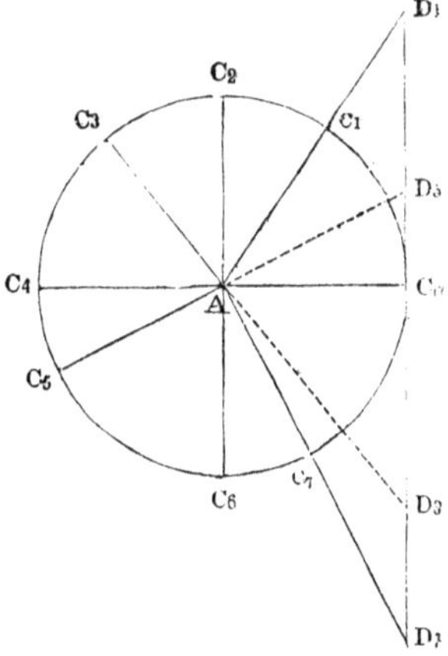


§ **10.** If from one extremity of an arc a tangent line be drawn, and limited by its intersection with the radius drawn through the other extremity of the arc, then will this portion of the tangent line thus included between the radius and the radius produced, be the trigonometrical tangent of the arc. And the radius thus produced, or simply the produced part, will be the secant of the arc. Thus, $C_0D_1$ is the tangent of the arc $C_0C_1$, and $C_0D_3$ is the tangent of the arc $C_0C_3$, in the same way $C_0D_5$, $C_0D_7$ are respectively the tangents of the arcs $C_0C_2C_5$, $C_0C_2C_4C_7$. For, by the former definition (§ 8), the tangent of the angle $C_0AD_1$, which is measured by the arc $C_0C_1$, is the quotient obtained by dividing $C_0D_1$ by $AC_0$, which quotient becomes $C_0D_1$, since $AC_0$, the radius, is a unit. Also the secant of this angle is the quotient obtained by dividing $AD_1$ by $AC_0$, which quotient becomes $AD_1$. When the arc terminates in the first and third quadrants, the tangents are counted on the line $D_7D_1$ from $C_0$ upwards, and are considered as positive. But in the second and fourth quadrants, the tangents are counted from $C_0$ downwards, and are considered as negative.

The secants of the arcs $C_0C_1$, $C_0C_3$, $C_0C_2C_5$, $C_0C_2C_4C_7$, are respectively AD , $AD_3$, $AD_5$, $AD_7$. By inspecting the diagram it will be seen that the secants represented by the dotted lines consist of the produced part of the radius, while those represented by the full lines consist of the radius together with the produced part; the latter are considered as positive and the former as negative. Hence in the first and fourth quadrants the secants are positive, and in the second and third quadrants they are negative.

§ **11.** These different lines may be clearly exhibited in the four quadrants as follows:

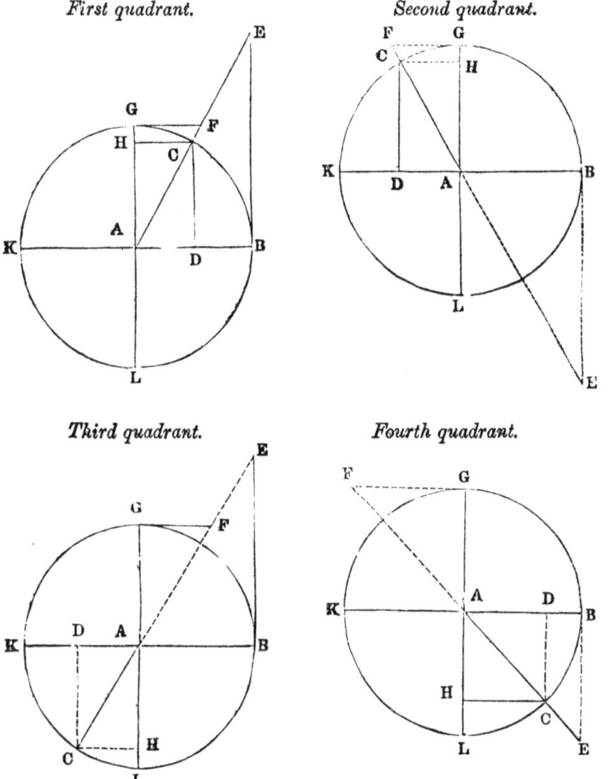


In the above diagrams the dotted lines are considered as negative and the full lines as positive.

The letters in the four diagrams are so arranged that in each case BC counted in the positive direction (§ 9) is the arc, CD the sine, CH the cosine, BE the tangent, GF the cotangent, AE the secant, and AF the cosecant.

The algebraic sines of these lines will be as follows:

| | In 1st quad. | In 2d quad. | In 3d quad. | In 4th quad. |
|---|---|---|---|---|
| Sine and cosecant, . | + | + | − | − |
| Cosine and secant, . | + | − | − | + |
| Tangent and cotangent, | + | − | + | − |

§ **12.** By carefully inspecting the diagrams of § 11, we see that denoting any arc when considered as positive by $a$, it will, when counted in an opposite direction, be denoted by $-a$, and the following relations will hold good for negative arcs:

$$\text{sin.}\ (-a) = -\text{sin.}\ a\ ;\ \text{cos.}\ (-a) = \text{cos.}\ a.$$
$$\text{tang.}\ (-a) = -\text{tang.}\ a\ ;\ \text{cotan.}\ (-a) = -\text{cotan.}\ a.$$
$$\text{sec.}\ (-a) = \text{sec.}\ a\ ;\ \text{cosec.}\ (-a) = -\text{cosec.}\ a.$$

§ **13.** If the arc KC′ is equal to the arc BC, then will the arc BC′ be the supplement (§ 7) of the arc BC. Denoting the arc BC by $a$, we have obviously the following relations:

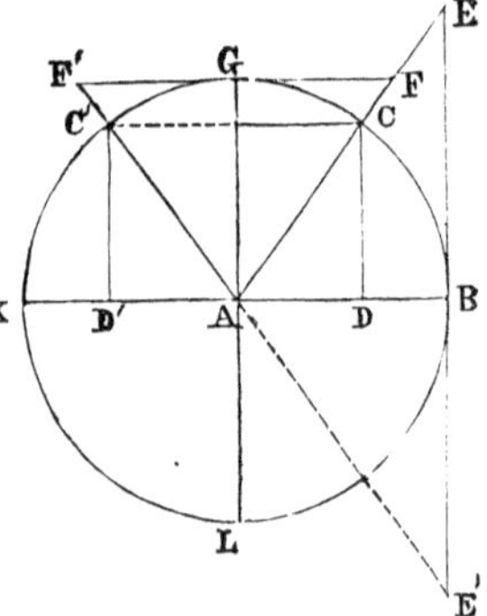


$$\text{sin.}\ a = \text{sin.}\ (180° - a)\ ;\ \text{cos.}\ a = -\text{cos.}\ (180° - a).$$
$$\text{tang.}\ a = -\text{tang.}\ (180° - a)\ ;\ \text{cotan.}\ a = -\text{cotan.}\ (180° - a).$$
$$\text{sec.}\ a = -\text{sec.}\ (180° - a)\ ;\ \text{cosec.}\ a = \text{cosec.}\ (180° - a).$$

§ **14.** The following particular values should be carefully noted by the student:

| | 0° | 90° | 180° | 270° | 360° same as 0° |
|---|---|---|---|---|---|
| Sine . . | 0 | +1 | 0 | −1 | 0 |
| Cosine . | +1 | 0 | −1 | 0 | +1 |
| Tangent . | 0 | ∞ | 0 | ∞ | 0 |
| Cotangent | ∞ | 0 | ∞ | 0 | ∞ |
| Secant . | +1 | ∞ | −1 | ∞ | −1 |
| Cosecant | ∞ | +1 | ∞ | −1 | ∞ |

In general, if $n$ denote any integral number, including also the value of $n=0$, we shall have,

$$
\begin{array}{ll}
\text{sin.} & 0^\circ = \text{sin. } 180^\circ = \text{sin. } 2n\times 90^\circ = 0 \\
\text{sin.} & 90^\circ = \text{sin. } (4n+1)\times 90^\circ = 1 \\
\text{sin.} & 270^\circ = \text{sin. } (4n+3)\times 90^\circ = -1 \\
\text{cos.} & 0^\circ = \text{cos. } 4n\times 90^\circ = 1 \\
\text{cos.} & 90^\circ = \text{cos. } 270^\circ = \text{cos. } (2n+1)\times 90^\circ = 0 \\
\text{cos.} & 180^\circ = \text{cos. } (4n+2)\times 90^\circ = -1 \\
\text{tang.} & 0^\circ = \text{tang. } 180^\circ = \text{tang. } 2n\times 90^\circ = 0 \\
\text{tang.} & 90^\circ = \text{tang. } 270^\circ = \text{tang. } (2n+1)\times 90^\circ = \infty \\
\text{cotan.} & 0^\circ = \text{cotan. } 180^\circ = \text{cotan. } 2n\times 90^\circ = \infty \\
\text{cotan.} & 90^\circ = \text{cotan. } 270^\circ = \text{cotan. } (2n+1)\times 90^\circ = 0 \\
\text{sec.} & 0^\circ = \text{sec. } 4n\times 90^\circ = 1 \\
\text{sec.} & 90^\circ = \text{sec. } 270^\circ = \text{sec. } (2n+1)\times 90^\circ = \infty \\
\text{sec.} & 180^\circ = \text{sec. } (4n+2)\times 90^\circ = -1 \\
\text{cosec.} & 0^\circ = \text{cosec. } 180^\circ = \text{cosec. } 2n\times 90^\circ = \infty \\
\text{cosec.} & 90^\circ = \text{cosec. } (4n+1)\times 90^\circ = 1 \\
\text{cosec.} & 270^\circ = \text{cosec. } (4n+3)\times 90^\circ = -1
\end{array}
$$

§ **15.** *To find the sine and cosine of the sum and difference of two arcs.*

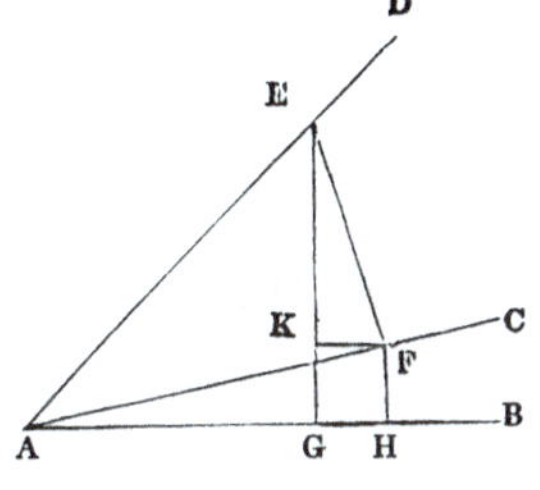


Let the angle BAC$=a$; that is, let $a$ denote the length of the arc which measures this angle; also let the angle CAD be denoted by $b$; then will BAD be denoted by $a+b$.

Draw EF perpendicular to AC, and from the points E and F draw EG and FH perpendicular to AB; also draw FK perpendicular to EG. Since the sides of the triangle EFK are respectively perpendicular to the sides of the triangle AFH, these triangles are similar. (Geom. B. III., T. VII.) Hence, the angle FEK is equal to FAH, equal to $a$.

$$\text{sin. }(a+b)=\frac{\text{EG}}{\text{AE}}=\frac{\text{HF}+\text{KE}}{\text{AE}}=\frac{\text{HF}}{\text{AE}}+\frac{\text{KE}}{\text{AE}}=$$

$$\frac{\text{HF}}{\text{AF}}\times\frac{\text{AF}}{\text{AE}}+\frac{\text{KE}}{\text{EF}}\times\frac{\text{EF}}{\text{AE}}.$$

By our definitions (§ 8), $\frac{HF}{AF} = \sin. a$; $\frac{AF}{AE} = \cos. b$; $\frac{KE}{EF} = \cos. a$; $\frac{EF}{AE} = \sin. b$. Hence, we have

$$\sin. (a+b) = \sin. a \cos. b + \cos. a \sin. b.$$

Again, we have

$$\cos. (a+b) = \frac{AG}{AE} = \frac{AH-KF}{AE} = \frac{AH}{AE} - \frac{KF}{AE} = \frac{AH}{AF} \times \frac{AF}{AE} - \frac{KF}{EF} \times \frac{EF}{AE} = \cos. a \cos. b - \sin. a \sin. b.$$

For the sine and cosine of $a-b$, let the angle BAC $= a$, and the angle CAD $= b$; then will BAD $= a-b$.

Then, as in the former diagram, we have the triangle EFK similar to the triangle AFH, and the angle KFE equal to the angle FAH, equal to $a$.

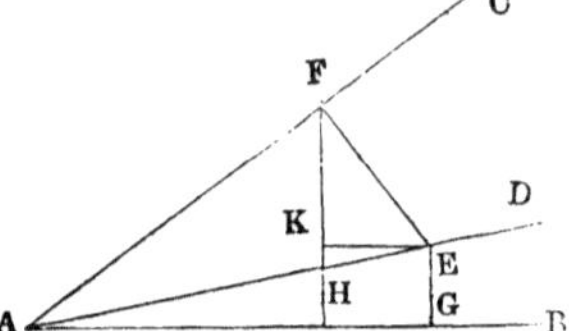


$$\sin. (a-b) = \frac{EG}{AE} = \frac{HF-KF}{AE} = \frac{HF}{AF} \times \frac{AF}{AE} - \frac{KF}{FE} \times \frac{FE}{AE} = \sin. a \cos. b - \cos. a \sin. b.$$

$$\cos. (a-b) = \frac{AG}{AE} = \frac{AH+KE}{AE} = \frac{AH}{AF} \times \frac{AF}{AE} + \frac{KE}{FE} \times \frac{FE}{AE} = \cos. a \cos. b + \sin. a \sin. b.$$

Collecting these results, we have

$$\sin. (a+b) = \sin. a \cos. b + \cos. a \sin. b. \quad . \quad . \quad (1.)$$
$$\sin. (a-b) = \sin. a \cos. b - \cos. a \sin. b. \quad . \quad . \quad (2.)$$
$$\cos. (a+b) = \cos. a \cos. b - \sin. a \sin. b. \quad . \quad . \quad (3.)$$
$$\cos. (a-b) = \cos. a \cos. b + \sin. a \sin. b. \quad . \quad . \quad (4.)$$

If we make $b = a$, equations (1) and (3) will give

$$\sin. 2a = 2 \sin. a \cos. a, \quad . \quad . \quad . \quad (5.)$$
$$\cos. 2a = \cos.^2 a - \sin.^2 a. \quad . \quad . \quad (6.)$$

From (5) of (B), § 8, we have

$$1 = \sin.^2 a + \cos.^2 a. \quad . \quad . \quad . \quad . \quad (7.)$$

The sum and difference of (6) and (7) give

$$1 + \cos.\ 2\,a = 2\cos.^2 a, \quad . \quad . \quad . \quad (8.)$$
$$1 - \cos.\ 2\,a = 2\sin.^2 a. \quad . \quad . \quad . \quad (9.)$$

Hence, $$\sin.\ a = \tfrac{1}{2}\sqrt{2 - 2\cos.\ 2\,a}, \quad . \quad . \quad (10.)$$

$$\cos.\ a = \tfrac{1}{2}\sqrt{2 + 2\cos.\ 2\,a}. \quad . \quad . \quad (11.)$$

Equations (5) and (6) make known the sine and cosine of a double arc in terms of the sine and cosine of the arc itself. Equations (10) and (11) give, conversely, the sine and cosine of half an arc in terms of the cosine of the arc itself.

Dividing (1) by (3), we have

$$\frac{\sin.\ (a+b)}{\cos.\ (a+b)} = \frac{\sin.\ a\cos.\ b + \cos.\ a\sin.\ b}{\cos.\ a\cos.\ b - \sin.\ a\sin.\ b}.$$

Dividing numerator and denominator of the right-hand member each by $\cos.\ a\cos.\ b$, we have

$$\frac{\sin.\ (a+b)}{\cos.\ (a+b)} = \frac{\dfrac{\sin.\ a}{\cos.\ a} + \dfrac{\sin.\ b}{\cos.\ b}}{1 - \dfrac{\sin.\ a}{\cos.\ a} \times \dfrac{\sin.\ b}{\cos.\ b}},$$

which becomes (§ 8),

$$\text{tang.}\ (a+b) = \frac{\text{tang.}\ a + \text{tang.}\ b}{1 - \text{tang.}\ a\ \text{tang.}\ b}. \quad . \quad . \quad (12.)$$

§ **16.** Equations (1), (2), (3), and (4) (§ 15), give as follows:

$$\sin.\ (a+b) + \sin.\ (a-b) = 2\sin.\ a\cos.\ b. \qquad (1.)$$
$$\sin.\ (a+b) - \sin.\ (a-b) = 2\cos.\ a\sin.\ b. \qquad (2.)$$
$$\cos.\ (a-b) + \cos.\ (a+b) = 2\cos.\ a\cos.\ b. \qquad (3.)$$
$$\cos.\ (a-b) - \cos.\ (a+b) = 2\sin.\ a\sin.\ b. \qquad (4.)$$

If in these results we substitute $p$ for $a+b$, and $q$ for $a-b$, which gives $a = \frac{1}{2}(p+q)$ and $b = \frac{1}{2}(p-q)$, they will become

$$\sin.\ p + \sin.\ q = 2\sin.\ \tfrac{1}{2}(p+q)\cos.\ \tfrac{1}{2}(p-q). \quad (5.)$$
$$\sin.\ p - \sin.\ q = 2\cos.\ \tfrac{1}{2}(p+q)\sin.\ \tfrac{1}{2}(p-q). \quad (6.)$$
$$\cos.\ q + \cos.\ p = 2\cos.\ \tfrac{1}{2}(p+q)\cos.\ \tfrac{1}{2}(p-q). \quad (7.)$$
$$\cos.\ q - \cos.\ p = 2\sin.\ \tfrac{1}{2}(p+q)\sin.\ \tfrac{1}{2}(p-q). \quad (8.)$$

Dividing (5) by (6), we have

$$\frac{\sin. p + \sin. q}{\sin. p - \sin. q} = \frac{\sin. \frac{1}{2}(p+q)}{\cos. \frac{1}{2}(p+q)} \times \frac{\cos. \frac{1}{2}(p-q)}{\sin. \frac{1}{2}(p-q)},$$

hence,

$$\frac{\sin. p + \sin. q}{\sin. p - \sin. q} = \frac{\sin. \frac{1}{2}(p+q)}{\cos. \frac{1}{2}(p+q)} \div \frac{\sin. \frac{1}{2}(p-q)}{\cos. \frac{1}{2}(p-q)} = \frac{\text{tang.} \frac{1}{2}(p+q)}{\text{tang.} \frac{1}{2}(p-q)}. \quad (9.)$$

### *Numerical Values of Sines, Tangents, &c.*

§ **17.** Having deduced a few of the most simple relations of the trigonometrical functions of angles, we will now proceed to determine their numerical values.

Sines of very small angles are obviously very nearly equal to the arcs which measure these angles. We have found (§ 6) that the length of an arc of $1'$ is equal to 0·0002908882, &c. If we regard this as the sine of $1'$, from which, as will be hereafter shown, it does not differ even in its eleventh decimal figure, we may find the cosine of $1'$ as follows (§ 8):

$$\cos. 1' = \sqrt{1 - \sin.^2 1'} = 0{\cdot}9999999577, \text{ \&c.}$$

Having thus found the sine and cosine of $1'$, we may continue our work for larger angles by the aid of equations (1) and (3), (§16), which give

$$\sin. (a+b) = 2 \sin. a \cos. b - \sin. (a-b).$$
$$\cos. (a+b) = 2 \cos. a \cos. b - \cos. (a-b).$$

Putting $b=1'$ and $a=1'$, $2'$, $3'$, $4'$, &c., successively, we shall obtain

$$\sin. 2' = 2 \sin. 1' \cos. 1' - \sin. 0'.$$
$$\sin. 3' = 2 \sin. 2' \cos. 1' - \sin. 1'.$$
$$\sin. 4' = 2 \sin. 3' \cos. 1' - \sin. 2'.$$
$$\sin. 5' = 2 \sin. 4' \cos. 1' - \sin. 3'.$$

&c. &c. &c.

$$\cos. 2' = 2 \cos. 1' \cos. 1' - \cos. 0'.$$
$$\cos. 3' = 2 \cos. 2' \cos. 1' - \cos. 1'.$$
$$\cos. 4' = 2 \cos. 3' \cos. 1' - \cos. 2'.$$
$$\cos. 5' = 2 \cos. 4' \cos. 1' - \cos. 3'.$$

&c. &c. &c.

Using the above value of sine and cosine of 1′, executing the work above indicated, and preserving only nine decimal places, we find

| | |
|---|---|
| sin. 1′=0·000290888. | cos. 1′=0·999999958. |
| sin. 2′=0·000581776. | cos. 2′=0·999999831. |
| sin. 3′=0·000872665. | cos. 3′=0·999999619. |
| sin. 4′=0·001163553. | cos. 4′=0·999999323. |
| sin. 5′=0·001454441. | cos. 5′=0·999998942. |
| &c. &c. | &c. &c. |

The work might be continued in this way to any desired extent.

§ **18.** We will now seek formulas which shall give the sine and cosine of any arc without our being obliged first to compute those of a smaller arc.

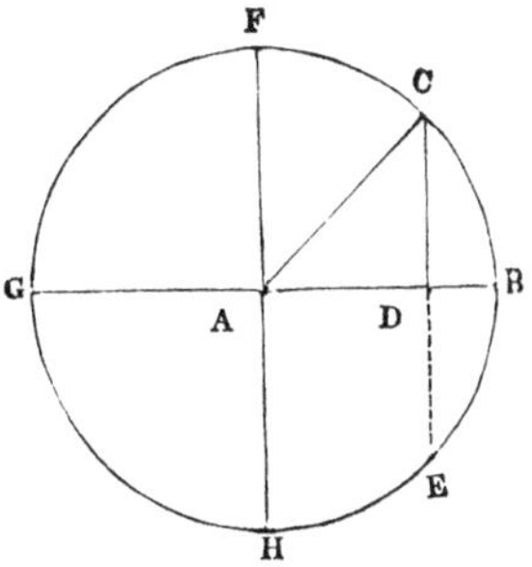

Denote the arc BC, which measures the angle BAC, by $x$, the radius being unity.

Then assume,

$$CD = \sin. x = a_0 + a_1 x + a_2 x^2 + a_3 x^3 + a_4 x^4 + a_5 x^5 +, \&c. \quad (1.)$$
$$DA = \cos. x = b_0 + b_1 x + b_2 x^2 + b_3 x^3 + b_4 x^4 + b_5 x^5 +, \&c. \quad (2.)$$

When $x=0$, the sine of $x$ becomes zero, which requires $a_0$ to $=0$. Since the sine of a negative arc (§ 12) is the same as *minus* the sine of the same arc taken positively, it follows that the expression (1) for the sine of $x$ must be of such a form as to change the algebraic signs of all its terms when $-x$ is written for $+x$, consequently it cannot contain any even powers of $x$. Therefore $a_2=0$; $a_4=0$, &c.

Again, when $x=0$, the cosine of $x$ becomes 1, so that $b_0=1$. And, since the cosine of a negative arc is precisely the same as the cosine of the same arc taken positively, it follows that the expression for the cosine of $x$ must be of such a form as not to change the algebraic signs of any of its terms when $-x$ is written for $+x$, consequently it cannot contain any odd powers of $x$. Therefore $b_1=0$; $b_3=0$; $b_5=0$, &c.

Equations (1) and (2) must therefore assume the following forms:

$$\sin. x = a_1 x + a_3 x^3 + a_5 x^5 + a_7 x^7 +, \text{\&c.} \quad . \quad . \quad (3.)$$
$$\cos. x = 1 + b_2 x^2 + b_4 x^4 + b_6 x^6 +, \text{\&c.} \quad . \quad . \quad . \quad (4.)$$

If we suppose the arc $x$ to be exceedingly small, we may obtain a very close approximate value for the sine of $x$, in (3); by retaining only the first power of $x$, we thus find

$$\sin. x = a_1 x, \text{ very nearly.}$$

But in very small arcs the sine and the arc are very nearly equal; hence, when $x$ is very small, we must have $\sin. x = x$, very nearly. Consequently $a_1 = 1$.

Using this value of $a_1$, equations (3) and (4) become

$$\sin. x = x + a_3 x^3 + a_5 x^5 + a_7 x^7 +, \text{\&c.} \quad . \quad . \quad (5.)$$
$$\cos. x = 1 + b_2 x^2 + b_4 x^4 + b_6 x^6 +, \text{\&c.} \quad . \quad . \quad (6.)$$

It now remains to find the other coefficients, $a_3$, $a_5$, $a_7$, &c., $b_2$, $b_4$, $b_6$, &c. For this purpose we proceed as follows:

In (5) write $2x$ for $x$, and we obtain

$$\sin. 2x = 2x + 8 a_3 x^3 + 32 a_5 x^5 + 128 a_7 x^7 +, \text{\&c.} \quad (7.)$$

Dividing (7) by 2, we have

$$\tfrac{1}{2} \sin. 2x = x + 4 a_3 x^3 + 16 a_5 x^5 + 64 a_7 x^7 +, \text{\&c.} \quad (8.)$$

Taking the product of (5) and (6), we find

$$\sin. x \cos. x = x + \left.\begin{matrix} a_3 \\ + b_2 \end{matrix}\right\} x^3 + \left.\begin{matrix} a_5 \\ + b_2 a_3 \\ + b_4 \end{matrix}\right\} x^5 + \left.\begin{matrix} a_7 \\ + b_2 a_5 \\ + b_4 a_3 \\ + b_6 \end{matrix}\right\} x^7 +, \text{\&c.} \quad (9.)$$

The left-hand members of (8) and (9) being equal [§ 15, eq. (5)], their right-hand members must also be equal; hence equating the coefficients of like powers of $x$, we obtain

$$\left.\begin{aligned} 4 a_3 &= a_3 + b_2 \\ 16 a_5 &= a_5 + b_2 a_3 + b_4 \\ 64 a_7 &= a_7 + b_2 a_5 + b_4 a_3 + b_6 \\ \text{\&c.} & \quad \text{\&c.} \end{aligned}\right\} \quad . \quad . \quad . \quad (10.)$$

Equations (5) and (6), when squared, become respectively

$$\sin.^2 x = x^2 + 2\,a_3\,x^4 + \left.\begin{matrix} a_3^2 \\ +2\,a_5 \end{matrix}\right\}\,x^6 +, \text{\&c.} \qquad (11.)$$

$$\cos.^2 x = 1 + 2\,b_2\,x^2 + \left.\begin{matrix} b_2^2 \\ +2\,b_4 \end{matrix}\right\}\,x^4 \left.\begin{matrix} +2\,b_6 \\ +2\,b_2\,b_4 \end{matrix}\right\}\,x^6 +, \text{\&c.} \qquad (12.)$$

If from the sum of (11) and (12) we subtract 1, we shall obtain

$$\sin.^2 x + \cos.^2 x - 1 = \left.\begin{matrix} +1 \\ +2\,b_2 \end{matrix}\right\}\,x^2 \left.\begin{matrix} +2\,a_3 \\ +b_2^2 \\ +2\,b_4 \end{matrix}\right\}\,x^4 \left.\begin{matrix} +a_3^2 \\ +2\,a_5 \\ +2\,b_6 \\ +2\,b_2\,b_4 \end{matrix}\right\}\,x^6 +, \text{\&c.} \quad (13.)$$

Since the left-hand member of (13) is equal to zero [§ 8, eq. 5 of B], the right-hand member must also equal zero; hence the coefficients of the different powers of $x$ must equal zero, which leads to the following conditions:

$$\left.\begin{matrix} 1 + 2\,b_2 = 0 \\ 2\,a_3 + b_2^2 + 2\,b_4 = 0 \\ a_3^2 + 2\,a_5 + 2\,b_6 + 2\,b_2\,b_4 = 0 \\ \text{\&c.} \qquad \text{\&c.} \end{matrix}\right\} \qquad (14.)$$

Equations (10) and (14) give immediately these results:

$$b_2 = -\tfrac{1}{2} = -\frac{1}{1\times 2}.$$

$$a_3 = -\tfrac{1}{6} = -\frac{1}{1\times 2\times 3}.$$

$$b_4 = \tfrac{1}{24} = \frac{1}{1\times 2\times 3\times 4}.$$

$$a_5 = \tfrac{1}{120} = \frac{1}{1\times 2\times 3\times 4\times 5}.$$

$$b_6 = -\tfrac{1}{720} = -\frac{1}{1\times 2\times 3\times 4\times 5\times 6}.$$

$$a_7 = -\tfrac{1}{5040} = -\frac{1}{1\times 2\times 3\times 4\times 5\times 6\times 7}.$$

&c. &c. &c.

Hence, equations (5) and (6) become

$$\sin.\,x = x - \frac{x^3}{1\times 2\times 3} + \frac{x^5}{1\times 2\times 3\times 4\times 5} - \frac{x^7}{1\times 2\times 3\times 4\times 5\times 6\times 7} +, \text{\&c.} \qquad (15.)$$

$$\cos.\,x = 1 - \frac{x^2}{1\times 2} + \frac{x^4}{1\times 2\times 3\times 4} - \frac{x^6}{1\times 2\times 3\times 4\times 5\times 6} +, \text{\&c.} \qquad (16.)$$

If we divide the right-hand member of (15) by the right-hand member of (16), we shall obtain an infinite series, giving the value of tang. $x$, as follows :

$$\text{tang. } x = x + \frac{x^3}{3} + \frac{2\,x^5}{15} + \frac{17\,x^7}{315} +, \text{ \&c.}$$

Other series might be found by combining (15) and (16) for the cotan. $x$, sec. $x$, and cosec. $x$.

The arc of 1′ has already been found [§ 6] to be equal to 0·0002908882086, &c. Putting this value for $x$, in the above series, we find

$$\frac{x^2}{1 \times 2} = 0{\cdot}0000000423079,$$

$$\frac{x^3}{1 \times 2 \times 3} = 0{\cdot}0000000000041.$$

If we confine ourselves to thirteen decimals, we shall obtain

$$\sin. 1' = x - \frac{x^3}{1 \times 2 \times 3} = 0{\cdot}0002908882045\,;$$

$$\cos. 1' = 1 - \frac{x^2}{1 \times 2} = 0{\cdot}9999999576921.$$

From this we see that the sine of 1′ does not differ from its corresponding arc until after we pass the 11th decimal place, as was asserted under § 17.

For ordinary purposes, seven decimal places is as accurate as is desired. Having already (§ 17) computed the sine and cosine of each minute as far as 5′, we will now, by the aid of equations (15) and (16), continue the computation, limiting ourselves to seven decimals, and shall obtain as follows :

| | | |
|---|---|---|
| arc of 6′ = 0·0017453, | sin. = 0·0017453, | cos. = 0·9999985. |
| " 7′ = 0·0020362, | " = 0·0020362, | " = 0·9999979. |
| " 8′ = 0·0023271, | " = 0·0023271, | " = 0·9999973. |
| " 9′ = 0·0026180, | " = 0·0026180, | " = 0·9999966. |
| " 10′ = 0·0029089, | " = 0·0029089, | " = 0·9999958. |
| " 11′ = 0·0031998, | " = 0·0031998, | " = 0·9999949. |
| " 12′ = 0·0034907, | " = 0·0034907, | " = 0·9999939. |
| " 13′ = 0·0037815, | " = 0·0037815, | " = 0·9999928. |
| " 14′ = 0·0040724, | " = 0·0040724, | " = 0·9999917. |
| " 15′ = 0·0043633, | " = 0·0043633, | " = 0·9999905. |

In the same way, for degrees, we find

| | | |
|---|---|---|
| arc of 1° = 0·0174533, | sin. = 0·0174524, | cos. = 0·9998477. |
| " 2° = 0·0349066, | " = 0·0348995, | " = 0·9993908. |
| " 3° = 0·0523599, | " = 0·0523360, | " = 0·9986295. |
| " 4° = 0·0698132, | " = 0·0697565, | " = 0·9975641. |
| " 5° = 0·0872665, | " = 0·0871557, | " = 0·9961947. |
| " 6° = 0·1047198, | " = 0·1045285, | " = 0·9945219. |
| " 7° = 0·1221730, | " = 0·1218693, | " = 0·9925462. |
| " 8° = 0·1396263, | " = 0·1391731, | " = 0·9902681. |
| " 9° = 0·1570796, | " = 0·1564345, | " = 0·9876883. |
| " 10° = 0·1745329, | " = 0·1736482, | " = 0·9848078. |

Having in this way computed the sines and cosines, we may [see equations (B), § 8] obtain the tangents by dividing the sines by their corresponding cosines; the cotangents will be obtained by dividing the cosines by the sines. The reciprocal of the cosines will give the secants, and the reciprocal of the sines will give the cosecants.

The values of the sines, cosines, tangents, and cotangents are given in Table III.

---

# CHAPTER II.

## EXPLANATION OF TABLES.

### TABLE I.

### LOGARITHMS.

**§ 19.** This table gives the logarithms of all numbers from 1 to 10000.

For the method of calculating these logarithms, the student is referred to my "Treatise on Algebra," chap. IX.

The common logarithm, or *Briggean* logarithm, as sometimes called, in honor of the inventor *Briggs*, has 10 for its base. Hence, for common logarithms, we have

| | | |
|---|---|---|
| $10^0=1$ | , that is, 0 is the logarithm of 1 | |
| $10^1=10$ | , " " 1 " " " " 10 | |
| $10^2=100$ | , " " 2 " " " " 100 | |
| $10^3=1000$ | , " " 3 " " " " 1000 | |
| $10^4=10000$ | , " " 4 " " " " 10000 | |
| $10^5=100000$, | " " 5 " " " " 100000 | |
| &c. | &c. | &c. |

If we use negative exponents, we shall have

| | | |
|---|---|---|
| $10^{-1}=0{\cdot}1$ | , that is, $-1$ is the logarithm of 0·1 | |
| $10^{-2}=0{\cdot}01$ | , " " $-2$ " " " " 0·01 | |
| $10^{-3}=0{\cdot}001$ | , " " $-3$ " " " " 0·001 | |
| $10^{-4}=0{\cdot}0001$ | , " " $-4$ " " " " 0·0001 | |
| $10^{-5}=0{\cdot}00001$, | " " $-5$ " " " " 0·00001 | |
| &c. | &c. | &c. |

From the above, we see that the logarithm of any number between 1 and 10 is a proper fraction between 0 and 1. The logarithm of any number between 10 and 100 is 1 plus a proper fraction. The logarithm of any number between 100 and 1000 is 2 plus a fraction.

We also see that the logarithm of any number between 1 and 0·1 is some number between 0 and $-1$; it may therefore be represented by $-1$ plus a fraction. For a similar reason, the logarithm of any number between 0·1 and 0·01 may be represented by $-2$ plus a fraction. The logarithm of any number between 0·01 and 0·001 is $-3$ plus a fraction.

Hence, we see that logarithms of all numbers greater than 10 or less than 1 consist of an integral part, positive or negative, plus a proper fraction. The integral part is called the *characteristic*, and may readily be assigned by the following

## RULE.

*The characteristic of the logarithm of any number greater than* 1 *is one less than the number of places of figures which express the integral part of the given number.*

*The characteristic of the logarithm of a decimal fraction is a negative number, and is equal to the number of places by which the first significant figure is removed from the units' place.*

Thus, we have as follows :

| *Number.* | *Characteristic.* | *Number.* | *Characteristic.* |
|---|---|---|---|
| 4 | 0 | 0·4 | −1 |
| 48 | 1 | 0·46 | −1 |
| 138 | 2 | 0·375 | −1 |
| 2304 | 3 | 0·05 | −2 |
| 4·7 | 0 | 0·0578 | −2 |
| 37·04 | 1 | 0·006 | −3 |
| 120·005 | 2 | 0·00675 | −3 |
| 2178·88 | 3 | 0·00035 | −4 |

The characteristic being so readily obtained by means of the above rule, has been omitted in the tables, except in the small table on page 1, where are found the logarithms of all numbers from 1 to 100, with their characteristics. But for all the other numbers of our table only the decimal part of the logarithms are given.

The decimal part of all logarithms are regarded as positive. The characteristic may, however, be considered as negative. When this is the case, it is usual to place the negative sign immediately over the characteristic. Thus the logarithm of 0·75 is $\bar{1}$·875061, or 0·875061−1.

§ **20.** *To find, in the table, the logarithm of any number from* 1 *to* 100.

Seek for the given number in one of the columns headed N, of the first page of the tables of logarithms, and against it on the right in the next column, will be found the logarithm. We thus find

log. 7 = 0·845098,
log. 39 = 1·591065.

§ **21.** *To find the logarithm of any number consisting of three places of figures.*

Seek the given number in one of the left-hand columns headed N, and against it in the column headed 0 will be found the decimal part of the logarithm, observing that if only four figures are given in column 0, they are the last four : the first two are to be taken from the number directly above which contains six places of figures. Thus, we find

$$\log. 617 = 2{\cdot}790285,$$
$$\log. 635 = 2{\cdot}802774.$$

§ **22.** *To find the logarithm of a number consisting of four places of figures.*

Seek the first three figures in column headed N, then passing horizontally across the page until you reach the column headed with the fourth figure of the given number, and you will have the last four decimal figures of the logarithm. The first two figures are to be taken from the column headed 0 ; observing, if the four figures found, stand opposite to a row of six figures in the column 0, to use the first two. But if the four figures found are opposite a line of only four figures, you are then to ascend the column till you come to a line of six figures, and then to use the two left-hand figures. The six figures thus obtained will be the decimal part of the logarithm. The characteristic is to be found by the Rule under § 19. Thus, we find

$$\log. 3657 = 3{\cdot}563125,$$
$$\log. 7641 = 3{\cdot}883150.$$

In some portions of the table, in the columns headed 1, 2, 3, 4, &c., the character ⋄ is to be found in place of a figure, which indicates that the two figures of the first column, which are to be prefixed, have changed, and are to be taken from the horizontal line directly below. This change of the two left-hand figures of the logarithm is indicated by an asterisk placed in the column 0. The place occupied by this character is to be supplied with a zero. The two figures from the column 0 must also be taken from the line below the asterisk, if the character ⋄ has been passed over while crossing the page horizontally. Thus, we find

$$\log. 2138 = 3{\cdot}330008,$$
$$\log. 2298 = 3{\cdot}361350,$$
$$\log. 3237 = 3{\cdot}510143.$$

§ **23.** *To find the logarithm of a number consisting of five or more places of figures.*

For example, let us seek the logarithm of 734582. We readily find as follows:

| | | |
|---|---|---|
| log. 734500 | = | 5·865992 |
| log. 734600 | = | 5·866051 |
| Diff. of numb. =100 | | Diff. of logs. =59 |

The difference between the first number 734500 and the given number is 734582−734500 = 82. Hence, if we suppose the difference of logarithms to be to each other as the difference of their corresponding numbers, which supposition is very nearly correct, we shall have

$$100 : 82 :: 59 : 48{\cdot}38,$$

for the difference between the logarithm of 734500 and the logarithm of 734582. Hence, we have

$$\begin{aligned} \text{log. } 734500 &= 5{\cdot}865992 \\ &\quad\; + 48{\cdot}38 \\ \hline \text{log. } 734582 &= 5{\cdot}866040 \end{aligned}$$

Had there been only five figures in the given number, that is, only one beyond the fourth place, the first term of the above proportion would have been 10 instead of 100; had there been three additional figures beyond the fourth place, the first term would have been 1000.

The difference between the logarithms of consecutive numbers of four places of figures is given in the table in a column headed D. Thus by turning to the table, we find 59, our third term of the above proportion, immediately opposite the logarithm of 7345.

From what has been done, we see that we may find the logarithm of a number consisting of more than four places of figures by the following method:

Consider all the figures after the fourth as zeros. Then find the decimal part of the logarithm of the number given by the first four figures, observing to give a characteristic for the whole number of figures by rule under § 19. Take from the column

D the number which is found directly opposite the logarithm already taken out, and multiply it by the figures which were regarded as zeros, pointing off in the product as though these figures were all decimals; add the result thus obtained to the logarithm already found, and it will give the logarithm of the given number.

In this way we find

$$\log. 365365 = 5{\cdot}562727.$$
$$\log. 635536 = 5{\cdot}803140.$$
$$\log. 704307 = 5{\cdot}847762.$$

NOTE.—Since the foregoing process of finding the logarithm of a number of more than four places of figures is founded on the supposition that the differences of logarithms of different numbers are to each other as the differences of the numbers, which supposition is not strictly true, it follows that this method can be used only to a limited extent. It ought never to be employed for a number consisting of more than six places of figures.

### § 24. *To find the logarithm of a Vulgar Fraction.*

Since a vulgar fraction is the quotient of the numerator divided by the denominator, we may obtain its logarithm by subtracting the logarithm of the denominator from the logarithm of the numerator. Thus, the logarithm of $\frac{37}{53}$ is $\log. 37 - \log. 53 = 1{\cdot}568202 - 1{\cdot}724276 = \bar{1}{\cdot}843926$.

In a similar manner we find

$$\log. \tfrac{365}{400} = \bar{1}{\cdot}960233.$$
$$\log. \tfrac{375}{463} = \bar{1}{\cdot}908450.$$

### § 25. *To find the logarithm of a Decimal Fraction.*

Since we have, by the property of logarithms,

$$\log. 4{\cdot}5 = \log. \tfrac{45}{10} = \log. 45 - \log. 10 = \log. 45 - 1;$$
$$\log. 3{\cdot}65 = \log. \tfrac{365}{100} = \log. 365 - \log. 100 = \log. 365 - 2;$$
$$\log. 3{\cdot}754 = \log. \tfrac{3754}{1000} = \log. 3754 - \log. 1000 = \log. 3754 - 3;$$

it follows that the decimal part of the logarithm of a decimal fraction is the same as though the number was wholly integral, the only difference between the logarithm of a decimal number and of the number considered as integral is in the characteristic. Hence, take out the logarithm as though the number were integral, and fix a characteristic according to rule under § 19.

We thus find,

log. 476 = 2·677607
log. 47·6 = 1·677607
log. 4·76 = 0·677607
log. 0·476 = $\bar{1}$·677607
log. 0·0476 = $\bar{2}$·677607
log. 0·00476 = $\bar{3}$·677607

§ **26.** *To find the natural number corresponding to any logarithm.*

Seek in the table, in the column 0, for the first two figures of the decimal part of the logarithm; the other four figures are to be sought for in the same column, or in any one of the columns 1, 2, 3, &c. If the decimal part of the logarithm is exactly found, then will the first three figures of the corresponding number be found in the column N, and the fourth figure will be found at the top of the page. This number must be made to correspond with the given characteristic of the given logarithm by annexing ciphers, or by pointing off decimals. Thus the

| | | | |
|---|---|---|---|
| logarithm | 5·311754 | gives | 205000, |
| " | 4·341237 | " | 21940, |
| " | 3·759970 | " | 5754, |
| " | 2·759970 | " | 575·4, |
| " | 1·759970 | " | 57·54. |

When the decimal part of the logarithm cannot be accurately found in the table, take out the four figures corresponding to the next less logarithm. Then for the additional figures, subtract this less logarithm from the given logarithm, and divide the remainder, with naughts annexed, by the corresponding number taken from column D. For example, let us seek the number whose logarithm is 1·234567. We find the next less number to the decimal 0·234567 to be 0·234517, which corresponds to 1716. We also find the number in column D to be 253. Hence

0·234567
0·234517
50; and 50 ÷ 253 = 0·198, nearly.

So that the number answering to the logarithm 1·234567 is 17·16198, nearly.

## ARITHMETICAL CALCULATIONS BY LOGARITHMS.

§ **27.** *Multiplication by Logarithms.*

Since the logarithm of the product of two or more factors is equal to the sum of their logarithms, we deduce, for multiplication by logarithms, this

RULE.

*Add the logarithms of the factors, and the sum will be the logarithm of the product.*

1. What is the product of 3·65 by 56·3 ?

$$\begin{aligned} \log.\ 3{\cdot}65 &= 0{\cdot}562293 \\ \log.\ 56{\cdot}3 &= 1{\cdot}750508 \\ \hline \log.\ \text{of product} &= 2{\cdot}312801, \end{aligned}$$

which gives 205·495 for the product.

2. What is the product of 7·8 by 35·3 ? *Ans.* 275·34
3. What is the product of 2·13 by 0·57 ? *Ans.* 1·2141.

NOTE.—When any of the characteristics of the logarithms are negative, we must observe the *algebraic rule* for their addition.

4. What is the continued product of 53·7, 0·12, and 0·004 ?

$$\begin{aligned} \log.\ 5{\cdot}37 &= 1{\cdot}729974 \\ \log.\ 0{\cdot}12 &= \bar{1}{\cdot}079181 \\ \log.\ 0{\cdot}004 &= \bar{3}{\cdot}602060 \\ \hline \text{product}=0{\cdot}025776, \text{ whose log.} &= \bar{2}{\cdot}411215 \end{aligned}$$

5. What is the square of 3·7 ; that is, what is the product of 3·7 by 3·7 ? *Ans.* 13·69.
6. What is the cube of 3·8 ; that is, what is the continued product of 3·8, 3·8, and 3·8 ? *Ans.* 54·872.

§ **28.** *For Division by Logarithms*, we obviously have this

RULE.

*Subtract the logarithm of the divisor from the logarithm of the dividend.*

EXAMPLES.

1. What is the quotient of 365 by 7·3?

$$\begin{aligned} \text{log. } 365 &= 2{\cdot}562293 \\ \text{log. } 7{\cdot}3 &= 0{\cdot}863323 \\ \hline \text{quotient is 5, its log.} &= 1{\cdot}698970 \end{aligned}$$

2. What is the quotient of 2·456 by 1·47?

*Ans.* 1·67075, nearly.

3. What is the quotient of 7·4 by 12·58? *Ans.* 0·588235.

NOTE.—When either or both of the characteristics of the logarithms are negative, we must observe the *algebraic rule* for the subtraction of the one from the other.

4. What is the quotient of 0·378 divided by 0·45?

$$\begin{aligned} \text{log. } 0{\cdot}378 &= \bar{1}{\cdot}577492 \\ \text{log. } 0{\cdot}45 &= \bar{1}{\cdot}653213 \\ \hline \text{quotient} = 0{\cdot}84\text{, whose log.} &= \bar{1}{\cdot}924279 \end{aligned}$$

5. What is the quotient of 0·10071 by 0·00373?

$$\begin{aligned} \text{log. } 0{\cdot}10071 &= \bar{1}{\cdot}003072 \\ \text{log. } 0{\cdot}00373 &= \bar{3}{\cdot}571709 \\ \hline \text{quotient} = 27\text{, whose log.} &= 1{\cdot}431363\text{, nearly.} \end{aligned}$$

## § 29. *Involution by Logarithms.*

Since the exponent denoting any power of any number expresses how many times this number is used as a factor to produce the given power, it follows that the logarithm of any power is equal to the logarithm of the number repeated as many times as there are units in the exponent. Hence we have this

RULE.

*Multiply the logarithm of the number by the exponent denoting the power.*

EXAMPLES.

1. What is the 5th power of 1·234?

$$\begin{aligned} \text{log. } 1{\cdot}234 &= 0{\cdot}091315 \\ &\quad\;\; 5 \text{ multiply.} \\ \hline \text{power} = 2{\cdot}86137\text{, whose log.} &= 0{\cdot}456575 \end{aligned}$$

2. What is the 7th power of 0·5?

$$\log. 0{\cdot}5 = \bar{1}{\cdot}698970$$
$$7 \text{ multiply.}$$

$$\text{power} = 0{\cdot}0078125, \text{ whose log.} = \bar{3}{\cdot}892790$$

NOTE.—It must be kept in mind that the decimal part of every logarithm is positive, so that, as in the last example, whatever is to carry to the product of the characteristic by the exponent is positive.

3. What is the 30th power of 1·07? *Ans.* 7·6123, nearly.
4. What is the 11th power of 0·11? *Ans.* 0·000000000285313.

## § 30. *Evolution by Logarithms.*

Since the exponent denoting a root indicates that the number is to be separated into as many equal factors as there are units in the exponents, we obviously have this

### RULE.

*Divide the logarithm of the number by the number denoting the root.*

#### EXAMPLES.

1. What is the 11th root of 11?

   log. 11 = 1·041393, which divided by 11 gives 0·094672 for the log. of the root. Hence the root = 1·24357, nearly.

NOTE.—When the characteristic is negative, and not divisible by the exponent, we must put with it a sufficiently large negative integer to make it divisible, and then connect with the decimal portion of the logarithm an equally large positive number. This will be best illustrated by the following example.

2. What is the 5th root of 0·00567?

   log. 0·00567 = $\bar{3}{\cdot}753583 = \bar{5} + 2{\cdot}753583$, which divided by 5 gives $\bar{1}{\cdot}550716$ for the logarithm of the root. Hence the root = 0·3554, nearly.

3. What is the 3d root of 0·365? *Ans.* 0·714657, nearly.
4. What is the 7th root of 7? *Ans.* 1·32047, nearly.
5. What is the 5th root of 0·5? *Ans.* 0·87055, nearly.

§ **31.** *Proportion by Logarithms.*

Since the fourth term of a proportion is found by dividing the product of the second and third terms by the first, we obviously have this

RULE.

*From the sum of the logarithms of the second and third terms, subtract the logarithm of the first term.*

What is the fourth term of a proportion of which the first, second, and third terms are respectively 0·0146, 4·5, and 1·07?

$$\begin{array}{rl} \text{log. } 4{\cdot}5 & = 0{\cdot}653213 \\ \text{log. } 1{\cdot}07 & = 0{\cdot}029384 \\ \hline & \phantom{=} 0{\cdot}682597 \\ \text{subtract log. } 0{\cdot}0146 & = \bar{2}{\cdot}164353 \\ \hline \text{log. of fourth term} & = 2{\cdot}518244 \end{array}$$

Consequently the fourth term = 329·794, nearly.

ARITHMETICAL COMPLEMENT.

§ **32.** The subtraction of a logarithm from the sum of two or more logarithms is usually made to depend upon addition, by making use of its *Arithmetical Complement*, which is the difference between 10 and the given logarithm, and is readily taken from the table, by beginning at the left hand and subtracting each figure from 9, except the last significant figure on the right, which must be taken from 10. Now it is obvious, that if instead of subtracting a given number from another, we first subtract the number from 10, and then add the result, we shall, after rejecting 10, obtain the true difference. Hence to work a proportion by logarithms, we may use this second

RULE.

*Take the arithmetical complement of the logarithm of the first term, and the logarithms of the second and third terms, and from the sum of the three reject* 10 *from the characteristic.*

EXAMPLES.

1. What is the fourth term of the proportion 6·5 : 30·5 : : 1·25?

| | | |
|---|---|---|
| Arith. comp. log. | 6·5 | = 9·187087 |
| log. | 30·5 | = 1·484300 |
| log. | 1·25 | = 0·096910 |

sum, after rejecting 10, = 0·768297, which is the log. of 5·86539, the fourth term.

2. What is the fourth term in the proportion 3456 : 10·5 : : 6543?

| | | |
|---|---|---|
| ar. co. log. | 3456 | = 6·461426 |
| log. | 10·5 | = 1·021189 |
| log. | 6543 | = 3·815777 |

fourth term = 19·8788, log. = 1·298392

3. In the proportion 1·23 : 2·34 : : 3·45, what is the fourth term? *Ans.* 6·56342, nearly.

## TABLE III.

### NATURAL SINES AND TANGENTS.

§ **33.** It is necessary to explain the arrangement of this table before explaining that of Table II., since the values of the latter have been obtained from those of the former. Table II., being logarithmic values, is with propriety placed immediately after the table of logarithms.

Table III. contains the natural sines and tangents to every minute of the first quadrant, and consequently those of the cosines and cotangents. The first 9 pages are devoted to sines and cosines, the remainder of this table gives the tangents and cotangents.

When the number of degrees does not exceed 45, they are to be found at the top of the page, and the additional minutes are given in the first or left-hand column of the page. But if the degrees exceed 45, they are to be found at the bottom of the page, and the additional minutes are then given in the last, or right-hand column of the page.

It will be seen by examining this table that the number of degrees at the top of any page added to the number at the bottom of the same, gives 89, to which adding the 60 intermediate minutes makes just 90 degrees. Thus, on page 70 we have 34° at the top and 55° at the bottom, the sum of which is 89°.

Since the cosine and cotangent (§ 8) of an angle is the same as the sine and tangent respectively of its complement, it follows that by this arrangement of the table those columns which are marked sine and tangent at the top of the page will be marked cosine and cotangent respectively, at the bottom of the page.

As the sine and cosine of any angle cannot exceed the radius, which is a unit, it follows that their values are always expressed in this table by a decimal. The tangent and cotangent may in some cases exceed a unit, and will then be expressed by a decimal joined to an integral part.

Thus, turning to this table, we find

sin. 17° 13′=0·29599; cos. 17° 13′=0·95519.
tang. 17° 13′=0·30987; cotan. 17° 13′=3·22715.
sin. 55° 43′=0·82626; cos. 55° 43′=0·56329.
tang. 55° 43′=1·46686; cotan. 55° 43′=0·68173.

## TABLE II.

### LOGARITHMIC SINES AND TANGENTS.

§ **34.** This table is constructed by taking the logarithms of the values given in Table III., and increasing each logarithm by 10. The arrangement is the same as given in the Table of Natural Sines, Tangents, &c.; that is, the degrees are to be found at the top of the page and minutes in the first column, when the angle is less than 45°. When the angle is greater than 45°, the degrees are to be found at the bottom of the page and the minutes in the last column. As in the last table, the column marked sine at top is cosine at the bottom; the column marked tangent at the top is cotangent at the bottom, and conversely.

Obtuse angles are also provided for in this table: when found at the top of the page they are at the right-hand side, and the minutes are then to be found in the right-hand column. But when found at the bottom of the page they are at the left-hand side, and the minutes are then to be found in the left-hand column.

As most cases of Trigonometry are solved by the assistance of proportions, which are most readily wrought by the help of logarithms, it has been found convenient to prepare beforehand the logarithms of the sines and tangents of all angles expressed by degrees and minutes within the limit of the first quadrant.

Since the sines, as well as tangents, of many arcs are less than a unit, the characteristic of their logarithms will be negative. By adding 10 to each logarithm, the characteristics will all be positive, and we shall not be as liable to commit errors in using them, as we should if some had positive characteristics, and others negative characteristics.

§ **35.** *To find, in the table, the logarithmic sine, cosine, tangent, or cotangent of any given angle.*

If the angle is less than 45°, we seek the degrees at the top of the page and the minutes in the first column on the left; then passing across the page horizontally until we come to the column having the appropriate heading of sine, cosine, tangent, or cotangent, as the case may require, we shall find the logarithm sought. Thus, on page 53, we find

$$\text{sin. } 35^\circ\ 53'=9{\cdot}767999\,;\ \text{cos. } 35^\circ\ 53'=9{\cdot}908599.$$
$$\text{tang. } 35^\circ\ 53'=9{\cdot}859400\,;\ \text{cotan. } 35^\circ\ 53'=10{\cdot}140600.$$

If the angle is greater than 45°, we seek for the degrees at the bottom of the page, and the minutes on the right-hand side of the page; then, as before, passing horizontally across the page till we reach the column having at the bottom the appropriate designation of sine, cosine, tangent, or cotangent, as the case may require, and the number thus given will be the one sought. If the angle is obtuse, we must make use of its supplement (§ 13).

§ **36.** When the angle is expressed in degrees, minutes, and seconds, we proceed as follows:

Let it be required, for example, to find the logarithmic

sine of 13° 14′ 17″. By what has already been said, we readily find

$$\text{sin. } 13° \ 14 \ = 9{\cdot}359678$$
$$\text{sin. } 13° \ 15' = 9{\cdot}360215$$
$$537 = \text{diff. for } 1'.$$

Hence, 537 ÷ 60 = 8·95 = diff. for 1″, and 895 = diff. for 100″. If we suppose the logarithmic sines to increase in the same ratio as the increase of their corresponding arcs, which supposition is very nearly correct, we shall be able to find the difference corresponding to 17″ by multiplying 8·95 by 17, which gives 152·15.

$$\text{sin. } 13° \ 14' = 9{\cdot}359678$$
$$152 \text{ add.}$$
$$9{\cdot}359830 = \text{sine of } 13° \ 14' \ 17''.$$

The column D, immediately at the right of the column of sines, gives the differences of the logarithmic sines for 100″. Thus, in the tables opposite the sine of 13° 14′ in the column D, we find 895, which we have already shown to be the increase for 100″. In the same way, the column D, immediately after the column of cosines, gives the difference of the logarithmic cosines for 100″. The column D, between the tangent and cotangent columns, answers for both, giving also the difference for 100″.

Hence, to find the logarithmic sine of an arc given in degrees, minutes, and seconds, we must first seek the value corresponding to the degrees and minutes; then multiply the corresponding tabular number in column D by the number of seconds, cutting off two places to the right for decimals; add the product to the value already found, and the result will be the logarithmic sine sought. Thus, find the sine of 37° 31′ 13″.

Sin. 37° 31′ = 9·784612. Tabular number D = 274, which multiplied by 13, the number of seconds, gives 274 × 13 = 3562; pointing off two decimals, we have 35·62, or 36 nearly, which added gives

$$\text{sin. } 37° \ 31' = 9{\cdot}784612$$
$$36$$
$$\text{sin. } 37° \ 31' \ 13'' = 9{\cdot}784648$$

The tangent of an arc where there are seconds is found by the same method as has been employed for finding the sine. In finding the values of the cosine and cotangent, we must bear in mind that the cosines and cotangents decrease as the arcs increase, consequently the corrections found by aid of columns D must be subtracted instead of being added.

Find the cosine of 23° 6′ 47″.

$$\begin{aligned} \cos.\ 23^\circ\ 6' &= 9{\cdot}963704 \\ &\quad \text{subtract } 42{\cdot}30 = \tfrac{47}{100} \text{ of } 90 \text{ (D.)} \\ \cos.\ 23^\circ\ 6'\ 47'' &= 9{\cdot}963662 \end{aligned}$$

Find tangent and cotangent of 30° 30′ 30″.

$$\begin{aligned} \text{tang. } 30^\circ\ 30' &= 9{\cdot}770148 \\ &\quad \text{add } 144 = \tfrac{30}{100} \text{ of } 481 \text{ (D.)} \\ \text{tang. } 30^\circ\ 30'\ 30'' &= 9{\cdot}770292 \end{aligned}$$

$$\begin{aligned} \text{cotan. } 30^\circ\ 30' &= 10{\cdot}229852 \\ &\quad \text{subtract } 144 = \tfrac{30}{100} \text{ of } 481 \text{ (D.)} \\ \text{cotan. } 30^\circ\ 30'\ 30'' &= 10{\cdot}229708 \end{aligned}$$

§ **37.** *To find the arc expressed in degrees, minutes, and seconds, when its logarithmic sine, cosine, tangent, or cotangent is given.*

Seek for the given value in the column having the proper designation of sine, cosine, tangent, or cotangent, as the case may require, keeping in mind (§ 34) that the columns designated by sine and cosine at the top of the page are marked cosine and sine respectively at the bottom of the page. Also, those marked tangent and cotangent at the top are respectively cotangent and tangent at the bottom.

If the exact value is found in the table, the arc will be accurately given in degrees and minutes. The degrees are to be taken from the top of the page, and the minutes from the left-hand column, when the proper precept of the column is found at the top. But when the precept is found at the bottom, the degrees are to be taken from the bottom of the page and the minutes from the right-hand column.

If the number cannot be exactly found in the table, then find, as above, the degrees and minutes corresponding to the nearest less logarithm. Divide the difference of these logarithms, with two ciphers annexed, by the tabular difference taken from column D, the quotient will give seconds, which seconds are to be added to the degrees and minutes already found, in the case of a sine or tangent, but to be subtracted in the case of a cosine or cotangent.

EXAMPLES.

1. Find the arc whose logarithmic sine is 9·365365.

9·365365 = given logarithmic sine.
sin. 13° 24′ = 9·365016

349 = difference,
and 34900 ÷ 883 = 39, the number of seconds.

Hence, the arc 13° 24′ 39″ corresponds with the logarithmic sine of 9·365365.

2. Find the arc whose cosine is 9·142857.

9·142857 = given logarithmic cosine.
cos. 82° 1′ = 9·142655

20200 ÷ 1500 = 13″.
subtract 13″

82° 0′ 47″ = arc required.

3. Find the arc whose tangent is 10·528017.

10·528017 = given logarithmic tangent.
tang. 73° 29′ = 10·527931

8600 ÷ 772 = 11″.
add 11″

73° 29′ 11″ = arc required.

4. Find the arc whose cotangent is 9·555555.

9·555555 = given logarithmic cotangent.
cotan. 70° 14′ = 9·555536

1900 ÷ 661 = 3″, nearly.
subtract 3″

70° 13′ 57″ = arc required.

§ **38.** The secants and cosecants are omitted in this table, since they are so readily found by the aid of the cosines and sines. By equation B, § 8, we have

$$\text{sec. A} = \frac{1}{\text{cos. A}}; \text{ cosec. A} = \frac{1}{\text{sin. A}}.$$

Clearing of fractions, we have

$$\text{sec. A} \times \text{cos. A} = 1; \text{ cosec. A} \times \text{sin. A} = 1.$$

Taking the logarithms, and observing to add 10 to each logarithm, we have

$$\text{log. sec. A} + \text{log. cos. A} = 20; \text{ log. cosec. A} + \text{log. sin. A} = 20.$$

Hence,
$$\text{log. sec. A} = 20 - \text{log. cos. A},$$
$$\text{log. cosec. A} = 20 - \text{log. sin. A}.$$

From this we see that

*The logarithmic secant is found by subtracting the logarithmic cosine from* 20; *and the logarithmic cosecant is found by subtracting the logarithmic sine from* 20.

EXAMPLES.

1. Find the logarithmic secant and cosecant of 41° 41′.

| | | |
|---|---|---|
| | 20· | 20· |
| sin. 41° 41′ = | 9·822830; cos. 41° 41′ = | 9·873223. |
| cosec. 41° 41′ = | 10·177170; sec. 41° 41′ = | 10·126777. |

2. What is the secant and cosecant of 15° 15′ 15″?

| | | |
|---|---|---|
| | 20· | 20· |
| sin. 15° 15′ 15″ = | 9·420123; cos. 15° 15′ 15″ = | 9·984423. |
| cosec. 15° 15′ 15″ = | 10·579877; sec. 15° 15′ 15″ = | 10·015577. |

§ **39.** Since we have (equation B, § 8) $\text{tang. A} = \frac{1}{\text{cotan. A}}$, it follows that

$$\text{tang. A} \times \text{cotan. A} = 1,$$

and
$$\text{log. tang. A} + \text{log. cotan. A} = 20.$$

*Hence, the logarithmic cotangent may be found by subtracting the logarithmic tangent from* 20; *and, conversely, the logarithmic tangent may be found by subtracting the logarithmic cotangent from* 20.

EXAMPLES.

1. The logarithmic tangent of an arc is 9·545454; what is the logarithmic cotangent of the same arc? *Ans.* 10·454546.

2. The logarithmic cotangent is 9·333444; what is that of the tangent of the same angle? *Ans.* 10·666556.

§ **40.** In any plane triangle there are three sides and three angles, making in all six parts to be considered. Any three of these six parts, provided one at least is a side, being given or known, the other three parts can be found. Thus, in Book I., Geometry, we have seen that when two triangles had three parts of the one respectively equal to the three corresponding parts of the other, provided the three parts compared were not all angles, the two triangles were identical, or equal in all respects.

The case in which three angles of a triangle are respectively equal to three angles of a second triangle, does not lead of necessity to an equality of these two triangles, but simply to their *similarity*. Hence, when in a triangle three parts, provided one at least is a side, are given, the remaining parts can be found.

§ **41.** In the case of right triangles, one angle, that is, the right angle, is always given; consequently two parts in addition to the right angle must be given, one of which must be a side, in order that we may find the remaining parts.

The geometrical or graphic method of solving a triangle when a sufficient number of parts are known, has already been exhibited in the Problems at the end of Book Second of Geometry. The method of calculating the numerical values of these parts—which we now proceed to explain—belongs to Trigonometry.

# CHAPTER III.

## SOLUTION OF RIGHT TRIANGLES.

### CASE I.

§ **42.** *Given, the hypotenuse and one of the angles, to find the other parts.*

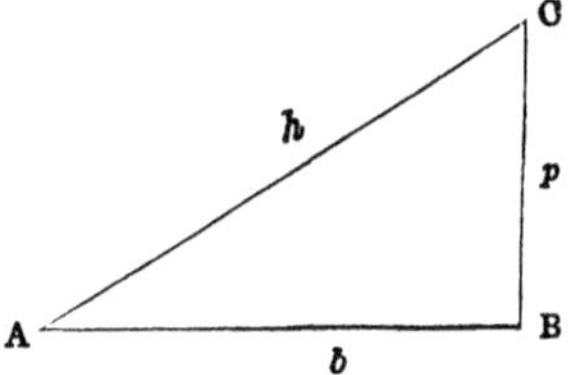


Suppose the given angle to be A. Then will the angle C be found by subtracting A from 90°.

By our definitions (§ 8), we have

$$\text{sin. A}=\frac{p}{h}; \ \text{cos. A}=\frac{b}{h},$$

which, cleared of fractions, give

$$p=h \text{ sin. A.} \quad . \quad . \quad . \quad (1.)$$
$$b=h \text{ cos. A.} \quad . \quad . \quad . \quad (2.)$$

By using logarithms, we have

$$\text{log. } p=\text{log. } h+\text{log. sin. A.} \quad . \quad (3.)$$
$$\text{log. } b=\text{log. } h+\text{log. cos. A.} \quad . \quad (4.)$$

As an example, suppose the hypotenuse to be 125, and the angle at A to be 37° 30′. Required the other parts.

$$h=125; \ \text{A}=37° \, 30'.$$

We find C = 90° − 37° 30′ = 52° 30′.

By Natural Numbers. (Table III.)

| sin. 37° 30′ = 0·60876; | cos. 37° 30′ = 0·79335 |
|---|---|
| 125 | 125 |
| 304380 | 396675 |
| 121752 | 158670 |
| 60876 | 79335 |
| 76·09500 = $p$. | 99·16875 = $b$. |

By Logarithms.

$$\begin{array}{lll} \log. 125 &= 2{\cdot}096910 \quad . \quad . \quad . & 2{\cdot}096910 \\ \log.\sin. 37^\circ\ 30' &= 9{\cdot}784447\ ;\ \log.\cos. 37^\circ\ 30' &= 9{\cdot}899467 \\ \hline \log. p &= 1{\cdot}881357 \qquad \log. b &= 1{\cdot}996377 \end{array}$$

Hence, $p = 76{\cdot}095$, and $b = 99{\cdot}169$.

Note.—In adding the log. sin. and log. cos. respectively to the log. 125, we rejected 10 from the sum to correct for the 10 which had been added to all logarithms of the sines, cosines, &c. (§ 34).

CASE II.

§ **43.** *Given, a side, that is, either the perpendicular or the base, and the angle opposite that side, to find the other parts.*

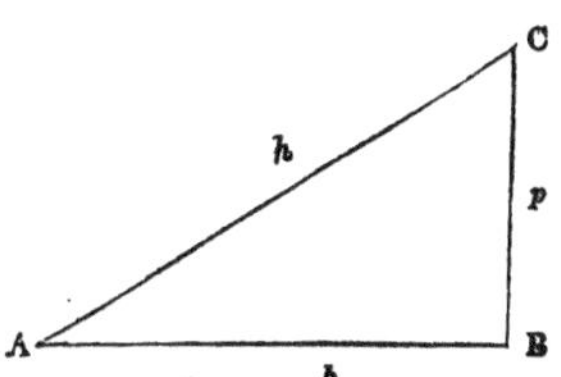


Suppose the given side to be $p$, and the given angle A. Then will the angle C be the complement of A.

By definitions (§ 8), we have

$$\sin. A = \frac{p}{h};\ \cot. A = \frac{b}{p},$$

which readily give

$$h = \frac{p}{\sin. A}. \quad . \quad . \quad . \quad . \quad (1.)$$

$$b = p \cot. A. \quad . \quad . \quad . \quad . \quad (2.)$$

By using logarithms, we have

$$\log. h = \log. p - \log. \sin. A. \quad . \quad (3.)$$

$$\log. b = \log. p + \log. \text{cotan}. A. \quad . \quad (4.)$$

For example, suppose the perpendicular to be 75, and the angle at A to be 39° 15′. Required the other parts.

$$p = 75\ ;\ A = 39^\circ\ 15'.$$

We find $C = 90^\circ - 39^\circ\ 15' = 50^\circ\ 45'$.

By Natural Numbers. (Table III.)

sin. 39° 15′ = 0·63271 ; cotan. 39° 15′ = 1·22394

```
0·63271 ) 75·00000 ( 118·53 = h.
          63271
          ------
          117290                 1·22394
           63271                      75
          ------                 -------
          540190                  611970
          506168                 856758
          ------                 --------
           340220                91·79550 = b.
           316355
           ------
            238650
            189813
            ------
             48837
```

By Logarithms.

log. 75 = 1·875061 . . . . . 1·875061
log. sin. 39° 15′ = 9·801201 ; log. cotan. 39° 15′ = 10·087760

log. h = 2·073860 log. b = 1·962821

Hence, $h = 118{\cdot}54$, nearly, and $b = 91{\cdot}795$, nearly.

Note.—In this example, before subtracting the log. of the sine of 39° 15′ from the log. of 75, it was necessary to increase this logarithm by 10, since the logarithm of the sine had been increased by the same quantity. In the second operation, after adding the log. of the cotangent to the log. of 75, we rejected 10 from the sum, for the reason assigned in the last note.

CASE III.

§ **44.** *Given, a side, and the angle adjacent that side, to find the other parts.*

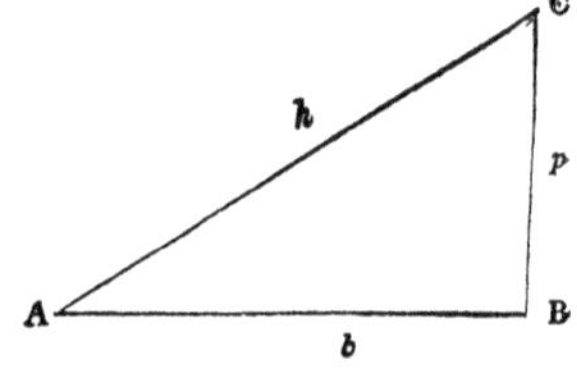


Suppose the given side to be $b$, and the given angle A. Then will the angle C be the complement of A.

By definitions (§ 8), we have

$$\cos. A = \frac{b}{h}; \ \text{tang.} A = \frac{p}{b},$$

which give

$$h = \frac{b}{\cos. A}. \quad . \quad . \quad . \quad . \qquad (1.)$$

$$p = b \ \text{tang.} A. \quad . \quad . \quad . \quad . \qquad (2.)$$

Using logarithms, we have

$$\log. h = \log. b - \log. \cos. A. \quad . \quad . \qquad (3.)$$

$$\log. p = \log. b + \log. \text{tang.} A. \quad . \quad . \qquad (4.)$$

For example, suppose the base to be 90·5, and the angle at A to be 50° 30′. Required the other parts.

$$b = 90{\cdot}5; \ A = 50° \, 30'.$$

We find $C = 90° - 50° \, 30' = 39° \, 30'$.

BY NATURAL NUMBERS. (TABLE III.)

cos. 50° 30′ = 0·63608 ; tang. 50° 30′ = 1·21310.

```
0·63608 ) 90·50000 ( 142·27 = h
          63608
          ------
          268920              1·2131
          254432                90·5
          ------             -------
           144880              60655
           127216            109179
           ------            -----------
            176640           109·78555 = p.
            127216
            ------
             494240
             445256
             ------
              48984
```

BY LOGARITHMS.

| | | | |
|---|---|---|---|
| log. 90·5 = | 1·956649 | . . . . . | 1·956649 |
| log. cos. 50° 30′ = | 9·803511 ; | log. tang. 50° 30′ = | 10·083896 |
| log. $h$ = | 2·153138 | log. $p$ = | 2·040545 |

Hence, $h = 142{\cdot}27$, and $p = 109{\cdot}78$.

### CASE IV.

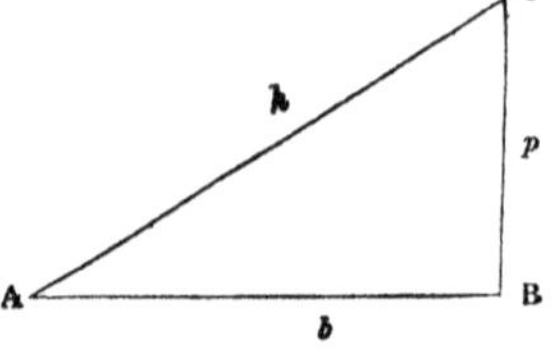


§ **45.** *Given, the hypotenuse and one of the sides, to find the other parts.*

Suppose the given side to be $b$, then (§ 8) we have

$$\sin. C = \cos. A = \frac{b}{h}, \quad . \quad . \quad . \quad . \qquad (1.)$$

or, using logarithms,

$$\log. \sin. C = \log. \cos. A = \log. b - \log. h. \quad . \qquad (2.)$$

Again, from the fundamental property of the right triangle, we have

$$b^2 + p^2 = h^2,$$

hence

$$p^2 = h^2 - b^2 = (h + b)\,(h - b),$$

$$p = \sqrt{(h+b)\,(h-b)}. \quad . \quad . \quad . \qquad (3.)$$

Using logarithms, we have

$$\log. p = \tfrac{1}{2}\,[\log. (h + b) + \log. (h - b)]. \quad . \qquad (4.)$$

For example, suppose the hypotenuse to be 112, and the base to be 97. Required the other parts.

$$h = 112 \,;\ b = 97.$$

112 ) 97·00000 ( 0·86607 = sin. C = cos. A.
896
———
740
672
———
680
672
———
800
784
———
16

Hence (by Table III.), C = 60° and A = 30°.
Again, $h + b = 209$, and $h - b = 15$.

*Extracting Square Root.*

| | | | |
|---|---|---|---|
| 209 | 5 | 3135 | (55·99 = $p$. |
| 15 | 105 | 25 | |
| 1045 | 1109 | 635 | |
| 209 | 11189 | 525 | |
| 3135 | | 11000 | |
| | | 9981 | |
| | | 101900 | |
| | | 100701 | |
| | | 199 | |

By Logarithms.

$$\begin{aligned} \log.\ 97 &= 1{\cdot}986772 \\ \log.\ 112 &= 2{\cdot}049218 \end{aligned}$$

$$9{\cdot}937554 = \log.\ \sin.\ C = \log.\ \cos.\ A.$$

Hence, $C = 60° \, 0' \, 19''$; $A = 29° \, 59' \, 41''$.

Again,

$$\begin{aligned} \log.\ 209 &= 2{\cdot}320146 \\ \log.\ 15 &= 1{\cdot}176091 \\ 2\,)\ &\ 3{\cdot}496237 \\ \log.\ p &= 1{\cdot}748118 \end{aligned}$$

Hence, $p = 55{\cdot}991$.

CASE V.

§ **46.** *Given, the two sides, to find the other parts.*

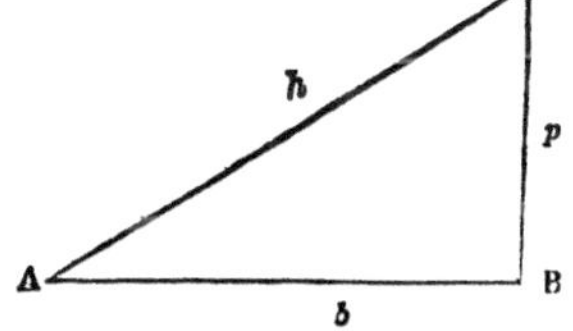


By § 8, we have

$$\text{tang. } A = \text{cotan. } C = \frac{p}{b}. \quad . \quad . \quad (1.)$$

or, using logarithms,

$$\log.\ \text{tang. } A = \log.\ \text{cotan. } C = \log. p - \log. b. \qquad (2).$$

The hypotenuse is immediately deduced from

$$h^2 = b^2 + p^2,$$

which gives

$$h = \sqrt{b^2 + p^2}. \quad . \quad . \quad . \qquad (3.)$$

Since this last formula is not well adapted to the use of logarithms, we will remark that after having found the angles, by means of equations (1) or (2), we may then determine the hypotenuse by Case II., equation (1), which gives

$$\text{log. } h = \text{log. } p - \text{log. sin. A.} \quad . \quad . \qquad (4.)$$

As an example, suppose the base to be 83·5 and the perpendicular to be 62·25. Required the other parts.

$$b = 83{\cdot}5\,;\ p = 62{\cdot}25.$$

BY NATURAL NUMBERS. (TABLE III.)

83·5 ) 62·25 ( 0·74550 = tang. A = cotan. C.
5845
3800
3340
4600
4175
4250
4175
750

Hence, A = 36° 42′
C = 53° 18′

83·5 = $b$
83·5
4175
2505
6680
6972·25 = $b^2$

62·25 = $p$
62·25
31125
12450
12450
37350
$p^2$ = 3875·0625
$b^2$ = 6972·25
10847·3125 = $b^2 + p^2$.

$$\sqrt{10847{\cdot}3125} = 104{\cdot}15 = h.$$

By Logarithms.

$$\log.\ 62{\cdot}25 = 1{\cdot}794139$$
$$\log.\ 83{\cdot}5 \ \ = 1{\cdot}921686$$

$$\log.\ \text{tang.}\ A = \log.\ \text{cotan.}\ C = 9{\cdot}872453$$

Hence, $A = 36^\circ\ 42'\ 18''$; $C = 53^\circ\ 17'\ 42''$.

Again,

$$\log.\ 62{\cdot}25 = 1{\cdot}794139$$
$$\log.\ \sin.\ 36^\circ\ 42'\ 18'' = 9{\cdot}776480$$

$$\log.\ h = 2{\cdot}017659$$

Hence, $h = 104{\cdot}15$.

## § 47. *Additional Examples of Right Triangles.*

1. Given, the hypotenuse, equal 365, and one of the acute angles, equal 33° 12′, to solve the triangle.

$$Ans. \begin{cases} \text{The other angle} = 56^\circ\ 48'. \\ \text{The sides} = \begin{cases} 199{\cdot}86. \\ 305{\cdot}42. \end{cases} \end{cases}$$

2. Given, one of the sides equal to 33·33, and the angle opposite this side = 83° 33′, to solve the triangle.

$$Ans. \begin{cases} \text{The other angle} = 6^\circ\ 27. \\ \text{The other side} \ \ = 3{\cdot}768. \\ \text{The hypotenuse} = 33{\cdot}542. \end{cases}$$

3. Given, one of the sides equal to 105·5, and the acute angle adjacent this side equal to 46° 3′, to solve the triangle.

$$Ans. \begin{cases} \text{The other angle} = 43^\circ\ 57'. \\ \text{The other side} \ \ = 109{\cdot}44. \\ \text{The hypotenuse} = 152{\cdot}01. \end{cases}$$

4. Given, the hypotenuse equal 111·1, and one of the sides equal 37·5, to solve the triangle.

$$Ans. \begin{cases} \text{The other side} = 104{\cdot}58. \\ \text{The angles} = \begin{cases} 19^\circ\ 43'\ 36''. \\ 70^\circ\ 16'\ 24''. \end{cases} \end{cases}$$

5. Given, the two sides equal 29·37 and 37·29, to solve the triangle.

$$Ans. \begin{cases} \text{The hypotenuse} = 47{\cdot}467. \\ \text{The angles} = \begin{cases} 38^\circ\ 13'\ 28''. \\ 51^\circ\ 46'\ 32''. \end{cases} \end{cases}$$

The above examples have all been wrought by the use of logarithms, which is, as a general thing, more simple than by the use of natural numbers only. It will be well, however, for the pupil to apply both methods, as was done in the examples which followed immediately after each case; by so doing, he will be the better prepared to comprehend the true nature of the trigonometrical values as well as their logarithmic values.

§ **48.** Hereafter, when we use the word *sine*, *cosine*, *tangent*, or *cotangent*, in connection with logarithmic calculations, we wish to have it understood as meaning the logarithms of those values, increased by 10, as given in Table II. In the usual analytical investigations of trigonometrical formulas, we make use of their real or natural values, as given in Table III.

---

# CHAPTER IV.

## SOLUTION OF OBLIQUE TRIANGLES.

§ **49.** All the different cases of oblique triangles may be solved by the aid of the following theorems:

### THEOREM I.

*The sides of any plane triangle are to each other as the sines of their opposite angles. And conversely, the sines of the angles of any plane triangle are to each other as their opposite sides.*

In the triangle ABC denote the sides opposite the angles A, B, C, respectively by the letters $a$, $b$, $c$. Draw CD perpendicular to AB, and we shall have (§ 8)

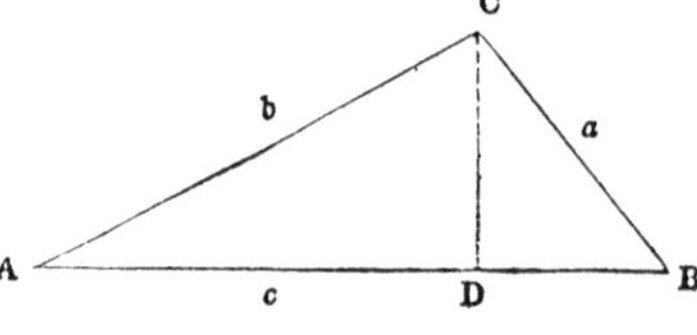


$$\sin. A = \frac{CD}{b},\ \sin. B = \frac{CD}{a};$$

dividing the first equation by the second, we have

$$\frac{\sin. A}{\sin. B} = \frac{a}{b}, \quad . \quad . \quad . \quad . \quad (1.)$$

which gives $a : b :: \sin. A : \sin. B$. . . (2.)

In the same way it may be shown that

$$a : c :: \sin. A : \sin. C. \quad . \quad . \quad (3.)$$

## THEOREM II.

*In any plane triangle, the sum of any two sides is to their difference as the tangent of half the sum of the opposite angles is to the tangent of half their difference.*

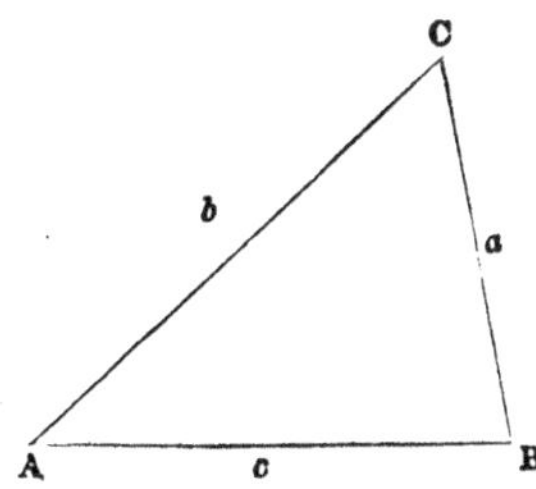


By Theorem I., we have

$$b : c :: \sin. B : \sin. C;$$

hence, by composition and division, we have

$$b+c : b-c :: \sin. B + \sin. C : \sin. B - \sin. C,$$

which gives

$$\frac{b+c}{b-c} = \frac{\sin. B + \sin. C}{\sin. B - \sin. C}. \quad . \quad . \quad . \quad (1.)$$

By equation (9), § 16, we have

$$\frac{\sin. B + \sin. C}{\sin. B - \sin. C} = \frac{\text{tang.} \frac{1}{2}(B+C)}{\text{tang.} \frac{1}{2}(B-C)}.$$

Hence, we have

$$\frac{b+c}{b-c} = \frac{\text{tang.} \frac{1}{2}(B+C)}{\text{tang.} \frac{1}{2}(B-C)}, \quad . \quad . \quad . \quad (2.)$$

or,

$$b+c : b-c :: \text{tang.} \tfrac{1}{2}(B+C) : \text{tang.} \tfrac{1}{2}(B-C). \quad (3.)$$

## THEOREM III.

*If from an angle of a plane triangle a perpendicular be drawn so as to meet the opposite side or base, the whole base will be to the sum of the other two sides as the difference of those sides is to the difference of the segments of the base.*

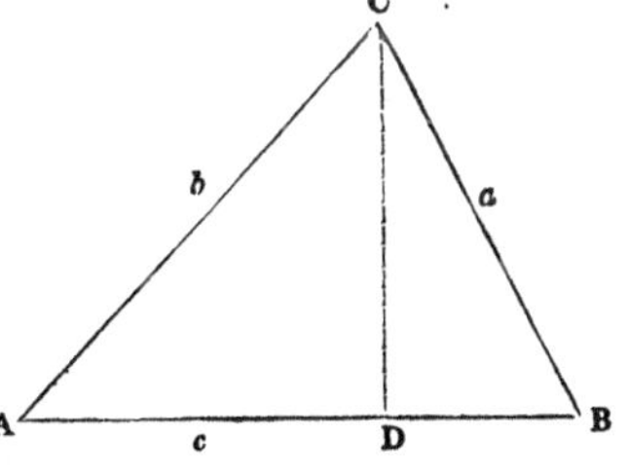


Drawing CD perpendicular to AB, we have

$$b^2 = AD^2 + CD^2 \quad . \quad . \quad . \quad (1.)$$

$$a^2 = BD^2 + CD^2. \quad . \quad . \quad . \quad (2.)$$

Subtracting (2) from (1), we have

$$b^2 - a^2 = \mathrm{AD}^2 - \mathrm{BD}^2,$$

or,

$$(b + a)(b - a) = (\mathrm{AD} + \mathrm{BD})(\mathrm{AD} - \mathrm{BD}), \quad . \qquad (3.)$$

which, converted into a proportion, will give, observing that $\mathrm{AD} + \mathrm{BD} = c$,

$$c : b + a :: b - a : \mathrm{AD} - \mathrm{BD}. \quad . \quad . \qquad (4.)$$

If the triangle has an obtuse angle, then the perpendicular should be drawn from this obtuse angle, otherwise it will not meet the opposite side. The theorem might be so modified as to embrace this case; but as it is used only in the solution of a triangle when the three sides are given, such modification is unnecessary.

§ **50.** The solution of all oblique triangles may be included in four cases, as follows:

CASE I.

*Given, a side and two angles, to find the other parts.*

Since the sum of the three angles of any plane triangle is equal to 180°, the third angle may be found by subtracting the sum of the two given angles from 180°. Having all the angles, the two remaining sides may be found by Theorem I. Thus, if we suppose the side $c$ to be given, we shall have

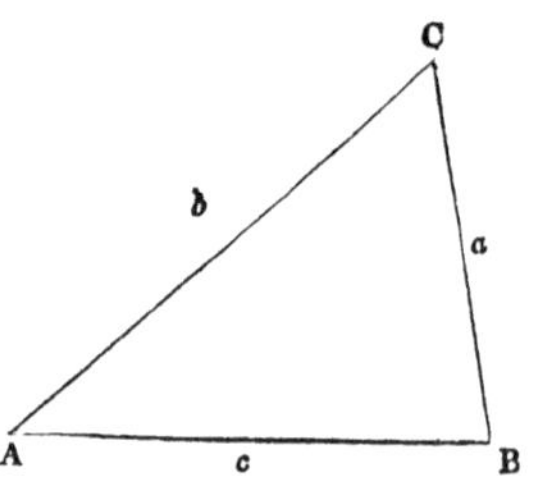


$$\sin. \mathrm{C} : \sin. \mathrm{A} :: c : a,$$
$$\sin. \mathrm{C} : \sin. \mathrm{B} :: c : b,$$

which give

$$a = \frac{c \sin. \mathrm{A}}{\sin. \mathrm{C}}. \quad . \quad . \quad . \quad . \qquad (1.)$$

$$b = \frac{c \sin. \mathrm{B}}{\sin. \mathrm{C}}. \quad . \quad . \quad . \quad . \qquad (2.)$$

Using logarithms, we have

$$\log. a = \log. c + \log. \sin. \mathrm{A} - \log. \sin. \mathrm{C}. \quad . \qquad (3.)$$
$$\log. b = \log. c + \log. \sin. \mathrm{B} - \log. \sin. \mathrm{C}. \quad . \qquad (4.)$$

By using arithmetical complements (§ 32), we shall have

$$\text{log. } a = \text{ar. co. log. sin. C} + \text{log. sin. A} + \text{log. } c \quad . \quad (5.)$$
$$\text{log. } b = \text{ar. co. log. sin. C} + \text{log. sin. B} + \text{log. } c \quad . \quad (6.)$$

As an example, suppose the angle A = 30° 20′, the angle B = 50° 10′, and consequently the angle C = 99° 30′. If the side $c$ is 93·37, what will be the lengths of the other sides?

Using equations (5) and (6), we have

| | | |
|---|---|---|
| ar. co. sin. 99° 30′ (80° 30′) = 0·005997 | . . | 0·005997 |
| sin. 30° 20′ = 9·703317; | sin. 50° 10′ = | 9·885311 |
| log. 93·37 = 1·970207 | . . | 1·970207 |
| log. $a$ = 1·679521 | log. $b$ = | 1·861515 |

Hence, $a = 47·81$; $b = 72·697$.

NOTE.—Since the angle C = 99° 30′ is an obtuse angle, we subtract it from 180, and obtain its supplement 80° 30′, the sine of which is 9·994003. Now subtracting this from 10, we have 0·005997 for the arithmetical complement of the logarithmic sine of 99° 30′.

The arithmetical complement may be readily taken from the table, by beginning at the left hand, and subtracting each figure from 9, except the last significant figure on the right, which must be subtracted from 10. (§ 32.)

CASE II.

*Given, two sides and an angle opposite one of them, to find the other parts.*

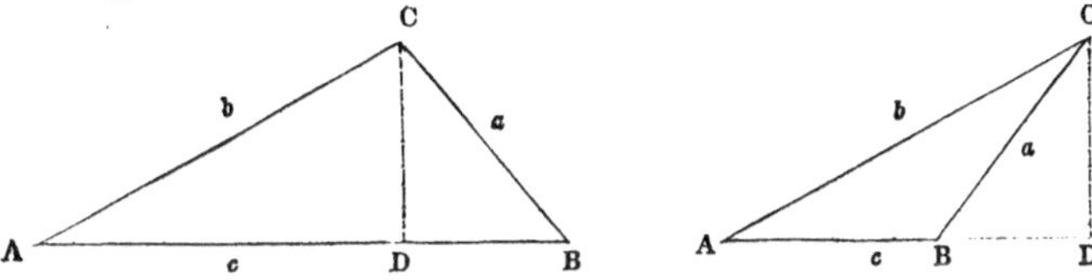


Suppose the sides $a$ and $b$ to be given, and the angle A opposite the side $a$.

By Theorem I., we have

$$a : b :: \text{sin. A} : \text{sin. B},$$

or,

$$\text{sin. B} = \frac{b \text{ sin. A}}{a}. \quad . \quad . \quad . \quad (1.)$$

Using logarithms, we have

$$\text{log. sin. B} = \text{ar. co. log. } a + \text{log. } b + \text{log. sin. A}. \quad (2.)$$

Having thus found the second angle, the third angle at C may be obtained by subtracting the sum of these two from 180°.

Then by Theorem I., we have

$$\sin. A : \sin. C :: a : c,$$

or, $$c = \frac{a \sin. C}{\sin. A}, \quad . \quad . \quad . \quad . \quad (3.)$$

which by logarithms, becomes

$$\log. c = \text{ar. co. log. sin. A} + \log. \sin. C + \log. a. \quad (4.)$$

SCHOLIUM. In equation (1), the numerator, $b$ sin. A, is the perpendicular CD; now if $a$ is less than this perpendicular, the value of the fraction $\frac{b \sin. A}{a}$, which gives the sine of B, will exceed a unit, which is *impossible*. But while $a$ is greater than this perpendicular, and does not at the same time exceed $b$, there will be two solutions, as represented in the two diagrams. When there are two solutions, the angle at B in the second triangle will be the supplement of the angle at B in the first.

As an example, suppose the side $a = 75{\cdot}5$ and the side $b = 98{\cdot}5$; and suppose the angle at A to be 37° 37′, required the other parts.

By equation (2), we have

$$\begin{aligned} \text{ar. co. log. } 75{\cdot}5 &= 8{\cdot}122053 \\ \log. 98{\cdot}5 &= 1{\cdot}993436 \\ \log. \sin. 37° 37 &= 9{\cdot}785597 \\ \hline \log. \sin. B &= 9{\cdot}901086 \end{aligned}$$

Hence, B = 52° 46′ 48″, consequently the angle at C = 89° 36′ 12″.

Now, by equation (4), we have

$$\begin{aligned} \text{ar. co. log. sin. } 37° 37' &= 0{\cdot}214403 \\ \log. \sin. 89° 36' 12'' &= 9{\cdot}999989 \\ \log. 75{\cdot}5 &= 1{\cdot}877947 \\ \hline \log. c &= 2{\cdot}092339 \\ \text{and } c &= 123{\cdot}69. \end{aligned}$$

The above solution corresponds with the first figure; if we take 127° 13′ 12″ for the angle at B, which is the supplement of 52° 46′ 48″, we shall then find 15° 9′ 48″ for the angle at C. Then, by equation (4), we have

$$\begin{aligned} \text{ar. co. log. sin. } 37° \, 37' &= 0{\cdot}214403 \\ \text{log. sin. } 15° \, 9' \, 48'' &= 9{\cdot}417590 \\ \text{log. } 75{\cdot}5 &= 1{\cdot}877947 \\ \hline \text{log. } c &= 1{\cdot}509940 \\ \text{and } c &= 32{\cdot}355. \end{aligned}$$

This second solution corresponds with the second figure.

CASE III.

*Given, two sides and the included angle, to find the other parts.*

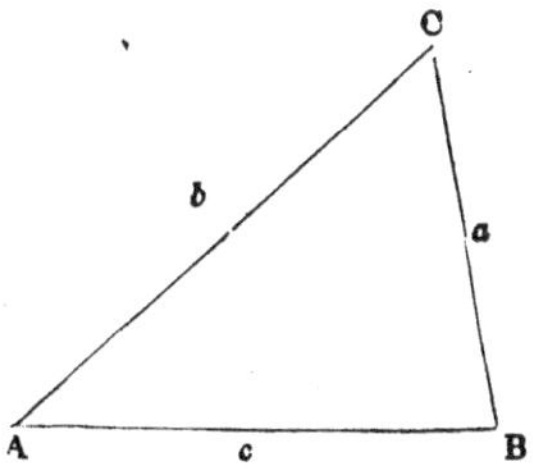

Suppose the sides $b$ and $c$, with the included angle A, to be given.

By subtracting the angle A from 180°, which is the sum of the three angles, we have

$$B + C = 180° - A.$$

and

$$\tfrac{1}{2}(B + C) = 90° - \tfrac{1}{2} A = \text{complement of } \tfrac{1}{2} A.$$

Then by Theorem II., we have

$$b + c : b - c :: \text{cotan. } \tfrac{1}{2} A : \text{tang. } \tfrac{1}{2}(B - C),$$

which gives

$$\text{tang. } \tfrac{1}{2}(B - C) = \frac{(b - c) \text{ cotan. } \tfrac{1}{2} A}{b + c}. \quad . \quad (1.)$$

Using logarithms, we have

$$\text{log. tang. } \tfrac{1}{2}(B - C) = \text{ar. co. log. } (b + c) + \text{log. } (b - c) + \text{log. cotan. } \tfrac{1}{2} A. \quad . \quad . \quad . \quad . \quad (2.)$$

Having thus found half the difference of the angles B and C, we add it to half their sum, and thus obtain the greater angle, which must be opposite the greater side. The smaller angle is found by subtracting half their difference from their half sum.

Again, having found all the angles, we have, by Theorem I.,

$$\text{sin. } B : \text{sin. } A :: b : a,$$

or,

$$a = \frac{b \text{ sin. } A}{\text{sin. } B}. \quad . \quad . \quad . \quad . \quad . \quad (3.)$$

Using logarithms, we have

$$\log. a = \text{ar. co. log. sin. B} + \text{log. sin. A} + \log. b. \qquad (4.)$$

As an example, suppose the side $b = 50{\cdot}24$, and the side $c = 43{\cdot}25$, and suppose the angle A to be 40° 15′, required the other parts.

$$A = 40^\circ\ 15' \ ; \ b = 50{\cdot}24 \ ; \ c = 43{\cdot}25,$$

consequently,

$$b + c = 93{\cdot}49 \ ; \ b - c = 6{\cdot}99 \ ; \ \tfrac{1}{2}A = 20^\circ\ 7'\ 30''.$$

Equation (2) gives

$$\begin{aligned} \text{ar. co. log. } 93{\cdot}49 &= \phantom{1}8{\cdot}029235 \\ \text{log. } 6{\cdot}99 &= \phantom{1}0{\cdot}844477 \\ \text{log. cotan. } 20^\circ\ 7'\ 30'' &= 10{\cdot}435993 \\ \hline \text{log. tang. } \tfrac{1}{2}(B-C) &= \phantom{1}9{\cdot}309705 \end{aligned}$$

$$\begin{aligned} \text{And } \tfrac{1}{2}(B-C) &= 11^\circ\ 31'\ 55'' \\ \tfrac{1}{2}(B+C) &= 69^\circ\ 52'\ 30'' \\ \hline B &= 81^\circ\ 24'\ 25'' = \text{sum.} \\ C &= 58^\circ\ 20'\ 35'' = \text{difference.} \end{aligned}$$

By equation (4), we have

$$\begin{aligned} \text{ar. co. log. sin. } 58^\circ\ 20'\ 35'' &= 0{\cdot}069965 \\ \text{log. sin. } 40^\circ\ 15' &= 9{\cdot}810316 \\ \text{log. } 43{\cdot}25 &= 1{\cdot}635986 \\ \hline \log. a &= 1{\cdot}516267 \end{aligned}$$

$$\text{And } a = 32{\cdot}829.$$

CASE IV.

*Given, the three sides, to find the angles.*

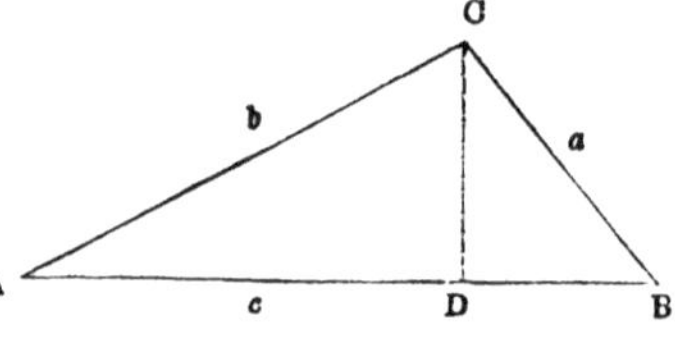


If we conceive CD to be drawn perpendicular to the side $c$, we shall have, by Theorem III.,

$$c : b + a :: b - a : AD - BD,$$

$$AD - BD = \frac{(b+a)\,(b-a)}{c}. \qquad (1.)$$

Using logarithms, we have

$$\log. (AD - BD) = \text{ar. co. log. } c + \log. (b + a) + \log. (b - a). \qquad (2.)$$

This gives half the difference of the segments AD and BD; if we add this half difference to the half sum, that is, add it to half of $c$, we shall obtain AD, the greater segment. The half difference subtracted from the half of $c$ will give BD, the shorter segment.

Then solving the two right triangles ADC and BDC, by Case IV. of Right Triangles, we have

$$\cos. A = \frac{AD}{b}; \quad \cos. B = \frac{BD}{a}.$$

Using logarithms,

$$\log. \cos. A = \log. AD - \log. b, \quad . \quad . \quad (3.)$$
$$\log. \cos. B = \log. BD - \log. a. \quad . \quad . \quad (4.)$$

Having found the angles A and B, subtract their sum from 180° for the third angle C.

For example, suppose we have $a = 50{\cdot}25$; $b = 60{\cdot}5$, and $c = 68{\cdot}4$. Required the angles.

$$b + a = 110{\cdot}75; \quad b - a = 10{\cdot}25.$$

By equation (2), we have

$$\begin{aligned} \text{ar. co. log. } 68{\cdot}4 &= 8{\cdot}164944 \\ \log. 110{\cdot}75 &= 2{\cdot}044344 \\ \log. 10{\cdot}25 &= 1{\cdot}010724 \\ \hline \log. (AD - BD) &= 1{\cdot}220012 \\ AD - BD &= 16{\cdot}596 \end{aligned}$$

$$\tfrac{1}{2} c + \tfrac{1}{2} (AD - BD) = 42{\cdot}498 = AD.$$
$$\tfrac{1}{2} c - \tfrac{1}{2} (AD - BD) = 25{\cdot}902 = BD.$$

By equations (3) and (4), we have

$$\log. 42{\cdot}498 = 1{\cdot}628369; \quad \log. 25{\cdot}902 = 1{\cdot}413333$$
$$\log. 60{\cdot}5 = 1{\cdot}781755; \quad \log. 50{\cdot}25 = 1{\cdot}701136$$
$$\log. \cos. A = 9{\cdot}846614; \quad \log. \cos. B = 9{\cdot}712197$$

Hence, $A = 45° \, 22' \, 35''$; $B = 58° \, 58' \, 18''$.

Consequently, $C = 75° \, 39' \, 7''$.

Perhaps a better method for this case would be as fol lows:—

If to equation (1), which is

$$AD - BD = \frac{b^2 - a^2}{c},$$

we add

$$AD + BD = c,$$

we shall have

$$2\ AD = \frac{b^2 + c^2 - a^2}{c};$$

hence,

$$\cos.\ A = \frac{AD}{b} = \frac{b^2 + c^2 - a^2}{2\,bc}. \quad . \quad . \quad (5.)$$

Adding 1 to each member of (5), we have

$$1 + \cos.\ A = 1 + \frac{b^2 + c^2 - a^2}{2\,bc} = \frac{b^2 + 2\,bc + c^2 - a^2}{2\,bc},$$

or,

$$1 + \cos.\ A = \frac{(b + c + a)\,(b + c - a)}{2\,bc}. \quad . \quad (6.)$$

Denoting half the sum of the three sides by $s$, we have

$$b + c + a = 2\,s;\ b + c - a = 2\,s - 2\,a = 2\,(s - a).$$

Hence equation (6) will become

$$1 + \cos.\ A = \frac{2\,s\,(s - a)}{bc}. \quad . \quad . \quad (7.)$$

By equation (8), § 15, we have

$$1 + \cos.\ A = 2\,(\cos.\ \tfrac{1}{2}\ A)^2.$$

Therefore equation (7) will become

$$2\,(\cos.\ \tfrac{1}{2}\ A)^2 = \frac{2\,s\,(s - a)}{bc},$$

or,

$$(\cos.\ \tfrac{1}{2}\ A)^2 = \frac{s\,(s - a)}{bc};$$

consequently,

$$\cos.\ \tfrac{1}{2}\ A = \sqrt{\frac{s\,(s - a)}{bc}}. \quad . \quad . \quad (8.)$$

In a similar manner, we find

$$\cos.\ \tfrac{1}{2}\ B = \sqrt{\frac{s\,(s - b)}{ac}}, \quad . \quad . \quad (9.)$$

$$\cos.\ \tfrac{1}{2}\ C = \sqrt{\frac{s\,(s - c)}{ab}}. \quad . \quad . \quad (10.)$$

By using logarithms, equations (8), (9), and (10) become

$$\left.\begin{array}{l}\log.\cos.\tfrac{1}{2}A=\tfrac{1}{2}[\text{ar. co. log.}\,b+\text{ar. co. log.}\,c+\log.\,s+\log.\,(s-a)]\\ \log.\cos.\tfrac{1}{2}B=\tfrac{1}{2}[\text{ar. co. log.}\,a+\text{ar. co. log.}\,c+\log.\,s+\log.\,(s-b)]\\ \log.\cos.\tfrac{1}{2}C=\tfrac{1}{2}[\text{ar. co. log.}\,a+\text{ar. co. log.}\,b+\log.\,s+\log.\,(s-c)]\end{array}\right\}(\text{A})$$

Applying equations (A) to the case already given, we have

$$a=50{\cdot}25\,;\ b=60{\cdot}5\,;\ c=68{\cdot}4\,;$$

consequently,

$$\tfrac{1}{2}\,(a+b+c)=s=89{\cdot}575,\text{ and }s-a=39{\cdot}325\,;\ s-b=29{\cdot}075\,;\ s-c=21{\cdot}175.$$

| | | | |
|---|---|---|---|
| ar. co. log. $b$ = | 8·218245 ; | ar. co. log. $a$ = | 8·298864 |
| ar. co. log. $c$ = | 8·164944 | ar. co. log. $c$ = | 8·164944 |
| log. $s$ = | 1·952187 | log. $s$ = | 1·952187 |
| log. $(s-a)$ = | 1·594669 | log. $(s-b)$ = | 1·463520 |
| | 2) 19·930045 | | 2) 19·879515 |
| log. cos. $\frac{1}{2}$ A = | 9·965022 | log. cos. $\frac{1}{2}$ B = | 9·939757 |

| | |
|---|---|
| ar. co. log. $a$ = | 8·298864 |
| ar. co. log. $b$ = | 8·218245 |
| log. $s$ = | 1·952187 |
| log. $(s-c)$ = | 1·325823 |
| | 2) 19·795119 |
| log. cos. $\frac{1}{2}$ C = | 9·897559 |

Hence $\frac{1}{2}$ A = 22° 41′ 17″ ; $\frac{1}{2}$ B = 29° 29′ 9″; $\frac{1}{2}$ C = 37° 49′ 34″, and A = 45° 22′ 34″ ; B = 58° 58′ 18″ ; C = 75° 39′ 8″.

When the sides of a triangle are expressed in small numbers, equation (5) may be advantageously employed: this equation, when the letters are properly permuted so as to apply to each angle, gives

$$\left.\begin{array}{l}\text{Nat. cos. A}=\dfrac{b^2+c^2-a^2}{2\,bc},\\[2ex] \text{Nat. cos. B}=\dfrac{a^2+c^2-b^2}{2\,ac},\\[2ex] \text{Nat. cos. C}=\dfrac{a^2+b^2-c^2}{2\,ab},\end{array}\right\}\quad\ldots\quad(\text{B.})$$

As an example, suppose the sides to be 5, 7, and 9. Required the angles.

$$a = 5\,;\ b = 7\,;\ c = 9.$$

$$\text{Nat. cos. A} = \frac{49 + 81 - 25}{2 \times 7 \times 9} = \frac{105}{126},$$

$$\text{Nat. cos. B} = \frac{25 + 81 - 49}{2 \times 5 \times 9} = \frac{57}{90},$$

$$\text{Nat. cos. C} = \frac{25 + 49 - 81}{2 \times 5 \times 7} = \frac{-7}{70}.$$

$$\begin{array}{ll} \log. 105 = 2{\cdot}021189 & \log. 57 = 1{\cdot}755875 \\ \log. 126 = \underline{2{\cdot}100371} & \log. 90 = \underline{1{\cdot}954243} \\ \cos. A = 9{\cdot}920818 & \cos. B = 9{\cdot}801632 \end{array}$$

$$\begin{array}{l} \log. -7 = 0{\cdot}845098n \\ \log.\ \ 70 = \underline{1{\cdot}845098} \\ \cos. C = 9{\cdot}000000n \end{array}$$

Hence, $A = 33°\ 33'\ 27''$; $B = 50°\ 42'\ 13''$; $C = 95°\ 44'\ 21''$.

NOTE.—In taking the logarithm of $-7$, we took it as though it had been $+7$, and at the right annexed the letter $n$ to indicate that the number is negative. So in the value of cos. C we write the $n$, thus indicating that the natural cosine of this angle is negative. We then seek in the tables for the angle as though its cosine was positive, and find 84° 15′ 39″, which taken from 180° gives the obtuse angle 95° 44′ 21″ for the true value of C.

## § 51. *Additional Examples of Oblique Triangles.*

1. Given, two angles of a triangle equal 47° 13′ and 63° 36′, and the side opposite the first angle equal 27·5, to solve the triangle.

*Ans.* The other angle = 69° 11′. The other sides = 33·562, 35·024.

2. Given, two sides of a triangle equal 47·13 and 63·36, and the angle opposite the first side equal 27° 50′, to solve the triangle.

*First Ans.* The other side = 92·719. The other angles = 38° 52′ 47″, 113° 17′ 13″.

*Second Ans.* The other side = 19·341. The other angles = 141° 7′ 13″, 11° 2′ 47″.

3. Given, two sides of a triangle equal 49·5 and 101·5, and the angle opposite the first side equal 30° 25′, to solve the triangle. *Ans.* The question is impossible.

4. Given, two sides of a triangle equal 630 and 800, and the angle opposite the first side equal 100°, to solve the triangle.
*Ans.* The question is impossible.

5. Given, two sides of a triangle, 77·5 and 90·25, and the included angle equal 83° 38′, to solve the triangle.

*Ans.* The other side = 112·25. The other angles = { 43° 19′ 38″. 53° 2′ 22″.

6. Given, the three sides of a triangle, 40, 50, and 60, to find the angles.

*Ans.* The angles = { 41° 24′ 35″. 55° 46′ 16″. 82° 49′ 9″.

7. Given, two sides of a triangle, 100 and 90, and the angle opposite the first side, 111° 15′, to solve the triangle.

*Ans.* The other side = 21·823. The other angles = { 11° 44′ 7″. 57° 0′ 53″.

8. In an obtuse-angled triangle one of the acute angles is 29° 15′, the other acute angle is 2° 45′; the side opposite the first angle is 6·3. What are the other parts?

*Ans.* The obtuse angle = 148°. The other sides = { 0·619. 6·832.

9. Given, two sides of a triangle, 332·21, 237·61, and their contained angle equal to 72° 29′ 48″, to find the other parts.

*Ans.* The other angles = { 40° 59′ 35″. 66° 30′ 37″. The other side = 345·46.

10. The three sides of a triangle are 10, 15, and 20. What are the angles?

*Ans.* { 28° 57′ 18″. 46° 34′ 3″. 104° 28′ 39″.

11. The three sides of a triangle are 123·48, 135·61, 140·91. What are the angles?

*Ans.* { 53° 0′ 10″. 61° 17′ 50″. 65° 42′ 0.

12. Two sides of a triangle are 59·34 and 12·13, and their included angle is 150° 38′. What are the other parts?

$$Ans. \begin{cases} \text{Angles} = \begin{cases} 24° \ 30' \ 12''. \\ 4° \ 51' \ 48''. \end{cases} \\ \text{Side} = 70·165. \end{cases}$$

---

# CHAPTER V.

## SPHERICAL TRIGONOMETRY.

§ **52.** Spherical Trigonometry treats of the methods of computing the unknown parts of a spherical triangle, when certain parts are given. But before proceeding to the investigation of the relation of the different parts of a spherical triangle, we will give some

### GENERAL TRIGONOMETRIC FORMULAS.

§ **53.** Under § 15 and § 16 we have already given several general and useful formulas which were needed in Plane Trigonometry. We now propose to continue these formulas to a greater extent, giving such additional ones as are needed in Spherical Trigonometry, as well as others which will be found convenient in the reduction and simplification of trigonometric expressions in general.

Dividing (10) by (11), § 15, we have

$$\frac{\sin. a}{\cos. a} = \tan. a = \frac{\sqrt{1 - \cos. 2a}}{\sqrt{1 + \cos. 2a}}. \quad . \quad . \quad (1.)$$

Multiplying both the numerator and denominator of the fraction constituting the right-hand member of (1) by the numerator, we have

$$\tan. a = \frac{1 - \cos. 2a}{\sqrt{1 - \cos.^2 2a}} = \frac{1 - \cos. 2a}{\sin. 2a}. \quad . \quad . \quad (2.)$$

If we multiply both the numerator and denominator of the right-hand member of (1) by the denominator, we shall obtain

$$\tan. a = \frac{\sqrt{1 - \cos.^2 2a}}{1 + \cos. 2a} = \frac{\sin. 2a}{1 + \cos. 2a}. \quad . \quad . \quad (3.)$$

Taking the reciprocals of (2) and (3), we have

$$\cot. a = \frac{\sin. 2a}{1 - \cos. 2a} = \frac{1 + \cos. 2a}{\sin. 2a}. \quad . \quad . \quad (4.)$$

We have already, § 15, deduced

$$\tan. (a + b) = \frac{\tan. a + \tan. b}{1 - \tan. a \tan. b}. \quad . \quad . \quad (5.)$$

By a similar process, we find

$$\tan. (a - b) = \frac{\tan. a - \tan. b}{1 + \tan. a \tan. b}. \quad . \quad . \quad (6.)$$

Reciprocating, we have

$$\cot. (a + b) = \frac{1 - \tan. a \tan. b}{\tan. a + \tan. b}. \quad . \quad . \quad (7.)$$

$$\cot. (a - b) = \frac{1 + \tan. a \tan. b}{\tan. a - \tan. b}. \quad . \quad . \quad (8.)$$

If we take $b = a$ in (5) and (7), we shall have

$$\tan. 2a = \frac{2 \tan. a}{1 - \tan.^2 a}; \quad . \quad . \quad (9.)$$

$$\cot. 2a = \frac{1 - \tan.^2 a}{2 \tan. a}. \quad . \quad . \quad (10.)$$

If we take $a = \frac{1}{2}(p + q)$, equation (5) of § 15 will become

$$\sin. (p + q) = 2 \sin. \tfrac{1}{2}(p + q) \cos. \tfrac{1}{2}(p + q). \quad (11.)$$

Using this in connection with (5), (6), (7), and (8) of § 16, we readily obtain, by division and reduction, as follows:

$$\frac{\sin. p + \sin. q}{\sin. p - \sin. q} = \frac{\tan. \frac{1}{2}(p + q)}{\tan. \frac{1}{2}(p - q)}; \quad . \quad . \quad (12.)$$

$$\frac{\cos. p + \cos. q}{\cos. q - \cos. p} = \frac{\cot. \frac{1}{2}(p + q)}{\tan. \frac{1}{2}(p - q)}; \quad . \quad . \quad (13.)$$

$$\frac{\sin. p + \sin. q}{\sin. (p + q)} = \frac{\cos. \frac{1}{2}(p - q)}{\cos. \frac{1}{2}(p + q)}; \quad . \quad . \quad (14.)$$

$$\frac{\sin. p - \sin. q}{\sin. (p + q)} = \frac{\sin. \frac{1}{2}(p - q)}{\sin. \frac{1}{2}(p + q)}; \quad . \quad . \quad (15.)$$

$$\frac{\sin. p + \sin. q}{\cos. p + \cos. q} = \tan. \tfrac{1}{2}(p + q); \quad . \quad . \quad (16.)$$

$$\frac{\sin. p + \sin. q}{\cos. q - \cos. p} = \cot. \tfrac{1}{2}(p - q); \quad . \quad . \quad (17.)$$

$$\frac{\sin. p - \sin. q}{\cos. p + \cos. q} = \tan. \tfrac{1}{2}(p - q); \quad . \quad . \quad (18.)$$

$$\frac{\sin. p - \sin. q}{\cos. q - \cos. p} = \cot. \tfrac{1}{2}(p + q). \quad . \quad . \quad (19.)$$

The following are readily obtained, and are of frequent use:

$$\frac{\sin. (p \pm q)}{\sin. p \sin. q} = \cot. q \pm \cot. p; \quad . \quad . \quad (20.)$$

$$\frac{\cos. (p \pm q)}{\sin. p \sin. q} = \cot. p \cot. q \mp 1; \quad . \quad . \quad (21.)$$

$$\frac{\sin. (p \pm q)}{\cos. p \cos. q} = \tan. p \pm \tan. q; \quad . \quad . \quad (22.)$$

$$\frac{\cos. (p \pm q)}{\cos. p \cos. q} = 1 \mp \tan. p \tan. q; \quad . \quad . \quad (23.)$$

$$\frac{\sin. (p \pm q)}{\sin. p \cos. q} = 1 \pm \cot. p \tan. q; \quad . \quad . \quad (24.)$$

$$\frac{\cos. (p \pm q)}{\sin. p \cos. q} = \cot. p \mp \tan. q; \quad . \quad (25.)$$

§ **54.** We now proceed to investigate the relation of the several parts of a spherical triangle. We shall confine our attention to such triangles only as have been treated of in Book VIII. of Geometry, namely, those whose sides and angles are each less than 180°.

§ **55.** *To express the cosine of an angle of a spherical triangle in terms of the sines and cosines of the sides.*

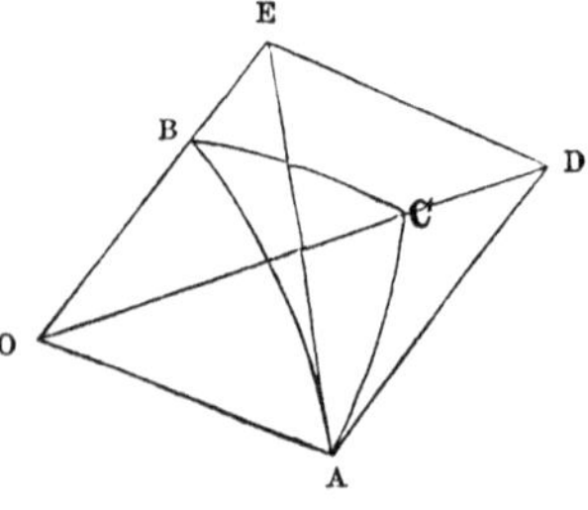


Let ABC be a spherical triangle described upon the surface of a sphere whose centre is at O.

We will denote the angles of this triangle by the large letters A, B, C, and the sides opposite them by the corresponding small letters $a$, $b$, $c$.

Draw AD tangent to the arc AC, and AE tangent to AB.

Then will the spherical angle A be equal to the angle DAE. (Geom., B. VIII., T. V.)

Draw the radius OC and produce it to meet AD in D; also, draw the radius OB and produce it to meet AE in E. The radius OA being drawn, will be perpendicular to AD, AE.

Now, by Plane Trigonometry, equation (5), Case IV., § 50, the triangle DAE gives

$$DE^2 = AD^2 + AE^2 - 2AD \times AE \times \cos. DAE. \qquad (1.)$$

The triangle DOE, in like manner, gives

$$DE^2 = OD^2 + OE^2 - 2OD \times OE \times \cos. DOE. \qquad (2.)$$

Equating the right-hand members of (1) and (2), and dividing each term by the square of the radius of the sphere, we have

$$\left.\begin{aligned} &\frac{AD^2}{OA^2} + \frac{AE^2}{OA^2} - 2 \times \frac{AD}{OA} \times \frac{AE}{OA} \times \cos. DAE \\ = &\frac{OD^2}{OA^2} + \frac{OE^2}{OA^2} - 2 \times \frac{OD}{OA} \times \frac{OE}{OA} \times \cos. DOE. \end{aligned}\right\} \quad . \quad (3.)$$

Now, since the triangles OAD, OAE are right-angled at A (Geom., B. II., T. V., C. I.), we have [(A), § 8]

$$\frac{AD}{OA} = \tan. AOD = \tan. b; \quad \frac{AE}{OA} = \tan. AOE = \tan. c;$$

$$\frac{OD}{OA} = \sec. AOD = \sec. b; \quad \frac{OE}{OA} = \sec. AOE = \sec. c.$$

Hence (3) gives, by observing that the angle DAE = A, and that the angle DOE = $a$,

$$\left.\begin{aligned} &\tan.^2 b + \tan.^2 c - 2 \tan. b \tan. c \cos. A \\ = &\sec.^2 b + \sec.^2 c - 2 \sec. b \sec. c \cos. a. \end{aligned}\right\} \quad . \quad (4.)$$

Transposing, and observing that $\sec.^2 b - \tan.^2 b$ and $\sec.^2 c - \tan.^2 c$ are each equal to 1, we have

$$2 + 2 \tan. b \tan. c \cos. A = 2 \sec. b \sec. c \cos. a. \qquad (5.)$$

Dividing by 2, and substituting for the tangents and secants their values in terms of sines and cosines [(B), § 8], we have

$$1 + \frac{\sin. b}{\cos. b} \times \frac{\sin. c}{\cos. c} \times \cos. A = \frac{1}{\cos. b} \times \frac{1}{\cos. c} \times \cos. a,$$

which, cleared of fractions, becomes

$$\cos. b \cos. c + \sin. b \sin. c \cos. A = \cos. a. \quad . \quad (6.)$$

Hence,

$$\cos. A = \frac{\cos. a - \cos. b \cos. c}{\sin. b \sin. c}.$$

By permuting,

$$\left.\begin{aligned} \cos. B &= \frac{\cos. b - \cos. c \cos. a}{\sin. c \sin. a}; \\ \cos. C &= \frac{\cos. c - \cos. a \cos. b}{\sin. a \sin. b}. \end{aligned}\right\} \quad . \quad . \quad (A.)$$

We shall not again have occasion to refer to a diagram for the purpose of obtaining any new relation of the respective parts of a spherical triangle, but shall be able to draw from equation (6) all other relations which may be required. Hence, (6) may be regarded as one of the fundamental relations of Spherical Trigonometry.

§ **56.** *To express the cosine of a side of a spherical triangle, in terms of the sines and cosines of the angles.*

Let A, B, C, $a$, $b$, $c$, be the angles and sides of a spherical triangle, and A′, B′, C′, $a'$, $b'$, $c'$ the corresponding quantities of its polar triangle.

Then, by (A), § 55, we have

$$\cos. A' = \frac{\cos. a' - \cos. b' \cos. c'}{\sin. b' \sin. c'}. \quad . \quad . \quad (7.)$$

Now we have these relations (Geom., B. VIII., T. VII.): $A' = 180° - a$; $B' = 180° - b$; $C' = 180° - c$; $a' = 180° - A$; $b' = 180° - B$; $c' = 180° - C$; hence, (7) becomes

$$\cos. (180° - a) = \frac{\cos. (180° - A) - \cos. (180° - B) \cos. (180° - C)}{\sin. (180° - B) \sin. (180° - C)},$$

which gives

$$\cos. a = \frac{\cos. A + \cos. B \cos. C}{\sin. B \sin. C};$$

and by permuting,

$$\left.\begin{aligned} \cos. b &= \frac{\cos. B + \cos. C \cos. A}{\sin. C \sin. A}; \\ \cos. c &= \frac{\cos. C + \cos. A \cos. B}{\sin. A \sin. B}. \end{aligned}\right\} \quad . \quad . \quad (B.)$$

By the application of this property of the *Polar Triangle*, any formula of a spherical triangle may be changed into a similar formula, where the sides will take the place of angles, and the angles will take the place of sides.

This principle of change may be thus stated: *For the sides write the supplements of the opposite angles, and for the angles write the supplements of the opposite sides.*

§ **57.** *To express the sine of an angle of a spherical triangle, in terms of the sines of the sides of the triangle.*

By (A) § 55, we have

$$\cos. A = \frac{\cos. a - \cos. b \cos c}{\sin. b \sin. c}.$$

Therefore

$$1 + \cos. A = \frac{\cos. a - \cos. b \cos. c + \sin. b \sin. c}{\sin. b \sin. c} = \frac{\cos. a - \cos. (b+c)}{\sin. b \sin. c}$$

$$= \frac{2 \sin. \tfrac{1}{2}(a+b+c) \sin. \tfrac{1}{2}(-a+b+c)}{\sin. b \sin. c.}. \quad \text{[See (8), § 16.]}$$

If $s = \tfrac{1}{2}(a+b+c)$ we shall have $s-a = \tfrac{1}{2}(-a+b+c)$;
$(s-b) = \tfrac{1}{2}(a-b+c)$; $s-c = \tfrac{1}{2}(a+b-c)$.

We therefore have

$$1 + \cos. A = \frac{2 \sin. s \sin. (s-a)}{\sin. b \sin. c}. \quad . \quad . \quad . \quad (1.)$$

By a similar process we find

$$1 - \cos. A = \frac{\cos. b \cos. c + \sin. b \sin. c - \cos. a}{\sin. b \sin. c} = \frac{\cos. (b-c) - \cos. a}{\sin. b \sin. c}$$

$$= \frac{2 \sin. \tfrac{1}{2}(a-b+c) \sin. \tfrac{1}{2}(a+b-c)}{\sin. b \sin. c} = \frac{2 \sin. (s-b) \sin. (s-c)}{\sin. b \sin. c}. \quad (2.)$$

Taking the product of (1) and (2) we have

$$1 - \cos.^2 A = \sin.^2 A = \frac{4 \sin. s \sin. (s-a) \sin. (s-b) \sin. (s-c)}{\sin.^2 b \sin.^2 c},$$

and

$$\sin. A = \frac{2}{\sin. b \sin. c} \left\{ \sin. s \sin. (s-a) \sin. (s-b) \sin. (s-c) \right\}^{\frac{1}{2}};$$

By permuting,

$$\sin. B = \frac{2}{\sin. c \sin. a} \left\{ \sin. s \sin. (s-a) \sin. (s-b) \sin. (s-c) \right\}^{\frac{1}{2}};$$

$$\sin. C = \frac{2}{\sin. a \sin. b} \left\{ \sin. s \sin. (s-a) \sin. (s-b) \sin. (s-c) \right\}^{\frac{1}{2}}. \quad \text{(C.)}$$

By (1) we have

$$1+\cos. A=2\cos.^2 \tfrac{1}{2} A=\frac{2\sin. s\sin. (s-a)}{\sin. b\sin. c};$$

consequently,

$$\cos. \tfrac{1}{2} A=\left\{\frac{\sin. s\sin. (s-a)}{\sin. b\sin. c}\right\}^{\frac{1}{2}};$$

And permuting,

$$\left.\begin{aligned}\cos. \tfrac{1}{2} B&=\left\{\frac{\sin. s\sin. (s-b)}{\sin. c\sin. a}\right\}^{\frac{1}{2}};\\ \cos. \tfrac{1}{2} C&=\left\{\frac{\sin. s\sin. (s-c)}{\sin. a\sin. b}\right\}^{\frac{1}{2}}.\end{aligned}\right\} \quad (C'.)$$

By (2) we have

$$1-\cos. A=2\sin.^2 \tfrac{1}{2} A=\frac{2\sin. (s-b\sin. (s-c)}{\sin. b\sin. c}.$$

Consequently,

$$\sin. \tfrac{1}{2} A=\left\{\frac{\sin. (s-b)\sin. (s-c)}{\sin. b\sin. c}\right\}^{\frac{1}{2}};$$

And permuting,

$$\left.\begin{aligned}\sin. \tfrac{1}{2} B&=\left\{\frac{\sin. (s-c)\sin. (s-a)}{\sin. c\sin. a}\right\}^{\frac{1}{2}};\\ \sin. \tfrac{1}{2} C&=\left\{\frac{\sin. (s-a)\sin. (s-b)}{\sin. a\sin. b}\right\}^{\frac{1}{2}}.\end{aligned}\right\} \quad (C''.)$$

Finally, dividing the expressions (C″) by those of (C′) we have

$$\left.\begin{aligned}\tan. \tfrac{1}{2} A&=\left\{\frac{\sin. (s-b)\sin. (s-c)}{\sin. s\sin. (s-a)}\right\}^{\frac{1}{2}};\\ \tan. \tfrac{1}{2} B&=\left\{\frac{\sin. (s-c)\sin. (s-a)}{\sin. s\sin. (s-b)}\right\}^{\frac{1}{2}};\\ \tan. \tfrac{1}{2} C&=\left\{\frac{\sin. (s-a)\sin. (s-b)}{\sin. s\sin. (s-c)}\right\}^{\frac{1}{2}}.\end{aligned}\right\} \quad (C'''.)$$

§ **58.** *To express the sine of a side of a spherical triangle, in terms of the sines and cosines of the angles.*

By (B) § 56, we have

$$\cos. a=\frac{\cos A+\cos. B\cos. C}{\sin. B\sin. C};$$

consequently,

$$1+\cos. a=\frac{\cos. A+\cos. B\cos. C+\sin. B\sin. C}{\sin. B\sin. C}=\frac{\cos. A+\cos.(B-C)}{\sin. B\sin. C}$$

$$=\frac{2\cos.\frac{1}{2}(A+B-C)\cos.\frac{1}{2}(A-B+C)}{\sin. B\sin. C}.$$

If $S=\frac{1}{2}(A+B+C)$ we shall have $S-A=\frac{1}{2}(-A+B+C)$; $S-B=\frac{1}{2}(A-B+C)$; $S-C=\frac{1}{2}(A+B-C)$. We therefore have

$$1+\cos. a=\frac{2\cos.(S-B)\cos.(S-C)}{\sin. B\sin. C}. \qquad (1.)$$

By a similar process we find

$$1-\cos. a=-\frac{\cos. B\cos. C-\sin. B\sin. C+\cos. A}{\sin. B\sin. C}$$

$$=-\frac{\cos.(B+C)+\cos. A}{\sin. B\sin. C}=-\frac{2\cos.\frac{1}{2}(A+B+C)\cos.\frac{1}{2}(-A+B+C)}{\sin. B\sin. C}$$

$$=-\frac{2\cos. S\cos.(S-A)}{\sin. B\sin. C}. \qquad (2.)$$

Taking the product of (1) and (2), we have

$$1-\cos.^2 a=\sin.^2 a=-\frac{4\cos. S\cos.(S-A)\cos.(S-B)\cos.(S-C)}{\sin.^2 B\sin.^2 C}.$$

And

$$\left.\begin{aligned}
&\sin. a=\frac{2}{\sin. B\sin. C}\left\{-\cos. S\cos.(S-A)\cos.(S-B)\cos.(S-C)\right\}^{\frac{1}{2}};\\
&\text{Permuting,}\\
&\sin. b=\frac{2}{\sin. C\sin. A}\left\{-\cos. S\cos.(S-A)\cos.(S-B)\cos.(S-C)\right\}^{\frac{1}{2}};\\
&\sin. c=\frac{2}{\sin. A\sin. B}\left\{-\cos. S\cos.(S-A)\cos.(S-B)\cos.(S-C)\right\}^{\frac{1}{2}}.
\end{aligned}\right\}(D.)$$

By (1) we have

$$1+\cos. a=2\cos.^2\tfrac{1}{2}a=\frac{2\cos.(S-B)\cos.(S-C)}{\sin. B\sin. C}.$$

Consequently,

$$\left.\begin{aligned}
\cos.\tfrac{1}{2}a&=\left\{\frac{\cos.(S-B)\cos.(S-C)}{\sin. B\sin. C}\right\}^{\frac{1}{2}};\\
\cos.\tfrac{1}{2}b&=\left\{\frac{\cos.(S-C)\cos.(S-A)}{\sin. C\sin. A}\right\}^{\frac{1}{2}};\\
\cos.\tfrac{1}{2}c&=\left\{\frac{\cos.(S-A)\cos.(S-B)}{\sin. A\sin. B}\right\}^{\frac{1}{2}}.
\end{aligned}\right\}\qquad (D'.)$$

By (2) we have

$$1 - \cos. a = 2 \sin.^2 \tfrac{1}{2} a = -\frac{2 \cos. S \cos. (S-A)}{\sin. B \sin. C}.$$

Consequently,

$$\left.\begin{aligned} \sin. \tfrac{1}{2} a &= \left\{ \frac{-\cos. S \cos. (S-A)}{\sin. B \sin. C} \right\}^{\frac{1}{2}}; \\ \sin. \tfrac{1}{2} b &= \left\{ \frac{-\cos. S \cos. (S-B)}{\sin. C \sin. A} \right\}^{\frac{1}{2}}; \\ \sin. \tfrac{1}{2} c &= \left\{ \frac{-\cos. S \cos. (S-C)}{\sin. A \sin. B} \right\}^{\frac{1}{2}}. \end{aligned}\right\} \quad (D''.)$$

Finally, dividing the expressions (D″) by those of (D′), we have

$$\left.\begin{aligned} \tan. \tfrac{1}{2} a &= \left\{ \frac{-\cos. S \cos. (S-A)}{\cos. (S-B) \cos. (S-C)} \right\}^{\frac{1}{2}}; \\ \tan. \tfrac{1}{2} b &= \left\{ \frac{-\cos. S \cos. (S-B)}{\cos. (S-C) \cos. (S-A)} \right\}^{\frac{1}{2}}; \\ \tan. \tfrac{1}{2} c &= \left\{ \frac{-\cos. S \cos. (S-C)}{\cos. (S-A) \cos. (S-B)} \right\}^{\frac{1}{2}}. \end{aligned}\right\} \quad (D'''.)$$

Groups marked (D), (D′), (D″), and (D‴) might have been immediately deduced from those marked respectively (C), (C′), (C″), and (C‴), by the application of the principle of the *Polar Triangle*, as indicated under § 56.

Since the negative sign precedes cos. S in the expressions under the radicals of (D), (D″), and (D‴) it might, at first view, be supposed that these values were under an impossible form. It is however easily shown that this is not the case. For (Geom. B. VIII., T. XV.) we know that the sum of all the angles of a spherical triangle is greater than two right angles, and less than six right angles.

Consequently,

$$A + B + C > 180° \text{ and } < 540°.$$

And

$$\tfrac{1}{2}(A + B + B) \text{ or } S > 90° \text{ and } < 270°.$$

Hence, the cosine of S is always negative, and $-\cos. S$ is therefore always positive.

Again, if $a'$, $b'$, $c'$ be the three sides of the polar triangle, since the sum of any two sides of a spherical triangle is greater than the third side (Geom. B. VIII., T. I), we have $b'+c'>a'$; that is,

$$180°-B+180°-C>180°-A;$$

consequently,

$$-A+B+C<180°,$$

and

$$\tfrac{1}{2}(-A+B+C)<90°;$$

hence, cos. (S−A) is always positive, and in like manner cos. (S−B), cos. (S−C), are always positive; hence the expressions (D), (D″), (D‴) are in every case possible.

§ **59.** *The sines of the angles of a spherical triangle are to each other as the sines of their opposite sides.*

Expressions (C) immediately give by division,

$$\left.\begin{aligned}\frac{\sin. A}{\sin. B}&=\frac{\sin. c\sin. a}{\sin. b\sin. c}=\frac{\sin. a}{\sin. b},\\ \frac{\sin. A}{\sin. C}&=\frac{\sin. a\sin. b}{\sin. b\sin. c}=\frac{\sin. a}{\sin. c};\\ \frac{\sin. B}{\sin. C}&=\frac{\sin. a\sin. b}{\sin. c\sin. a}=\frac{\sin. b}{\sin. c}.\end{aligned}\right\}\quad . \quad . \quad (E.)$$

§ **60.** *To express the tangent of the sum and difference of two angles of a spherical triangle, in terms of the sides opposite to these angles, and the third angle of the triangle.*

By (A) § 55, we have from the third expression,

$$\cos. c=\cos. a\cos. b+\sin. a\sin. b\cos. C.\quad . \quad (1.)$$

Substituting this value of cos. $c$ in the first expression of (A), we find

$$\cos. A=\frac{\cos. a-\cos. a\cos.^2 b-\cos. b\sin. a\sin. b\cos. C}{\sin. b\sin. c}$$

$$=\frac{\cos. a(1-\cos.^2 b)-\cos. b\sin. a\sin. b\cos. C}{\sin. b\sin. c}$$

$$=\frac{\cos. a\sin. b-\cos. b\sin. a\cos. C}{\sin. c}.\quad . \quad (2.)$$

If we substitute the value of cos. $c$, given by (1), in the second expression of (A), we shall obtain

$$\cos. B=\frac{\cos. b\sin. a-\cos. a\sin. b\cos. C}{\sin. c}.\quad . \quad (3.)$$

Adding (2) and (3), we have

$$\cos. A+\cos. B=\frac{\sin. a \cos. b+\cos. a \sin. b-(\sin. a \cos. b+\cos. a \sin. b) \cos. C}{\sin. c}$$

$$=\frac{\sin. (a+b)-\sin. (a+b) \cos. C}{\sin. c}=\frac{\sin. (a+b)(1-\cos. C)}{\sin. c}. \quad (4.)$$

Again, we have, by the first equation of (E),

$$\frac{\sin. A}{\sin. B}=\frac{\sin. a}{\sin. b};$$

consequently,

$$\frac{\sin. A \pm \sin. B}{\sin. B}=\frac{\sin. a \pm \sin. b}{\sin. b}.$$

Hence,

$$\sin. A \pm \sin. B=(\sin. a \pm \sin. b)\frac{\sin. B}{\sin. b}=(\sin. a \pm \sin. b)\frac{\sin. C}{\sin. c}. \quad (5.)$$

Dividing (5) by (4), using first the positive sign, we have

$$\frac{\sin. A+\sin. B}{\cos. A+\cos. B}=\frac{\sin. a+\sin. b}{\sin. (a+b)} \times \frac{\sin. C}{1-\cos. C};$$

hence, by equations (16), (14), and (4) of § 53, this becomes

$$\tan. \tfrac{1}{2}(A+B)=\frac{\cos. \frac{1}{2}(a-b)}{\cos. \frac{1}{2}(a+b)} \times \cot. \tfrac{1}{2} C. \quad . \quad (6.)$$

In a similar manner, by dividing (5), with the negative sign, by (4), we find

$$\frac{\sin. A-\sin. B}{\cos. A+\cos. B}=\frac{\sin. a-\sin. b}{\sin. (a+b)} \times \frac{\sin. C}{1-\cos. C},$$

which, by equations (18), (15), and (4) of § 53, becomes

$$\tan. \tfrac{1}{2}(A-B)=\frac{\sin. \frac{1}{2}(a-b)}{\sin. \frac{1}{2}(a+b)} \times \cot. \tfrac{1}{2} C. \quad . \quad (7.)$$

Equations (6) and (7) and their permuted values give as follows:

$$\left.\begin{array}{l}\left\{\begin{array}{l}\tan.\tfrac{1}{2}(\mathrm{A}+\mathrm{B})=\dfrac{\cos.\frac{1}{2}(a-b)}{\cos.\frac{1}{2}(a+b)}\times\cot.\tfrac{1}{2}\mathrm{C};\\[2ex] \tan.\tfrac{1}{2}(\mathrm{A}-\mathrm{B})=\dfrac{\sin.\frac{1}{2}(a-b)}{\sin.\frac{1}{2}(a+b)}\times\cot.\tfrac{1}{2}\mathrm{C}.\end{array}\right.\\[4ex] \left\{\begin{array}{l}\tan.\tfrac{1}{2}(\mathrm{B}+\mathrm{C})=\dfrac{\cos.\frac{1}{2}(b-c)}{\cos.\frac{1}{2}(b+c)}\times\cot.\tfrac{1}{2}\mathrm{A};\\[2ex] \tan.\tfrac{1}{2}(\mathrm{B}-\mathrm{C})=\dfrac{\sin.\frac{1}{2}(b-c)}{\sin.\frac{1}{2}(b+c)}\times\cot.\tfrac{1}{2}\mathrm{A}.\end{array}\right.\\[4ex] \left\{\begin{array}{l}\tan.\tfrac{1}{2}(\mathrm{C}+\mathrm{A})=\dfrac{\cos.\frac{1}{2}(c-a)}{\cos.\frac{1}{2}(c+a)}\times\cot.\tfrac{1}{2}\mathrm{B};\\[2ex] \tan.\tfrac{1}{2}(\mathrm{C}-\mathrm{A})=\dfrac{\sin.\frac{1}{2}(c-a)}{\sin.\frac{1}{2}(c+a)}\times\cot.\tfrac{1}{2}\mathrm{B}.\end{array}\right.\end{array}\right\}\quad(\mathrm{F}.)$$

If we convert equations (1) and (2) of (F) into proportions, we have

$$\cos.\tfrac{1}{2}(a+b):\cos.\tfrac{1}{2}(a-b)::\cot.\tfrac{1}{2}\mathrm{C}:\tan.\tfrac{1}{2}(\mathrm{A}+\mathrm{B}),$$
$$\sin.\tfrac{1}{2}(a+b):\sin.\tfrac{1}{2}(a-b)::\cot.\tfrac{1}{2}\mathrm{C}:\tan.\tfrac{1}{2}(\mathrm{A}-\mathrm{B});$$

which proportions are known as *Napier's first and second Analogies*. They are used in solving a triangle when two sides and the included angle are given.

§ **61.** *To express the tangent of the sum and difference of two sides of a spherical triangle, in terms of the angles opposite to them, and the third side of the triangle.*

Let A, B, C, $a$, $b$, $c$ be the angles and sides of a spherical triangle, A′, B′, C′, $a'$, $b'$, $c'$, the corresponding parts of the polar triangle; then by (F), we have

$$\tan.\tfrac{1}{2}(\mathrm{A}'+\mathrm{B}')=\frac{\cos.\frac{1}{2}(a'-b')}{\cos.\frac{1}{2}(a'+b')}\times\cot.\tfrac{1}{2}\mathrm{C}'.$$

Therefore,

$$\tan.\tfrac{1}{2}[(180^\circ-a)+(180^\circ-b)]=\frac{\cos.\frac{1}{2}[(180^\circ-\mathrm{A})-(180^\circ-\mathrm{B})]}{\cos.\frac{1}{2}[(180^\circ-\mathrm{A})+(180^\circ-\mathrm{B})]}\times\cot.\tfrac{1}{2}(180^\circ-c).$$

This becomes

$$\tan.\tfrac{1}{2}(a+b)=\frac{\cos.\frac{1}{2}(\mathrm{A}-\mathrm{B})}{\cos.\frac{1}{2}(\mathrm{A}+\mathrm{B})}\times\tan.\tfrac{1}{2}c.\qquad(1.)$$

Similarly we have

$$\tan.\tfrac{1}{2}(A'-B')=\frac{\sin.\tfrac{1}{2}(a'-b')}{\sin.\tfrac{1}{2}(a'+b')}\times\cot.\tfrac{1}{2}C'.$$

Therefore,

$$\tan.\tfrac{1}{2}[(180^\circ-a)-(180^\circ-b)]=\frac{\sin.\tfrac{1}{2}[(180^\circ-A)-(180^\circ-B)]}{\sin.\tfrac{1}{2}[(180^\circ-A)+(180^\circ-B)]}\times\cot.\tfrac{1}{2}(180^\circ-c),$$

which becomes

$$\tan.\tfrac{1}{2}(a-b)=\frac{\sin.\tfrac{1}{2}(A-B)}{\sin.\tfrac{1}{2}(A+B)}\times\tan.\tfrac{1}{2}c. \quad . \quad (2.)$$

Equations (1) and (2), together with their permuted values, give as follows:

$$\left.\begin{array}{l}
\left\{\begin{array}{l}
\tan.\tfrac{1}{2}(a+b)=\dfrac{\cos.\tfrac{1}{2}(A-B)}{\cos.\tfrac{1}{2}(A+B)}\times\tan.\tfrac{1}{2}c; \\[2ex]
\tan.\tfrac{1}{2}(a-b)=\dfrac{\sin.\tfrac{1}{2}(A-B)}{\sin.\tfrac{1}{2}(A+B)}\times\tan.\tfrac{1}{2}c.
\end{array}\right. \\[5ex]
\left\{\begin{array}{l}
\tan.\tfrac{1}{2}(b+c)=\dfrac{\cos.\tfrac{1}{2}(B-C)}{\cos.\tfrac{1}{2}(B+C)}\times\tan.\tfrac{1}{2}a; \\[2ex]
\tan.\tfrac{1}{2}(b-c)=\dfrac{\cos.\tfrac{1}{2}(B-C)}{\sin.\tfrac{1}{2}(B+C)}\times\tan.\tfrac{1}{2}a.
\end{array}\right. \\[5ex]
\left\{\begin{array}{l}
\tan.\tfrac{1}{2}(c+a)=\dfrac{\cos.\tfrac{1}{2}(C-A)}{\cos.\tfrac{1}{2}(C+A)}\times\tan.\tfrac{1}{2}b; \\[2ex]
\tan.\tfrac{1}{2}(c-a)=\dfrac{\sin.\tfrac{1}{2}(C-A)}{\sin.\tfrac{1}{2}(C+A)}\times\tan.\tfrac{1}{2}b.
\end{array}\right.
\end{array}\right\} \quad (G.)$$

The first and second conditions of (G) being converted into proportions, gives *Napier's third and fourth Analogies*, as follows:

$$\cos.\tfrac{1}{2}(A+B):\cos.\tfrac{1}{2}(A-B)::\tan.\tfrac{1}{2}c:\tan.\tfrac{1}{2}(a+b),$$

$$\sin.\tfrac{1}{2}(A+B):\sin.\tfrac{1}{2}(A-B)::\tan.\tfrac{1}{2}c:\tan.\tfrac{1}{2}(a-b).$$

These analogies are employed in the solution of a triangle when two angles and the interjacent side are given.

§ **62.** *To express the cotangent of an angle of a spherical triangle, in terms of the side opposite, one of the other sides, and the angle included between these two sides.*

Equation (2) of § 60, when cleared of fractions, becomes

$$\cos.A\sin.c=\cos.a\sin.b-\cos.b\sin.a\cos.C. \quad . \quad (1.)$$

By (E) we have

$$\sin. A \sin. c = \sin. C \sin. a. \quad . \quad . \quad (2.)$$

Dividing (1) by (2), we find

$$\cot. A = \cot. a \sin. b \operatorname{cosec.} C - \cos. b \cot. C. \quad . \quad (3.)$$

By interchanging B and C, $b$ and $c$, (3) will give

$$\cot. A = \cot. a \sin. c \operatorname{cosec.} B - \cos. c \cot. B. \quad . \quad (4.)$$

Proceeding in like manner for the other angles, we shall obtain the following group:

$$\left.\begin{aligned} \cot. A &= \cot. a \sin. b \operatorname{cosec.} C - \cos. b \cot. C \\ &= \cot. a \sin. c \operatorname{cosec.} B - \cos. c \cot. B; \\ \cot. B &= \cot. b \sin. c \operatorname{cosec.} A - \cos. c \cot. A \\ &= \cot. b \sin. a \operatorname{cosec.} C - \cos. a \cot. C; \\ \cot. C &= \cot. c \sin. a \operatorname{cosec.} B - \cos. a \cot. B \\ &= \cot. c \sin. b \operatorname{cosec.} A - \cos. b \cot. A. \end{aligned}\right\} \quad (H.)$$

These six equations are of such a nature that they do not yield any new relations by the application of the *principle of the Polar Triangle* (§ 56).

§ **63.** By aid of the eight groups of formulas designated by (A), (B), (C), (D), (E), (F), (G), (H), we shall be enabled to solve all the cases of spherical triangles, whether right-angled or oblique-angled. We shall, in the next chapter, proceed to apply these formulas.

§ **64.** Before, however, passing to the solution of spherical triangles, we will deduce some other general formulas of *spherics*, which will frequently be found useful in Astronomy.

Equation (6) of § 55, becomes, by permuting,

$$\cos. c \cos. a + \sin. c \sin. a \cos. B = \cos. b. \quad . \quad . \quad (1.)$$

Multiplying (6) of § 55, by $\cos. c$, it becomes

$$\cos. c \cos. a = \cos. b \cos.^2 c + \sin. b \sin. c \cos. c \cos. A. \quad (2.)$$

Subtracting (2) from (1) and substituting $\sin.^2 c$ for $1 - \cos.^2 c$, we have

$$\sin. c \sin. a \cos. B = \cos. b \sin.^2 c - \sin. b \sin. c \cos. c \cos. A;$$

which, divided by $\sin. c$, becomes

$$\sin. a \cos. B = \cos. b \sin. c - \sin. b \cos. c \cos. A. \quad (3.)$$

Interchanging B, C, and $b$, $c$, (3) becomes

$$\sin. a \cos. C = \cos. c \sin. b - \sin. c \cos. b \cos. A. \qquad (4.)$$

Equations (3) and (4) with their permutations, give

$$\left.\begin{array}{l} \left\{\begin{array}{l} \sin. a \cos. B = \cos. b \sin. c - \sin. b \cos. c \cos. A; \\ \sin. a \cos. C = \cos. c \sin. b - \sin. c \cos. b \cos. A. \end{array}\right. \\ \left\{\begin{array}{l} \sin. b \cos. C = \cos. c \sin. a - \sin. c \cos. a \cos. B; \\ \sin. b \cos. A = \cos. a \sin. c - \sin. a \cos. c \cos. B. \end{array}\right. \\ \left\{\begin{array}{l} \sin. c \cos. A = \cos. a \sin. b - \sin. a \cos. b \cos. C; \\ \sin. c \cos. B = \cos. b \sin. a - \sin. b \cos. a \cos. C. \end{array}\right. \end{array}\right\} \quad (I.)$$

If we apply the principle of the Polar Triangle to equations (I) they become

$$\left.\begin{array}{l} \left\{\begin{array}{l} \sin. A \cos. b = \cos. B \sin. C + \sin. B \cos. C \cos. a; \\ \sin. A \cos. c = \cos. C \sin. B + \sin. C \cos. B \cos. a. \end{array}\right. \\ \left\{\begin{array}{l} \sin. B \cos. c = \cos. C \sin. A + \sin. C \cos. A \cos. b; \\ \sin. B \cos. a = \cos. A \sin. C + \sin. A \cos. C \cos. b. \end{array}\right. \\ \left\{\begin{array}{l} \sin. C \cos. a = \cos. A \sin. B + \sin. A \cos. B \cos. c; \\ \sin. C \cos. b = \cos. B \sin. A + \sin. B \cos. A \cos. c. \end{array}\right. \end{array}\right\} \quad (J.)$$

---

# CHAPTER VI.

## SOLUTION OF SPHERICAL RIGHT TRIANGLES.

§ **65.** We shall confine ourselves to such triangles as have only one right angle; those that have two or three right angles will be considered hereafter. (See § 74.)

A spherical triangle consists of 6 parts, 3 sides and 3 angles; and any three of these being known the others may be found. In the present case, one of the angles being a right angle, it follows that any other two parts being known the other three may be found.

The number of combinations of 5 things taken 3 and 3 at a time, is $\frac{5 \times 4 \times 3}{1 \times 2 \times 3} = 10$; therefore, ten different cases present themselves in the solution of right-angled spherical triangles.

§ **66.** The solution of these ten cases may all be comprised in two rules first given by *Napier*, and known under the name of *Napier's Rules for Circular Parts.*

OF THE CIRCULAR PARTS.

The right angle is not taken into consideration. The two sides, the complements of the two angles, and the complement of the hypotenuse constitute the five circular parts.

Let ABC be a spherical triangle, right-angled at C. If we arrange the five circulating parts upon the circumference of its circumscribed circle, we shall readily discover which are *adjacent*, and which are *opposite*, when either of the five parts is chosen as the *middle* part.

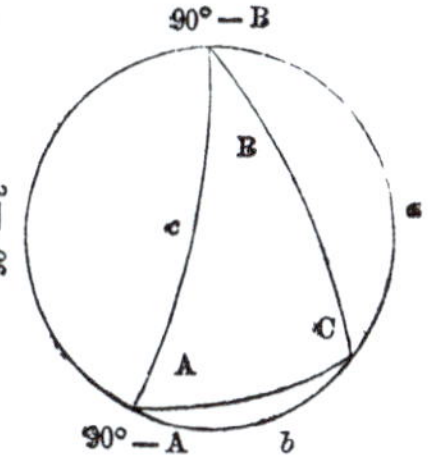


NAPIER'S RULES.

I. *The sine of the middle part is equal to the product of the tangents of the adjacent parts.*

II. *The sine of the middle part is equal to the product of the cosines of the opposite parts.*

If now we take in turn each of the five parts as the middle part, and apply these Rules, we shall obtain ten formulas, as follows.

*First.* Let $a$ be the middle part, then will $b$ and $(90° - B)$ be the adjacent parts, and $(90° - c)$ and $(90° - A)$ the opposite parts.

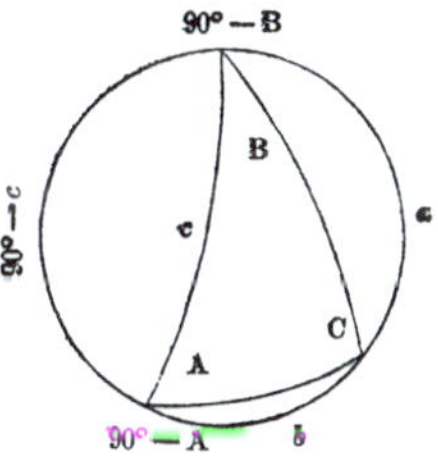


RULE I. $\sin. a = \tan. b \tan. (90° - B) = \tan. b \cot. B.$ (1.)

RULE II. $\sin. a = \cos. (90° - c) \cos. (90° - A) = \sin. c \sin. A.$ (2.)

*Secondly.* Let $b$ be the middle part, then will $a$ and $(90° - A)$ be the adjacent parts, and $(90° - c)$ and $(90° - B)$ the opposite parts.

RULE I. $\sin. b = \tan. a \tan. (90° - A) = \tan. a \cot. A.$ (3.)

RULE II. $\sin. b = \cos. (90° - c) \cos. (90° - B) = \sin. c \sin. B.$ (4.)

*Thirdly*. Let $(90° - c)$ be the middle part, then will $(90° - A)$ and $(90° - B)$ be the adjacent parts, and $a$ and $b$ the opposite parts.

RULE I. $\sin. (90° - c) = \tan. (90° - A) \tan. (90° - B)$,

or $\cos. c = \cot. A \cot. B$. . . . . (5.)

RULE II. $\sin. (90° - c) = \cos. a \cos. b$, or $\cos. c = \cos. a \cos. b$. (6.)

*Fourthly*. Let $(90° - A)$ be the middle part, then will $b$ and $(90° - c)$ be the adjacent parts, and $a$ and $(90° - B)$ the opposite parts.

RULE I. $\sin. (90° - A) = \tan. b \tan. (90° - c)$,

or $\cos. A = \tan. b \cot. c$. . . . (7.)

RULE II. $\sin. (90° - A) = \cos. a \cos. (90° - B)$,

or $\cos A = \cos. a \sin. B$. . . . (8.)

*Fifthly*. Let $(90° - B)$ be the middle part, then will $a$ and $(90° - c)$ be the adjacent parts, and $b$ and $(90° - A)$ the opposite parts.

RULE I. $\sin. (90° - B) = \tan. a \tan. (90° - c)$,

or $\cos B = \tan. a \cot. c$. . . . (9.)

RULE II. $\sin. (90° - B) = \cos. b \cos. (90° - A)$,

or $\cos. B = \cos. b \sin. A$. . . . (10.)

Collecting these ten results, we have

$$\left.\begin{aligned}
\sin. a &= \tan. b \cot. B, && (1.)\\
&= \sin. c \sin. A. && (2.)\\
\sin. b &= \tan. a \cot. A, && (3.)\\
&= \sin. c \sin. B. && (4.)\\
\cos. c &= \cot. A \cot. B, && (5.)\\
&= \cos. a \cos. b. && (6.)\\
\cos. A &= \tan. b \cot. c, && (7.)\\
&= \cos. a \sin. B. && (8.)\\
\cos. B &= \tan. a \cot. c, && (9.)\\
&= \cos. b \sin. A. && (10.)
\end{aligned}\right\} \quad (K.)$$

Since the acute angles A and B, as well as their opposite sides $a$ and $b$, admit of being interchanged, it follows that there are in reality only six distinct formulas in the group (K). Thus (1) and (2) are immediately changed into (3) and (4) respectively. Also (7) and (8) give, by this change, (9) and (10) respectively.

Hence, (1), (2), (5), (6,) (7), and (8) give all that is required for the solution of all spherical right triangles.

That these six formulas, as drawn from *Napier's Rules*, are correct, may be shown as follows:

Equation (4) of (H) § 62, gives, when $C=90°$,

$$\cot. B = \cot. b \sin. a, \text{ which immediately reduces to}$$
$$\sin. a = \tan. b \cot. B, \text{ which is (1) of (K).}$$

Equation (2) of (E) § 59, gives, when $C=90°$,

$$\sin. A = \frac{\sin. a}{\sin. c}, \text{ or } \sin. a = \sin. c \sin. A, \text{ which is (2) of (K).}$$

Equation (3) of (B) § 56, gives, when $C=90°$,

$$\cos. c = \frac{\cos. A \cos. B}{\sin. A \sin. B}, \text{ or } \cos. c = \cot. A \cot. B, \text{ which is (5) of (K).}$$

Equation (3) of (A) § 55, gives, when $C=90°$,

$$0 = \frac{\cos. c - \cos. a \cos. b}{\sin. a \sin. b}, \text{ or } \cos. c = \cos. a \cos. b, \text{ which is (6) of (K).}$$

Equation (6) of (H) § 62, gives, when $C=90°$,

$$0 = \cot. c \sin. b \operatorname{cosec}. A - \cos. b \cot. A, \text{ or } \cos. A = \tan. b \cot. c,$$
which is (7) of (K).

Equation (1) of (B) § 56, gives, when $C=90°$,

$$\cos. a = \frac{\cos. A}{\sin. B}, \text{ or } \cos. A = \cos. a \sin. B, \text{ which is (8) of (K).}$$

Having proved the formulas (K), which were derived from *Napier's two Rules*, we are at liberty to use them in solving the six cases of spherical right triangles, or we may in all cases make the direct application of those Rules.

**§ 67.** That we may, in the foregoing formulas, distinguish the trigonometric functions of parts less than 90° from those greater than 90°, we must pay particular attention to the rule for the algebraic signs.

Since the sine of an angle and the sine of its supplement are identical, it follows that there will be two solutions when the required part is determined by means of its sine, unless the ambiguity can be removed by the application of the following propositions:

PROPOSITION I.

*In a spheric right triangle, an angle and its opposite side are always both less than* 90°, *or else both greater than* 90°.

To establish this, take equation (8) of (K),

$$\cos. A = \cos. a \sin. B,$$

which immediately gives

$$\sin B = \frac{\cos. A}{\cos. a}. \quad . \quad . \quad . \quad (1.)$$

Now, since B is less than 180°, its sine must be positive, hence the fraction which is equal to sin. B must also be positive; therefore the numerator and denominator must be either both positive or else both negative—that is, A and $a$ must be either both less than 90°, or else both greater than 90°.

PROPOSITION II.

*When the two sides of a spherical triangle including the right angle are both less or both greater than* 90°, *the hypotenuse will be less than* 90°. *But when one side is less than* 90° *and the other greater than* 90°, *the hypotenuse will be greater than* 90°.

This may be established by equation (6) of (K), which is

$$\cos. c = \cos. a \cos. b.$$

If $a$ and $b$ are both less than 90°, their cosines will be positive; if they are both greater than 90°, their cosines will be negative, and in both cases the product of their cosines, which gives cos. $c$, will be positive; consequently $c$ will be less than 90°.

If $a$ and $b$ are in different quadrants—that is, if one is less than 90° and the other greater than 90°, the product of their cosines will be negative; consequently $c$ will be greater than 90°.

CASE I.

§ **68.** *Given the hypotenuse and one angle.*

For example, suppose $c$ and A given.

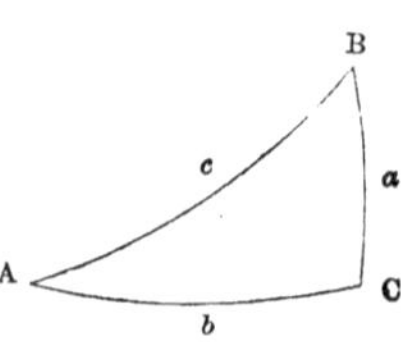


*To find a.* Equation (2) of (K) immediately gives

$$\sin. a = \sin. c \sin A.$$

There will be two values of $a$ corresponding to the same sine. We must, § 67, use the one which is in the same quadrant as A (Proposition I.).

*To find b.* Equation (7) of (K) readily gives

$$\tan. b = \tan. c \cos. \mathrm{A}.$$

*To find* B. Equation (5) of (K) readily gives

$$\cot. \mathrm{B} = \cos. c \tan. \mathrm{A}.$$

The quadrants in which $b$ and B are situated are determined by the algebraic sign of tan. $b$ and cot. B, § 67.

EXAMPLES.

1. Given $c = 104° \; 30'$, $\mathrm{A} = 75° \; 15'$. To solve the triangle,

$$\begin{aligned} \log. \sin. c &= 9{\cdot}985942 \\ \log. \sin. \mathrm{A} &= 9{\cdot}985447 \\ \hline \log. \sin. a &= 9{\cdot}971389 \end{aligned}$$

Therefore $a = 69° \; 25' \; 48''$, or $110° \; 34' \; 12''$.

By Prop. I., under § 67, it follows that $a$ must be in the same quadrant with A; consequently we must take $69° \; 25' \; 48''$ as the value of $a$. The supplement of this is not applicable to this triangle.

$$\begin{aligned} \log. \tan. c &= 10{\cdot}587342\,n \\ \log. \cos. \mathrm{A} &= 9{\cdot}405862 \\ \hline \log. \tan. b &= 9{\cdot}993204\,n \\ b &= 135° \; 26' \; 54''. \end{aligned}$$

$$\begin{aligned} \log. \cos. c &= 9{\cdot}398600\,n \\ \log. \tan. \mathrm{A} &= 10{\cdot}579585 \\ \hline \log. \cot. \mathrm{B} &= 9{\cdot}978185\,n \\ \mathrm{B} &= 133° \; 33' \; 42''. \end{aligned}$$

In the above work, since $c$ is an arc greater than a quadrant, its tangent and cosine are both negative, hence at the right of their logarithms we have placed the letter $n$. (See Note under Case IV., § 50.)

Now, since the addition of logarithms corresponds to multiplication, and the subtraction of one logarithm from another corresponds to division, it follows that when both logarithms, in the case of addition or of subtraction, are derived from negative values—that is, when both logarithms have the letter $n$—the result must be positive, and its logarithm will not be marked with $n$. But when one of the logarithms is distinguished with $n$ and the other is not, the resulting logarithm is also to be marked with $n$, and the corresponding numerical value will be negative.

Our results for the logarithms of tan. $b$ and cot. B above are both marked with $n$, consequently their numerical values are both negative. We seek in the Table in the usual manner for the arcs whose logarithmic tangent and cotangent are respectively 9·993204 and 9·978185, and find them to be $44° \; 33' \; 6''$ and $46° \; 26' \; 18''$; we then take the supplements of those arcs for the values sought. Or, we may at once take out of the Tables the obtuse angles, since our Tables have been so arranged as to give the supplementary angles in all cases. By this arrangement, the logarithmic work with obtuse angles is performed with the same ease as in the case of acute angles.

2. Given $c = 65° \, 5'$, $A = 48° \, 12'$, to solve the triangle.

$$Ans. \begin{cases} a = 42° \, 32' \, 20''. \\ b = 55° \, 7' \, 32''. \\ B = 64° \, 46' \, 13''. \end{cases}$$

CASE II.

§ **69.** *Given the hypotenuse and a side:* for example, $c$ and $a$.

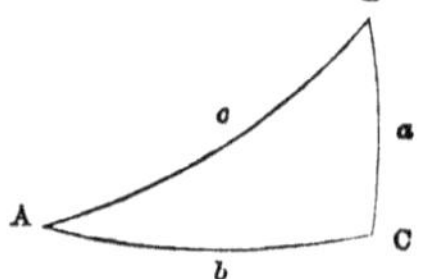


*To find* $b$. Equation (6) of (K) gives

$$\cos. b = \frac{\cos. c}{\cos. a}.$$

*To find* A. Equation (2) of (K) gives

$$\sin. A = \frac{\sin. a}{\sin. c}.$$

We must take A in the same quadrant as $a$, § 67, Prop. I.

*To find* B. Equation (9) of (K) is

$$\cos. B = \tan. a \cot. c.$$

EXAMPLES.

1. Given $c = 94° \, 5'$, $a = 100° \, 45'$, to solve the triangle.

| | | |
|---|---|---|
| $\cos. c = 8{\cdot}852525\,n$ | $\sin. a = 9{\cdot}992311$ | $\tan. a = 10{\cdot}721576\,n$ |
| $\cos. a = 9{\cdot}270735\,n$ | $\sin. c = 9{\cdot}998896$ | $\cot. c = 8{\cdot}853628\,n$ |
| $\cos. b = 9{\cdot}581790$ | $\sin. A = 9{\cdot}993415$ | $\cos. B = 9{\cdot}575204$ |
| $b = 67° \, 33' \, 26''$, | $A = 99° \, 57' \, 8''$, | $B = 67° \, 54' \, 47''$. |

In this example, since $c$ and $a$ are each greater than 90°, their cosines, as used in finding the side $b$, are negative; this we have indicated by the letter $n$ at the right of their logarithms. (See Note under Case IV., § 50.) Also their tangents as well as cotangents are negative, and the letter $n$ indicates this in the work for finding B.

Since the sines of all angles not exceeding 180° are positive, it follows that the logarithms of the sines of $a$ and $c$, as used in finding A, do not require to be distinguished by the letter $n$. But the value of A being given in a sine leads to two arcs 80° 2′ 52″, 99° 57′ 8″, supplements of each other. Prop. I., § 67, requires us to use the obtuse angle for the value of A.

2. Given $c = 66° \, 32'$, $a = 37° \, 48'$, to solve the triangle.

$$Ans. \begin{cases} b = 59° \, 44' \, 13''. \\ A = 41° \, 55' \, 34''. \\ B = 70° \, 18' \, 42''. \end{cases}$$

CASE III.

§ **70.** *Given one side and its opposite angle:* for example, $a$ and A.

Equations (3), (2), and (8) of (K) readily give

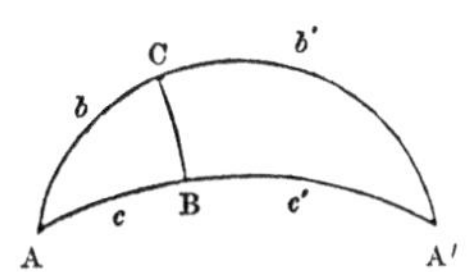


$$\sin. b = \tan. a \cot. A,$$

$$\sin. c = \frac{\sin. a}{\sin. A}.$$

$$\sin. B = \frac{\cos. A}{\cos. a}.$$

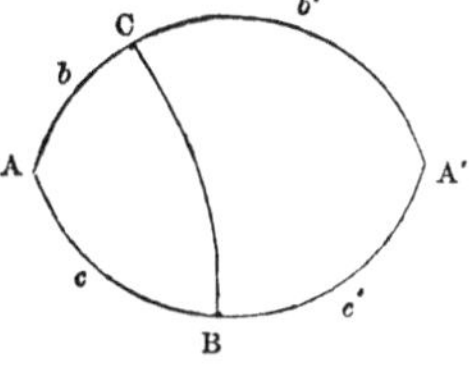


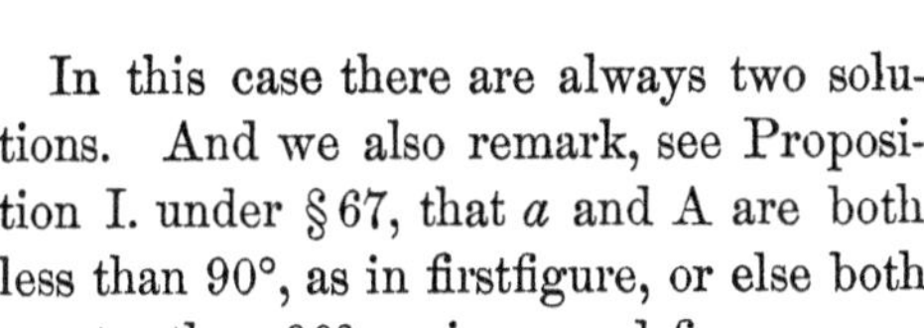

In this case there are always two solutions. And we also remark, see Proposition I. under § 67, that $a$ and A are both less than 90°, as in firstfigure, or else both greater than 90°, as in second figure.

If AB and AC be produced to meet at A′, we shall have ABA′ and ACA′ each a semi-circumference, and the angle A = A′. The two triangles ABC, A′BC both have the same given parts $a$ and A = A′, but the required parts, $b'$, $c'$, B′, of the second triangle are respectively the supplements of $b$, $c$, and B of the first triangle.

EXAMPLES.

1. Given $a = 110° \, 4$, A = 98° 30′, to solve the triangle.

$$\begin{array}{lll} \tan. a = 10{\cdot}437364 \, n & \sin. a = 9{\cdot}972802 & \cos. A = 9{\cdot}169702 \, n \\ \cot. A = 9{\cdot}174499 \, n & \sin. A = 9{\cdot}995203 & \cos. a = 9{\cdot}535438 \, n \\ \hline \sin. b = 9{\cdot}611863 & \sin. c = 9{\cdot}977599 & \sin. B = 9{\cdot}634264 \end{array}$$

$$b = \begin{cases} 24° \; 9' \; 1'', \\ 155° \, 50' \, 59''. \end{cases} \quad c = \begin{cases} 71° \, 45' \, 19'', \\ 108° \, 14' \, 41''. \end{cases} \quad B = \begin{cases} 25° \, 31' \; 3'', \\ 154° \, 28' \, 57''. \end{cases}$$

By paying attention to the Propositions under § 67, we shall have for the two solutions as follows:

$$Ans. \begin{cases} b = 24°\ 9'\ 1'' \\ c = 108°\ 14'\ 41'' \\ B = 25°\ 31'\ 3'' \end{cases} \text{ or } \begin{cases} b = 155°\ 50'\ 59'' \\ c = 71°\ 45'\ 19'' \\ B = 154°\ 28'\ 57'' \end{cases}$$

2. Given $a = 50°\ 12'$, $A = 75°\ 30'$, to solve the triangle.

$$Ans. \begin{cases} b = 18°\ 5'\ 0'' \\ c = 52°\ 31'\ 10'' \\ B = 23°\ 1'\ 35'' \end{cases} \text{ or } \begin{cases} b = 161°\ 55'\ 0'' \\ c = 127°\ 28'\ 50'' \\ B = 156°\ 58'\ 25'' \end{cases}$$

CASE IV.

**§ 71.** *Given one side and its adjacent angle:* for example, $a$ and B.

Equations (1), (9), and (8) of (K) readily give

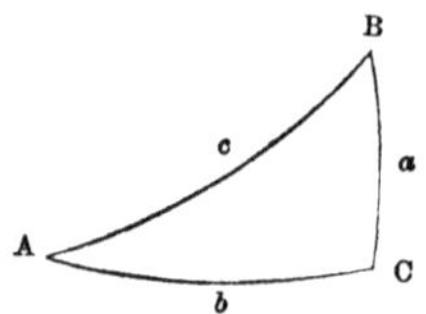


$$\tan. b = \sin. a \tan. B,$$
$$\cot. c = \cot. a \cos. B,$$
$$\cos. A = \cos. a \sin. B.$$

EXAMPLES.

1. Given $a = 101°\ 30'$, $B = 42°\ 15'$, to solve the triangle.

| | | |
|---|---|---|
| $\sin. a = 9{\cdot}991193$ | $\cot. a = 9{\cdot}308463\,n$ | $\cos. a = 9{\cdot}299655\,n$ |
| $\tan. B = 9{\cdot}958247$ | $\cos. B = 9{\cdot}869360$ | $\sin. B = 9{\cdot}827606$ |
| $\tan. b = 9{\cdot}949440$ | $\cot. c = 9{\cdot}177823\,n$ | $\cos. A = 9{\cdot}127261\,n$ |
| $b = 41°\ 40'\ 21''$, | $c = 98°\ 33'\ 52''$, | $A = 97°\ 42'\ 13''$. |

Since the logarithms of cot. $c$ and cos. A have the letter $n$, indicating that the cot. $c$ and cos. A are negative (see remarks under Case II.), it follows that we must take from the Tables the arcs which are greater than 90°. We thus find

$$Ans. \begin{cases} b = 41°\ 40'\ 21''. \\ c = 98°\ 33'\ 52''. \\ A = 97°\ 42'\ 13''. \end{cases}$$

2. Given $a = 48°\ 30'$, $B = 40°\ 20'$, to solve the triangle.

$$Ans. \begin{cases} b = 32°\ 27'\ 10''. \\ c = 56°\ 0'\ 13''. \\ A = 64°\ 36'\ 15''. \end{cases}$$

CASE V.

§ **72.** *Given the two sides*, $a$ and $b$.

Equations (6), (3), and (1) of (K) readily give

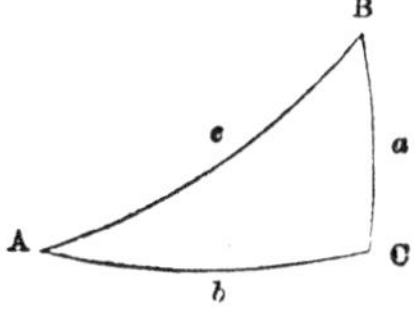


$$\cos. c = \cos. a \cos. b,$$

$$\cot. A = \cot. a \sin. b,$$

$$\cot. B = \sin. a \cot. b.$$

EXAMPLES.

1. Given $a = 75° 15'$, $b = 120° 15'$, to solve the triangle.

| | | |
|---|---|---|
| $\cos. a = 9{\cdot}405862$ | $\cot. a = 9{\cdot}420415$ | $\sin. a = 9{\cdot}985447$ |
| $\cos. b = 9{\cdot}702236\,n$ | $\sin. b = 9{\cdot}936431$ | $\cot. b = 9{\cdot}765805\,n$ |
| $\cos. c = 9{\cdot}108098\,n$ | $\cot. A = 9{\cdot}356846$ | $\cot. B = 9.751252\,n$ |
| $c = 97° 22' 9''$, | $A = 77° 11' 14''$, | $B = 119° 25' 17''$. |

2. Given $a = 120°$, $b = 30°$, to solve the triangle.

*Ans.* $\begin{cases} c = 115° 39' 32''. \\ A = 106° \ 6' \ 8''. \\ B = 33° 42' 36''. \end{cases}$

CASE VI.

§ **73.** *Given the two angles*, A and B.

Equations (8), (10), and (5) of (K) readily give

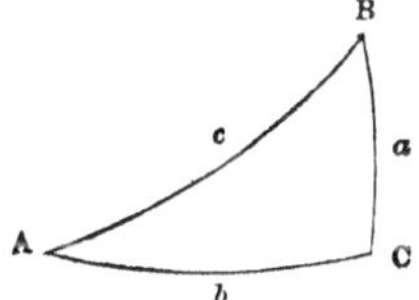


$$\cos. a = \frac{\cos. A}{\sin. B},$$

$$\cos. b = \frac{\cos. B}{\sin. A},$$

$$\cos. c = \cot. A \cot. B.$$

EXAMPLES.

1. Given $A = 62° 15'$, $B = 56° 30'$, to solve the triangle.

| | | |
|---|---|---|
| $\cos. A = 9{\cdot}668027$ | $\cos. B = 9{\cdot}741889$ | $\cot. A = 9{\cdot}721089$ |
| $\sin. B = 9{\cdot}921107$ | $\sin. A = 9{\cdot}946937$ | $\cot. B = 9{\cdot}820783$ |
| $\cos. a = 9.746920$ | $\cos. b = 9.794952$ | $\cos. c = 9{\cdot}541872$ |
| $a = 56° 3' 25''$, | $b = 51° 24' 56''$, | $c = 69° 37' 14''$. |

2. Given $A = 99° 30'$, $B = 150° 15'$, to solve the triangle.

*Ans.* $\begin{cases} a = 109° 25' 39''. \\ b = 151° 40' 30''. \\ c = 72° 58' 30''. \end{cases}$

§ **74.** When a spherical triangle has two or even three of its angles right, it is then isosceles, and may be divided into two equal right triangles, each having only one right angle (except the unique case of three right angles in the given triangle), and the solution may be obtained by the foregoing cases. When all the three angles are right, the sides are all equal, and each a quadrant.

§ **75.** The foregoing relations of spheric right triangles may be associated with the corresponding relations of plane right triangles as follows:

| *In plane right triangles.* | *In spheric right triangles.* |
|---|---|
| | 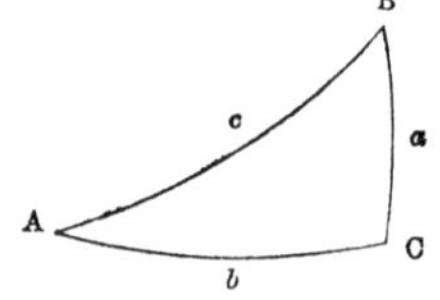  |
| $\text{Sin A} = \frac{a}{c}$; $\sin. B = \frac{b}{c}$. | $\text{Sin. A} = \frac{\sin. a}{\sin. c}$; $\sin. B = \frac{\sin. b}{\sin. c}$. |
| $\text{Cos. A} = \frac{b}{c}$; $\cos. B = \frac{a}{c}$ | $\text{Cos. A} = \frac{\tan. b}{\tan. c}$; $\cos. B = \frac{\tan. a}{\tan. c}$. |
| $\text{Tan. A} = \frac{a}{b}$; $\tan. B = \frac{b}{a}$. | $\text{Tan. A} = \frac{\tan. a}{\sin. b}$; $\tan. B = \frac{\tan. b}{\sin. a}$. |
| $\text{Sin. A} = \cos. B$; $\sin. B = \cos. A$. | $\text{Sin. A} = \frac{\cos. B}{\cos. b}$; $\sin. B = \frac{\cos. A}{\cos. a}$. |
| $c^2 = a^2 + b^2$. | $\left\{ \begin{matrix} \text{Cos. } c = \cos. a \cos. b, \text{ or} \\ \log. \cos. c = \log. \cos. a + \log. \cos. b \end{matrix} \right\}$ |
| $1 = \cot. A \cot. B$. | $\text{Cos. } c = \cot. A \cot. B$. |

# CHAPTER VII.

## SOLUTION OF SPHERICAL OBLIQUE TRIANGLES.

### CASE I.

§ **76.** *When two sides and the included angle are given.*

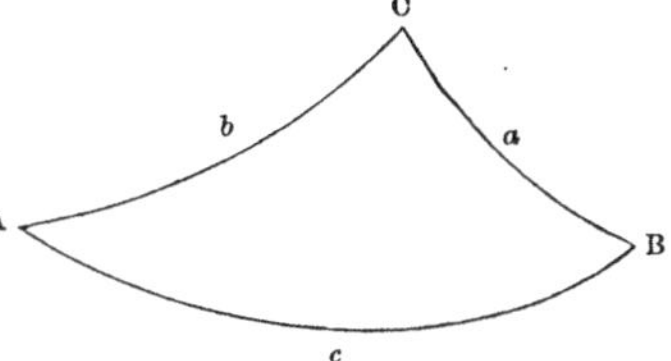


For example, suppose $a$, $b$, and C to be given.

Using formula (F), and writing $B-A$ for $A-B$, and $b-a$ for $a-b$, since $b$ is greater than $a$, and consequently $B>A$, the greater angle being opposite the greater side, we have

$$\tan.\tfrac{1}{2}(B+A)=\frac{\cos.\tfrac{1}{2}(b-a)}{\cos.\tfrac{1}{2}(b+a)}\cot.\tfrac{1}{2}C;$$

$$\tan.\tfrac{1}{2}(B-A)=\frac{\sin.\tfrac{1}{2}(b-a)}{\sin.\tfrac{1}{2}(b+a)}\cot.\tfrac{1}{2}C.$$

Knowing the tangents of $\frac{1}{2}(B+A)$ and $\frac{1}{2}(B-A)$, we readily find from the Tables $\frac{1}{2}(B+A)$ and $\frac{1}{2}(B-A)$, and consequently, by addition and subtraction, we have B and A.

The first equation of (G) gives

$$\tan.\tfrac{1}{2}c=\frac{\cos.\tfrac{1}{2}(B+A)}{\cos.\tfrac{1}{2}(B-A)}\tan.\tfrac{1}{2}(b+a).$$

And, in like manner, if any two sides and the included angle be given, the remaining parts may be found.

#### EXAMPLES.

1. Given $a=70°\ 35'$; $b=120°\ 15'$; $C=54°\ 40$, to find the remaining parts.

$$\tfrac{1}{2}(b+a)=95°\ 25';\ \tfrac{1}{2}(b-a)=24°\ 50';\ \tfrac{1}{2}C=27°\ 20'.$$

| | | | |
|---|---|---|---|
| $\cos.\frac{1}{2}(b-a)=$ | 9·957863 | $\sin.\frac{1}{2}(b-a)=$ | 9·623229 |
| ar. co. $\cos.\frac{1}{2}(b+a)=$ | 1·025038$n$ | ar. co. $\sin.\frac{1}{2}(b+a)=$ | 0·001944 |
| $\cot.\frac{1}{2}C$ | =10·286614 | . . . . . | 10·286614 |
| $\tan.\frac{1}{2}(B+A)=$ | 11·269515$n$ | $\tan.\frac{1}{2}(B-A)=$ | 9·911787 |
| $\frac{1}{2}(B+A)=$ | 93° 4′ 39″. | $\frac{1}{2}(B-A)=$ | 39° 13′ 14″. |

Consequently, $A=53° \ 51' \ 25''$; $B=132° \ 17' \ 53''$, observing that the greater angle is opposite the greater side (Geom., B. VIII., T. XIV.).

$$\begin{aligned} \cos.\tfrac{1}{2}(B+A) &= 8{\cdot}729865\,n \\ \text{ar. co. } \cos.\tfrac{1}{2}(B-A) &= 0{\cdot}110856 \\ \tan.\tfrac{1}{2}(b+a) &= 11{\cdot}023094\,n \\ \hline \tan.\tfrac{1}{2}c \quad &= 9{\cdot}863815 \end{aligned}$$

$\tfrac{1}{2}c=36° \ 9' \ 37''$, and $c=72° \ 19' \ 14''$.

Collecting results, we have

$$Ans. \begin{cases} A=\ 53° \ 51' \ 25''. \\ B=132° \ 17' \ 53''. \\ c=\ 72° \ 19' \ 14''. \end{cases}$$

In this case there can be no ambiguity—only one solution can be obtained.

2. Given $a=100° \ 50'$; $b=99° \ 30'$; $C=68° \ 40'$, to find the remaining parts.

$$Ans. \begin{cases} A=97° \ 51' \ 59''. \\ B=95° \ 52' \ 59''. \\ c=67° \ 27' \ 10''. \end{cases}$$

CASE II.

§ **77.** *When two angles and the interjacent side are given.*

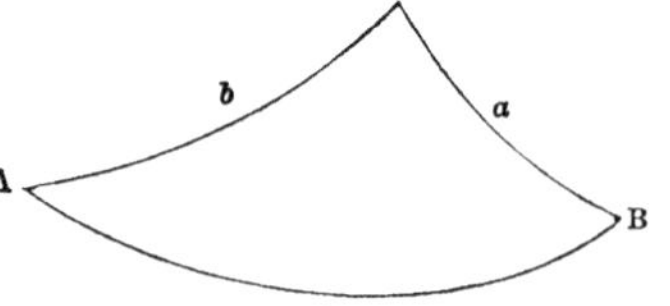


For example, suppose A, B, $c$ to be given.

By formula (G) we have

$$\tan.\tfrac{1}{2}(b+a)=\frac{\cos.\tfrac{1}{2}(B-A)}{\cos.\tfrac{1}{2}(B+A)}\tan.\tfrac{1}{2}c,$$

$$\tan.\tfrac{1}{2}(b-a)=\frac{\sin.\tfrac{1}{2}(B-A)}{\sin.\tfrac{1}{2}(B+A)}\tan.\tfrac{1}{2}c.$$

Having found $a$ and $b$ by the above, we use the first equation of formula (F), which readily gives

$$\cot.\tfrac{1}{2}C=\frac{\cos.\tfrac{1}{2}(b+a)}{\cos.\tfrac{1}{2}(b-a)}\tan.\tfrac{1}{2}(B+A).$$

And, in like manner, if any two angles and the interjacent side be given, the remaining parts may be found.

EXAMPLES.

1. Given $c=110°$; $A=84°$; $B=88°$, to determine the remaining parts.

$$\tfrac{1}{2}(B+A)=86°;\ \tfrac{1}{2}(B-A)=2°;\ \tfrac{1}{2}c=55°.$$

| | | | |
|---|---|---|---|
| $\cos.\tfrac{1}{2}(B-A)=$ | 9·999735 | $\sin.\tfrac{1}{2}(B-A)=$ | 8·542819 |
| ar. co. $\cos.\tfrac{1}{2}(B+A)=$ | 1·156415 | ar. co. $\sin.\tfrac{1}{2}(B+A)=$ | 0·001059 |
| $\tan.\tfrac{1}{2}c=$ | 10·154773 | . . . . $=$ | 10·154773 |
| $\tan.\tfrac{1}{2}(b+a)=$ | 11·310923 | $\tan.\tfrac{1}{2}(b-a)=$ | 8·698651 |

$$\tfrac{1}{2}(b+a)=87°\ 12'\ 7''. \qquad \tfrac{1}{2}(b-a)=2°\ 51'\ 37''.$$

Consequently, $a=84°\ 20'\ 30''$; $b=90°\ 3'\ 44''$.

Observing that the greater side is opposite the greater angle (Geom., B. VIII., T. XIV.)

| | |
|---|---|
| $\cos.\tfrac{1}{2}(b+a)=$ | 8·688563 |
| ar. co. $\cos.\tfrac{1}{2}(b-a)=$ | 0·000541 |
| $\tan.\tfrac{1}{2}(B+A)=$ | 11·155356 |
| $\cot.\tfrac{1}{2}C=$ | 9·844460 |

$$\tfrac{1}{2}C=55°\ 2'\ 51'',\ \text{and}\ C=110°\ 5'\ 42''.$$

Collecting, we have

$$Ans.\begin{cases} a=84°\ 20'\ 30''. \\ b=90°\ 3'\ 44''. \\ C=110°\ 5'\ 42''. \end{cases}$$

This case can never admit of more than one solution.

2. Given $c=70°$; $A=140°$; $B=50°$, to find the other parts.

$$Ans.\begin{cases} a=126°\ 24'\ 41''. \\ b=73°\ 33'\ 19''. \\ C=48°\ 38'\ 18''. \end{cases}$$

CASE III.

§ **78.** *When two sides and angle opposite one of them are given.*

For example, suppose $a$, $b$, and A to be given.

We first find B by (E),

$$\frac{\sin. A}{\sin. B} = \frac{\sin. a}{\sin. b},$$

which gives

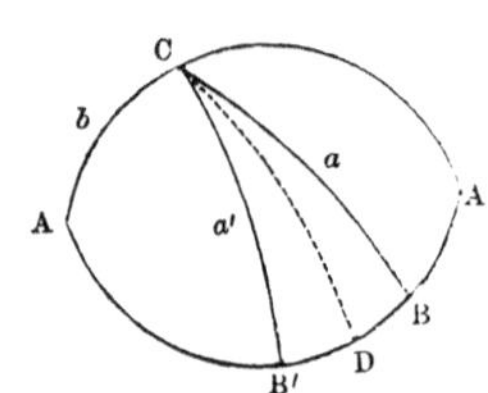


$$\sin. B = \frac{\sin. b \sin. A}{\sin. a}. \qquad (1.)$$

Having found B, we proceed to find C and $c$ by means of the following, drawn from (F) and (G):

$$\cot. \tfrac{1}{2} C = \frac{\cos. \frac{1}{2}(a+b) \tan. \frac{1}{2}(A+B)}{\cos. \frac{1}{2}(a-b)}. \qquad (2.)$$

$$\tan. \tfrac{1}{2} c = \frac{\cos. \frac{1}{2}(A+B) \tan. \frac{1}{2}(a+b)}{\cos. \frac{1}{2}(A-B)}. \qquad (3.)$$

Equation (1) will give two values for B supplementary to each other. If these two values of B, when used in (2) and (3), do not cause either to become negative, there will be two solutions. Since the arcs C and $c$ are each less than 180°, $\frac{1}{2}$ C and $\frac{1}{2} c$ must each be less than 90°, and their tangents and cotangents must be positive. Hence their values given by the right-hand members of (2) and (3) must also be positive.

If we draw the perpendicular CD, we see that it will be the shortest arc that can be drawn from C, in case the angle A (see first figure) is acute; but this perpendicular will be the longest arc that can be drawn from C, when the angle A is obtuse (see second figure). From this we see that $a$ may frequently take two positions, as indicated in the diagrams, on opposites sides of this perpendicular, thus giving two triangles ABC, AB′C, both fulfilling the required conditions.

EXAMPLES.

1. Given $a=40°$; $b=118°$; $A=30°$, to solve the triangle.

$$\begin{aligned} \sin. b &= 9{\cdot}945935 \\ \text{ar. co. } \sin. a &= 0{\cdot}191933 \\ \sin. A &= 9{\cdot}698970 \\ \hline \sin. B &= 9{\cdot}836838 \end{aligned}$$

$B=43°\ 22'\ 42''$, or $136°\ 37'\ 18''$.

$B=43°\ 22'\ 42''$, *first value of* B.
$A=30°$.

$\frac{1}{2}(B+A)=36°\ 41'\ 21''$; $\frac{1}{2}(B-A)=6°\ 41'\ 21''$.
$\frac{1}{2}(b+a)=79°$; $\frac{1}{2}(b-a)=39°$.

$$\begin{aligned} \cos. \tfrac{1}{2}(b+a) &= 9{\cdot}280599 \\ \text{ar. co. } \cos. \tfrac{1}{2}(b-a) &= 0{\cdot}109497 \\ \tan. \tfrac{1}{2}(B+A) &= 9{\cdot}872204 \\ \hline \cot. \tfrac{1}{2}C &= 9{\cdot}262300 \end{aligned}$$

$\frac{1}{2}C=79°\ 37'\ 59''$,
and $C=159°\ 15'\ 58''$;

$$\begin{aligned} \cos. \tfrac{1}{2}(B+A) &= 9{\cdot}904114 \\ \text{ar. co. } \cos. \tfrac{1}{2}(B-A) &= 0{\cdot}002966 \\ \tan. \tfrac{1}{2}(b+a) &= 10{\cdot}711348 \\ \hline \tan. \tfrac{1}{2}c &= 10{\cdot}618428 \end{aligned}$$

$\frac{1}{2}c=76°\ 27'\ 48''$,
and $c=152°\ 55'\ 36''$.

$B=136°\ 37'\ 18''$, *second value of* B.
$A=30°$.

$\frac{1}{2}(B+A)=83°\ 18'\ 39''$; $\frac{1}{2}(B-A)=53°\ 18'\ 39''$.

$$\begin{aligned} \cos. \tfrac{1}{2}(b+a) &= 9{\cdot}280599 \\ \text{ar. co. } \cos. \tfrac{1}{2}(b-a) &= 0{\cdot}109497 \\ \tan. \tfrac{1}{2}(B+A) &= 10{\cdot}930770 \\ \hline \cot. \tfrac{1}{2}C &= 10{\cdot}320866 \end{aligned}$$

$\frac{1}{2}C=25°\ 31'\ 58''$,
and $C=51°\ 3'\ 56''$;

$$\begin{aligned} \cos. \tfrac{1}{2}(B+A) &= 9{\cdot}066264 \\ \text{ar. co. } \cos. \tfrac{1}{2}(B-A) &= 0{\cdot}223681 \\ \tan. \tfrac{1}{2}(b+a) &= 10{\cdot}711348 \\ \hline \tan. \tfrac{1}{2}c &= 10{\cdot}001293 \end{aligned}$$

$\frac{1}{2}c=45°\ 5'\ 7''$,
and $c=90°\ 10'\ 14''$.

Collecting results, we have these two solutions:

$$\textit{Ans.} \left\{ \begin{aligned} B &= 43°\ 22'\ 42'' \\ C &= 159°\ 15'\ 58'' \\ c &= 152°\ 55'\ 36'' \end{aligned} \right\} \text{ or } \left\{ \begin{aligned} B &= 136°\ 37'\ 18''. \\ C &= 51°\ 3'\ 56''. \\ c &= 90°\ 10'\ 14''. \end{aligned} \right.$$

2. Given $a = 100°$; $b = 64° 30'$; $A = 95° 30'$, to find the other parts.

There is only one solution, as follows:

$$Ans. \begin{cases} B = 65° 49' 25''. \\ C = 98° 32' 24''. \\ c = 101° 55' 58''. \end{cases}$$

CASE IV.

§ **79.** *When two angles and the side opposite one of them are given.*

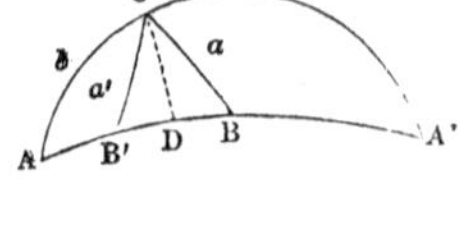


For example, suppose A, B, and $a$ to be given.

We first find $b$ by (E),

$$\frac{\sin. A}{\sin. B} = \frac{\sin. a}{\sin. b},$$

which gives

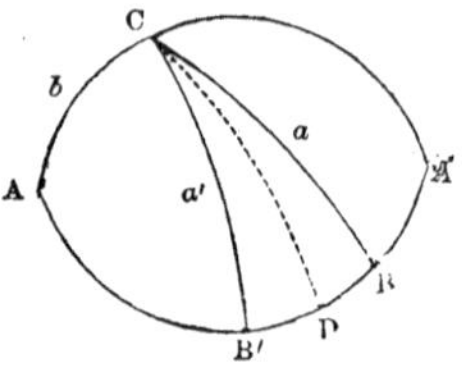


$$\sin. b = \frac{\sin. a \sin. B}{\sin. A}. \qquad (1.)$$

Having found $b$, we proceed to find C and $c$, by using precisely the same formulas as in Case III., which are

$$\cot. \tfrac{1}{2} C = \frac{\cos. \frac{1}{2}(a+b) \tan. \frac{1}{2}(A+B)}{\cos. \frac{1}{2}(a-b)}. \quad . \quad . \quad (2.)$$

$$\tan. \tfrac{1}{2} c = \frac{\cos. \frac{1}{2}(A+B) \tan. \frac{1}{2}(a+b)}{\cos. \frac{1}{2}(A-B)}. \quad . \quad . \quad (3.)$$

As in the last case, equation (1) will give two values for $b$ supplementary to each other.

If these two values of $b$, when used in (2) and (3), do not cause either to become negative, there will be two solutions. These two triangles, which thus satisfy the conditions of the problem, will be indicated in the above diagrams by ABC, A'B'C.

EXAMPLES.

1. Given $A = 140°$; $B = 60°$; $a = 150°$, to solve the triangle.

$$\begin{aligned} \sin. a &= 9{\cdot}698970 \\ \text{ar. co. } \sin. A &= 0{\cdot}191933 \\ \sin. B &= 9{\cdot}937531 \\ \hline \sin. b &= 9{\cdot}828434 \end{aligned}$$

$$b = 42° \ 20' \ 58'' \text{ or } 137° \ 39' \ 2''.$$

$$\begin{aligned} a &= 150° \\ b &= \ 42° \ 20' \ 58'', \textit{first value of } b. \end{aligned}$$

$$\tfrac{1}{2}(a+b) = \ 96° \ 10' \ 29''; \ \tfrac{1}{2}(a-b) = 53° \ 49' \ 31''.$$
$$\tfrac{1}{2}(A+B) = 100°; \ \tfrac{1}{2}(A-B) = 40°.$$

$$\begin{aligned} \cos. \tfrac{1}{2}(a+b) &= \ 9{\cdot}031654n \\ \text{ar.co.cos.} \tfrac{1}{2}(a-b) &= \ 0{\cdot}228964 \\ \tan. \tfrac{1}{2}(A+B) &= 10{\cdot}753681n \\ \hline \cot. \tfrac{1}{2} C &= 10{\cdot}014299 \end{aligned}$$

$$\tfrac{1}{2} C = 44° \ 3' \ 25'', \text{ and } C = 88° \ 6' \ 50''.$$

$$\begin{aligned} \cos. \tfrac{1}{2}(A+B) &= \ 9{\cdot}239670n \\ \text{ar.co.cos.} \tfrac{1}{2}(A-B) &= \ 0{\cdot}115746 \\ \tan. \tfrac{1}{2}(a+b) &= 10{\cdot}965820n \\ \hline \tan. \tfrac{1}{2} c &= 10{\cdot}321236 \end{aligned}$$

$$\tfrac{1}{2} c = \ 64° \ 29' \ 11'', \text{ and } c = 128° \ 58' \ 22''.$$

$$\begin{aligned} a &= 150° \\ b &= 137° \ 39' \ \ 2'', \textit{second value of } b. \end{aligned}$$

$$\tfrac{1}{2}(a+b) = 143° \ 49' \ 31''; \ \tfrac{1}{2}(a-b) = 6° \ 10' \ 29''.$$

$$\begin{aligned} \cos. \tfrac{1}{2}(a+b) &= \ 9{\cdot}906993n \\ \text{ar.co.cos.} \tfrac{1}{2}(a-b) &= \ 0{\cdot}002525 \\ \tan. \tfrac{1}{2}(A+B) &= 10{\cdot}753681n \\ \hline \cot. \tfrac{1}{2} C &= 10{\cdot}663199 \end{aligned}$$

$$\tfrac{1}{2} C = 12° \ 15' \ 10'', \text{ and } C = 24° \ 30' \ 20''.$$

$$\begin{aligned} \cos. \tfrac{1}{2}(A+B) &= 9{\cdot}239670n \\ \text{ar. co. cos.} \tfrac{1}{2}(A-B) &= 0{\cdot}115746 \\ \tan. \tfrac{1}{2}(a+b) &= 9{\cdot}864043n \\ \hline \tan. \tfrac{1}{2} c &= 9{\cdot}219459 \end{aligned}$$

$$\tfrac{1}{2} c = \ 9° \ 24' \ 41'', \text{ and } c = 18° \ 49' \ 22''.$$

Collecting results, we have these two solutions:

$$\textit{Ans.} \left\{ \begin{aligned} b &= \ 42° \ 20' \ 58'' \\ c &= 128° \ 58' \ 22'' \\ C &= \ 88° \ \ 6' \ 50'' \end{aligned} \right\} \text{ or } \left\{ \begin{aligned} b &= 137° \ 39' \ \ 2''. \\ c &= \ 18° \ 49' \ 22''. \\ C &= \ 24° \ 30' \ 20''. \end{aligned} \right.$$

2. Given $A=80^\circ\ 20'$; $B=115^\circ\ 30'$; $a=84^\circ\ 20'$, to find the other parts.

There is only one solution, as follows:

$$Ans.\begin{cases} b = 114^\circ\ 20'\ 26''. \\ c = \ \ 82^\circ\ 39'\ 34''. \\ C = \ \ 79^\circ\ 17'\ \ 8''. \end{cases}$$

CASE V.

§ **80.** *When the three sides are given.*

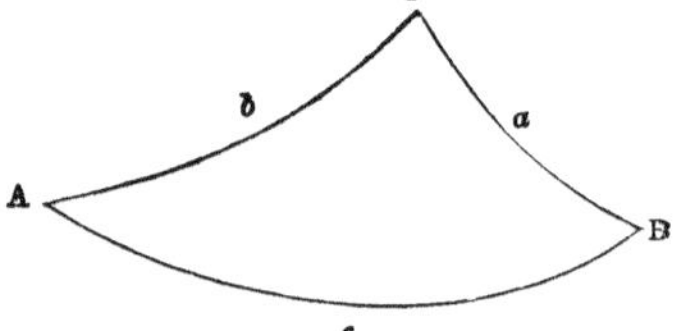


Either of the groups (C′), (C″), (C‴) will give the angles.

If we use (C‴) we have

$$\tan.\tfrac{1}{2}A=\left\{\frac{\sin.(s-b)\sin.(s-c)}{\sin.s\sin.(s-a)}\right\}^{\frac{1}{2}}; \quad . \quad . \quad (1.)$$

$$\tan.\tfrac{1}{2}B=\left\{\frac{\sin.(s-c)\sin.(s-a)}{\sin.s\sin.(s-b)}\right\}^{\frac{1}{2}}; \quad . \quad . \quad (2.)$$

$$\tan.\tfrac{1}{2}C=\left\{\frac{\sin.(s-a)\sin.(s-b)}{\sin.s\sin.(s-c)}\right\}^{\frac{1}{2}}; \quad . \quad . \quad (3.)$$

where $s=\frac{1}{2}(a+b+c)$.

Since the values of the half angles are given by these formulas, there will be no ambiguity in this case.

EXAMPLES.

1. Given $a=100^\circ$; $b=80^\circ$; $c=75^\circ$, to find the angles.

$$\begin{aligned} a &= 100^\circ \\ b &= \ \ 80^\circ \\ c &= \ \ 75^\circ \end{aligned}$$

$\frac{1}{2}(a+b+c)=s=127^\circ\ 30'$; $s-a=27^\circ\ 30'$; $s-b=47^\circ\ 30'$, $s-c=52^\circ\ 30'$.

$$
\begin{array}{rl}
\sin.(s-b) = & 9{\cdot}867631 \\
\sin.(s-c) = & 9{\cdot}899467 \\
\text{ar. co.} \sin.(s-a) = & 0{\cdot}335594 \\
\text{ar. co.} \sin.\ s \quad = & 0{\cdot}100533 \\
\hline
& 2)\,20{\cdot}203225 \\
\hline
\tan.\tfrac{1}{2}\,\mathrm{A} = & 10{\cdot}101612
\end{array}
$$

$\frac{1}{2}\,\mathrm{A} = 38^\circ\ 21'\ 27''$,
and $\mathrm{A} = 76^\circ\ 42'\ 54''$.

$$
\begin{array}{rl}
\sin.(s-c) = & 9{\cdot}899467 \\
\sin.(s-a) = & 9{\cdot}664406 \\
\text{ar. co.} \sin.(s-b) = & 0{\cdot}132369 \\
\text{ar. co.} \sin.\ s \quad = & 0{\cdot}100533 \\
\hline
& 2)\,19{\cdot}796775 \\
\hline
\tan.\tfrac{1}{2}\,\mathrm{B} = & 9{\cdot}898387
\end{array}
$$

$\frac{1}{2}\,\mathrm{B} = 38^\circ\ 21'\ 27''$,
and $\mathrm{B} = 76^\circ\ 42'\ 54''$.

$$
\begin{array}{rl}
\sin.(s-a) = & 9{\cdot}664406 \\
\sin.(s-b) = & 9{\cdot}867631 \\
\text{ar. co.} \sin.(s-c) = & 0{\cdot}100533 \\
\text{ar. co.} \sin.\ s \quad = & 0{\cdot}100533 \\
\hline
& 2)\,19{\cdot}733103 \\
\hline
\tan.\tfrac{1}{2}\,\mathrm{C} = & 9{\cdot}866551
\end{array}
$$

$\frac{1}{2}\,\mathrm{C} = 36^\circ\ 19'\ 57''$ and $\mathrm{C} = 72^\circ\ 39'\ 54''$.

2. Given $a = 49^\circ\ 8'$; $b = 57^\circ\ 16'$; $c = 96^\circ\ 12'$, to find the angles.

$$
Ans. \left\{
\begin{array}{l}
\mathrm{A} = \ \ 31^\circ\ 32'\ 42''. \\
\mathrm{B} = \ \ 35^\circ\ 35'\ 15''. \\
\mathrm{C} = 136^\circ\ 32'\ 48''.
\end{array}
\right.
$$

CASE VI.

§ **81.** *When the three angles are given.*

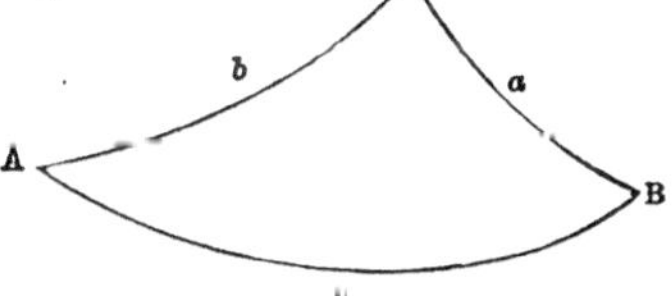


As in last case, we have three groups (D′), (D″), (D‴), either of which will give the sides.

If we use (D‴), we have

$$\tan.\tfrac{1}{2}\,a = \left\{\frac{-\cos.\mathrm{S}\cos.(\mathrm{S}-\mathrm{A})}{\cos.(\mathrm{S}-\mathrm{B})\cos.(\mathrm{S}-\mathrm{C})}\right\}^{\frac{1}{2}}; \quad . \quad . \quad (1.)$$

$$\tan.\tfrac{1}{2}\,b = \left\{\frac{-\cos.\mathrm{S}\cos.(\mathrm{S}-\mathrm{B})}{\cos.(\mathrm{S}-\mathrm{C})\cos.(\mathrm{S}-\mathrm{A})}\right\}^{\frac{1}{2}}; \quad . \quad . \quad (2.)$$

$$\tan.\tfrac{1}{2}\,c = \left\{\frac{-\cos.\mathrm{S}\cos.(\mathrm{S}-\mathrm{C})}{\cos.(\mathrm{S}-\mathrm{A})\cos.(\mathrm{S}-\mathrm{B})}\right\}^{\frac{1}{2}}; \quad . \quad . \quad (3.)$$

where $\mathrm{S} = \frac{1}{2}(\mathrm{A}+\mathrm{B}+\mathrm{C})$.

As in last case there can be no ambiguity in the values of $a$, $b$, $c$ as here found.

EXAMPLES.

1. Given $A=110°$; $B=100°$; $C=90°$, to find the sides.

$$\begin{aligned} A &= 110° \\ B &= 100° \\ C &= 90° \end{aligned}$$

$\frac{1}{2}(A+B+C)=S=150°$; $S-A=40°$; $S-B=50°$; $S-C=60°$.

$$\begin{array}{rlrl} -\cos. S & = 9{\cdot}937531 & \ldots\ldots & = 9{\cdot}937531 \\ \cos.(S-A) & = 9{\cdot}884254 & \cos.(S-B) & = 9{\cdot}808067 \\ \text{ar. co. } \cos.(S-B) & = 0{\cdot}191933 & \text{ar. co. } \cos.(S-C) & = 0{\cdot}301030 \\ \text{ar. co. } \cos.(S-C) & = 0{\cdot}301030 & \text{ar. co. } \cos.(S-A) & = 0{\cdot}115746 \\ \hline & 2)20{\cdot}314748 & & 2)20{\cdot}162374 \\ \hline \tan.\frac{1}{2}a & = 10{\cdot}157374 & \tan.\frac{1}{2}b & = 10{\cdot}081187 \end{array}$$

$$\begin{array}{rlrl} \frac{1}{2}a & = 55°\ 9'\ 40'', & \frac{1}{2}b & = 50°\ 19'\ 28'', \\ \text{and } a & = 110°\ 19'\ 20''. & \text{and } b & = 100°\ 38'\ 56''. \end{array}$$

$$\begin{array}{rl} -\cos. S & = 9{\cdot}937531 \\ \cos.(S-C) & = 9{\cdot}698970 \\ \text{ar. co. } \cos.(S-A) & = 0{\cdot}115746 \\ \text{ar. co. } \cos.(S-B) & = 0{\cdot}191933 \\ \hline & 2)19.944180 \\ \hline \tan.\frac{1}{2}c & = 9{\cdot}972090 \end{array}$$

$\frac{1}{2}c=43°\ 9'\ 37''$ and $c=86°\ 19'\ 14''$.

Since the angle C is right, the solution might have been obtained by Case VI. of Right Triangles. (See Ex. 6, § 82.)

2. Given $A=85°$; $B=60°$; $C=50°$, to find the sides.

$$\textit{Ans.} \begin{cases} a = 51°\ 59'\ 16''. \\ b = 43°\ 13'\ 48''. \\ c = 37°\ 17'\ 26''. \end{cases}$$

§ **82.** The methods which we have given for the solution of the six cases of spherical oblique triangles, as well as for the six cases of spherical right triangles, are all direct and always applicable; but frequently other methods may be used for particular cases, which would lead to simpler methods of solution. We can, however, hardly expect any thing more simple or more easy to be retained by the memory, in the case of right triangles, than the *Rules of Napier;* but in oblique triangles we can frequently obtain simpler solutions than by the methods already given. We may obviously divide any oblique triangle into two right triangles, and thus make all the cases of oblique triangles depend, for their solution, upon the principles of *Napier's Rules.*

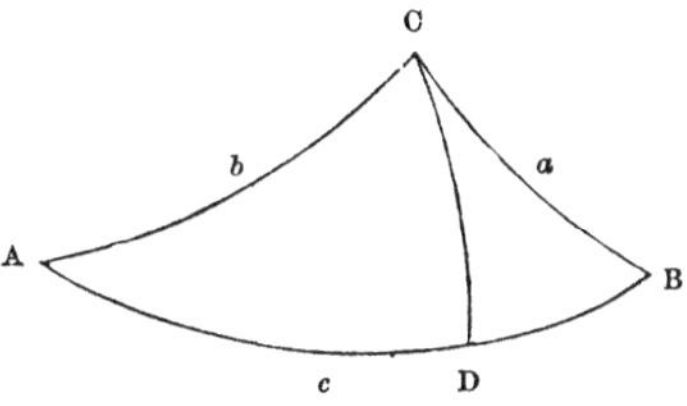


As an example of this method, let us suppose in the triangle ABC, that we have the arcs AB, AC, and the angle A given, which corresponds with Case I. If we draw the perpendicular arc CD, we shall have, by applying *Napier's Rules* to the right triangles ADC, BDC, as follows:

$$\cos. b = \cos. \text{AD} \cos. \text{CD}. \quad . \quad . \quad . \quad (1.)$$
$$\cos. a = \cos. \text{BD} \cos. \text{CD}. \quad . \quad . \quad . \quad (2.)$$

Dividing (1) by (2), we have

$$\frac{\cos. \text{AD}}{\cos. \text{BD}} = \frac{\cos. b}{\cos. a}, \text{ or } \cos. \text{AD} : \cos. \text{BD} :: \cos. b : \cos. a. \quad (3.)$$

That is, *if from an angle of a spheric oblique triangle an arc be drawn perpendicular to the opposite side, dividing it into two segments, we shall have the cosines of those segments to each other as the cosines of their adjacent sides.*

Returning to the triangle ADC, we have

$$\tan. \text{AD} = \tan. b \cos. \text{A}. \quad . \quad . \quad . \quad (4.)$$

Thus we know AD, which subtracted from AB $= c$, gives BD.

The segments are then known, and condition (3) gives at once $\cos. a$.

It is obvious that, by the aid of the great variety of formulas which we have given, the method of solving many of the cases may be varied to almost any extent we please. We will leave this kind of exercise wholly to the student, remarking that in our general solutions we have given those methods which, under all circumstances, seemed to be the safest and best.

## § 83. Examples for Practice.

1. Given, in a right spherical triangle, $A = 91° 11'$; $B = 111° 11'$, to find the remaining parts.

$$Ans. \begin{cases} a = 91° 16' 8''. \\ b = 111° 48' 43''. \\ c = 89° 32' 28''. \end{cases}$$

2. Given, in a right spheric triangle, $a=35° 44'$; $A=37° 28'$, to find the other parts.

$$Ans. \begin{cases} b = 69° 50' 24'' \\ c = 73° 45' 15'' \\ B = 77° 54' 0'' \end{cases} \text{ or } \begin{cases} b = 110° 9' 36''. \\ c = 106° 14' 45''. \\ B = 102° 6' 0''. \end{cases}$$

3. Given, in a spheric oblique triangle, $a=138°$; $A=95°$; $C=104°$, to find the other parts.

$$Ans. \begin{cases} b = 16° 34' 19'' \\ c = 139° 19' 40'' \\ B = 25° 7' 38'' \end{cases} \text{ or } \begin{cases} b = 7° 0' 21''. \\ c = 40° 40' 20''. \\ B = 10° 27' 42''. \end{cases}$$

4. Given $a=81° 17'$; $b=114° 3'$; $c=59° 12'$, to find the angles.

$$Ans. \begin{cases} A = 62° 39' 43''. \\ B = 124° 50' 50''. \\ C = 50° 31' 43''. \end{cases}$$

5. Given, in a right spheric triangle, $a=118° 54'$; $B=12° 19'$, to find the other parts.

$$Ans. \begin{cases} b = 10° 49' 17''. \\ c = 118° 20' 20''. \\ A = 95° 55' 2''. \end{cases}$$

6. Given, in a right spheric triangle, $A = 110°$; $B = 100°$, to find the other part. (This is the same as first example under Case VI. of Oblique Spherical Trigonometry.)

$$Ans. \begin{cases} a = 110° 19' 20''. \\ b = 100° 38' 56''. \\ c = 86° 19' 14''. \end{cases}$$

# CHAPTER VIII.

## MENSURATION OF SURFACES.

§ **84.** THE *area* of a surface is found by comparing it with some known and fixed surface, which is taken as the *unit surface*, which unit surface is usually in the form of a *square;* as, a square inch, a square foot, a square yard, &c.

The *superficial* unit receives its name from that of the *linear* unit, which measures its side. Thus, a square whose side is one inch is called a square inch; one whose side is one foot is called a square foot; one whose side is one yard is called a square yard.

There are some superficial units which have no corresponding linear unit, such as the *acre*, in land measure, and its fourth part, called a *rood.* There is no such linear measure as an acre or a rood. It would be absurd to speak of an acre long or a rood long.

The usual units of square measure are given by the following

TABLE OF SQUARE MEASURE.

| sq. inches. | | sq. feet. | | sq. yd. | | sq. rd. | | sq. ch's. | | Acres. | | S. M. |
|---|---|---|---|---|---|---|---|---|---|---|---|---|
| 144 | = | 1 | | | | | | | | | | |
| 1296 | = | 9 | = | 1 | | | | | | | | |
| 39204 | = | $272\frac{1}{4}$ | = | $30\frac{1}{4}$ | = | 1 | | | | | | |
| 627264 | = | 4356 | = | 484 | = | 16 | = | 1 | | | | |
| 6272640 | = | 43560 | = | 4840 | = | 160 | = | 10 | = | 1 | | |
| 4014489600 | = | 27878400 | = | 3097600 | = | 102400 | = | 6400 | = | 640 | = | 1 |

§ **85.** In measuring land, Gunter's Chain is most commonly employed. Its length is 4 rods, or 66 feet, and is divided into 100 links, so that we have as follows:

| | | |
|---|---|---|
| $7\frac{92}{100}$ | inches, or 7·92 inches = . . . . . | 1 link. |
| 100 | links=792 inches=66 feet=22 yards=4 rods= | 1 chain. |
| 80 | chains = . . . . . . . . . . | 1 mile. |
| 10000 | *sq.* links = 16 *sq.* rods = . . . . . | 1 *sq.* chain. |
| 100000 | *sq.* links = 10 *sq.* chains = 160 *sq.* rods = . | 1 acre. |

The unit of measure for land is the *acre.* It cannot be given in an exact square, since it is made to consist of 160 square rods, and 160 is not a square number, which proves that the side of a square containing 160 *sq.* rods cannot be accurately expressed in rods ; and if not in rods, then not in any multiple or submultiple of rods.

One-fourth of an acre is called a *rood*, and the square rod in land measure is frequently called a *perch.*

If links be multiplied by links, the product will be square links, which may be reduced to acres by dividing by 100000, or by simply pointing off five decimal places. If chains be multiplied by chains, the product will be square chains, which may be reduced to acres by dividing by 10, or pointing off one decimal figure. If chains are multiplied by links, the product will be converted into acres by pointing off three decimals.

We will also add, that to convert square feet into acres, we must divide by 43560. To convert square yards into acres, we must divide by 4840; and to convert square rods into acres, we must divide by 160.

### PROBLEM I.

*To find the area of a square or rectangular piece of land.* The obvious rule in this case will be as follows :

*Take the product of two adjacent sides for the area.* [Geom., B. III., T. XXI., Scholium.]

#### EXAMPLES.

1. How many acres in a piece of ground in the form of a square, each of whose sides is 13 chains, 25 links, or 13·25 chains?

$$13{\cdot}25 \times 13{\cdot}25 = 175{\cdot}5625 \ sq. \text{ chains.}$$

Or, since 13·25 chains = 1325 links, we have

$$1325 \times 1325 = 1755625 \text{ square links,}$$
$$\text{and } 17{\cdot}55625 = \text{the number of acres.}$$

Converting this decimal of an acre into roods and perches, we have

$$\begin{array}{r} 0{\cdot}55625 \\ 4 \\ \hline \text{roods} = 2{\cdot}22500 \\ 40 \\ \hline \text{perches} = 9{\cdot}00000 \end{array} \qquad \textit{Ans. } 17A.\ 2R.\ 9P.$$

2. How many acres in a rectangular piece of land whose adjacent sides are 30·25 chains and 25·21 chains?

$$2521 \times 3025 = 7626025 \ sq. \text{ links} = 76A.\ 1R.\ 1\tfrac{64}{100}P.$$

3. How many rods in a rectangular piece of ground 282 feet wide and 325 feet long? *Ans.* 336·639 *sq.* rods.

PROBLEM II.

*To find the area of a parallelogram.*

FIRST RULE.

*When the altitude is given, multiply it by the base, for the area.* [Geom., B. III., T. XXII.]

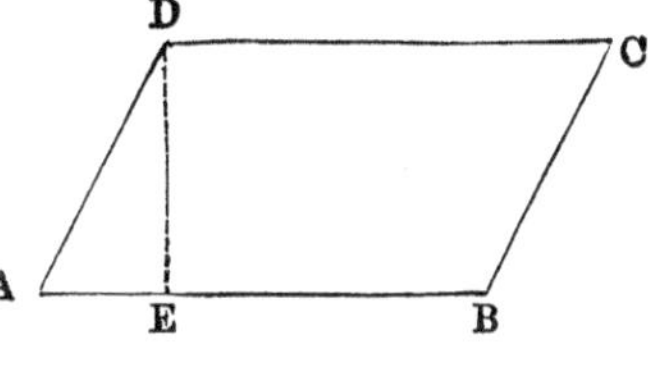


If we multiply AD by the sine of the angle at A, we shall obtain DE. (Case I., Right Triangles.) That is,

$$DE = AD \times \sin. A,$$

$$\text{and } area \text{ ABCD} = DE \times AB = AD \times AB \times \sin. A. \quad (1.)$$

Using logarithms,

$$\log. area = \log. AD + \log. AB + \log. \sin. A - 10. \quad (2.)$$

We subtract 10 to correct for the 10 added to log. sin. of A.

Hence we have this

SECOND RULE.

*Multiply the product of two adjacent sides by the natural sine of the included angle.*

### OR BY LOGARITHMS.

*To the sum of the logarithms of two adjacent sides add the logarithmic sine of the included angle, and subtract* 10, *and the result will be the logarithm of the area.*

#### EXAMPLES.

1. How many acres in a field in the form of a parallelogram, having a base of 13 chains 14 links, and an altitude of 10 chains 37 links? $1314 \times 1037 = 1362618$ *sq.* links $= 13A.\ 2R.\ 20P.$

2. How many acres in a field in the form of a parallelogram, the adjacent sides being 25·17 chains and 30·25 chains, and having 70° 30′ for the included angle?

### BY LOGARITHMS.

$$
\begin{aligned}
\text{log. } 25{\cdot}17 &= 1{\cdot}400883 \\
\text{log. } 30{\cdot}25 &= 1{\cdot}480725 \\
\text{sin. } 70^\circ\ 30' &= 9{\cdot}974347 \\
\hline
\text{log. } \textit{area} &= 2{\cdot}855955
\end{aligned}
$$

Hence, area = 717·72 *sq.* chains, which is

*Ans.* $71A.\ 3R.\ 3\tfrac{1}{2}P.$

3. A field, in the form of a parallelogram, has 10·21 chains, 12·12 chains for the adjacent sides, and 30° 45′ for the included angle. How many acres does it contain?

*Ans.* 63·27 *sq.* chains.

### PROBLEM III.

*To find the area of a triangle.*

#### FIRST RULE.

*When the altitude is given, multiply its half by the base.* [Geom., B. III. T. XXIII.]

*Or multiply the altitude by half the base, or take half the product of the altitude and base.*

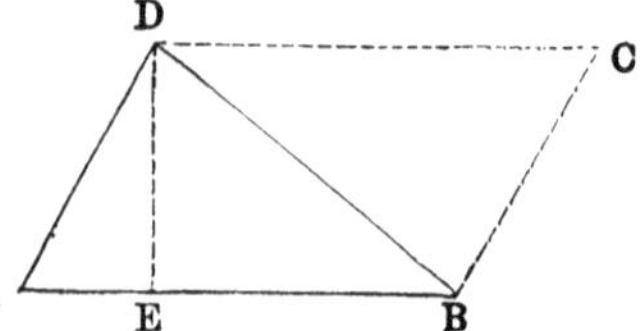


Since the triangle ABD is obviously just one-half of the parallelogram ABCD, whose measure is given by equation (1) of Prob. II., we obviously have

$$\text{area ABD} = \frac{\text{DE} \times \text{AB}}{2} = \frac{\text{AD} \times \text{AB} \times \sin. \text{A}}{2}. \qquad (1.)$$

Or, using logarithms,

$$\log.\ \text{area} = \log.\ \text{AD} + \log.\ \text{AB} + \log.\ \sin.\ \text{A} - 10{\cdot}301030. \qquad (2.)$$

The log. of 2, the denominator of (1), being 0·301030, is to be subtracted in addition to the 10.

Hence, we have this

### SECOND RULE.

*Multiply the product of any two adjacent sides by the sine of the included angle, and one-half this last product will be the area.*

### BY LOGARITHMS.

*To the sum of the logarithms of any two adjacent sides add the logarithmic sine of the included angle, and subtract* 10·301030; *and the remainder will be the logarithm of the area.*

We have also given [Geom., B. VIII., P. IV.] a demonstration to this

### THIRD RULE.

*Take half the sum of the three sides, and from this half sum subtract each side separately; then take the square root of the continued product of the half sum and the three remainders, and it will be the area.*

### BY LOGARITHMS.

*Find the half sum of the three sides, and the three remainders as before; take half the sum of their logarithms, and it will be the logarithm of the area.*

### EXAMPLES.

1. How many square feet in a triangle whose base is 110·5 feet and altitude 97·25 feet?

*Ans.* $\frac{1}{2}$ of 110·5 × 97·25 = 5373·0625 *sq.* feet.

2. If an angle of a triangle is 33° 33′, included by sides whose

lengths are 45·25 feet and 37·5 feet respectively, what will be its area in square rods?

$$\begin{aligned} \log. 45{\cdot}25 &= 1{\cdot}655619 \\ \log. 37{\cdot}5 &= 1{\cdot}574031 \\ \sin. 33^\circ 33' &= 9{\cdot}742462 \\ \hline & \phantom{=}\ 12{\cdot}972112 \\ & \phantom{=}\ 10{\cdot}301030 \\ \hline & \phantom{=}\ 2{\cdot}671082 = \text{log. of } sq. \text{ feet.} \\ \text{subtract } & \phantom{=}\ 2{\cdot}434968 = \text{log. of } 272\tfrac{1}{4}, sq. ft. \text{ in a } sq. rd. \\ \hline & \phantom{=}\ 0{\cdot}236114 = \text{log. of } sq. \text{ rods.} \end{aligned}$$

*Ans.* 1·7223 *sq.* rods.

3. How many acres in a triangular field whose sides are 70, 110, and 120 chains, respectively?

$$\begin{aligned} \tfrac{1}{2} \text{ of } (70+110+120) &= 150, \log. = 2{\cdot}176091 \\ 150-\ \ 70 &= \ \ 80, \log. = 1{\cdot}903090 \\ 150-110 &= \ \ 40, \log. = 1{\cdot}602060 \\ 150-120 &= \ \ 30, \log. = 1{\cdot}477121 \\ \hline & \quad 2\,)\,7{\cdot}158362 \\ \hline & \text{log. area} = 3{\cdot}579181 \end{aligned}$$

*Ans.* 3794·7 *sq.* chains = 379·47 acres.

4. How many *sq.* feet in a triangle, each of whose sides is denoted by $a$?

*Ans.* $\sqrt{\tfrac{3}{2}a \times \tfrac{1}{2}a \times \tfrac{1}{2}a \times \tfrac{1}{2}a} = \dfrac{a^2}{4}\sqrt{3}.$

5. Referring to the last example, how many *sq.* feet in an equilateral triangle whose side is 10 feet?

*Ans.* 43·301 *sq.* feet.

## PROBLEM IV

*To find the area of a trapezoid.*

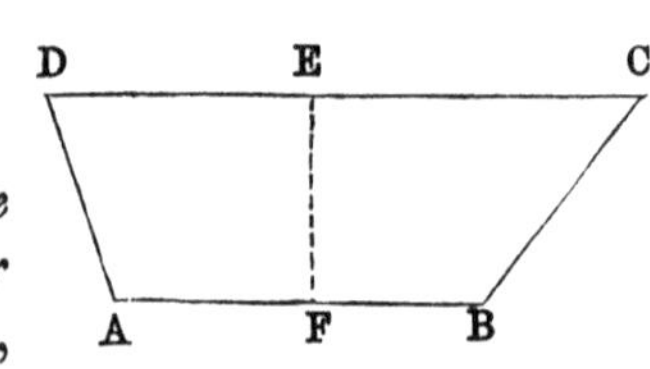


### RULE.

*Multiply half the sum of the parallel sides by the perpendicular distance between them.* [Geom., B. III., T. XXIII., Cor. II.]

EXAMPLES.

1. What is the area of a trapezoid, whose parallel sides are 20 and 30 chains, and which are perpendicularly distant 12·5 chains?

$$\tfrac{1}{2} \text{ of } (20 + 30) = 25;$$
$$25 \times 12{\cdot}5 = 312{\cdot}5 \textit{ sq.} \text{ chains, or}$$
$$31\tfrac{1}{4} \text{ acres.}$$

2. If the parallel sides of a trapezoid, whose lengths are 25 and 35 feet, are at the distance of 12 feet from each other, what will be its area? *Ans.* 360 *sq.* feet.

3. What is the area of a trapezoid, whose parallel sides are 3 and 4·90 chains, and the perpendicular distance between them 1·66 chains? *Ans.* 6·557 *sq.* chains.

4. How many square feet in a trapezoidal table whose parallel sides are 5·5 feet and 4·25 feet, the width being 3·5 feet? *Ans.* 17·0625 *sq.* feet.

SCHOLIUM.—In land surveying, we frequently have occasion to measure long irregular portions of a field, which are bounded on one side by a straight line.

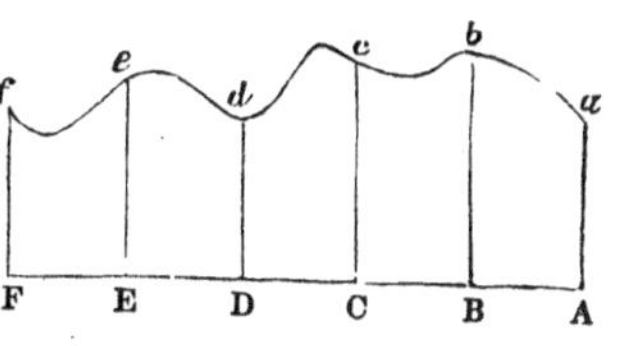


At equal intervals along the straight line AF, as at B, C, D, and E, measure the perpendiculars B$b$, C$c$, D$d$, &c.; also measure the extreme perpendiculars A$a$ and F$f$. When the equal intervals AB, BC, CD, &c., are short, the figures AB$ba$, BC$cb$, CD$dc$, &c., may, without much error, be regarded as trapezoids; hence, calling $s$ one of the equal distances, AB, BC, CD, &c., we have

$$\text{area AB}ba = \frac{\text{A}a + \text{B}b}{2} \times s,$$

$$\text{area BC}cb = \frac{\text{B}b + \text{C}c}{2} \times s,$$

$$\text{area CD}dc = \frac{\text{C}c + \text{D}d}{2} \times s,$$

$$\text{area DE}ed = \frac{\text{D}d + \text{E}e}{2} \times s,$$

$$\text{area EF}fe = \frac{\text{E}e + \text{F}f}{2} \times s.$$

The total area is $(\tfrac{1}{2}\,\text{A}a + \text{B}b + \text{C}c + \text{D}d + \text{E}e + \tfrac{1}{2}\,\text{F}f) \times s$.

Hence this

RULE.

*To half the sum of the extreme breadths add all the intermediate breadths; multiply this sum by one of the equal distances, and the product will be the area.*

As an example, suppose seven breadths of an irregular field, taken at equal intervals, to be 32 links, 47 links, 63 links, 13 links, 17 links, 5 links, and 2 links; if one of the equal intervals is 10 links, what will be the area?

$$\frac{32+2}{2}+47+63+13+17+5=162,$$

and $162 \times 10 = 1620$ for the number of square links in the area.

PROBLEM V.

*To find the area of any irregular polygonal figure.*

RULE.

*Draw a sufficient number of diagonals to divide the whole figure into triangles. Then compute each triangle separately, and take their sum.*

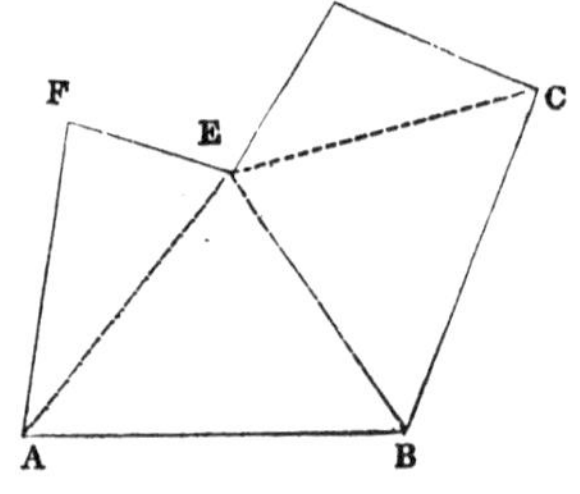


As an illustration, suppose ABCDEF to be the polygon. Draw the diagonals EA, EB, EC, and suppose by actual measurement we find AB = 12·5; BC = 11·25; CD = 7; DE = 6·75; EF = 5·75; FA = 10·5; EA = 10·75; EB = 10; EC = 9·75.

Now computing each triangle by the *Third Rule* under Problem III., we shall find

| | |
|---|---|
| Area EFA = | 29·3795 |
| " EAB = | 51·6642 |
| " EBC = | 45·6431 |
| " ECD = | 23·6247 |
| area ABCDEF = | 150·3115 |

PROBLEM VI.

*To find the area of a quadrilateral in terms of its diagonals and their included angle.*

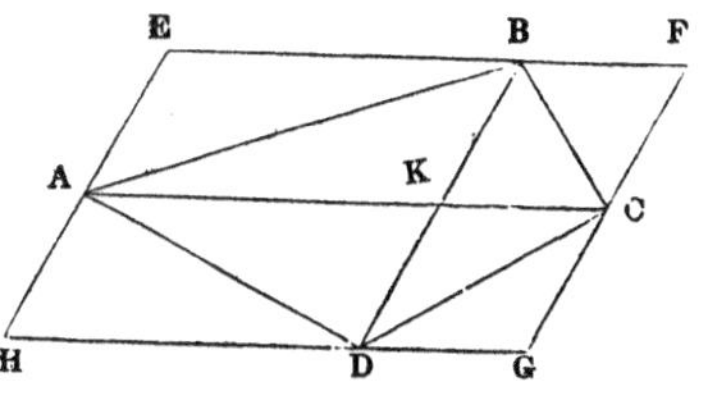


Let ABCD be the quadrilateral figure, of which the diagonals AC, BD, and their angle K of intersection are given.

Through A, B, C, D, draw lines parallel to the diagonals, which will together form the parallelogram EFGH, whose angles are equal to the angles at K; and whose area is obviously double that of the quadrilateral ABCD.

But, Prob. II., EFGH $=$ HE $\times$ HG $\times$ sin. H $=$ BD $\times$ AC $\times$ sin. K; and we have ABCD $= \frac{1}{2}$ BD $\times$ AC $\times$ sin. K.

Hence we have this

RULE.

*Multiply the product of the two diagonals by the sine of their included angle, and one-half this last product will be the area.*

BY LOGARITHMS.

*To the sum of the logarithms of the two diagonals add the logarithmic sine of their included angle, and subtract* 10·301030; *and the remainder will give the logarithm of the area.*

NOTE.—This rule will agree with the second rule of Prob. III., if the word *diagonals* is changed into *sides.*

EXAMPLES.

1. The diagonals of a quadrilateral are 37·5 and 46·25, and they make with each other an angle of 47° 47′. What is the area?

$$
\begin{aligned}
\text{log. } 37{\cdot}5 &= 1{\cdot}574031 \\
\text{log. } 46{\cdot}25 &= 1{\cdot}665112 \\
\text{sin. } 47^\circ\ 47' &= 9{\cdot}869589 \\
\hline
&\phantom{=}\ 13{\cdot}108732 \\
&\phantom{=}\ 10{\cdot}301030 \\
\hline
\text{log. area} &= 2{\cdot}807702 \\
\text{and } \textit{area} &= 642{\cdot}247.
\end{aligned}
$$

2. The diagonals of a quadrilateral are 90 and 100, and make an angle of 75° with each other. What is the area?

*Ans.* 4346·68.

## PROBLEM VII.

*To find the area of any regular polygon.*

### FIRST RULE.

*Take one-half the product of the perimeter and perpendicular drawn from its centre to one of its sides.* [Geom., B. IV., T. V.]

When the perpendicular is not given, it may be found as follows: the angle ACB will be found by dividing 360° by the number of sides in the polygon; and the angle BCD, which is one-half of ACB, can be found by dividing 180° by the number of sides. Hence if we denote the number of sides by $n$, we shall have $\frac{180°}{n}$, for the angle BCD.

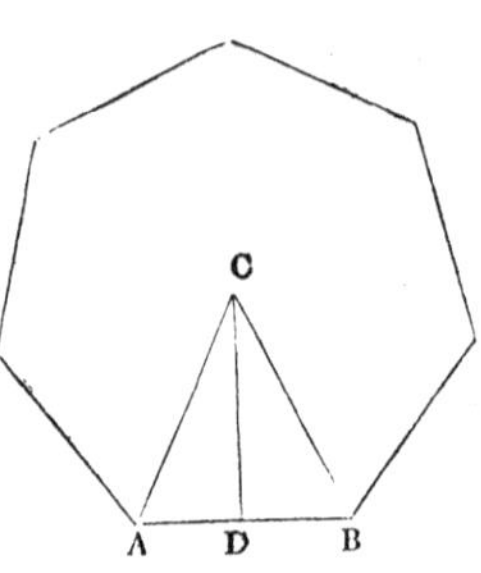


Now BD, which is one-half of a side of the polygon, being multiplied by cotan. BCD, will give CD. If, then, $s$ denote a side and $p$ the perpendicular, we shall have

$$p = \tfrac{1}{2} \text{ of } s \times \text{cotan. } \frac{180°}{n}. \quad . \quad . \quad . \quad (1.)$$

And, the perimeter being $n \times s$, we shall obtain

$$\text{area} = \tfrac{1}{2} \text{ of } p \times n \times s = \tfrac{1}{4} \text{ of } s^2 \times n \times \text{cotan. } \frac{180°}{n}. \quad (2.)$$

If the side of the polygon is taken as a unit, then $s = 1$, and

$$\text{area} = \tfrac{1}{4} \text{ of } n \times \text{cotan. } \frac{180°}{n}. \quad . \quad . \quad . \quad (3.)$$

If, in formula (3), we take successively $n = 3, 4, 5, 6, 7$, &c., we shall obtain the areas of regular polygons, whose sides are unity, as given in the following

TABLE.

| *Names.* | *Sides.* | *Areas.* | *Logarithms.* |
|---|---|---|---|
| Triangle | 3 | 0·4330127 | $\bar{1}$·6365007 |
| Square | 4 | 1·0000000 | 0·0000000 |
| Pentagon | 5 | 1·7204774 | 0·2356490 |
| Hexagon | 6 | 2·5980762 | 0·4146519 |
| Heptagon | 7 | 3·6339124 | 0·5603744 |
| Octagon | 8 | 4·8284271 | 0·6838057 |
| Nonagon | 9 | 6·1818242 | 0·7911166 |
| Decagon | 10 | 7·6942088 | 0·8861640 |
| Undecagon | 11 | 9·3656399 | 0·9715375 |
| Dodecagon | 12 | 11·1961524 | 1·0490687 |

Since the areas of similar figures are to each other as the squares of their homologous sides, we may find the area of any regular polygon whose number of sides is not greater than 12 by this

SECOND RULE.

*Multiply the square of the length of a side of the polygon by the area of a similar polygon whose side is unity, as found in the preceding table.*

BY LOGARITHMS.

*To twice the logarithm of a side of the given polygon add the logarithm of the corresponding tabular number.*

EXAMPLES.

1. What is the area of a regular dodecagon whose side is 7·75?

log. 7·75 = 0·889302
log. 7·75 = 0·889302
tabular log. = 1·049069

log. area = 2·827673
Area = 672·469.

2. What is the area of an octagonal room, each side of which is 10·5 feet? *Ans.* 532·334 *sq.* feet.

3. What is the number of square inches in a pentagon, each of whose sides is 10 inches? *Ans.* 172·048 *sq.* inches.

4. What is the area of an equilateral triangle, having 100 for a side? *Ans.* 4330·127.

### PROBLEM VIII.

*To find the circumference of a circle when its diameter is known, and conversely.*

If the diameter of a circle is a unit, its circumference will be 3·141592653589793, &c. (§ 5). Denote this ratio by $\pi$, and put

$$R = \text{radius},$$
$$D = \text{diameter},$$
$$C = \text{circumference}.$$

Now since circumferences are to each other as their diameters, and consequently as their radii, we shall have

$$C = D \times \pi = D \times 3{\cdot}14159265, \quad . \quad . \quad (1.)$$
$$C = R \times 2\pi = R \times 6{\cdot}28318531, \quad . \quad . \quad (2.)$$
$$D = C \times \frac{1}{\pi} = C \times 0{\cdot}31830989, \quad . \quad . \quad (3.)$$
$$R = C \times \frac{1}{2\pi} = C \times 0{\cdot}15915494. \quad . \quad . \quad (4.)$$

Using logarithms, we have

$$\log. C = 0{\cdot}4971499 + \log. D, \quad . \quad . \quad . \quad (5.)$$
$$\log. C = 0{\cdot}7981799 + \log. R, \quad . \quad . \quad . \quad (6.)$$
$$\log. D = \bar{1}{\cdot}5028501 + \log. C, \quad . \quad . \quad . \quad (7.)$$
$$\log. R = \bar{1}{\cdot}2018201 + \log. C. \quad . \quad . \quad . \quad (8.)$$

#### EXAMPLES.

1. What is the circumference of a circle whose diameter is 12·5 feet? *Ans.* 3·1416 × 12·5 = 39·27 feet, nearly.

2. If the radius of the earth is 3956 miles, what is its circumference? *Ans.* 24856·281 miles, nearly.

3. If the circumference of a circle is 100 feet, what is its diameter? *Ans.* 0·31830989 × 100 = 31·83 feet, nearly.

4. If the circumference of a wheel is 15·5 feet, what is its radius? *Ans.* 2·4669 feet.

5. What is the circumference of a circle whose diameter is 100 feet? *Ans.* 314·159 feet.

### PROBLEM IX.

*To find the area of a circle when its diameter is known, and conversely. Also when the circumference is known.*

Denoting the area by A and continuing the notation as under Prob. VIII., we have [Geom., B. IV., T. XIII.]

$$A = \tfrac{1}{4} \text{ of } D \times C.$$

Substituting the value of C as already given, we have

$$A = D^2 \times \frac{\pi}{4} = D^2 \times 0{\cdot}78539816, \quad . \quad . \quad (1.)$$

$$A = R^2 \times \pi = R^2 \times 3{\cdot}14159265, \quad . \quad . \quad (2.)$$

$$A = C^2 \times \frac{1}{4\pi} = C^2 \times 0{\cdot}07957747, \quad . \quad . \quad (3.)$$

$$D = \sqrt{A} \times 2\sqrt{\frac{1}{\pi}} = \sqrt{A} \times 1{\cdot}12837917, \quad . \quad (4.)$$

$$R = \sqrt{A} \times \sqrt{\frac{1}{\pi}} = \sqrt{A} \times 0{\cdot}56418958, \quad . \quad (5.)$$

$$C = \sqrt{A} \times 2\sqrt{\pi} = \sqrt{A} \times 3{\cdot}54490770. \quad . \quad (6.)$$

Using logarithms, we have

$$\log. A = \bar{1}{\cdot}8950899 + 2 \times \log. D, \quad . \quad (7.)$$

$$\log. A = 0{\cdot}4971499 + 2 \times \log. R, \quad . \quad (8.)$$

$$\log. A = \bar{2}{\cdot}9007901 + 2 \times \log. C, \quad . \quad (9.)$$

$$\log. D = 0{\cdot}0524551 + \tfrac{1}{2} \times \log. A, \quad . \quad (10.)$$

$$\log. R = \bar{1}{\cdot}7514251 + \tfrac{1}{2} \times \log. A, \quad . \quad (11.)$$

$$\log. C = 0{\cdot}5496049 + \tfrac{1}{2} \times \log. A. \quad . \quad (12.)$$

#### EXAMPLES.

1. What is the area of a circle, whose diameter is 125 feet?

By equation (7), we have

$$\begin{array}{rl} & \bar{1}{\cdot}895090 \\ 2 \times \log. 125 = & 4{\cdot}193820 \\ \hline \log. A = & 4{\cdot}088910 \\ A = & 12271{\cdot}8. \end{array}$$

If more decimals are required, they may be obtained by using equation (1), which gives $0{\cdot}78539816 \times 125 \times 125 = 12271{\cdot}84625$.

2. What is the number of square inches in a circular stand-top, whose radius is 15 inches?

*Ans.* $3{\cdot}1416 \times 15 \times 15 = 706{\cdot}86$ *sq.* inches.

3. What is the area of a circular fish-pond, whose circumference is 25 rods?

*Ans.* $0{\cdot}07957747 \times 25 \times 25 = 49{\cdot}734$ *sq.* rods.

4. What is the number of rods in the diameter, the radius, and the circumference of a circle, which shall contain just 1 acre?

Using equations (10), (11), and (12), we have

$$\begin{array}{rccc} & 0{\cdot}052455 & \bar{1}{\cdot}751425 & 0{\cdot}549605 \\ \tfrac{1}{2} \text{ of log. } 160 = & 1{\cdot}102060 \;\ldots & 1{\cdot}102060 \;\ldots & 1{\cdot}102060 \\ \hline & 1{\cdot}154515 & 0{\cdot}853485 & 1{\cdot}651665 \\ & D = 14{\cdot}273. & R = 7{\cdot}1365. & C = 44{\cdot}84. \end{array}$$

NOTE.—This last example would have required far more labor had we used equations (4), (5), and (6), which are independent of logarithms.

5. How many chains must the diameter of a circular farm be which shall contain just 100 acres? *Ans.* 35·68 chains.

### PROBLEM X.

*To determine formulas for the area and for the arc of a sector when the angle at the centre is known, and conversely.*

Denote the number of degrees which measures the angle at the centre by $n$, the length of the corresponding arc by $c$, and the area of corresponding sector by $a$. Then, since arcs of the same circle are to each other as their corresponding angles at the centre; also, since sectors are to each other in the same ratio, we shall have

$$360 : n :: C : c,$$
$$360 : n :: A : a.$$

And

$$c = \frac{n}{360} \times C, \quad \ldots \quad (1.)$$

$$a = \frac{n}{360} \times A. \quad \ldots \quad (2.)$$

### EXAMPLES.

1. What is the length of an arc of 15° of the earth, its entire circumference being 24856·28 miles?

*Ans.* $\frac{15}{360}$ of $24856{\cdot}28 = 1035{\cdot}68$ miles, nearly.

2. What is the area of a sector of 10 degrees of a circle, whose diameter is 10 feet?

If, in equation (2), we substitute the value of A as given by equation (1) of Prob. IX., it will become

$$a = \frac{n}{360} \times \frac{\pi}{4} \times D^2 = \tfrac{10}{360} \times \frac{\pi}{4} \times 100 = 2{\cdot}18 \; sq. \text{ feet, nearly.}$$

3. What is the length of an arc of 1° of the moon, its diameter being 2160 miles?

In equation (1), substitute the value of C as given by (1) of Prob. VIII., and we have

$$c = \frac{n}{360} \times \pi \times D = \tfrac{1}{360} \times \pi \times 2160 = 18{\cdot}849 \text{ miles, nearly.}$$

4. An arc on the earth of 500 miles corresponds to how great an angle at the centre?

Equation (1) gives immediately [see equation (1), Prob. IX.]

$$n = \frac{c}{C} \times 360° = \frac{500}{24856{\cdot}28} \times 360° = 7°{\cdot}242 = 7° \; 14' \; 31''.$$

5. A sector whose area is 10 square feet corresponds to how great an angle at the centre, provided the diameter of the circle is 10 feet?

Equation (1) gives immediately [see equation (1), Prob. IX.]

$$n = \frac{a}{A} \times 360° = \frac{a}{D^2} \times \frac{4}{\pi} \times 360° = \tfrac{10}{100} \times \frac{4}{\pi} \times 360° = 45°{\cdot}8366 = 45° \; 50' \; 12'.$$

### PROBLEM XI.

*To find the area of a segment of a circle.*

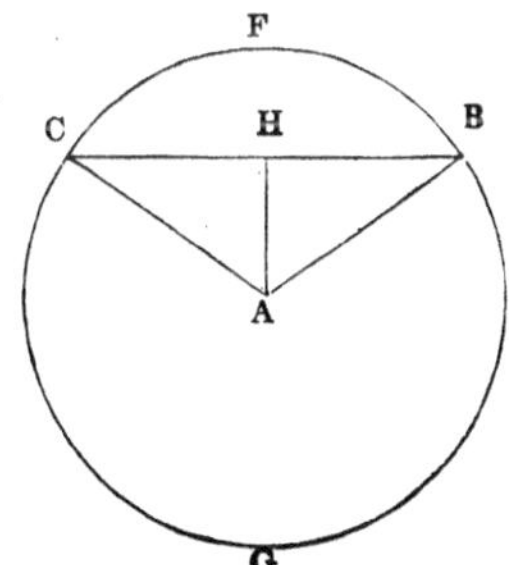


If $n$ denote the number of degrees in the angle BAC, we shall have [Prob. X.]

$$\text{area of sector ABFC} = \frac{n}{360} \times R^2 \times \pi,$$

since $A = R^2 \times \pi$, by equation (5), Prob. IX.,

$$\text{area of triangle ABC} = \tfrac{1}{2} R^2 \times \sin. n.$$

[Prob. III., Second Rule.]

Hence the area of the segment BHCF, being the difference between the sector and triangle, is

$$\left(\frac{n}{360} \times \pi - \tfrac{1}{2}\ \text{sin.}\ n\right) \times R^2. \quad . \quad . \quad (1.)$$

And the area of the segment BHCG, which is greater than a semicircle, is the sum of the sector CBG and triangle ABC. Hence, to determine the area of a segment, we have this

RULE.

*Find the area of a sector which has the same arc as the segment; also, the area of the triangle formed by the chord of the segment and the radii of the sector.*

*Then take the sum of these areas when the segment exceeds the semicircle, and their difference when it is less.*

EXAMPLES.

1. In a circle whose radius is 10 feet, find the area of a segment whose arc corresponds with an angle of 10° at the centre.

Area of sector $= \frac{n}{360} \times \pi \times R^2 = \frac{10}{360} \times 3{\cdot}14159 \times 10^2 = 8{\cdot}727.$

" " triangle $= \frac{1}{2}\ \text{sin.}\ 10° \times 10^2 = \frac{1}{2}$ of $0{\cdot}17365 \times 10^2 = 8{\cdot}683.$

The difference = area of segment = 0·044 of a *sq.* foot.

2. In a circle whose radius is 15 rods, it is required to find the area of a segment whose arc corresponds to a central angle of 200°.

Area of sector = 392·699
" of triangle = 38·477
" of segment 431·176 *sq.* rods.

PROBLEM XII.

*To find the area of an ellipse.*

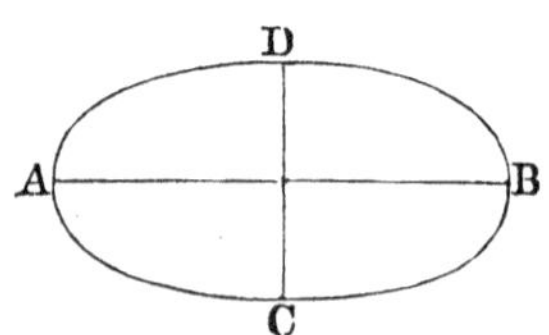


A line drawn through the *centre* of an ellipse is called its diameter. The longest diameter is called the *transverse* diameter; the shortest is called the *conjugate* diameter. Thus AB is the transverse diameter, and CD is the conjugate diameter. These diameters mutually bisect each other at right angles.

The area of an ellipse may be found by this

RULE.

*Multiply the product of the transverse and conjugate diameters by* $0·7854 = \frac{1}{4}\pi$.

A demonstration of this rule cannot be given by the elementary principles of Geometry.

EXAMPLES.

1. How many square feet in the surface of an elliptical pond, whose transverse diameter is 100 feet, and conjugate diameter 60 feet? *Ans.* $100 \times 60 \times 0·7854 = 4712·4$ square feet.

2. How many square inches is an elliptical table, whose transverse diameter is 5 feet 3 inches, and conjugate diameter 3 feet 6 inches? And how many square feet?

*Ans.* { 2078·1684 square inches.
14·4317 square feet.

---

# CHAPTER IX.

## MENSURATION OF SOLIDS,

*Which includes the measure of their surfaces as well as that of their volumes.*

§ **86.** The common unit for measuring solids is a cube; as, a cubic inch, a cubic foot, a cubic yard, &c. The values of the usual units of solid measure are given in the following

TABLE.

| | | | |
|---|---|---|---|
| 1728 | cubic inches | = | 1 cubic foot. |
| 27 | cubic feet | = | 1 cubic yard. |
| $4492\frac{1}{8}$ | cubic feet | = | 1 cubic rod. |
| 128 | cubic feet | = | 1 cord of wood. |
| 231 | cubic inches | = | 1 gallon (liquid measure). |
| $268\frac{4}{5}$ | cubic inches | = | 1 gallon (dry measure). |
| $2150\frac{2}{5}$ | cubic inches | = | 1 bushel. |

§ **87.** By an act of Parliament of Great Britain, which took effect on the 1st of January, 1826, the Imperial gallon of 277·274 cubic inches was adopted as the only gallon. This value was found to be the measure of 10 pounds, avoirdupois, of distilled water. Estimating 8 gallons to the bushel, we have for the Imperial bushel 2218·192 cubic inches.

New York as well as the other States of the Union still continue to use the old English gallons as given in the foregoing table.

### PROBLEM I.

*To find the surface of a right prism, or of a right cylinder.*

#### RULE.

*Multiply the perimeter of the base by the altitude, for the convex surface. To which add the areas of the two bases if the entire surface is required.* [See Geom., B. VI., T. I. Also B. VII., T. I.]

#### EXAMPLES.

1. What is the entire surface of an octagonal prism, whose height is 25 feet, and each side of its base being 1 foot 3 inches?

$1{\cdot}25 \times 8 \times 25 =$ convex surface 250 *sq.* feet.
area of bases 7·544

257·544 *sq.* feet.

2. What is the convex surface of a cylinder, whose length is 40 feet, and its diameter 2 feet? *Ans.* 251·328 *sq.* feet.

### PROBLEM II.

*To find the volume of a right prism, or of a right cylinder.*

#### RULE.

*Multiply the area of the base by the altitude.* [See Geom., B. VI., T. XII., and B. VII., T. II.]

#### EXAMPLES.

1. What is the volume of a triangular prism, whose height is 20 feet, and each side of whose base is 2 feet?

*Ans.* 34·641 cubic feet.

2. How many Imperial gallons in a cylindrical oil-can, the diameter being 20 inches, and depth 40 inches?

Denoting the diameter, in inches, by $d$, and height by $h$, and putting $g$ for the number of inches in a gallon, we have

$$\text{Number of gallons} = \frac{\pi}{4\,g} \times d^2 \times h. \quad . \quad . \quad (1.)$$

For Imperial gallons, $g = 277{\cdot}274$; for United States, or old English gallons, $g = 231$. Hence

$$\text{Imperial gallons} = 0{\cdot}0028326 \times d^2 \times h, \quad (2.)$$
$$\text{United States gallons} = 0{\cdot}0034 \times d^2 \times h. \quad (3.)$$

USING LOGARITHMS.

$$\text{log. [Imp. galls.]} = \bar{3}{\cdot}4521808 + 2 \times \text{log. } d + \text{log. } h, \quad (4.)$$
$$\text{log. [U. S. galls.]} = \bar{3}{\cdot}5314779 + 2 \times \text{log. } d + \text{log. } h. \quad (5.)$$

Now, by equation (2), we find

*Ans.* $0{\cdot}0028326 \times 20^2 \times 40 = 45{\cdot}32$ Imp. gallons.

3. How many cubic feet in a cylindric log 14 feet long, and 14 inches in diameter? *Ans.* 14·966 cubic feet.

4. How many cubic inches in a cylindric measure of $18\frac{1}{2}$ inches diameter, and 8 inches deep?

*Ans.* 2150·425 cubic inches.

NOTE.—The dimensions of the measure as given in the last example are those of the old English bushel, usually called the Winchester bushel; so named from the place where it was deposited for safe keeping.

5. How many United States gallons of oil in a can of 18 inches diameter, the oil being only 13 inches deep?

*Ans.* 14·32 U. S. gallons.

PROBLEM III.

*To find the surface of a regular pyramid, or of a conc.*

RULE.

*Multiply the perimeter of the base by half the slant height. Add the surface of base when the entire surface is required.*

[See Geom., B. VI., T. XV. Also B. VII., T. III.]

EXAMPLES.

1. What is the convex surface of a cone whose slant height is 20 feet, and the diameter of the base 6 inches?

*Ans.* 15·708 *sq.* feet.

2. What is the convex surface of an octagonal pyramid, its base being 10 inches on each side, and the slant height being 35 feet?

*Ans.* $116\frac{2}{3}$ *sq.* feet.

### PROBLEM IV.

*To find the volume of a regular pyramid, or of a cone.*

RULE.

*Multiply the area of the base by one-third of the altitude.* [See Geom., B. VI., T. XVIII., and B. VII., T. V.]

EXAMPLES.

1. How many cubic feet in a conical stick of timber 40 feet long, and 20 inches diameter at the base?

*Ans.* 29·0888 cubic feet.

2. How many cubic yards in a conical stack of hay, which is 18 feet in diameter and 27 feet high?

*Ans.* 84·823 cubic yards.

3. How many cubic feet in a regular octagonal pyramid of marble, whose height is 30 feet, and a side of the base 2 feet?

*Ans.* 193·136 cubic feet.

### PROBLEM V.

*To find the surface of a frustum of a regular pyramid, or of a frustum of a cone.*

RULE.

*Multiply half the sum of the perimeters of the two bases by the slant height. When the entire surface is required, add the areas of the two bases.*

[See Geom., B. VI., T. XV., Cor. II. Also B. VII., T. IV.]

EXAMPLES.

1. What is the entire surface of a frustum of a cone, the slant height of which is 15 feet, and the diameter of one base 2 feet 6 inches, and that of the other 1 foot 3 inches?

*Ans.* 94·4934 *sq.* feet.

2. What is the convex surface of a frustum of a regular hexagonal pyramid, the slant height being 12 feet, each side of the larger base being 4 feet, and each side of the smaller base 3 feet?

*Ans.* 252 *sq.* feet.

PROBLEM VI.

*To find the volume of a frustum of a regular pyramid, or of a cone.*

If we denote the volume of the frustum by V, we shall have (Geom. B. VI., T. XIX.), (See also B. VII., T. V., C. II.)

$$V=(A+a+\sqrt{A\times a})\times\tfrac{1}{3}h=A\times\tfrac{1}{3}h+a\times\tfrac{1}{3}h+\sqrt{A\times a}\times\tfrac{1}{3}h.$$

Hence the following

RULE.

*Add together the areas of the two bases, and a mean proportional between them, and multiply this sum by one third of the altitude.*

In the case of the cone, if D and $d$ denote the diameters of the bases, we shall have

$$A=D^2\times\frac{\pi}{4};\ a=d^2\times\frac{\pi}{4},\text{ and }A^{\frac{1}{2}}a^{\frac{1}{2}}=Dd\times\frac{\pi}{4}.$$

And $$V=(D^2+d^2+Dd)\times h\times 0{\cdot}26179939. \quad (1.)$$

If D, $d$, and $h$ are estimated in inches, we shall have the number of United States gallons, which is the same as the number of old English gallons, in this volume, by dividing by 231. If we divide by 277·274, we shall obtain the number of Imperial gallons. Thus,

$$\text{U.S. gallons}=(D^2+d^2+Dd)\times h\times 0{\cdot}0011333, \quad (2.)$$
$$\text{Imp. gallons}=(D^2+d^2+Dd)\times h\times 0{\cdot}0009442. \quad (3.)$$

Using logarithms, we have

$$\log.[\text{U.S. galls.}]=\bar{3}{\cdot}0543566+\log.(D^2+d^2+Dd)+\log.h, \quad (4.)$$
$$\log.[\text{Imp. galls.}]=\bar{4}{\cdot}9750595+\log.(D^2+d^2+Dd)+\log.h. \quad (5.)$$

### EXAMPLES.

1. A block of marble is in the form of a frustum of a square pyramid: the larger base is 4 feet each way, and the smaller base is 3 feet on a side; the height is 6 feet. How many cubic feet does it contain?

$$A = 4^2 = 16; \quad a = 3^2 = 9; \quad \text{and } A^{\frac{1}{2}} a^{\frac{1}{2}} = 12.$$

$$V = (A + a + A^{\frac{1}{2}} a^{\frac{1}{2}}) \times \tfrac{1}{3} h = (16 + 9 + 12) \times 2 = 74 \text{ cubic ft.}$$

2. A cistern in form of a frustum of a cone is 9 feet deep, having for diameters 8 feet and 10 feet. How many U. S. gallons of 231 cubic inches will it contain?

Using equation (4), we have

$$D^2 + d^2 + Dd = 35136; \quad h = 108.$$

$$\begin{array}{rl} & \bar{3}\cdot054357 \\ \text{log. } 35136 = & 4\cdot545752 \\ \text{log. } 108 = & 2\cdot033424 \\ \hline \text{log. [U. S. galls.]} = & 3\cdot633533 \end{array}$$

*Ans.* 4300·6 U. S. galls.

3. How many Imperial gallons in a cistern in the form of a frustum of a cone, the bottom diameter being 12 feet and top diameter being 10 feet, the depth also 10 feet?

Using equation (5), we have

$$D^2 + d^2 + Dd = 52416; \quad h = 120.$$

$$\begin{array}{rl} & \bar{4}\cdot975060 \\ \text{log. } 52416 = & 4\cdot719464 \\ \text{log. } 120 = & 2\cdot079181 \\ \hline \text{log. [Imp. galls.]} = & 3\cdot773705 \end{array}$$

*Ans.* 5938·9 Imp. galls.

4. How many cubic feet in a block of marble in the form of the frustum of a regular hexagonal pyramid, each side of the larger base being 4 feet, and each side of the smaller base being 3 feet, and the altitude being 9 feet?

*Ans.* 288·386 cubic feet.

PROBLEM VII.

*To find the volume of a wedge.*

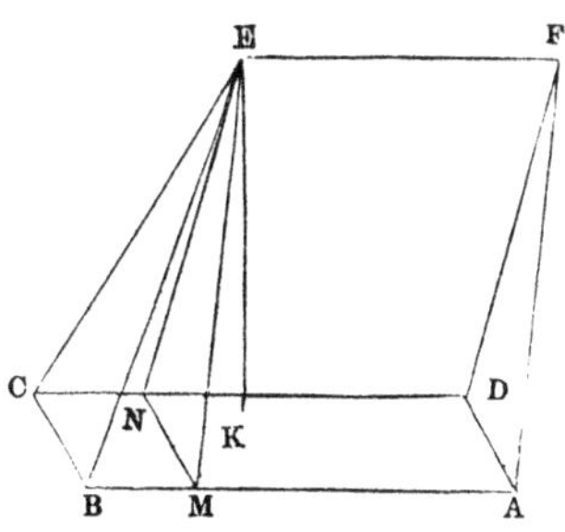


*Definition.*—A *wedge* is a solid bounded by five planes; a rectangle ABCD, called the base of the wedge; two trapezoids ABEF, DCEF, called the sides of the wedge, and which intersect in the line EF, parallel to the base, which is called the edge of the wedge; and two triangles ADF, BCE, which are called the ends of the wedge. If through E we pass the plane EMN parallel to the end FAD, the portion EMBCN thus cut off is a pyramid, and the remaining wedge having AMND for its base is evidently a triangular prism, and has for its measure one-half of a parallelopipedon, having the same base and same altitude.

Hence if $L = AB$, the length of base, $l = EF$, the length of the edge $b = AD$, the breadth of base, and $h = EK$, the altitude, then we shall have

Volume of the wedge whose base is AMND

$$= \tfrac{1}{2} \text{ of } AM \times AD \times EK = \tfrac{1}{2} lbh. \quad . \quad . \quad (1.)$$

Volume of pyramid EMBCN $= \tfrac{1}{3}$ of $MB \times BC \times EK$

$$= \tfrac{1}{3}(L - l)\, bh = \tfrac{1}{3} Lbh - \tfrac{1}{3} lbh. \quad . \quad . \quad . \quad (2.)$$

Taking the sum, we have for the volume of the entire wedge

$$\tfrac{1}{3} Lbh + \tfrac{1}{6} lbh = \tfrac{1}{6}(2\,L + l)\, bh. \quad . \quad . \quad (3.)$$

If the length of the wedge is less than the length of the edge, the volume sought will be the difference between the prism and pyramid. And in this case the volume of the pyramid will be $\tfrac{1}{3}(l - L)\, bh = \tfrac{1}{3} lbh - \tfrac{1}{3} Lbh$, which subtracted from $\tfrac{1}{2} lbh$, the volume of the prism, we have

$$\tfrac{1}{6} lbh + \tfrac{1}{3} Lbh = \tfrac{1}{6}(2\,L + l)\, bh, \quad . \quad . \quad (4.)$$

which is precisely the same as expression (3).

Hence the following

RULE.

*To twice the length of the base add the edge, multiply the sum by the breadth of the base, and that product by $\frac{1}{6}$ of the height of the wedge.*

EXAMPLES.

1. What is the solidity of a wedge, whose base is 10 inches long, 4 inches wide, whose height is 8 inches, and whose edge is 9 inches? *Ans.* $154\frac{2}{3}$ cubic inches.

2. The length and breadth of the base of a wedge is 70 inches and 30 inches, the edge is 110 inches, and the altitude 42 inches. What is the solidity? *Ans.* 30·3819 cubic feet.

PROBLEM VIII.

*To find the volume of a rectangular prismoid.*

*Definition.*—A rectangular prismoid takes its name from its semblance to a prism. It may also be said to resemble a frustum of a quadrangular pyramid. The upper and lower bases are rectangles, having their corresponding sides parallel, and the convex surface consists of four trapezoids. The altitude of the prismoid is the perpendicular distance between the bases.

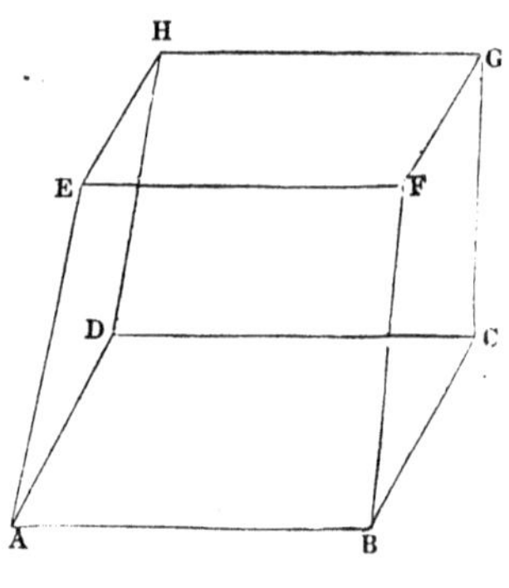


Put L = AB, the length of lower base.
B = AD, the breadth of " "
$l$ = EF, the length of upper base.
$b$ = EH, the breadth of " "
$h$ = the altitude of the prismoid.

If we pass a plane through EF and DC, the prismoid will be divided into two wedges, whose bases are the bases of the prismoid, and whose common altitude is the same as the altitude of the prismoid.

The solidity of these wedges will be (Problem VII.)

$$\tfrac{1}{6}(2\,\mathrm{L}+l)\,\mathrm{B}h,\ \tfrac{1}{6}(2\,l+\mathrm{L})\,bh.$$

Adding, we have for solidity of prismoid,

$$\tfrac{1}{6}\,h\,(2\,\mathrm{BL}+2\,bl+\mathrm{B}l+b\mathrm{L}). \quad . \quad . \quad . \quad (1.)$$

If we put $p=\tfrac{1}{2}(\mathrm{L}+l)$; $q=\tfrac{1}{2}(\mathrm{B}+b)$, we shall have

$$\text{solidity of prismoid} = \tfrac{1}{6}\,h\,(\mathrm{BL}+bl+4\,pq). \quad . \quad . \quad (2.)$$

In this last expression $pq$ is the area of a parallel section equally distant from the two bases. Hence the following

### RULE.

*To the sum of the two bases add four times the area of a parallel section equally distant from the bases, and multiply the sum by $\tfrac{1}{6}$ of the height.*

### EXAMPLES.

1. How many cubic feet in a block of marble in the form of a rectangular prismoid, the lower base being 4 feet by 3, the upper base 3 feet by 2, and the height being 5 feet?

*Ans.* $44\tfrac{1}{6}$ cubic feet.

2. How many cubic feet in a stick 36 feet long, in the form of a rectangular prismoid, the one end being 20 inches by 16 inches, and the other end being 14 inches by 12 inches?

*Ans.* 60 cubic feet.

### PROBLEM IX.

*To find the inclination of two adjacent faces of a regular polyedron, and the radii of the inscribed and circumscribed spheres; also, the area of its surface, and its volume.*

Let AB be the edge common to two adjacent faces, C and E the centres of those faces. Draw CO and EO perpendicular to these faces, meeting at O; also draw CD, ED perpendicular to the common edge AB. The angle CDE will be the inclination of these faces.

Let I = inclination of the faces = angle CDE;

$a$ = one of the edges = AB;

$r$ = radius of the inscribed sphere = OC = OE;

R = radius of the circumscribed sphere = OA = OB;

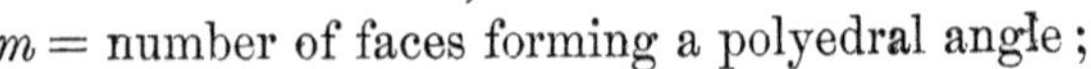

$m$ = number of faces forming a polyedral angle;

$n$ = number of sides in each face;

$f$ = whole number of faces in polyedral;

S = whole surface of polyedron;

V = the volume.

If, now, a sphere be described about O as a centre with a radius equal to unity, its intersection with the planes OAC, OAD, ODC will form a spherical triangle $acd$, right-angled at $d$. By *Napier's Rules*, this triangle gives

$$\cos. cd = \frac{\cos. cad}{\sin. acd}; \quad . \quad . \quad . \quad (1.)$$

$$\cos. ac = \cot. cad \cot. acd. \quad . \quad . \quad (2.)$$

Since $cd$ measures the angle COD, and COD $= 90° - \frac{1}{2}$I, we have $\cos. cd = \sin. \frac{1}{2}$I. We also have

$$\text{angle } cad = \frac{\pi}{m}; \quad \text{angle } acd = \frac{\pi}{n};$$

hence (1) becomes

$$\sin. \tfrac{1}{2}\text{I} = \frac{\cos. \frac{\pi}{m}}{\sin. \frac{\pi}{n}}. \quad . \quad . \quad . \quad (3.)$$

The triangle ACD gives

$$\text{CD} = \tfrac{1}{2} a \cot. \frac{\pi}{n},$$

but triangle COD gives $r$ = CD tan. $\frac{1}{2}$ I; hence we have

$$r = \tfrac{1}{2} a \tan. \tfrac{1}{2}\text{I} \cot. \frac{\pi}{n}. \quad . \quad . \quad . \quad (4.)$$

Triangle AOC gives $R \cos. AOC = R \cos. ac = r$; using the value of $\cos. ac$ given by (2), we find

$$R = r \tan. \frac{\pi}{m} \tan. \frac{\pi}{n} = \tfrac{1}{2} a \tan. \tfrac{1}{2} I \tan. \frac{\pi}{m}. \quad . \quad (5.)$$

The area of one of the faces is

$$n \times AD \times CD = n \times \tfrac{1}{2} a \times \tfrac{1}{2} a \cot. \frac{\pi}{n} = a^2 \times \frac{n}{4} \cot. \frac{\pi}{n};$$

consequently we have

$$S = a^2 \times \frac{nf}{4} \cot. \frac{\pi}{n}. \quad . \quad . \quad . \quad (6.)$$

Since the volume is equal to $S \times \frac{1}{3} r$, we find

$$V = a^3 \times \frac{nf}{24} \tan. \tfrac{1}{2} I \cot.^2 \frac{\pi}{n}. \quad . \quad . \quad (7.)$$

These equations are general and apply to the five *Platonic* bodies. The values of $m$ and $n$ are as follows:

Tetraedron, $m = 3$, $n = 3$. Hexaedron, $m = 3$, $n = 4$.
Octaedron, $m = 4$, $n = 3$. Dodecaedron, $m = 3$, $n = 5$.
Icosaedron, $m = 5$, $n = 3$.

Using these values in (3), we find as follows:

Tetraedron, $\sin. \tfrac{1}{2} I = \dfrac{\cos. 60°}{\sin. 60°} = \tfrac{1}{3} \sqrt{3}.$

Hexaedron, $\sin. \tfrac{1}{2} I = \dfrac{\cos. 60°}{\sin. 45°} = \tfrac{1}{2} \sqrt{2}.$

Octaedron, $\sin. \tfrac{1}{2} I = \dfrac{\cos. 45°}{\sin. 60°} = \tfrac{1}{3} \sqrt{6}.$

Dodecaedron, $\sin. \tfrac{1}{2} I = \dfrac{\cos. 60°}{\sin. 36°} = \tfrac{1}{10} \sqrt{50 + 10 \sqrt{5}}.$

Icosaedron, $\sin. \tfrac{1}{2} I = \dfrac{\cos. 36°}{\sin. 60°} = \tfrac{1}{10} (5 + \sqrt{5}).$

Having thus found $\sin. \frac{1}{2} I$, we can find $\tan. \frac{1}{2} I$ by the formula

$$\tan. \tfrac{1}{2} I = \frac{\sin. \frac{1}{2} I}{\sqrt{1 - \sin.^2 \frac{1}{2} I}};$$

and then equations (4), (5), (6), and (7) will give, after considerable reduction, as follows:

Tetraedron, $\begin{cases} r = a \times \frac{1}{12}\sqrt{6}; \quad R = a \times \frac{1}{4}\sqrt{6}; \\ S = a^2 \times \sqrt{3}; \quad V = a^3 \times \frac{1}{12}\sqrt{2}. \end{cases}$

Hexaedron, $\begin{cases} r = a \times \frac{1}{2}; \quad R = a \times \frac{1}{2}\sqrt{2}; \\ S = a^2 \times 6; \quad V = a^3 \times 1. \end{cases}$

Octaedron, $\begin{cases} r = a \times \frac{1}{6}\sqrt{6}; \quad R = a \times \frac{1}{2}\sqrt{2}; \\ S = a^2 \times 2\sqrt{3}; \quad V = a^3 \times \frac{1}{3}\sqrt{2}. \end{cases}$

Dodecaedron, $\begin{cases} r = a \times \frac{1}{20}\sqrt{250 + 110\sqrt{5}}; \quad R = a \times \frac{1}{4}(\sqrt{15} + \sqrt{3}); \\ S = a^2 \times 3\sqrt{25 + 10\sqrt{5}}; \quad V = a^3 \times \frac{1}{4}(15 + 7\sqrt{5}). \end{cases}$

Icosaedron, $\begin{cases} r = a \times \frac{1}{12}(3\sqrt{3} + \sqrt{15}); \quad R = a \times \frac{1}{4}\sqrt{10 + 2\sqrt{5}}; \\ S = a^2 \times 5\sqrt{3}; \quad V = a^3 \times \frac{5}{12}(3 + \sqrt{5}). \end{cases}$

We give, in the next table, to seven decimal places, the values, and their logarithms, of the surfaces and volumes of the five polyedrons, having unity for the edges of each:

TABLE.

| *Names.* | *No. of Faces.* | *Surfaces.* | *Volumes.* |
|---|---|---|---|
| Tetraedron . . | 4 | 1·7320508 | 0·1178513 |
| | | log. [0·2385607] | log. [$\bar{1}$·0713344] |
| Hexaedron . . | 6 | 6·0000000 | 1·0000000 |
| | | log. [0·7781513] | log. [0·0000000] |
| Octaedron . . | 8 | 3·4641016 | 0·4714045 |
| | | log. [0·5395907] | log. [$\bar{1}$·6733937] |
| Dodecaedron . | 12 | 20·6457288 | 7·6631189 |
| | | log. [1·3148302] | log. [0·8844056] |
| Icosaedron . . | 20 | 8·6602540 | 2·1816950 |
| | | log. [0·9375307] | log. [0·3387940] |

The numbers included within the brackets are the logarithms of those numbers immediately above them.

Since surfaces of similar solids are to each other as the squares of their like dimensions, we may find the surface of any regular polyedron by this

RULE.

*Multiply the square of one of the edges by the tabular surface of the similar unit solid, as given in the foregoing table.*

Or using logarithms.

*To twice the logarithm of the edge add the tabular logarithm of the similar solid, as given in the table, and the sum will be the logarithm of the surface.*

Again, since the volumes of similar solids are to each other as the cubes of their like dimensions, we may find the volume of any regular polyedron by this

RULE.

*Multiply the cube of one of the edges by the tabular volume of the similar unit solid, as given in the table.*

Or using logarithms.

*To three times the logarithm of the edge add the tabular logarithm of the similar solid, as given in the table, and the sum will be the logarithm of the volume.*

EXAMPLES.

1. What is the surface and the volume of a regular icosaedron, each edge of which is 10 inches?

*Ans.* { Surface = 866·025 *sq.* inches.<br>Volume = 2181·695 cubic inches.

2. What is the surface and the volume of a regular dodecaedron, each edge of which is 3 feet?

*Ans.* { Surface = 185·811 *sq.* feet.<br>Volume = 206·902 cubic feet.

3. What is the surface and the volume of a regular tetraedron, each edge being 4 inches?

*Ans.* { Surface = 27·712 *sq.* inches.<br>Volume = 7·542 cubic inches.

PROBLEM X.

*To find the surface and volume of any parallelopipedon.*

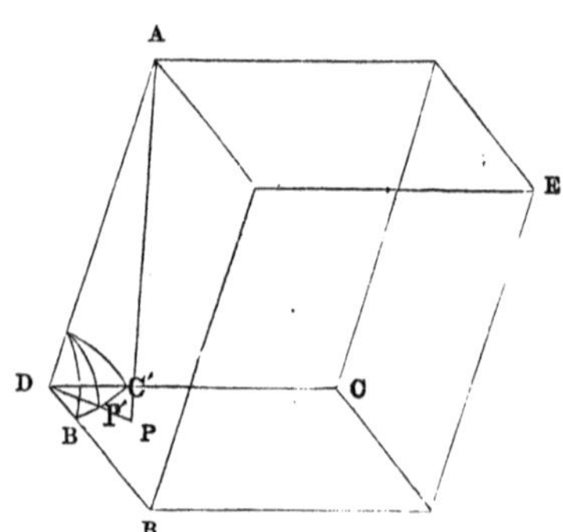


Let DE be a parallelopipedon, having the edges

$$DB = a,$$
$$DC = b,$$
$$DA = c,$$

and their inclinations

$$BDC = \alpha,$$
$$CDA = \beta,$$
$$ADB = \gamma, \text{ given.}$$

The area of any face, as the face BC, $= ab \sin. \alpha$; hence, for the whole surface, we have

$$S = (ab \sin. \alpha + bc \sin. \beta + ca \sin. \gamma). \quad . \quad (1.)$$

To find the volume, draw AP perpendicular to BC, and pass a plane through AD and AP, intersecting BC in DP. Conceive the surface of a sphere whose radius is unity to be described about D as a centre, and the arcs formed by its intersection with the planes passing through the centre D will be as follows:

$$B'C' = \alpha, \quad C'A' = \beta, \quad A'B' = \gamma.$$

The arc A′P′ measures the angle ADP.
The altitude $AP = AD \sin. ADP = c \sin. A'P'$; this, multiplied by the base $BC = ab \sin. \alpha$, gives, for the volume,

$$V = abc \sin. \alpha \sin. A'P'. \quad . \quad . \quad . \quad (2.)$$

The right spherical triangle A′B′P′ gives

$$\sin. A'P' = \sin. \gamma \sin. A'B'P';$$

hence (2) becomes

$$V = abc \sin. \alpha \sin. \gamma \sin. A'B'P'. \quad . \quad . \quad (3.)$$

In the spherical triangle A′B′C′ we have, by formula (C), § 57, if $s = \frac{1}{2}(\alpha + \beta + \gamma)$,

$$\sin. A'B'P' = \frac{2}{\sin. \alpha \sin. \gamma} \sqrt{\sin. s \sin. (s - \alpha) \sin. (s - \beta) \sin. (s - \gamma)};$$

hence (3) becomes

$$V = 2abc \sqrt{\sin. s \sin. (s - \alpha) \sin. (s - \beta) \sin. (s - \gamma)}. \quad (4.)$$

Since the volume of the tetraedron formed by passing a plane through the three points ABC is one sixth of the parallelopipedon, it follows, that if $a$, $b$, $c$ denote the three edges of a polyedral angle of a tetraedron, and $\alpha$, $\beta$, $\gamma$ are their inclinations, the volume of the tetraedron will be

$$V = \tfrac{1}{3} abc \sqrt{\sin. s \sin. (s-\alpha) \sin. (s-\beta) \sin. (s-\gamma)}. \quad (5.)$$

PROBLEM XI.

*To find the surface of a sphere.*

RULE.

*Multiply the square of the diameter by* $\pi = 3{\cdot}1415926$. [See Geom., B. VII., T. XII., S.]

EXAMPLES.

1. Required the number of square inches in the surface of a ball 6 inches in diameter. *Ans.* 113·0976 *sq.* inches.
2. How many square miles on the surface of the earth, it being 7912 miles in diameter. *Ans.* 196662896.

PROBLEM XII.

*To find the volume of a sphere.*

RULE.

*Multiply the cube of the diameter by* $\tfrac{1}{6}\pi = 0{\cdot}5235988$. [See Geom., B. VII., T. XII., S.]

EXAMPLES.

1. What is the volume of a sphere 6 inches in diameter? *Ans.* 113·0976 cubic inches.
2. How many cubic miles in the earth, its diameter being 7912 miles? *Ans.* 259332805350 cubic miles.

PROBLEM XIII.

*To find the area of a portion of the surface of a sphere included between three arcs of great circles. That is, to find the area of a spherical triangle.*

RULE.

*From the sum of the three angles of the triangle subtract* 180°, *divide the remainder by* 90° *and multiply the quotient by one-eighth of the entire surface of the sphere of which the triangle forms a part.* [Geom. B. VIII., T. XIX., S.]

EXAMPLES.

1. How many square inches in a spherical triangle, described on a sphere 6 inches in diameter, and whose angles are 90°, 100°, 110°? *Ans.* 47·124 *sq.* inches.

2. What part of the entire surface of a sphere is the spherical triangle, each of whose angles is 120°? *Ans.* $\frac{1}{4}$.

§ **88.** The following multipliers with their logarithms will be very convenient for reference.

TABLE OF MULTIPLIERS.

| No. | Multiplier | *Log. of multipliers.* |
|---|---|---|
| 1. | Radius of a circle × 6·28318531 = Circumference | 0·7981799 |
| 2. | Square of the radius of a circle × 3·14159265 = Area | 0·4971499 |
| 3. | Diameter of a circle × 3·14159265 = Circumference | 0·4971499 |
| 4. | Square of the diameter of a circle × 0·78539816 = Area | $\bar{1}$·8950899 |
| 5. | Circumference of a circle × 0·15915494 = Radius | $\bar{1}$·2018201 |
| 6. | Circumference of a circle × 0·31830989 = Diameter | $\bar{1}$·5028501 |
| 7. | Square root of area of a circle × 0·56418958 = Radius | $\bar{1}$·7514251 |
| 8. | Square root of area of a circle × 1·12837917 = Diameter | 0·0524551 |
| 9. | Radius of circle × 1·73205081 = Side of inscribed equilateral triangle | 0·2385606 |
| 10. | Side of inscribed equilateral triangle × 0·57735027 = Radius of circle | $\bar{1}$·7614394 |
| 11. | Radius of a circle × 1·41421356 = Side of inscribed square | 0·1505150 |
| 12. | Side of inscribed square × 0·70710678 = Radius | $\bar{1}$·8494850 |
| 13. | Square of radius of a sphere × 12·56637061 = Surface | 1·0992099 |
| 14. | Cube of radius of a sphere × 4·18879020 = Volume | 0·6220886 |
| 15. | Square of diameter of a sphere × 3·14159265 = Surface | 0·4971499 |
| 16. | Cube of diameter of a sphere × 0·52359878 = Volume | $\bar{1}$·7189986 |
| 17. | Square of circumference of a sphere × 0·31830989 = Surface | $\bar{1}$·5028501 |
| 18. | Cube of circumference of a sphere × 0·01688686 = Volume | $\bar{2}$·2275490 |
| 19. | Square root of surface of a sphere × 0·28209479 = Radius | $\bar{1}$·4503951 |
| 20. | Square root of surface of a sphere × 0·56418958 = Diameter | $\bar{1}$·7514251 |
| 21. | Square root of surface of a sphere × 1·77245385 = Circumference | 0·2485748 |
| 22. | Cube root of volume of a sphere × 0·62035049 = Radius | $\bar{1}$·7926371 |
| 23. | Cube root of volume of a sphere × 1·24070098 = Diameter | 0·0936671 |
| 24. | Cube root of volume of a sphere × 3·89777707 = Circumference | 0·5908170 |
| 25. | Radius of a sphere × 1·15470054 = Side of inscribed cube | 0·0624694 |
| 26. | Side of inscribed cube × 0·86602540 = Radius | $\bar{1}$·9375306 |

# TABLE I.,

OF

# LOGARITHMS OF NUMBERS

FROM

## 1 TO 10000.

| N. | Log. | N. | Log. | N. | Log. | N. | Log. |
|---|---|---|---|---|---|---|---|
| 1 | 0·000000 | 26 | 1·414973 | 51 | 1·707570 | 76 | 1·880814 |
| 2 | 0·301030 | 27 | 1·431364 | 52 | 1·716003 | 77 | 1·886491 |
| 3 | 0·477121 | 28 | 1·447158 | 53 | 1·724276 | 78 | 1·892095 |
| 4 | 0·602060 | 29 | 1·462398 | 54 | 1·732394 | 79 | 1·897627 |
| 5 | 0·698970 | 30 | 1·477121 | 55 | 1·740363 | 80 | 1·903090 |
| 6 | 0·778151 | 31 | 1·491362 | 56 | 1·748188 | 81 | 1·908485 |
| 7 | 0·845098 | 32 | 1·505150 | 57 | 1·755875 | 82 | 1·913814 |
| 8 | 0·903090 | 33 | 1·518514 | 58 | 1·763428 | 83 | 1·919078 |
| 9 | 0·954243 | 34 | 1·531479 | 59 | 1·770852 | 84 | 1·924279 |
| 10 | 1·000000 | 35 | 1·544068 | 60 | 1·778151 | 85 | 1·929419 |
| 11 | 1·041393 | 36 | 1·556303 | 61 | 1·785330 | 86 | 1·934498 |
| 12 | 1·079181 | 37 | 1·568202 | 62 | 1·792392 | 87 | 1·939519 |
| 13 | 1·113943 | 38 | 1·579784 | 63 | 1·799341 | 88 | 1·944483 |
| 14 | 1·146128 | 39 | 1·591065 | 64 | 1·806180 | 89 | 1·949390 |
| 15 | 1·176091 | 40 | 1·602060 | 65 | 1·812913 | 90 | 1·954243 |
| 16 | 1·204120 | 41 | 1·612784 | 66 | 1·819544 | 91 | 1·959041 |
| 17 | 1·230449 | 42 | 1·623249 | 67 | 1·826075 | 92 | 1·963788 |
| 18 | 1·255273 | 43 | 1·633468 | 68 | 1·832509 | 93 | 1·968483 |
| 19 | 1·278754 | 44 | 1·643453 | 69 | 1·838849 | 94 | 1·973128 |
| 20 | 1·301030 | 45 | 1·653213 | 70 | 1·845098 | 95 | 1·977724 |
| 21 | 1·322219 | 46 | 1·662758 | 71 | 1·851258 | 96 | 1·982271 |
| 22 | 1·342423 | 47 | 1·672098 | 72 | 1·857333 | 97 | 1·986772 |
| 23 | 1·361728 | 48 | 1·681241 | 73 | 1·863323 | 98 | 1·991226 |
| 24 | 1·380211 | 49 | 1·690196 | 74 | 1·869232 | 99 | 1·995635 |
| 25 | 1·397940 | 50 | 1·698970 | 75 | 1·875061 | 100 | 2·000000 |

N. B. In the following table, in the last nine columns of each page, where the first or leading figures change from 9's to 0's, the character ✦ is introduced instead of the 0's, to catch the eye, and to indicate that from thence the annexed first two figures of the Logarithm in the second column stand in the next lower line, directly under the *asterisk.*

| N. | 0 | 1 | 2 | 3 | 4 | 5 | 6 | 7 | 8 | 9 | D. |
|---|---|---|---|---|---|---|---|---|---|---|---|
| 100 | 00 0000 | 0434 | 0868 | 1301 | 1734 | 2166 | 2598 | 3029 | 3461 | 3891 | 432 |
| 101 | 4321 | 4751 | 5181 | 5609 | 6038 | 6466 | 6894 | 7321 | 7748 | 8174 | 428 |
| 102 | *8600 | 9026 | 9451 | 9876 | ✦300 | 0724 | 1147 | 1570 | 1993 | 2415 | 424 |
| 103 | 01 2837 | 3259 | 3680 | 4100 | 4521 | 4940 | 5360 | 5779 | 6197 | 6616 | 419 |
| 104 | *7033 | 7451 | 7868 | 8284 | 8700 | 9116 | 9532 | 9947 | ✦361 | 0775 | 416 |
| 105 | 02 1189 | 1603 | 2016 | 2428 | 2841 | 3252 | 3664 | 4075 | 4486 | 4896 | 412 |
| 106 | 5306 | 5715 | 6125 | 6533 | 6942 | 7350 | 7757 | 8164 | 8571 | 8978 | 408 |
| 107 | *9384 | 9789 | ✦195 | 0600 | 1004 | 1408 | 1812 | 2216 | 2619 | 3021 | 404 |
| 108 | 03 3424 | 3826 | 4227 | 4628 | 5029 | 5430 | 5830 | 6230 | 6629 | 7028 | 400 |
| 109 | *7426 | 7825 | 8223 | 8620 | 9017 | 9414 | 9811 | ✦207 | 0602 | 0998 | 396 |
| 110 | 04 1393 | 1787 | 2182 | 2576 | 2969 | 3362 | 3755 | 4148 | 4540 | 4932 | 393 |
| 111 | 5323 | 5714 | 6105 | 6495 | 6885 | 7275 | 7664 | 8053 | 8442 | 8830 | 389 |
| 112 | *9218 | 9606 | 9993 | ✦380 | 0766 | 1153 | 1538 | 1924 | 2309 | 2694 | 386 |
| 113 | 05 3078 | 3463 | 3846 | 4230 | 4613 | 4996 | 5378 | 5760 | 6142 | 6524 | 382 |
| 114 | *6905 | 7286 | 7666 | 8046 | 8426 | 8805 | 9185 | 9563 | 9942 | ✦320 | 379 |
| 115 | 06 0698 | 1075 | 1452 | 1829 | 2206 | 2582 | 2958 | 3333 | 3709 | 4083 | 376 |
| 116 | 4458 | 4833 | 5206 | 5580 | 5953 | 6326 | 6699 | 7071 | 7443 | 7815 | 372 |
| 117 | *8186 | 8557 | 8928 | 9298 | 9668 | ✦038 | 0407 | 0776 | 1145 | 1514 | 369 |
| 118 | 07 1882 | 2250 | 2617 | 2985 | 3352 | 3718 | 4085 | 4451 | 4816 | 5182 | 366 |
| 119 | 5547 | 5912 | 6276 | 6640 | 7004 | 7368 | 7731 | 8094 | 8457 | 8819 | 363 |
| 120 | *9181 | 9543 | 9904 | ✦266 | 0626 | 0987 | 1347 | 1707 | 2067 | 2426 | 360 |
| 121 | 08 2785 | 3144 | 3503 | 3861 | 4219 | 4576 | 4934 | 5291 | 5647 | 6004 | 357 |
| 122 | 6360 | 6716 | 7071 | 7426 | 7781 | 8136 | 8490 | 8845 | 9198 | 9552 | 355 |
| 123 | *9905 | ✦258 | 0611 | 0963 | 1315 | 1667 | 2018 | 2370 | 2721 | 3071 | 351 |
| 124 | 09 3422 | 3772 | 4122 | 4471 | 4820 | 5169 | 5518 | 5866 | 6215 | 6562 | 349 |
| 125 | *6910 | 7257 | 7604 | 7951 | 8298 | 8644 | 8990 | 9335 | 9681 | ✦026 | 346 |
| 126 | 10 0371 | 0715 | 1059 | 1403 | 1747 | 2091 | 2434 | 2777 | 3119 | 3462 | 343 |
| 127 | 3804 | 4146 | 4487 | 4828 | 5169 | 5510 | 5851 | 6191 | 6531 | 6871 | 340 |
| 128 | *7210 | 7549 | 7888 | 8227 | 8565 | 8903 | 9241 | 9579 | 9916 | ✦253 | 338 |
| 129 | 11 0590 | 0926 | 1263 | 1599 | 1934 | 2270 | 2605 | 2940 | 3275 | 3609 | 335 |
| 130 | 3943 | 4277 | 4611 | 4944 | 5278 | 5611 | 5943 | 6276 | 6608 | 6940 | 333 |
| 131 | *7271 | 7603 | 7934 | 8265 | 8595 | 8926 | 9256 | 9586 | 9915 | ✦245 | 330 |
| 132 | 12 0574 | 0903 | 1231 | 1560 | 1888 | 2216 | 2544 | 2871 | 3198 | 3525 | 328 |
| 133 | 3852 | 4178 | 4504 | 4830 | 5156 | 5481 | 5806 | 6131 | 6456 | 6781 | 325 |
| 134 | *7105 | 7429 | 7753 | 8076 | 8399 | 8722 | 9045 | 9368 | 9690 | ✦012 | 323 |
| 135 | 13 0334 | 0655 | 0977 | 1298 | 1619 | 1939 | 2260 | 2580 | 2900 | 3219 | 321 |
| 136 | 3539 | 3858 | 4177 | 4496 | 4814 | 5133 | 5451 | 5769 | 6086 | 6403 | 318 |
| 137 | 6721 | 7037 | 7354 | 7671 | 7987 | 8303 | 8618 | 8934 | 9249 | 9564 | 315 |
| 138 | *9879 | ✦194 | 0508 | 0822 | 1136 | 1450 | 1763 | 2076 | 2389 | 2702 | 314 |
| 139 | 14 3015 | 3327 | 3639 | 3951 | 4263 | 4574 | 4885 | 5196 | 5507 | 5818 | 311 |
| 140 | 6128 | 6438 | 6748 | 7058 | 7367 | 7676 | 7985 | 8294 | 8603 | 8911 | 309 |
| 141 | *9219 | 9527 | 9835 | ✦142 | 0449 | 0756 | 1063 | 1370 | 1676 | 1982 | 307 |
| 142 | 15 2288 | 2594 | 2900 | 3205 | 3510 | 3815 | 4120 | 4424 | 4728 | 5032 | 305 |
| 143 | 5336 | 5640 | 5943 | 6246 | 6549 | 6852 | 7154 | 7457 | 7759 | 8061 | 303 |
| 144 | *8362 | 8664 | 8965 | 9266 | 9567 | 9868 | ✦168 | 0469 | 0769 | 1068 | 301 |
| 145 | 16 1368 | 1667 | 1967 | 2266 | 2564 | 2863 | 3161 | 3460 | 3758 | 4055 | 299 |
| 146 | 4353 | 4650 | 4947 | 5244 | 5541 | 5838 | 6134 | 6430 | 6726 | 7022 | 297 |
| 147 | 7317 | 7613 | 7908 | 8203 | 8497 | 8792 | 9086 | 9380 | 9674 | 9968 | 295 |
| 148 | 17 0262 | 0555 | 0848 | 1141 | 1434 | 1726 | 2019 | 2311 | 2603 | 2895 | 293 |
| 149 | 3186 | 3478 | 3769 | 4060 | 4351 | 4641 | 4932 | 5222 | 5512 | 5802 | 291 |
| 150 | 6091 | 6381 | 6670 | 6959 | 7248 | 7536 | 7825 | 8113 | 8401 | 8689 | 289 |
| 151 | *8977 | 9264 | 9552 | 9839 | ✦126 | 0413 | 0699 | 0985 | 1272 | 1558 | 287 |
| 152 | 18 1844 | 2129 | 2415 | 2700 | 2985 | 3270 | 3555 | 3839 | 4123 | 4407 | 285 |
| 153 | 4691 | 4975 | 5259 | 5542 | 5825 | 6108 | 6391 | 6674 | 6956 | 7239 | 283 |
| 154 | *7521 | 7803 | 8084 | 8366 | 8647 | 8928 | 9209 | 9490 | 9771 | ✦051 | 281 |
| 155 | 19 0332 | 0612 | 0892 | 1171 | 1451 | 1730 | 2010 | 2289 | 2567 | 2846 | 279 |
| 156 | 3125 | 3403 | 3681 | 3959 | 4237 | 4514 | 4792 | 5069 | 5346 | 5623 | 278 |
| 157 | 5900 | 6176 | 6453 | 6729 | 7005 | 7281 | 7556 | 7832 | 8107 | 8382 | 276 |
| 158 | *8657 | 8932 | 9206 | 9481 | 9755 | ✦029 | 0303 | 0577 | 0850 | 1124 | 274 |
| 159 | 20 1397 | 1670 | 1943 | 2216 | 2488 | 2761 | 3033 | 3305 | 3577 | 3848 | 272 |
| N. | 0 | 1 | 2 | 3 | 4 | 5 | 6 | 7 | 8 | 9 | D. |

| N. | 0 | 1 | 2 | 3 | 4 | 5 | 6 | 7 | 8 | 9 | D. |
|---|---|---|---|---|---|---|---|---|---|---|---|
| 160 | 20 4120 | 4391 | 4663 | 4934 | 5204 | 5475 | 5746 | 6016 | 6286 | 6556 | 271 |
| 161 | 6826 | 7096 | 7365 | 7634 | 7904 | 8173 | 8441 | 8710 | 8979 | 9247 | 269 |
| 162 | *9515 | 9783 | ◆051 | 0319 | 0586 | 0853 | 1121 | 1388 | 1654 | 1921 | 267 |
| 163 | 21 2188 | 2454 | 2720 | 2986 | 3252 | 3518 | 3783 | 4049 | 4314 | 4579 | 266 |
| 164 | 4844 | 5109 | 5373 | 5638 | 5902 | 6166 | 6430 | 6694 | 6957 | 7221 | 264 |
| 165 | 7484 | 7747 | 8010 | 8273 | 8536 | 8798 | 9060 | 9323 | 9585 | 9846 | 262 |
| 166 | 22 0108 | 0370 | 0631 | 0892 | 1153 | 1414 | 1675 | 1936 | 2196 | 2456 | 261 |
| 167 | 2716 | 2976 | 3236 | 3496 | 3755 | 4015 | 4274 | 4533 | 4792 | 5051 | 259 |
| 168 | 5309 | 5568 | 5826 | 6084 | 6342 | 6600 | 6858 | 7115 | 7372 | 7630 | 258 |
| 169 | *7887 | 8144 | 8400 | 8657 | 8913 | 9170 | 9426 | 9682 | 9938 | ◆193 | 256 |
| 170 | 23 0449 | 0704 | 0960 | 1215 | 1470 | 1724 | 1979 | 2234 | 2488 | 2742 | 254 |
| 171 | 2996 | 3250 | 3504 | 3757 | 4011 | 4264 | 4517 | 4770 | 5023 | 5276 | 253 |
| 172 | 5528 | 5781 | 6033 | 6285 | 6537 | 6789 | 7041 | 7292 | 7544 | 7795 | 252 |
| 173 | *8046 | 8297 | 8548 | 8799 | 9049 | 9299 | 9550 | 9800 | ◆050 | 0300 | 250 |
| 174 | 24 0549 | 0799 | 1048 | 1297 | 1546 | 1795 | 2044 | 2293 | 2541 | 2790 | 249 |
| 175 | 3038 | 3286 | 3534 | 3782 | 4030 | 4277 | 4525 | 4772 | 5019 | 5266 | 248 |
| 176 | 5513 | 5759 | 6006 | 6252 | 6499 | 6745 | 6991 | 7237 | 7482 | 7728 | 246 |
| 177 | *7973 | 8219 | 8464 | 8709 | 8954 | 9198 | 9443 | 9687 | 9932 | ◆176 | 245 |
| 178 | 25 0420 | 0664 | 0908 | 1151 | 1395 | 1638 | 1881 | 2125 | 2368 | 2610 | 243 |
| 179 | 2853 | 3096 | 3338 | 3580 | 3822 | 4064 | 4306 | 4548 | 4790 | 5031 | 242 |
| 180 | 5273 | 5514 | 5755 | 5996 | 6237 | 6477 | 6718 | 6958 | 7198 | 7439 | 241 |
| 181 | 7679 | 7918 | 8158 | 8398 | 8637 | 8877 | 9116 | 9355 | 9594 | 9833 | 239 |
| 182 | 26 0071 | 0310 | 0548 | 0787 | 1025 | 1263 | 1501 | 1739 | 1976 | 2214 | 238 |
| 183 | 2451 | 2688 | 2925 | 3162 | 3399 | 3636 | 3873 | 4109 | 4346 | 4582 | 237 |
| 184 | 4818 | 5054 | 5290 | 5525 | 5761 | 5996 | 6232 | 6467 | 6702 | 6937 | 235 |
| 185 | 7172 | 7406 | 7641 | 7875 | 8110 | 8344 | 8578 | 8812 | 9046 | 9279 | 234 |
| 186 | *9513 | 9746 | 9980 | ◆213 | 0446 | 0679 | 0912 | 1144 | 1377 | 1609 | 233 |
| 187 | 27 1842 | 2074 | 2306 | 2538 | 2770 | 3001 | 3233 | 3464 | 3696 | 3927 | 232 |
| 188 | 4158 | 4389 | 4620 | 4850 | 5081 | 5311 | 5542 | 5772 | 6002 | 6232 | 230 |
| 189 | 6462 | 6692 | 6921 | 7151 | 7380 | 7609 | 7838 | 8067 | 8296 | 8525 | 229 |
| 190 | *8754 | 8982 | 9211 | 9439 | 9667 | 9895 | ◆123 | 0351 | 0578 | 0806 | 228 |
| 191 | 28 1033 | 1261 | 1488 | 1715 | 1942 | 2169 | 2396 | 2622 | 2849 | 3075 | 227 |
| 192 | 3301 | 3527 | 3753 | 3979 | 4205 | 4431 | 4656 | 4882 | 5107 | 5332 | 226 |
| 193 | 5557 | 5782 | 6007 | 6232 | 6456 | 6681 | 6905 | 7130 | 7354 | 7578 | 225 |
| 194 | 7802 | 8026 | 8249 | 8473 | 8696 | 8920 | 9143 | 9366 | 9589 | 9812 | 223 |
| 195 | 29 0035 | 0257 | 0480 | 0702 | 0925 | 1147 | 1369 | 1591 | 1813 | 2034 | 222 |
| 196 | 2256 | 2478 | 2699 | 2920 | 3141 | 3363 | 3584 | 3804 | 4025 | 4246 | 221 |
| 197 | 4466 | 4687 | 4907 | 5127 | 5347 | 5567 | 5787 | 6007 | 6226 | 6446 | 220 |
| 198 | 6665 | 6884 | 7104 | 7323 | 7542 | 7761 | 7979 | 8198 | 8416 | 8635 | 219 |
| 199 | *8853 | 9071 | 9289 | 9507 | 9725 | 9943 | ◆161 | 0378 | 0595 | 0813 | 218 |
| 200 | 30 1030 | 1247 | 1464 | 1681 | 1898 | 2114 | 2331 | 2547 | 2764 | 2980 | 217 |
| 201 | 3196 | 3412 | 3628 | 3844 | 4059 | 4275 | 4491 | 4706 | 4921 | 5136 | 216 |
| 202 | 5351 | 5566 | 5781 | 5996 | 6211 | 6425 | 6639 | 6854 | 7068 | 7282 | 215 |
| 203 | 7496 | 7710 | 7924 | 8137 | 8351 | 8564 | 8778 | 8991 | 9204 | 9417 | 213 |
| 204 | *9630 | 9843 | ◆056 | 0268 | 0481 | 0693 | 0906 | 1118 | 1330 | 1542 | 212 |
| 205 | 31 1754 | 1966 | 2177 | 2389 | 2600 | 2812 | 3023 | 3234 | 3445 | 3656 | 211 |
| 206 | 3867 | 4078 | 4289 | 4499 | 4710 | 4920 | 5130 | 5340 | 5551 | 5760 | 210 |
| 207 | 5970 | 6180 | 6390 | 6599 | 6809 | 7018 | 7227 | 7436 | 7646 | 7854 | 209 |
| 208 | 8063 | 8272 | 8481 | 8689 | 8898 | 9106 | 9314 | 9522 | 9730 | 9938 | 208 |
| 209 | 32 0146 | 0354 | 0562 | 0769 | 0977 | 1184 | 1391 | 1598 | 1805 | 2012 | 207 |
| 210 | 2219 | 2426 | 2633 | 2839 | 3046 | 3252 | 3458 | 3665 | 3871 | 4077 | 206 |
| 211 | 4282 | 4488 | 4694 | 4899 | 5105 | 5310 | 5516 | 5721 | 5926 | 6131 | 205 |
| 212 | 6336 | 6541 | 6745 | 6950 | 7155 | 7359 | 7563 | 7767 | 7972 | 8176 | 204 |
| 213 | *8380 | 8583 | 8787 | 8991 | 9194 | 9398 | 9601 | 9805 | ◆008 | 0211 | 203 |
| 214 | 33 0414 | 0617 | 0819 | 1022 | 1225 | 1427 | 1630 | 1832 | 2034 | 2236 | 202 |
| 215 | 2438 | 2640 | 2842 | 3044 | 3246 | 3447 | 3649 | 3850 | 4051 | 4253 | 202 |
| 216 | 4454 | 4655 | 4856 | 5057 | 5257 | 5458 | 5658 | 5859 | 6059 | 6260 | 201 |
| 217 | 6460 | 6660 | 6860 | 7060 | 7260 | 7459 | 7659 | 7858 | 8058 | 8257 | 200 |
| 218 | *8456 | 8656 | 8855 | 9054 | 9253 | 9451 | 9650 | 9849 | ◆047 | 0246 | 199 |
| 219 | 34 0444 | 0642 | 0841 | 1039 | 1237 | 1435 | 1632 | 1830 | 2028 | 2225 | 198 |
| N. | 0 | 1 | 2 | 3 | 4 | 5 | 6 | 7 | 8 | 9 | D. |

| N. | 0 | 1 | 2 | 3 | 4 | 5 | 6 | 7 | 8 | 9 | D. |
|---|---|---|---|---|---|---|---|---|---|---|---|
| 220 | 34 2423 | 2620 | 2817 | 3014 | 3212 | 3409 | 3606 | 3802 | 3999 | 4196 | 197 |
| 221 | 4392 | 4589 | 4785 | 4981 | 5178 | 5374 | 5570 | 5766 | 5962 | 6157 | 196 |
| 222 | 6353 | 6549 | 6744 | 6939 | 7135 | 7330 | 7525 | 7720 | 7915 | 8110 | 195 |
| 223 | *8305 | 8500 | 8694 | 8889 | 9083 | 9278 | 9472 | 9666 | 9860 | ♦054 | 194 |
| 224 | 35 0248 | 0442 | 0636 | 0829 | 1023 | 1216 | 1410 | 1603 | 1796 | 1989 | 193 |
| 225 | 2183 | 2375 | 2568 | 2761 | 2954 | 3147 | 3339 | 3532 | 3724 | 3916 | 193 |
| 226 | 4108 | 4301 | 4493 | 4685 | 4876 | 5068 | 5260 | 5452 | 5643 | 5834 | 192 |
| 227 | 6026 | 6217 | 6408 | 6599 | 6790 | 6981 | 7172 | 7363 | 7554 | 7744 | 191 |
| 228 | 7935 | 8125 | 8316 | 8506 | 8696 | 8886 | 9076 | 9266 | 9456 | 9646 | 190 |
| 229 | *9835 | ♦025 | 0215 | 0404 | 0593 | 0783 | 0972 | 1161 | 1350 | 1539 | 189 |
| 230 | 36 1728 | 1917 | 2105 | 2294 | 2482 | 2671 | 2859 | 3048 | 3236 | 3424 | 188 |
| 231 | 3612 | 3800 | 3988 | 4176 | 4363 | 4551 | 4739 | 4926 | 5113 | 5301 | 188 |
| 232 | 5488 | 5675 | 5862 | 6049 | 6236 | 6423 | 6610 | 6796 | 6983 | 7169 | 187 |
| 233 | 7356 | 7542 | 7729 | 7915 | 8101 | 8287 | 8473 | 8659 | 8845 | 9030 | 186 |
| 234 | *9216 | 9401 | 9587 | 9772 | 9958 | ♦143 | 0328 | 0513 | 0698 | 0883 | 185 |
| 235 | 37 1068 | 1253 | 1437 | 1622 | 1806 | 1991 | 2175 | 2360 | 2544 | 2728 | 184 |
| 236 | 2912 | 3096 | 3280 | 3464 | 3647 | 3831 | 4015 | 4198 | 4382 | 4565 | 184 |
| 237 | 4748 | 4932 | 5115 | 5298 | 5481 | 5664 | 5846 | 6029 | 6212 | 6394 | 183 |
| 238 | 6577 | 6759 | 6942 | 7124 | 7306 | 7488 | 7670 | 7852 | 8034 | 8216 | 182 |
| 239 | *8398 | 8580 | 8761 | 8943 | 9124 | 9306 | 9487 | 9668 | 9849 | ♦030 | 181 |
| 240 | 38 0211 | 0392 | 0573 | 0754 | 0934 | 1115 | 1296 | 1476 | 1656 | 1837 | 181 |
| 241 | 2017 | 2197 | 2377 | 2557 | 2737 | 2917 | 3097 | 3277 | 3456 | 3636 | 180 |
| 242 | 3815 | 3995 | 4174 | 4353 | 4533 | 4712 | 4891 | 5070 | 5249 | 5428 | 179 |
| 243 | 5606 | 5785 | 5964 | 6142 | 6321 | 6499 | 6677 | 6856 | 7034 | 7212 | 178 |
| 244 | 7390 | 7568 | 7746 | 7923 | 8101 | 8279 | 8456 | 8634 | 8811 | 8989 | 178 |
| 245 | *9166 | 9343 | 9520 | 9698 | 9875 | ♦051 | 0228 | 0405 | 0582 | 0759 | 177 |
| 246 | 39 0935 | 1112 | 1288 | 1464 | 1641 | 1817 | 1993 | 2169 | 2345 | 2521 | 176 |
| 247 | 2697 | 2873 | 3048 | 3224 | 3400 | 3575 | 3751 | 3926 | 4101 | 4277 | 176 |
| 248 | 4452 | 4627 | 4802 | 4977 | 5152 | 5326 | 5501 | 5676 | 5850 | 6025 | 175 |
| 249 | 6199 | 6374 | 6548 | 6722 | 6896 | 7071 | 7245 | 7419 | 7592 | 7766 | 174 |
| 250 | 7940 | 8114 | 8287 | 8461 | 8634 | 8808 | 8981 | 9154 | 9328 | 9501 | 173 |
| 251 | *9674 | 9847 | ♦020 | 0192 | 0365 | 0538 | 0711 | 0883 | 1056 | 1228 | 173 |
| 252 | 40 1401 | 1573 | 1745 | 1917 | 2089 | 2261 | 2433 | 2605 | 2777 | 2949 | 172 |
| 253 | 3121 | 3292 | 3464 | 3635 | 3807 | 3978 | 4149 | 4320 | 4492 | 4663 | 171 |
| 254 | 4834 | 5005 | 5176 | 5346 | 5517 | 5688 | 5858 | 6029 | 6199 | 6370 | 171 |
| 255 | 6540 | 6710 | 6881 | 7051 | 7221 | 7391 | 7561 | 7731 | 7901 | 8070 | 170 |
| 256 | 8240 | 8410 | 8579 | 8749 | 8918 | 9087 | 9257 | 9426 | 9595 | 9764 | 169 |
| 257 | *9933 | ♦102 | 0271 | 0440 | 0609 | 0777 | 0946 | 1114 | 1283 | 1451 | 169 |
| 258 | 41 1620 | 1788 | 1956 | 2124 | 2293 | 2461 | 2629 | 2796 | 2964 | 3132 | 168 |
| 259 | 3300 | 3467 | 3635 | 3803 | 3970 | 4137 | 4305 | 4472 | 4639 | 4806 | 167 |
| 260 | 4973 | 5140 | 5307 | 5474 | 5641 | 5808 | 5974 | 6141 | 6308 | 6474 | 167 |
| 261 | 6641 | 6807 | 6973 | 7139 | 7306 | 7472 | 7638 | 7804 | 7970 | 8135 | 166 |
| 262 | 8301 | 8467 | 8633 | 8798 | 8964 | 9129 | 9295 | 9460 | 9625 | 9791 | 165 |
| 263 | *9956 | ♦121 | 0286 | 0451 | 0616 | 0781 | 0945 | 1110 | 1275 | 1439 | 165 |
| 264 | 42 1604 | 1768 | 1933 | 2097 | 2261 | 2426 | 2590 | 2754 | 2918 | 3082 | 164 |
| 265 | 3246 | 3410 | 3574 | 3737 | 3901 | 4065 | 4228 | 4392 | 4555 | 4718 | 164 |
| 266 | 4882 | 5045 | 5208 | 5371 | 5534 | 5697 | 5860 | 6023 | 6186 | 6349 | 163 |
| 267 | 6511 | 6674 | 6836 | 6999 | 7161 | 7324 | 7486 | 7648 | 7811 | 7973 | 162 |
| 268 | 8135 | 8297 | 8459 | 8621 | 8783 | 8944 | 9106 | 9268 | 9429 | 9591 | 162 |
| 269 | *9752 | 9914 | ♦075 | 0236 | 0398 | 0559 | 0720 | 0881 | 1042 | 1203 | 161 |
| 270 | 43 1364 | 1525 | 1685 | 1846 | 2007 | 2167 | 2328 | 2488 | 2649 | 2809 | 161 |
| 271 | 2969 | 3130 | 3290 | 3450 | 3610 | 3770 | 3930 | 4090 | 4249 | 4409 | 160 |
| 272 | 4569 | 4729 | 4888 | 5048 | 5207 | 5367 | 5526 | 5685 | 5844 | 6004 | 159 |
| 273 | 6163 | 6322 | 6481 | 6640 | 6799 | 6957 | 7116 | 7275 | 7433 | 7592 | 159 |
| 274 | 7751 | 7909 | 8067 | 8226 | 8384 | 8542 | 8701 | 8859 | 9017 | 9175 | 158 |
| 275 | *9333 | 9491 | 9648 | 9806 | 9964 | ♦122 | 0279 | 0437 | 0594 | 0752 | 158 |
| 276 | 44 0909 | 1066 | 1224 | 1381 | 1538 | 1695 | 1852 | 2009 | 2166 | 2323 | 157 |
| 277 | 2480 | 2637 | 2793 | 2950 | 3106 | 3263 | 3419 | 3576 | 3732 | 3889 | 157 |
| 278 | 4045 | 4201 | 4357 | 4513 | 4669 | 4825 | 4981 | 5137 | 5293 | 5449 | 156 |
| 279 | 5604 | 5760 | 5915 | 6071 | 6226 | 6382 | 6537 | 6692 | 6848 | 7003 | 155 |
| N. | 0 | 1 | 2 | 3 | 4 | 5 | 6 | 7 | 8 | 9 | D. |

| N. | 0 | 1 | 2 | 3 | 4 | 5 | 6 | 7 | 8 | 9 | D. |
|---|---|---|---|---|---|---|---|---|---|---|---|
| 280 | 44 7158 | 7313 | 7468 | 7623 | 7778 | 7933 | 8088 | 8242 | 8397 | 8552 | 155 |
| 281 | *8706 | 8861 | 9015 | 9170 | 9324 | 9478 | 9633 | 9787 | 9941 | ♦095 | 154 |
| 282 | 45 0249 | 0403 | 0557 | 0711 | 0865 | 1018 | 1172 | 1326 | 1479 | 1633 | 154 |
| 283 | 1786 | 1940 | 2093 | 2247 | 2400 | 2553 | 2706 | 2859 | 3012 | 3165 | 153 |
| 284 | 3318 | 3471 | 3624 | 3777 | 3930 | 4082 | 4235 | 4387 | 4540 | 4692 | 153 |
| 285 | 4845 | 4997 | 5150 | 5302 | 5454 | 5606 | 5758 | 5910 | 6062 | 6214 | 152 |
| 286 | 6366 | 6518 | 6670 | 6821 | 6973 | 7125 | 7276 | 7428 | 7579 | 7731 | 152 |
| 287 | 7882 | 8033 | 8184 | 8336 | 8487 | 8638 | 8789 | 8940 | 9091 | 9242 | 151 |
| 288 | *9392 | 9543 | 9694 | 9845 | 9995 | ♦146 | 0296 | 0447 | 0597 | 0748 | 151 |
| 289 | 46 0898 | 1048 | 1198 | 1348 | 1499 | 1649 | 1799 | 1948 | 2098 | 2248 | 150 |
| 290 | 2398 | 2548 | 2697 | 2847 | 2997 | 3146 | 3296 | 3445 | 3594 | 3744 | 150 |
| 291 | 3893 | 4042 | 4191 | 4340 | 4490 | 4639 | 4788 | 4936 | 5085 | 5234 | 149 |
| 292 | 5383 | 5532 | 5680 | 5829 | 5977 | 6126 | 6274 | 6423 | 6571 | 6719 | 149 |
| 293 | 6868 | 7016 | 7164 | 7312 | 7460 | 7608 | 7756 | 7904 | 8052 | 8200 | 148 |
| 294 | 8347 | 8495 | 8643 | 8790 | 8938 | 9085 | 9233 | 9380 | 9527 | 9675 | 148 |
| 295 | *9822 | 9969 | ♦116 | 0263 | 0410 | 0557 | 0704 | 0851 | 0998 | 1145 | 147 |
| 296 | 47 1292 | 1438 | 1585 | 1732 | 1878 | 2025 | 2171 | 2318 | 2464 | 2610 | 146 |
| 297 | 2756 | 2903 | 3049 | 3195 | 3341 | 3487 | 3633 | 3779 | 3925 | 4071 | 146 |
| 298 | 4216 | 4362 | 4508 | 4653 | 4799 | 4944 | 5090 | 5235 | 5381 | 5526 | 146 |
| 299 | 5671 | 5816 | 5962 | 6107 | 6252 | 6397 | 6542 | 6687 | 6832 | 6976 | 145 |
| 300 | 7121 | 7266 | 7411 | 7555 | 7700 | 7844 | 7989 | 8133 | 8278 | 8422 | 145 |
| 301 | 8566 | 8711 | 8855 | 8999 | 9143 | 9287 | 9431 | 9575 | 9719 | 9863 | 144 |
| 302 | 48 0007 | 0151 | 0294 | 0438 | 0582 | 0725 | 0869 | 1012 | 1156 | 1299 | 144 |
| 303 | 1443 | 1586 | 1729 | 1872 | 2016 | 2159 | 2302 | 2445 | 2588 | 2731 | 143 |
| 304 | 2874 | 3016 | 3159 | 3302 | 3445 | 3587 | 3730 | 3872 | 4015 | 4157 | 143 |
| 305 | 4300 | 4442 | 4585 | 4727 | 4869 | 5011 | 5153 | 5295 | 5437 | 5579 | 142 |
| 306 | 5721 | 5863 | 6005 | 6147 | 6289 | 6430 | 6572 | 6714 | 6855 | 6997 | 142 |
| 307 | 7138 | 7280 | 7421 | 7563 | 7704 | 7845 | 7986 | 8127 | 8269 | 8410 | 141 |
| 308 | 8551 | 8692 | 8833 | 8974 | 9114 | 9255 | 9396 | 9537 | 9677 | 9818 | 141 |
| 309 | *9958 | ♦099 | 0239 | 0380 | 0520 | 0661 | 0801 | 0941 | 1081 | 1222 | 140 |
| 310 | 49 1362 | 1502 | 1642 | 1782 | 1922 | 2062 | 2201 | 2341 | 2481 | 2621 | 140 |
| 311 | 2760 | 2900 | 3040 | 3179 | 3319 | 3458 | 3597 | 3737 | 3876 | 4015 | 139 |
| 312 | 4155 | 4294 | 4433 | 4572 | 4711 | 4850 | 4989 | 5128 | 5267 | 5406 | 139 |
| 313 | 5544 | 5683 | 5822 | 5960 | 6099 | 6238 | 6376 | 6515 | 6653 | 6791 | 139 |
| 314 | 6930 | 7068 | 7206 | 7344 | 7483 | 7621 | 7759 | 7897 | 8035 | 8173 | 138 |
| 315 | 8311 | 8448 | 8586 | 8724 | 8862 | 8999 | 9137 | 9275 | 9412 | 9550 | 138 |
| 316 | *9687 | 9824 | 9962 | ♦099 | 0236 | 0374 | 0511 | 0648 | 0785 | 0922 | 137 |
| 317 | 50 1059 | 1196 | 1333 | 1470 | 1607 | 1744 | 1880 | 2017 | 2154 | 2291 | 137 |
| 318 | 2427 | 2564 | 2700 | 2837 | 2973 | 3109 | 3246 | 3382 | 3518 | 3655 | 136 |
| 319 | 3791 | 3927 | 4063 | 4199 | 4335 | 4471 | 4607 | 4743 | 4878 | 5014 | 136 |
| 320 | 5150 | 5286 | 5421 | 5557 | 5693 | 5828 | 5964 | 6099 | 6234 | 6370 | 136 |
| 321 | 6505 | 6640 | 6776 | 6911 | 7046 | 7181 | 7316 | 7451 | 7586 | 7721 | 135 |
| 322 | 7856 | 7991 | 8126 | 8260 | 8395 | 8530 | 8664 | 8799 | 8934 | 9068 | 135 |
| 323 | *9203 | 9337 | 9471 | 9606 | 9740 | 9874 | ♦009 | 0143 | 0277 | 0411 | 134 |
| 324 | 51 0545 | 0679 | 0813 | 0947 | 1081 | 1215 | 1349 | 1482 | 1616 | 1750 | 134 |
| 325 | 1883 | 2017 | 2151 | 2284 | 2418 | 2551 | 2684 | 2818 | 2951 | 3084 | 133 |
| 326 | 3218 | 3351 | 3484 | 3617 | 3750 | 3883 | 4016 | 4149 | 4282 | 4414 | 133 |
| 327 | 4548 | 4681 | 4813 | 4946 | 5079 | 5211 | 5344 | 5476 | 5609 | 5741 | 133 |
| 328 | 5874 | 6006 | 6139 | 6271 | 6403 | 6535 | 6668 | 6800 | 6932 | 7064 | 132 |
| 329 | 7196 | 7328 | 7460 | 7592 | 7724 | 7855 | 7987 | 8119 | 8251 | 8382 | 132 |
| 330 | 8514 | 8646 | 8777 | 8909 | 9040 | 9171 | 9303 | 9434 | 9566 | 9697 | 131 |
| 331 | *9828 | 9959 | ♦090 | 0221 | 0353 | 0484 | 0615 | 0745 | 0876 | 1007 | 131 |
| 332 | 52 1138 | 1269 | 1400 | 1530 | 1661 | 1792 | 1922 | 2053 | 2183 | 2314 | 131 |
| 333 | 2444 | 2575 | 2705 | 2835 | 2966 | 3096 | 3226 | 3356 | 3486 | 3616 | 130 |
| 334 | 3746 | 3876 | 4006 | 4136 | 4266 | 4396 | 4526 | 4656 | 4785 | 4915 | 130 |
| 335 | 5045 | 5174 | 5304 | 5434 | 5563 | 5693 | 5822 | 5951 | 6081 | 6210 | 129 |
| 336 | 6339 | 6469 | 6598 | 6727 | 6856 | 6985 | 7114 | 7243 | 7372 | 7501 | 129 |
| 337 | 7630 | 7759 | 7888 | 8016 | 8145 | 8274 | 8402 | 8531 | 8660 | 8788 | 129 |
| 338 | *8917 | 9045 | 9174 | 9302 | 9430 | 9559 | 9687 | 9815 | 9943 | ♦072 | 128 |
| 339 | 53 0200 | 0328 | 0456 | 0584 | 0712 | 0840 | 0968 | 1096 | 1223 | 1351 | 128 |
| N. | 0 | 1 | 2 | 3 | 4 | 5 | 6 | 7 | 8 | 9 | D. |

| N. | 0 | 1 | 2 | 3 | 4 | 5 | 6 | 7 | 8 | 9 | D. |
|---|---|---|---|---|---|---|---|---|---|---|---|
| 340 | 53 1479 | 1607 | 1734 | 1862 | 1990 | 2117 | 2245 | 2372 | 2500 | 2627 | 128 |
| 341 | 2754 | 2882 | 3009 | 3136 | 3264 | 3391 | 3518 | 3645 | 3772 | 3899 | 127 |
| 342 | 4026 | 4153 | 4280 | 4407 | 4534 | 4661 | 4787 | 4914 | 5041 | 5167 | 127 |
| 343 | 5294 | 5421 | 5547 | 5674 | 5800 | 5927 | 6053 | 6180 | 6306 | 6432 | 126 |
| 344 | 6558 | 6685 | 6811 | 6937 | 7063 | 7189 | 7315 | 7441 | 7567 | 7693 | 126 |
| 345 | 7819 | 7945 | 8071 | 8197 | 8322 | 8448 | 8574 | 8699 | 8825 | 8951 | 126 |
| 346 | *9076 | 9202 | 9327 | 9452 | 9578 | 9703 | 9829 | 9954 | ✦079 | 0204 | 125 |
| 347 | 54 0329 | 0455 | 0580 | 0705 | 0830 | 0955 | 1080 | 1205 | 1330 | 1454 | 125 |
| 348 | 1579 | 1704 | 1829 | 1953 | 2078 | 2203 | 2327 | 2452 | 2576 | 2701 | 125 |
| 349 | 2825 | 2950 | 3074 | 3199 | 3323 | 3447 | 3571 | 3696 | 3820 | 3944 | 124 |
| 350 | 4068 | 4192 | 4316 | 4440 | 4564 | 4688 | 4812 | 4936 | 5060 | 5183 | 124 |
| 351 | 5307 | 5431 | 5555 | 5678 | 5802 | 5925 | 6049 | 6172 | 6296 | 6419 | 124 |
| 352 | 6543 | 6666 | 6789 | 6913 | 7036 | 7159 | 7282 | 7405 | 7529 | 7652 | 123 |
| 353 | 7775 | 7898 | 8021 | 8144 | 8267 | 8389 | 8512 | 8635 | 8758 | 8881 | 123 |
| 354 | *9003 | 9126 | 9249 | 9371 | 9494 | 9616 | 9739 | 9861 | 9984 | ✦106 | 123 |
| 355 | 55 0228 | 0351 | 0473 | 0595 | 0717 | 0840 | 0962 | 1084 | 1206 | 1328 | 122 |
| 356 | 1450 | 1572 | 1694 | 1816 | 1938 | 2060 | 2181 | 2303 | 2425 | 2547 | 122 |
| 357 | 2668 | 2790 | 2911 | 3033 | 3155 | 3276 | 3398 | 3519 | 3640 | 3762 | 121 |
| 358 | 3883 | 4004 | 4126 | 4247 | 4368 | 4489 | 4610 | 4731 | 4852 | 4973 | 121 |
| 359 | 5094 | 5215 | 5336 | 5457 | 5578 | 5699 | 5820 | 5940 | 6061 | 6182 | 121 |
| 360 | 6303 | 6423 | 6544 | 6664 | 6785 | 6905 | 7026 | 7146 | 7267 | 7387 | 120 |
| 361 | 7507 | 7627 | 7748 | 7868 | 7988 | 8108 | 8228 | 8349 | 8469 | 8589 | 120 |
| 362 | 8709 | 8829 | 8948 | 9068 | 9188 | 9308 | 9428 | 9548 | 9667 | 9787 | 120 |
| 363 | *9907 | ✦026 | 0146 | 0265 | 0385 | 0504 | 0624 | 0743 | 0863 | 0982 | 119 |
| 364 | 56 1101 | 1221 | 1340 | 1459 | 1578 | 1698 | 1817 | 1936 | 2055 | 2174 | 119 |
| 365 | 2293 | 2412 | 2531 | 2650 | 2769 | 2887 | 3006 | 3125 | 3244 | 3362 | 119 |
| 366 | 3481 | 3600 | 3718 | 3837 | 3955 | 4074 | 4192 | 4311 | 4429 | 4548 | 119 |
| 367 | 4666 | 4784 | 4903 | 5021 | 5139 | 5257 | 5376 | 5494 | 5612 | 5730 | 118 |
| 368 | 5848 | 5966 | 6084 | 6202 | 6320 | 6437 | 6555 | 6673 | 6791 | 6909 | 118 |
| 369 | 7026 | 7144 | 7262 | 7379 | 7497 | 7614 | 7732 | 7849 | 7967 | 8084 | 118 |
| 370 | 8202 | 8319 | 8436 | 8554 | 8671 | 8788 | 8905 | 9023 | 9140 | 9257 | 117 |
| 371 | *9374 | 9491 | 9608 | 9725 | 9842 | 9959 | ✦076 | 0193 | 0309 | 0426 | 117 |
| 372 | 57 0543 | 0660 | 0776 | 0893 | 1010 | 1126 | 1243 | 1359 | 1476 | 1592 | 117 |
| 373 | 1709 | 1825 | 1942 | 2058 | 2174 | 2291 | 2407 | 2523 | 2639 | 2755 | 116 |
| 374 | 2872 | 2988 | 3104 | 3220 | 3336 | 3452 | 3568 | 3684 | 3800 | 3915 | 116 |
| 375 | 4031 | 4147 | 4263 | 4379 | 4494 | 4610 | 4726 | 4841 | 4957 | 5072 | 116 |
| 376 | 5188 | 5303 | 5419 | 5534 | 5650 | 5765 | 5880 | 5996 | 6111 | 6226 | 115 |
| 377 | 6341 | 6457 | 6572 | 6687 | 6802 | 6917 | 7032 | 7147 | 7262 | 7377 | 115 |
| 378 | 7492 | 7607 | 7722 | 7836 | 7951 | 8066 | 8181 | 8295 | 8410 | 8525 | 115 |
| 379 | 8639 | 8754 | 8868 | 8983 | 9097 | 9212 | 9326 | 9441 | 9555 | 9669 | 114 |
| 380 | *9784 | 9898 | ✦012 | 0126 | 0241 | 0355 | 0469 | 0583 | 0697 | 0811 | 114 |
| 381 | 58 0925 | 1039 | 1153 | 1267 | 1381 | 1495 | 1608 | 1722 | 1836 | 1950 | 114 |
| 382 | 2063 | 2177 | 2291 | 2404 | 2518 | 2631 | 2745 | 2858 | 2972 | 3085 | 114 |
| 383 | 3199 | 3312 | 3426 | 3539 | 3652 | 3765 | 3879 | 3992 | 4105 | 4218 | 113 |
| 384 | 4331 | 4444 | 4557 | 4670 | 4783 | 4896 | 5009 | 5122 | 5235 | 5348 | 113 |
| 385 | 5461 | 5574 | 5686 | 5799 | 5912 | 6024 | 6137 | 6250 | 6362 | 6475 | 113 |
| 386 | 6587 | 6700 | 6812 | 6925 | 7037 | 7149 | 7262 | 7374 | 7486 | 7599 | 112 |
| 387 | 7711 | 7823 | 7935 | 8047 | 8160 | 8272 | 8384 | 8496 | 8608 | 8720 | 112 |
| 388 | 8832 | 8944 | 9056 | 9167 | 9279 | 9391 | 9503 | 9615 | 9726 | 9838 | 112 |
| 389 | *9950 | ✦061 | 0173 | 0284 | 0396 | 0507 | 0619 | 0730 | 0842 | 0953 | 112 |
| 390 | 59 1065 | 1176 | 1287 | 1399 | 1510 | 1621 | 1732 | 1843 | 1955 | 2066 | 111 |
| 391 | 2177 | 2288 | 2399 | 2510 | 2621 | 2732 | 2843 | 2954 | 3064 | 3175 | 111 |
| 392 | 3286 | 3397 | 3508 | 3618 | 3729 | 3840 | 3950 | 4061 | 4171 | 4282 | 111 |
| 393 | 4393 | 4503 | 4614 | 4724 | 4834 | 4945 | 5055 | 5165 | 5276 | 5386 | 110 |
| 394 | 5496 | 5606 | 5717 | 5827 | 5937 | 6047 | 6157 | 6267 | 6377 | 6487 | 110 |
| 395 | 6597 | 6707 | 6817 | 6927 | 7037 | 7146 | 7256 | 7366 | 7476 | 7586 | 110 |
| 396 | 7695 | 7805 | 7914 | 8024 | 8134 | 8243 | 8353 | 8462 | 8572 | 8681 | 110 |
| 397 | 8791 | 8900 | 9009 | 9119 | 9228 | 9337 | 9446 | 9556 | 9665 | 9774 | 109 |
| 398 | *9883 | 9992 | ✦101 | 0210 | 0319 | 0428 | 0537 | 0646 | 0755 | 0864 | 109 |
| 399 | 60 0973 | 1082 | 1191 | 1299 | 1408 | 1517 | 1625 | 1734 | 1843 | 1951 | 109 |
| N. | 0 | 1 | 2 | 3 | 4 | 5 | 6 | 7 | 8 | 9 | D. |

| N. | 0 | 1 | 2 | 3 | 4 | 5 | 6 | 7 | 8 | 9 | D. |
|---|---|---|---|---|---|---|---|---|---|---|---|
| 400 | 60 2060 | 2169 | 2277 | 2386 | 2494 | 2603 | 2711 | 2819 | 2928 | 3036 | 108 |
| 401 | 3144 | 3253 | 3361 | 3469 | 3577 | 3686 | 3794 | 3902 | 4010 | 4118 | 108 |
| 402 | 4226 | 4334 | 4442 | 4550 | 4658 | 4766 | 4874 | 4982 | 5089 | 5197 | 108 |
| 403 | 5305 | 5413 | 5521 | 5628 | 5736 | 5844 | 5951 | 6059 | 6166 | 6274 | 108 |
| 404 | 6381 | 6489 | 6596 | 6704 | 6811 | 6919 | 7026 | 7133 | 7241 | 7348 | 107 |
| 405 | 7455 | 7562 | 7669 | 7777 | 7884 | 7991 | 8098 | 8205 | 8312 | 8419 | 107 |
| 406 | 8526 | 8633 | 8740 | 8847 | 8954 | 9061 | 9167 | 9274 | 9381 | 9488 | 107 |
| 407 | *9594 | 9701 | 9808 | 9914 | ♦021 | 0128 | 0234 | 0341 | 0447 | 0554 | 107 |
| 408 | 61 0660 | 0767 | 0873 | 0979 | 1086 | 1192 | 1298 | 1405 | 1511 | 1617 | 106 |
| 409 | 1723 | 1829 | 1936 | 2042 | 2148 | 2254 | 2360 | 2466 | 2572 | 2678 | 106 |
| 410 | 2784 | 2890 | 2996 | 3102 | 3207 | 3313 | 3419 | 3525 | 3630 | 3736 | 106 |
| 411 | 3842 | 3947 | 4053 | 4159 | 4264 | 4370 | 4475 | 4581 | 4686 | 4792 | 106 |
| 412 | 4897 | 5003 | 5108 | 5213 | 5319 | 5424 | 5529 | 5634 | 5740 | 5845 | 105 |
| 413 | 5950 | 6055 | 6160 | 6265 | 6370 | 6476 | 6581 | 6686 | 6790 | 6895 | 105 |
| 414 | 7000 | 7105 | 7210 | 7315 | 7420 | 7525 | 7629 | 7734 | 7839 | 7943 | 105 |
| 415 | 8048 | 8153 | 8257 | 8362 | 8466 | 8571 | 8676 | 8780 | 8884 | 8989 | 105 |
| 416 | *9093 | 9198 | 9302 | 9406 | 9511 | 9615 | 9719 | 9824 | 9928 | ♦032 | 104 |
| 417 | 62 0136 | 0240 | 0344 | 0448 | 0552 | 0656 | 0760 | 0864 | 0968 | 1072 | 104 |
| 418 | 1176 | 1280 | 1384 | 1488 | 1592 | 1695 | 1799 | 1903 | 2007 | 2110 | 104 |
| 419 | 2214 | 2318 | 2421 | 2525 | 2628 | 2732 | 2835 | 2939 | 3042 | 3146 | 104 |
| 420 | 3249 | 3353 | 3456 | 3559 | 3663 | 3766 | 3869 | 3973 | 4076 | 4179 | 103 |
| 421 | 4282 | 4385 | 4488 | 4591 | 4695 | 4798 | 4901 | 5004 | 5107 | 5210 | 103 |
| 422 | 5312 | 5415 | 5518 | 5621 | 5724 | 5827 | 5929 | 6032 | 6135 | 6238 | 103 |
| 423 | 6340 | 6443 | 6546 | 6648 | 6751 | 6853 | 6956 | 7058 | 7161 | 7263 | 103 |
| 424 | 7366 | 7468 | 7571 | 7673 | 7775 | 7878 | 7980 | 8082 | 8185 | 8287 | 102 |
| 425 | 8389 | 8491 | 8593 | 8695 | 8797 | 8900 | 9002 | 9104 | 9206 | 9308 | 102 |
| 426 | *9410 | 9512 | 9613 | 9715 | 9817 | 9919 | ♦021 | 0123 | 0224 | 0326 | 102 |
| 427 | 63 0428 | 0530 | 0631 | 0733 | 0835 | 0936 | 1038 | 1139 | 1241 | 1342 | 102 |
| 428 | 1444 | 1545 | 1647 | 1748 | 1849 | 1951 | 2052 | 2153 | 2255 | 2356 | 101 |
| 429 | 2457 | 2559 | 2660 | 2761 | 2862 | 2963 | 3064 | 3165 | 3266 | 3367 | 101 |
| 430 | 3468 | 3569 | 3670 | 3771 | 3872 | 3973 | 4074 | 4175 | 4276 | 4376 | 100 |
| 431 | 4477 | 4578 | 4679 | 4779 | 4880 | 4981 | 5081 | 5182 | 5283 | 5383 | 100 |
| 432 | 5484 | 5584 | 5685 | 5785 | 5886 | 5986 | 6087 | 6187 | 6287 | 6388 | 100 |
| 433 | 6488 | 6588 | 6688 | 6789 | 6889 | 6989 | 7089 | 7189 | 7290 | 7390 | 100 |
| 434 | 7490 | 7590 | 7690 | 7790 | 7890 | 7990 | 8090 | 8190 | 8290 | 8389 | 99 |
| 435 | 8489 | 8589 | 8689 | 8789 | 8888 | 8988 | 9088 | 9188 | 9287 | 9387 | 99 |
| 436 | *9486 | 9586 | 9686 | 9785 | 9885 | 9984 | ♦084 | 0183 | 0283 | 0382 | 99 |
| 437 | 64 0481 | 0581 | 0680 | 0779 | 0879 | 0978 | 1077 | 1177 | 1276 | 1375 | 99 |
| 438 | 1474 | 1573 | 1672 | 1771 | 1871 | 1970 | 2069 | 2168 | 2267 | 2366 | 99 |
| 439 | 2465 | 2563 | 2662 | 2761 | 2860 | 2959 | 3058 | 3156 | 3255 | 3354 | 99 |
| 440 | 3453 | 3551 | 3650 | 3749 | 3847 | 3946 | 4044 | 4143 | 4242 | 4340 | 98 |
| 441 | 4439 | 4537 | 4636 | 4734 | 4832 | 4931 | 5029 | 5127 | 5226 | 5324 | 98 |
| 442 | 5422 | 5521 | 5619 | 5717 | 5815 | 5913 | 6011 | 6110 | 6208 | 6306 | 98 |
| 443 | 6404 | 6502 | 6600 | 6698 | 6796 | 6894 | 6992 | 7089 | 7187 | 7285 | 98 |
| 444 | 7383 | 7481 | 7579 | 7676 | 7774 | 7872 | 7969 | 8067 | 8165 | 8262 | 98 |
| 445 | 8360 | 8458 | 8555 | 8653 | 8750 | 8848 | 8945 | 9043 | 9140 | 9237 | 97 |
| 446 | *9335 | 9432 | 9530 | 9627 | 9724 | 9821 | 9919 | ♦016 | 0113 | 0210 | 97 |
| 447 | 65 0308 | 0405 | 0502 | 0599 | 0696 | 0793 | 0890 | 0987 | 1084 | 1181 | 97 |
| 448 | 1278 | 1375 | 1472 | 1569 | 1666 | 1762 | 1859 | 1956 | 2053 | 2150 | 97 |
| 449 | 2246 | 2343 | 2440 | 2536 | 2633 | 2730 | 2826 | 2923 | 3019 | 3116 | 97 |
| 450 | 3213 | 3309 | 3405 | 3502 | 3598 | 3695 | 3791 | 3888 | 3984 | 4080 | 96 |
| 451 | 4177 | 4273 | 4369 | 4465 | 4562 | 4658 | 4754 | 4850 | 4946 | 5042 | 96 |
| 452 | 5138 | 5235 | 5331 | 5427 | 5523 | 5619 | 5715 | 5810 | 5906 | 6002 | 96 |
| 453 | 6098 | 6194 | 6290 | 6386 | 6482 | 6577 | 6673 | 6769 | 6864 | 6960 | 96 |
| 454 | 7056 | 7152 | 7247 | 7343 | 7438 | 7534 | 7629 | 7725 | 7820 | 7916 | 96 |
| 455 | 8011 | 8107 | 8202 | 8298 | 8393 | 8488 | 8584 | 8679 | 8774 | 8870 | 95 |
| 456 | 8965 | 9060 | 9155 | 9250 | 9346 | 9441 | 9536 | 9631 | 9726 | 9821 | 95 |
| 457 | *9916 | ♦011 | 0106 | 0201 | 0296 | 0391 | 0486 | 0581 | 0676 | 0771 | 95 |
| 458 | 66 0865 | 0960 | 1055 | 1150 | 1245 | 1339 | 1434 | 1529 | 1623 | 1718 | 95 |
| 459 | 1813 | 1907 | 2002 | 2096 | 2191 | 2286 | 2380 | 2475 | 2569 | 2663 | 95 |
| N. | 0 | 1 | 2 | 3 | 4 | 5 | 6 | 7 | 8 | 9 | D. |

| N. | 0 | 1 | 2 | 3 | 4 | 5 | 6 | 7 | 8 | 9 | D. |
|---|---|---|---|---|---|---|---|---|---|---|---|
| 460 | 66 2758 | 2852 | 2947 | 3041 | 3135 | 3230 | 3324 | 3418 | 3512 | 3607 | 94 |
| 461 | 3701 | 3795 | 3889 | 3983 | 4078 | 4172 | 4266 | 4360 | 4454 | 4548 | 94 |
| 462 | 4642 | 4736 | 4830 | 4924 | 5018 | 5112 | 5206 | 5299 | 5393 | 5487 | 94 |
| 463 | 5581 | 5675 | 5769 | 5862 | 5956 | 6050 | 6143 | 6237 | 6331 | 6424 | 94 |
| 464 | 6518 | 6612 | 6705 | 6799 | 6892 | 6986 | 7079 | 7173 | 7266 | 7360 | 94 |
| 465 | 7453 | 7546 | 7640 | 7733 | 7826 | 7920 | 8013 | 8106 | 8199 | 8293 | 93 |
| 466 | 8386 | 8479 | 8572 | 8665 | 8759 | 8852 | 8945 | 9038 | 9131 | 9224 | 93 |
| 467 | *9317 | 9410 | 9503 | 9596 | 9689 | 9782 | 9875 | 9967 | •060 | 0153 | 93 |
| 468 | 67 0246 | 0339 | 0431 | 0524 | 0617 | 0710 | 0802 | 0895 | 0988 | 1080 | 93 |
| 469 | 1173 | 1265 | 1358 | 1451 | 1543 | 1636 | 1728 | 1821 | 1913 | 2005 | 93 |
| 470 | 2098 | 2190 | 2283 | 2375 | 2467 | 2560 | 2652 | 2744 | 2836 | 2929 | 92 |
| 471 | 3021 | 3113 | 3205 | 3297 | 3390 | 3482 | 3574 | 3666 | 3758 | 3850 | 92 |
| 472 | 3942 | 4034 | 4126 | 4218 | 4310 | 4402 | 4494 | 4586 | 4677 | 4769 | 92 |
| 473 | 4861 | 4953 | 5045 | 5137 | 5228 | 5320 | 5412 | 5503 | 5595 | 5687 | 92 |
| 474 | 5778 | 5870 | 5962 | 6053 | 6145 | 6236 | 6328 | 6419 | 6511 | 6602 | 92 |
| 475 | 6694 | 6785 | 6876 | 6968 | 7059 | 7151 | 7242 | 7333 | 7424 | 7516 | 91 |
| 476 | 7607 | 7698 | 7789 | 7881 | 7972 | 8063 | 8154 | 8245 | 8336 | 8427 | 91 |
| 477 | 8518 | 8609 | 8700 | 8791 | 8882 | 8973 | 9064 | 9155 | 9246 | 9337 | 91 |
| 478 | *9428 | 9519 | 9610 | 9700 | 9791 | 9882 | 9973 | •063 | 0154 | 0245 | 91 |
| 479 | 68 0336 | 0426 | 0517 | 0607 | 0698 | 0789 | 0879 | 0970 | 1060 | 1151 | 91 |
| 480 | 1241 | 1332 | 1422 | 1513 | 1603 | 1693 | 1784 | 1874 | 1964 | 2055 | 90 |
| 481 | 2145 | 2235 | 2326 | 2416 | 2506 | 2596 | 2686 | 2777 | 2867 | 2957 | 90 |
| 482 | 3047 | 3137 | 3227 | 3317 | 3407 | 3497 | 3587 | 3677 | 3767 | 3857 | 90 |
| 483 | 3947 | 4037 | 4127 | 4217 | 4307 | 4396 | 4486 | 4576 | 4666 | 4756 | 90 |
| 484 | 4845 | 4935 | 5025 | 5114 | 5204 | 5294 | 5383 | 5473 | 5563 | 5652 | 90 |
| 485 | 5742 | 5831 | 5921 | 6010 | 6100 | 6189 | 6279 | 6368 | 6458 | 6547 | 89 |
| 486 | 6636 | 6726 | 6815 | 6904 | 6994 | 7083 | 7172 | 7261 | 7351 | 7440 | 89 |
| 487 | 7529 | 7618 | 7707 | 7796 | 7886 | 7975 | 8064 | 8153 | 8242 | 8331 | 89 |
| 488 | 8420 | 8509 | 8598 | 8687 | 8776 | 8865 | 8953 | 9042 | 9131 | 9220 | 89 |
| 489 | *9309 | 9398 | 9486 | 9575 | 9664 | 9753 | 9841 | 9930 | •019 | 0107 | 89 |
| 490 | 69 0196 | 0285 | 0373 | 0462 | 0550 | 0639 | 0728 | 0816 | 0905 | 0993 | 89 |
| 491 | 1081 | 1170 | 1258 | 1347 | 1435 | 1524 | 1612 | 1700 | 1789 | 1877 | 88 |
| 492 | 1965 | 2053 | 2142 | 2230 | 2318 | 2406 | 2494 | 2583 | 2671 | 2759 | 88 |
| 493 | 2847 | 2935 | 3023 | 3111 | 3199 | 3287 | 3375 | 3463 | 3551 | 3639 | 88 |
| 494 | 3727 | 3815 | 3903 | 3991 | 4078 | 4166 | 4254 | 4342 | 4430 | 4517 | 88 |
| 495 | 4605 | 4693 | 4781 | 4868 | 4956 | 5044 | 5131 | 5219 | 5307 | 5394 | 88 |
| 496 | 5482 | 5569 | 5657 | 5744 | 5832 | 5919 | 6007 | 6094 | 6182 | 6269 | 87 |
| 497 | 6356 | 6444 | 6531 | 6618 | 6706 | 6793 | 6880 | 6968 | 7055 | 7142 | 87 |
| 498 | 7229 | 7317 | 7404 | 7491 | 7578 | 7665 | 7752 | 7839 | 7926 | 8014 | 87 |
| 499 | 8101 | 8188 | 8275 | 8362 | 8449 | 8535 | 8622 | 8709 | 8796 | 8883 | 87 |
| 500 | 8970 | 9057 | 9144 | 9231 | 9317 | 9404 | 9491 | 9578 | 9664 | 9751 | 87 |
| 501 | *9838 | 9924 | •011 | 0098 | 0184 | 0271 | 0358 | 0444 | 0531 | 0617 | 87 |
| 502 | 70 0704 | 0790 | 0877 | 0963 | 1050 | 1136 | 1222 | 1309 | 1395 | 1482 | 86 |
| 503 | 1568 | 1654 | 1741 | 1827 | 1913 | 1999 | 2086 | 2172 | 2258 | 2344 | 86 |
| 504 | 2431 | 2517 | 2603 | 2689 | 2775 | 2861 | 2947 | 3033 | 3119 | 3205 | 86 |
| 505 | 3291 | 3377 | 3463 | 3549 | 3635 | 3721 | 3807 | 3895 | 3979 | 4065 | 86 |
| 506 | 4151 | 4236 | 4322 | 4408 | 4494 | 4579 | 4665 | 4751 | 4837 | 4922 | 86 |
| 507 | 5008 | 5094 | 5179 | 5265 | 5350 | 5436 | 5522 | 5607 | 5693 | 5778 | 86 |
| 508 | 5864 | 5949 | 6035 | 6120 | 6206 | 6291 | 6376 | 6462 | 6547 | 6632 | 85 |
| 509 | 6718 | 6803 | 6888 | 6974 | 7059 | 7144 | 7229 | 7315 | 7400 | 7485 | 85 |
| 510 | 7570 | 7655 | 7740 | 7826 | 7911 | 7996 | 8081 | 8166 | 8251 | 8336 | 85 |
| 511 | 8421 | 8506 | 8591 | 8676 | 8761 | 8846 | 8931 | 9015 | 9100 | 9185 | 85 |
| 512 | *9270 | 9355 | 9440 | 9524 | 9609 | 9694 | 9779 | 9863 | 9948 | •033 | 85 |
| 513 | 71 0117 | 0202 | 0287 | 0371 | 0456 | 0540 | 0625 | 0710 | 0794 | 0879 | 85 |
| 514 | 0963 | 1048 | 1132 | 1217 | 1301 | 1385 | 1470 | 1554 | 1639 | 1723 | 84 |
| 515 | 1807 | 1892 | 1976 | 2060 | 2144 | 2229 | 2313 | 2397 | 2481 | 2566 | 84 |
| 516 | 2650 | 2734 | 2818 | 2902 | 2986 | 3070 | 3154 | 3238 | 3323 | 3407 | 84 |
| 517 | 3491 | 3575 | 3650 | 3742 | 3826 | 3910 | 3994 | 4078 | 4162 | 4246 | 84 |
| 518 | 4330 | 4414 | 4497 | 4581 | 4665 | 4749 | 4833 | 4916 | 5000 | 5084 | 84 |
| 519 | 5167 | 5251 | 5335 | 5418 | 5502 | 5586 | 5669 | 5753 | 5836 | 5920 | 84 |
| N. | 0 | 1 | 2 | 3 | 4 | 5 | 6 | 7 | 8 | 9 | D. |

| N. | 0 | 1 | 2 | 3 | 4 | 5 | 6 | 7 | 8 | 9 | D. |
|---|---|---|---|---|---|---|---|---|---|---|---|
| 520 | 71 6003 | 6087 | 6170 | 6254 | 6337 | 6421 | 6504 | 6588 | 6671 | 6754 | 83 |
| 521 | 6838 | 6921 | 7004 | 7088 | 7171 | 7254 | 7338 | 7421 | 7504 | 7587 | 83 |
| 522 | 7671 | 7754 | 7837 | 7920 | 8003 | 8086 | 8169 | 8253 | 8336 | 8419 | 83 |
| 523 | 8502 | 8585 | 8668 | 8751 | 8834 | 8917 | 9000 | 9083 | 9165 | 9248 | 83 |
| 524 | *9331 | 9414 | 9497 | 9580 | 9663 | 9745 | 9828 | 9911 | 9994 | *077 | 83 |
| 525 | 72 0159 | 0242 | 0325 | 0407 | 0490 | 0573 | 0655 | 0738 | 0821 | 0903 | 83 |
| 526 | 0986 | 1068 | 1151 | 1233 | 1316 | 1398 | 1481 | 1563 | 1646 | 1728 | 82 |
| 527 | 1811 | 1893 | 1975 | 2058 | 2140 | 2222 | 2305 | 2387 | 2469 | 2552 | 82 |
| 528 | 2634 | 2716 | 2798 | 2881 | 2963 | 3045 | 3127 | 3209 | 3291 | 3374 | 82 |
| 529 | 3456 | 3538 | 3620 | 3702 | 3784 | 3866 | 3948 | 4030 | 4112 | 4194 | 82 |
| 530 | 4276 | 4358 | 4440 | 4522 | 4604 | 4685 | 4767 | 4849 | 4931 | 5013 | 82 |
| 531 | 5095 | 5176 | 5258 | 5340 | 5422 | 5503 | 5585 | 5667 | 5748 | 5830 | 82 |
| 532 | 5912 | 5993 | 6075 | 6156 | 6238 | 6320 | 6401 | 6483 | 6564 | 6646 | 82 |
| 533 | 6727 | 6809 | 6890 | 6972 | 7053 | 7134 | 7216 | 7297 | 7379 | 7460 | 81 |
| 534 | 7541 | 7623 | 7704 | 7785 | 7866 | 7948 | 8029 | 8110 | 8191 | 8273 | 81 |
| 535 | 8354 | 8435 | 8516 | 8597 | 8678 | 8759 | 8841 | 8922 | 9003 | 9084 | 81 |
| 536 | 9165 | 9246 | 9327 | 9408 | 9489 | 9570 | 9651 | 9732 | 9813 | 9893 | 81 |
| 537 | *9974 | *055 | 0136 | 0217 | 0298 | 0378 | 0459 | 0540 | 0621 | 0702 | 81 |
| 538 | 73 0782 | 0863 | 0944 | 1024 | 1105 | 1186 | 1266 | 1347 | 1428 | 1508 | 81 |
| 539 | 1589 | 1669 | 1750 | 1830 | 1911 | 1991 | 2072 | 2152 | 2233 | 2313 | 81 |
| 540 | 2394 | 2474 | 2555 | 2635 | 2715 | 2796 | 2876 | 2956 | 3037 | 3117 | 80 |
| 541 | 3197 | 3278 | 3358 | 3438 | 3518 | 3598 | 3679 | 3759 | 3839 | 3919 | 80 |
| 542 | 3999 | 4079 | 4160 | 4240 | 4320 | 4400 | 4480 | 4560 | 4640 | 4720 | 80 |
| 543 | 4800 | 4880 | 4960 | 5040 | 5120 | 5200 | 5279 | 5359 | 5439 | 5519 | 80 |
| 544 | 5599 | 5679 | 5759 | 5838 | 5918 | 5998 | 6078 | 6157 | 6237 | 6317 | 80 |
| 545 | 6397 | 6476 | 6556 | 6635 | 6715 | 6795 | 6874 | 6954 | 7034 | 7113 | 80 |
| 546 | 7193 | 7272 | 7352 | 7431 | 7511 | 7590 | 7670 | 7749 | 7829 | 7908 | 79 |
| 547 | 7987 | 8067 | 8146 | 8225 | 8305 | 8384 | 8463 | 8543 | 8622 | 8701 | 79 |
| 548 | 8781 | 8860 | 8939 | 9018 | 9097 | 9177 | 9256 | 9335 | 9414 | 9493 | 79 |
| 549 | *9572 | 9651 | 9731 | 9810 | 9889 | 9968 | *047 | 0126 | 0205 | 0284 | 79 |
| 550 | 74 0363 | 0442 | 0521 | 0600 | 0678 | 0757 | 0836 | 0915 | 0994 | 1073 | 79 |
| 551 | 1152 | 1230 | 1309 | 1388 | 1467 | 1546 | 1624 | 1703 | 1782 | 1860 | 79 |
| 552 | 1939 | 2018 | 2096 | 2175 | 2254 | 2332 | 2411 | 2489 | 2568 | 2646 | 79 |
| 553 | 2725 | 2804 | 2882 | 2961 | 3039 | 3118 | 3196 | 3275 | 3353 | 3431 | 78 |
| 554 | 3510 | 3588 | 3667 | 3745 | 3823 | 3902 | 3980 | 4058 | 4136 | 4215 | 78 |
| 555 | 4293 | 4371 | 4449 | 4528 | 4606 | 4684 | 4762 | 4840 | 4919 | 4997 | 78 |
| 556 | 5075 | 5153 | 5231 | 5309 | 5387 | 5465 | 5543 | 5621 | 5699 | 5777 | 78 |
| 557 | 5855 | 5933 | 6011 | 6089 | 6167 | 6245 | 6323 | 6401 | 6479 | 6556 | 78 |
| 558 | 6634 | 6712 | 6790 | 6868 | 6945 | 7023 | 7101 | 7179 | 7256 | 7334 | 78 |
| 559 | 7412 | 7489 | 7567 | 7645 | 7722 | 7800 | 7878 | 7955 | 8033 | 8110 | 78 |
| 560 | 8188 | 8266 | 8343 | 8421 | 8498 | 8576 | 8653 | 8731 | 8808 | 8885 | 77 |
| 561 | 8963 | 9040 | 9118 | 9195 | 9272 | 9350 | 9427 | 9504 | 9582 | 9659 | 77 |
| 562 | *9736 | 9814 | 9891 | 9968 | *045 | 0123 | 0200 | 0277 | 0354 | 0431 | 77 |
| 563 | 75 0508 | 0586 | 0663 | 0740 | 0817 | 0894 | 0971 | 1048 | 1125 | 1202 | 77 |
| 564 | 1279 | 1356 | 1433 | 1510 | 1587 | 1664 | 1741 | 1818 | 1895 | 1972 | 77 |
| 565 | 2048 | 2125 | 2202 | 2279 | 2356 | 2433 | 2509 | 2586 | 2663 | 2740 | 77 |
| 566 | 2816 | 2893 | 2970 | 3047 | 3123 | 3200 | 3277 | 3353 | 3430 | 3506 | 77 |
| 567 | 3583 | 3660 | 3736 | 3813 | 3889 | 3966 | 4042 | 4119 | 4195 | 4272 | 77 |
| 568 | 4348 | 4425 | 4501 | 4578 | 4654 | 4730 | 4807 | 4883 | 4960 | 5036 | 76 |
| 569 | 5112 | 5189 | 5265 | 5341 | 5417 | 5494 | 5570 | 5646 | 5722 | 5799 | 76 |
| 570 | 5875 | 5951 | 6027 | 6103 | 6180 | 6256 | 6332 | 6408 | 6484 | 6560 | 76 |
| 571 | 6636 | 6712 | 6788 | 6864 | 6940 | 7016 | 7092 | 7168 | 7244 | 7320 | 76 |
| 572 | 7396 | 7472 | 7548 | 7624 | 7700 | 7775 | 7851 | 7927 | 8003 | 8079 | 76 |
| 573 | 8155 | 8230 | 8306 | 8382 | 8458 | 8533 | 8609 | 8685 | 8761 | 8836 | 76 |
| 574 | 8912 | 8988 | 9063 | 9139 | 9214 | 9290 | 9366 | 9441 | 9517 | 9592 | 76 |
| 575 | *9668 | 9743 | 9819 | 9894 | 9970 | *045 | 0121 | 0196 | 0272 | 0347 | 75 |
| 576 | 76 0422 | 0498 | 0573 | 0649 | 0724 | 0799 | 0875 | 0950 | 1025 | 1101 | 75 |
| 577 | 1176 | 1251 | 1326 | 1402 | 1477 | 1552 | 1627 | 1702 | 1778 | 1853 | 75 |
| 578 | 1928 | 2003 | 2078 | 2153 | 2228 | 2303 | 2378 | 2453 | 2529 | 2604 | 75 |
| 579 | 2679 | 2754 | 2829 | 2904 | 2978 | 3053 | 3128 | 3203 | 3278 | 3353 | 75 |
| N. | 0 | 1 | 2 | 3 | 4 | 5 | 6 | 7 | 8 | 9 | D. |

| N. | 0 | 1 | 2 | 3 | 4 | 5 | 6 | 7 | 8 | 9 | D. |
|---|---|---|---|---|---|---|---|---|---|---|---|
| 580 | 76 3428 | 3503 | 3578 | 3653 | 3727 | 3802 | 3877 | 3952 | 4027 | 4101 | 75 |
| 581 | 4176 | 4251 | 4326 | 4400 | 4475 | 4550 | 4624 | 4699 | 4774 | 4848 | 75 |
| 582 | 4923 | 4998 | 5072 | 5147 | 5221 | 5296 | 5370 | 5445 | 5520 | 5594 | 75 |
| 583 | 5669 | 5743 | 5818 | 5892 | 5966 | 6041 | 6115 | 6190 | 6264 | 6338 | 74 |
| 584 | 6413 | 6487 | 6562 | 6636 | 6710 | 6785 | 6859 | 6933 | 7007 | 7082 | 74 |
| 585 | 7156 | 7230 | 7304 | 7379 | 7453 | 7527 | 7601 | 7675 | 7749 | 7823 | 74 |
| 586 | 7898 | 7972 | 8046 | 8120 | 8194 | 8268 | 8342 | 8416 | 8490 | 8564 | 74 |
| 587 | 8638 | 8712 | 8786 | 8860 | 8934 | 9008 | 9082 | 9156 | 9230 | 9303 | 74 |
| 588 | *9377 | 9451 | 9525 | 9599 | 9673 | 9746 | 9820 | 9894 | 9968 | •042 | 74 |
| 589 | 77 0115 | 0189 | 0263 | 0336 | 0410 | 0484 | 0557 | 0631 | 0705 | 0778 | 74 |
| 590 | 0852 | 0926 | 0999 | 1073 | 1146 | 1220 | 1293 | 1367 | 1440 | 1514 | 74 |
| 591 | 1587 | 1661 | 1734 | 1808 | 1881 | 1955 | 2028 | 2102 | 2175 | 2248 | 73 |
| 592 | 2322 | 2395 | 2468 | 2542 | 2615 | 2688 | 2762 | 2835 | 2908 | 2981 | 73 |
| 593 | 3055 | 3128 | 3201 | 3274 | 3348 | 3421 | 3494 | 3567 | 3640 | 3713 | 73 |
| 594 | 3786 | 3860 | 3933 | 4006 | 4079 | 4152 | 4225 | 4298 | 4371 | 4444 | 73 |
| 595 | 4517 | 4590 | 4663 | 4736 | 4809 | 4882 | 4955 | 5028 | 5100 | 5173 | 73 |
| 596 | 5246 | 5319 | 5392 | 5465 | 5538 | 5610 | 5683 | 5756 | 5829 | 5902 | 73 |
| 597 | 5974 | 6047 | 6120 | 6193 | 6265 | 6338 | 6411 | 6483 | 6556 | 6629 | 73 |
| 598 | 6701 | 6774 | 6846 | 6919 | 6992 | 7064 | 7137 | 7209 | 7282 | 7354 | 73 |
| 599 | 7427 | 7499 | 7572 | 7644 | 7717 | 7789 | 7862 | 7934 | 8006 | 8079 | 72 |
| 600 | 8151 | 8224 | 8296 | 8368 | 8441 | 8513 | 8585 | 8658 | 8730 | 8802 | 72 |
| 601 | 8874 | 8947 | 9019 | 9091 | 9163 | 9236 | 9308 | 9380 | 9452 | 9524 | 72 |
| 602 | *9596 | 9669 | 9741 | 9813 | 9885 | 9957 | •029 | 0101 | 0173 | 0245 | 72 |
| 603 | 78 0317 | 0389 | 0461 | 0533 | 0605 | 0677 | 0749 | 0821 | 0893 | 0965 | 72 |
| 604 | 1037 | 1109 | 1181 | 1253 | 1324 | 1396 | 1468 | 1540 | 1612 | 1684 | 72 |
| 605 | 1755 | 1827 | 1899 | 1971 | 2042 | 2114 | 2186 | 2258 | 2329 | 2401 | 72 |
| 606 | 2473 | 2544 | 2616 | 2688 | 2759 | 2831 | 2902 | 2974 | 3046 | 3117 | 72 |
| 607 | 3189 | 3260 | 3332 | 3403 | 3475 | 3546 | 3618 | 3689 | 3761 | 3832 | 71 |
| 608 | 3904 | 3975 | 4046 | 4118 | 4189 | 4261 | 4332 | 4403 | 4475 | 4546 | 71 |
| 609 | 4617 | 4689 | 4760 | 4831 | 4902 | 4974 | 5045 | 5116 | 5187 | 5259 | 71 |
| 610 | 5330 | 5401 | 5472 | 5543 | 5615 | 5686 | 5757 | 5828 | 5899 | 5970 | 71 |
| 611 | 6041 | 6112 | 6183 | 6254 | 6325 | 6396 | 6467 | 6538 | 6609 | 6680 | 71 |
| 612 | 6751 | 6822 | 6893 | 6964 | 7035 | 7106 | 7177 | 7248 | 7319 | 7390 | 71 |
| 613 | 7460 | 7531 | 7602 | 7673 | 7744 | 7815 | 7885 | 7956 | 8027 | 8098 | 71 |
| 614 | 8168 | 8239 | 8310 | 8381 | 8451 | 8522 | 8593 | 8663 | 8734 | 8804 | 71 |
| 615 | 8875 | 8946 | 9016 | 9087 | 9157 | 9228 | 9299 | 9369 | 9440 | 9510 | 71 |
| 616 | *9581 | 9651 | 9722 | 9792 | 9863 | 9933 | •004 | 0074 | 0144 | 0215 | 70 |
| 617 | 79 0285 | 0356 | 0426 | 0496 | 0567 | 0637 | 0707 | 0778 | 0848 | 0918 | 70 |
| 618 | 0988 | 1059 | 1129 | 1199 | 1269 | 1340 | 1410 | 1480 | 1550 | 1620 | 70 |
| 619 | 1691 | 1761 | 1831 | 1901 | 1971 | 2041 | 2111 | 2181 | 2252 | 2322 | 70 |
| 620 | 2392 | 2462 | 2532 | 2602 | 2672 | 2742 | 2812 | 2882 | 2952 | 3022 | 70 |
| 621 | 3092 | 3162 | 3231 | 3301 | 3371 | 3441 | 3511 | 3581 | 3651 | 3721 | 70 |
| 622 | 3790 | 3860 | 3930 | 4000 | 4070 | 4139 | 4209 | 4279 | 4349 | 4418 | 70 |
| 623 | 4488 | 4558 | 4627 | 4697 | 4767 | 4836 | 4906 | 4976 | 5045 | 5115 | 70 |
| 624 | 5185 | 5254 | 5324 | 5393 | 5463 | 5532 | 5602 | 5672 | 5741 | 5811 | 70 |
| 625 | 5880 | 5949 | 6019 | 6088 | 6158 | 6227 | 6297 | 6366 | 6436 | 6505 | 69 |
| 626 | 6574 | 6644 | 6713 | 6782 | 6852 | 6921 | 6990 | 7060 | 7129 | 7198 | 69 |
| 627 | 7268 | 7337 | 7406 | 7475 | 7545 | 7614 | 7683 | 7752 | 7821 | 7890 | 69 |
| 628 | 7960 | 8029 | 8098 | 8167 | 8236 | 8305 | 8374 | 8443 | 8513 | 8582 | 69 |
| 629 | 8651 | 8720 | 8789 | 8858 | 8927 | 8996 | 9065 | 9134 | 9203 | 9272 | 69 |
| 630 | 9341 | 9409 | 9478 | 9547 | 9616 | 9685 | 9754 | 9823 | 9892 | 9961 | 69 |
| 631 | 80 0029 | 0098 | 0167 | 0236 | 0305 | 0373 | 0442 | 0511 | 0580 | 0648 | 69 |
| 632 | 0717 | 0786 | 0854 | 0923 | 0992 | 1061 | 1129 | 1198 | 1266 | 1335 | 69 |
| 633 | 1404 | 1472 | 1541 | 1609 | 1678 | 1747 | 1815 | 1884 | 1952 | 2021 | 69 |
| 634 | 2089 | 2158 | 2226 | 2295 | 2363 | 2432 | 2500 | 2568 | 2637 | 2705 | 69 |
| 635 | 2774 | 2842 | 2910 | 2979 | 3047 | 3116 | 3184 | 3252 | 3321 | 3389 | 68 |
| 636 | 3457 | 3525 | 3594 | 3662 | 3730 | 3798 | 3867 | 3935 | 4003 | 4071 | 68 |
| 637 | 4139 | 4208 | 4276 | 4344 | 4412 | 4480 | 4548 | 4616 | 4685 | 4753 | 68 |
| 638 | 4821 | 4889 | 4957 | 5025 | 5093 | 5161 | 5229 | 5297 | 5365 | 5433 | 68 |
| 639 | 5501 | 5569 | 5637 | 5705 | 5773 | 5841 | 5908 | 5976 | 6044 | 6112 | 68 |
| N. | 0 | 1 | 2 | 3 | 4 | 5 | 6 | 7 | 8 | 9 | D. |

| N. | 0 | 1 | 2 | 3 | 4 | 5 | 6 | 7 | 8 | 9 | D. |
|---|---|---|---|---|---|---|---|---|---|---|---|
| 640 | 80 6180 | 6248 | 6316 | 6384 | 6451 | 6519 | 6587 | 6655 | 6723 | 6790 | 68 |
| 641 | 6858 | 6926 | 6994 | 7061 | 7129 | 7197 | 7264 | 7332 | 7400 | 7467 | 68 |
| 642 | 7535 | 7603 | 7670 | 7738 | 7806 | 7873 | 7941 | 8008 | 8076 | 8143 | 68 |
| 643 | 8211 | 8279 | 8346 | 8414 | 8481 | 8549 | 8616 | 8684 | 8751 | 8818 | 67 |
| 644 | 8886 | 8953 | 9021 | 9088 | 9156 | 9223 | 9290 | 9358 | 9425 | 9492 | 67 |
| 645 | *9560 | 9627 | 9694 | 9762 | 9829 | 9896 | 9964 | ◆031 | 0098 | 0165 | 67 |
| 646 | 81 0233 | 0300 | 0367 | 0434 | 0501 | 0569 | 0636 | 0703 | 0770 | 0837 | 67 |
| 647 | 0904 | 0971 | 1039 | 1106 | 1173 | 1240 | 1307 | 1374 | 1441 | 1508 | 67 |
| 648 | 1575 | 1642 | 1709 | 1776 | 1843 | 1910 | 1977 | 2044 | 2111 | 2178 | 67 |
| 649 | 2245 | 2312 | 2379 | 2445 | 2512 | 2579 | 2646 | 2713 | 2780 | 2847 | 67 |
| 650 | 2913 | 2980 | 3047 | 3114 | 3181 | 3247 | 3314 | 3381 | 3448 | 3514 | 67 |
| 651 | 3581 | 3648 | 3714 | 3781 | 3848 | 3914 | 3981 | 4048 | 4114 | 4181 | 67 |
| 652 | 4248 | 4314 | 4381 | 4447 | 4514 | 4581 | 4647 | 4714 | 4780 | 4847 | 67 |
| 653 | 4913 | 4980 | 5046 | 5113 | 5179 | 5246 | 5312 | 5378 | 5445 | 5511 | 66 |
| 654 | 5578 | 5644 | 5711 | 5777 | 5843 | 5910 | 5976 | 6042 | 6109 | 6175 | 66 |
| 655 | 6241 | 6308 | 6374 | 6440 | 6506 | 6573 | 6639 | 6705 | 6771 | 6838 | 66 |
| 656 | 6904 | 6970 | 7036 | 7102 | 7169 | 7235 | 7301 | 7367 | 7433 | 7499 | 66 |
| 657 | 7565 | 7631 | 7698 | 7764 | 7830 | 7896 | 7962 | 8028 | 8094 | 8160 | 66 |
| 658 | 8226 | 8292 | 8358 | 8424 | 8490 | 8556 | 8622 | 8688 | 8754 | 8820 | 66 |
| 659 | 8885 | 8951 | 9017 | 9083 | 9149 | 9215 | 9281 | 9346 | 9412 | 9478 | 66 |
| 660 | *9544 | 9610 | 9676 | 9741 | 9807 | 9873 | 9939 | ◆004 | 0070 | 0136 | 66 |
| 661 | 82 0201 | 0267 | 0333 | 0399 | 0464 | 0530 | 0595 | 0661 | 0727 | 0792 | 66 |
| 662 | 0858 | 0924 | 0989 | 1055 | 1120 | 1186 | 1251 | 1317 | 1382 | 1448 | 66 |
| 663 | 1514 | 1579 | 1645 | 1710 | 1775 | 1841 | 1906 | 1972 | 2037 | 2103 | 65 |
| 664 | 2168 | 2233 | 2299 | 2364 | 2430 | 2495 | 2560 | 2626 | 2691 | 2756 | 65 |
| 665 | 2822 | 2887 | 2952 | 3018 | 3083 | 3148 | 3213 | 3279 | 3344 | 3409 | 65 |
| 666 | 3474 | 3539 | 3605 | 3670 | 3735 | 3800 | 3865 | 3930 | 3996 | 4061 | 65 |
| 667 | 4126 | 4191 | 4256 | 4321 | 4386 | 4451 | 4516 | 4581 | 4646 | 4711 | 65 |
| 668 | 4776 | 4841 | 4906 | 4971 | 5036 | 5101 | 5166 | 5231 | 5296 | 5361 | 65 |
| 669 | 5426 | 5491 | 5556 | 5621 | 5686 | 5751 | 5815 | 5880 | 5945 | 6010 | 65 |
| 670 | 6075 | 6140 | 6204 | 6269 | 6334 | 6399 | 6464 | 6528 | 6593 | 6658 | 65 |
| 671 | 6723 | 6787 | 6852 | 6917 | 6981 | 7046 | 7111 | 7175 | 7240 | 7305 | 65 |
| 672 | 7369 | 7434 | 7499 | 7563 | 7628 | 7692 | 7757 | 7821 | 7886 | 7951 | 65 |
| 673 | 8015 | 8080 | 8144 | 8209 | 8273 | 8338 | 8402 | 8467 | 8531 | 8595 | 64 |
| 674 | 8660 | 8724 | 8789 | 8853 | 8918 | 8982 | 9046 | 9111 | 9175 | 9239 | 64 |
| 675 | 9304 | 9368 | 9432 | 9497 | 9561 | 9625 | 9690 | 9754 | 9818 | 9882 | 64 |
| 676 | *9947 | ◆011 | 0075 | 0139 | 0204 | 0268 | 0332 | 0396 | 0460 | 0525 | 64 |
| 677 | 83 0589 | 0653 | 0717 | 0781 | 0845 | 0909 | 0973 | 1037 | 1102 | 1166 | 64 |
| 678 | 1230 | 1294 | 1358 | 1422 | 1486 | 1550 | 1614 | 1678 | 1742 | 1806 | 64 |
| 679 | 1870 | 1934 | 1998 | 2062 | 2126 | 2189 | 2253 | 2317 | 2381 | 2445 | 64 |
| 680 | 2509 | 2573 | 2637 | 2700 | 2764 | 2828 | 2892 | 2956 | 3020 | 3083 | 64 |
| 681 | 3147 | 3211 | 3275 | 3338 | 3402 | 3466 | 3530 | 3593 | 3657 | 3721 | 64 |
| 682 | 3784 | 3848 | 3912 | 3975 | 4039 | 4103 | 4166 | 4230 | 4294 | 4357 | 64 |
| 683 | 4421 | 4484 | 4548 | 4611 | 4675 | 4739 | 4802 | 4866 | 4929 | 4993 | 64 |
| 684 | 5056 | 5120 | 5183 | 5247 | 5310 | 5373 | 5437 | 5500 | 5564 | 5627 | 63 |
| 685 | 5691 | 5754 | 5817 | 5881 | 5944 | 6007 | 6071 | 6134 | 6197 | 6261 | 63 |
| 686 | 6324 | 6387 | 6451 | 6514 | 6577 | 6641 | 6704 | 6767 | 6830 | 6894 | 63 |
| 687 | 6957 | 7020 | 7083 | 7146 | 7210 | 7273 | 7336 | 7399 | 7462 | 7525 | 63 |
| 688 | 7588 | 7652 | 7715 | 7778 | 7841 | 7904 | 7967 | 8030 | 8093 | 8156 | 63 |
| 689 | 8219 | 8282 | 8345 | 8408 | 8471 | 8534 | 8597 | 8660 | 8723 | 8786 | 63 |
| 690 | 8849 | [illegible] | [illegible] | [illegible] | 9101 | 9164 | [illegible] | [illegible] | [illegible] | 9415 | 63 |
| 691 | *9478 | 9541 | 9604 | 9667 | 9729 | 9792 | 9855 | 9918 | 9981 | ◆043 | 63 |
| 692 | 84 0106 | 0169 | 0232 | 0294 | 0357 | 0420 | 0482 | 0545 | 0608 | 0671 | 63 |
| 693 | 0733 | 0796 | 0859 | 0921 | 0984 | 1046 | 1109 | 1172 | 1234 | 1297 | 63 |
| 694 | 1359 | 1422 | 1485 | 1547 | 1610 | 1672 | 1735 | 1797 | 1860 | 1922 | 63 |
| 695 | 1985 | 2047 | 2110 | 2172 | 2235 | 2297 | 2360 | 2422 | 2484 | 2547 | 62 |
| 696 | 2609 | 2672 | 2734 | 2796 | 2859 | 2921 | 2983 | 3046 | 3108 | 3170 | 62 |
| 697 | 3233 | 3295 | 3357 | 3420 | 3482 | 3544 | 3606 | 3669 | 3731 | 3793 | 62 |
| 698 | 3855 | 3918 | 3980 | 4042 | 4104 | 4166 | 4229 | 4291 | 4353 | 4415 | 62 |
| 699 | 4477 | 4539 | 4601 | 4664 | 4726 | 4788 | 4850 | 4912 | 4974 | 5036 | 62 |
| N. | 0 | 1 | 2 | 3 | 4 | 5 | 6 | 7 | 8 | 9 | D. |

| N. | 0 | 1 | 2 | 3 | 4 | 5 | 6 | 7 | 8 | 9 | D. |
|---|---|---|---|---|---|---|---|---|---|---|---|
| 700 | 84 5098 | 5160 | 5222 | 5284 | 5346 | 5408 | 5470 | 5532 | 5594 | 5656 | 62 |
| 701 | 5718 | 5780 | 5842 | 5904 | 5966 | 6028 | 6090 | 6151 | 6213 | 6275 | 62 |
| 702 | 6337 | 6399 | 6461 | 6523 | 6585 | 6646 | 6708 | 6770 | 6832 | 6894 | 62 |
| 703 | 6955 | 7017 | 7079 | 7141 | 7202 | 7264 | 7326 | 7388 | 7449 | 7511 | 62 |
| 704 | 7573 | 7634 | 7696 | 7758 | 7819 | 7881 | 7943 | 8004 | 8066 | 8128 | 62 |
| 705 | 8189 | 8251 | 8312 | 8374 | 8435 | 8497 | 8559 | 8620 | 8682 | 8743 | 62 |
| 706 | 8805 | 8866 | 8928 | 8989 | 9051 | 9112 | 9174 | 9235 | 9297 | 9358 | 61 |
| 707 | 9419 | 9481 | 9542 | 9604 | 9665 | 9726 | 9788 | 9849 | 9911 | 9972 | 61 |
| 708 | 85 0033 | 0095 | 0156 | 0217 | 0279 | 0340 | 0401 | 0462 | 0524 | 0585 | 61 |
| 709 | 0646 | 0707 | 0769 | 0830 | 0891 | 0952 | 1014 | 1075 | 1136 | 1197 | 61 |
| 710 | 1258 | 1320 | 1381 | 1442 | 1503 | 1564 | 1625 | 1686 | 1747 | 1809 | 61 |
| 711 | 1870 | 1931 | 1992 | 2053 | 2114 | 2175 | 2236 | 2297 | 2358 | 2419 | 61 |
| 712 | 2480 | 2541 | 2602 | 2663 | 2724 | 2785 | 2846 | 2907 | 2968 | 3029 | 61 |
| 713 | 3090 | 3150 | 3211 | 3272 | 3333 | 3394 | 3455 | 3516 | 3577 | 3637 | 61 |
| 714 | 3698 | 3759 | 3820 | 3881 | 3941 | 4002 | 4063 | 4124 | 4185 | 4245 | 61 |
| 715 | 4306 | 4367 | 4428 | 4488 | 4549 | 4610 | 4670 | 4731 | 4792 | 4852 | 61 |
| 716 | 4913 | 4974 | 5034 | 5095 | 5156 | 5216 | 5277 | 5337 | 5398 | 5459 | 61 |
| 717 | 5519 | 5580 | 5640 | 5701 | 5761 | 5822 | 5882 | 5943 | 6003 | 6064 | 61 |
| 718 | 6124 | 6185 | 6245 | 6306 | 6366 | 6427 | 6487 | 6548 | 6608 | 6668 | 60 |
| 719 | 6729 | 6789 | 6850 | 6910 | 6970 | 7031 | 7091 | 7152 | 7212 | 7272 | 60 |
| 720 | 7332 | 7393 | 7453 | 7513 | 7574 | 7634 | 7694 | 7755 | 7815 | 7875 | 60 |
| 721 | 7935 | 7995 | 8056 | 8116 | 8176 | 8236 | 8297 | 8357 | 8417 | 8477 | 60 |
| 722 | 8537 | 8597 | 8657 | 8718 | 8778 | 8838 | 8898 | 8958 | 9018 | 9078 | 60 |
| 723 | 9138 | 9198 | 9258 | 9318 | 9379 | 9439 | 9499 | 9559 | 9619 | 9679 | 60 |
| 724 | *9739 | 9799 | 9859 | 9918 | 9978 | *038 | 0098 | 0158 | 0218 | 0278 | 60 |
| 725 | 86 0338 | 0398 | 0458 | 0518 | 0578 | 0637 | 0697 | 0757 | 0817 | 0877 | 60 |
| 726 | 0937 | 0996 | 1056 | 1116 | 1176 | 1236 | 1295 | 1355 | 1415 | 1475 | 60 |
| 727 | 1534 | 1594 | 1654 | 1714 | 1773 | 1833 | 1893 | 1952 | 2012 | 2072 | 60 |
| 728 | 2131 | 2191 | 2251 | 2310 | 2370 | 2430 | 2489 | 2549 | 2608 | 2668 | 60 |
| 729 | 2728 | 2787 | 2847 | 2906 | 2966 | 3025 | 3085 | 3144 | 3204 | 3263 | 60 |
| 730 | 3323 | 3382 | 3442 | 3501 | 3561 | 3620 | 3680 | 3739 | 3799 | 3858 | 59 |
| 731 | 3917 | 3977 | 4036 | 4096 | 4155 | 4214 | 4274 | 4333 | 4392 | 4452 | 59 |
| 732 | 4511 | 4570 | 4630 | 4689 | 4748 | 4808 | 4867 | 4926 | 4985 | 5045 | 59 |
| 733 | 5104 | 5163 | 5222 | 5282 | 5341 | 5400 | 5459 | 5519 | 5578 | 5637 | 59 |
| 734 | 5696 | 5755 | 5814 | 5874 | 5933 | 5992 | 6051 | 6110 | 6169 | 6228 | 59 |
| 735 | 6287 | 6346 | 6405 | 6465 | 6524 | 6583 | 6642 | 6701 | 6760 | 6819 | 59 |
| 736 | 6878 | 6937 | 6996 | 7055 | 7114 | 7173 | 7232 | 7291 | 7350 | 7409 | 59 |
| 737 | 7467 | 7526 | 7585 | 7644 | 7703 | 7762 | 7821 | 7880 | 7939 | 7998 | 59 |
| 738 | 8056 | 8115 | 8174 | 8233 | 8292 | 8350 | 8409 | 8468 | 8527 | 8586 | 59 |
| 739 | 8644 | 8703 | 8762 | 8821 | 8879 | 8938 | 8997 | 9056 | 9114 | 9173 | 59 |
| 740 | 9232 | 9290 | 9349 | 9408 | 9466 | 9525 | 9584 | 9642 | 9701 | 9760 | 59 |
| 741 | *9818 | 9877 | 9935 | 9994 | *053 | 0111 | 0170 | 0228 | 0287 | 0345 | 59 |
| 742 | 87 0404 | 0462 | 0521 | 0579 | 0638 | 0696 | 0755 | 0813 | 0872 | 0930 | 58 |
| 743 | 0989 | 1047 | 1106 | 1164 | 1223 | 1281 | 1339 | 1398 | 1456 | 1515 | 58 |
| 744 | 1573 | 1631 | 1690 | 1748 | 1806 | 1865 | 1923 | 1981 | 2040 | 2098 | 58 |
| 745 | 2156 | 2215 | 2273 | 2331 | 2389 | 2448 | 2506 | 2564 | 2622 | 2681 | 58 |
| 746 | 2739 | 2797 | 2855 | 2913 | 2972 | 3030 | 3088 | 3146 | 3204 | 3262 | 58 |
| 747 | 3321 | 3379 | 3437 | 3495 | 3553 | 3611 | 3669 | 3727 | 3785 | 3844 | 58 |
| 748 | 3902 | 3960 | 4018 | 4076 | 4134 | 4192 | 4250 | 4308 | 4366 | 4424 | 58 |
| 749 | 4482 | 4540 | 4598 | 4656 | 4714 | 4772 | 4830 | 4888 | 4945 | 5003 | 58 |
| 750 | 5061 | 5119 | 5177 | 5235 | 5293 | 5351 | 5409 | 5466 | 5524 | 5582 | 58 |
| 751 | 5640 | 5698 | 5756 | 5813 | 5871 | 5929 | 5987 | 6045 | 6102 | 6160 | 58 |
| 752 | 6218 | 6276 | 6333 | 6391 | 6449 | 6507 | 6564 | 6622 | 6680 | 6737 | 58 |
| 753 | 6795 | 6853 | 6910 | 6968 | 7026 | 7083 | 7141 | 7199 | 7256 | 7314 | 58 |
| 754 | 7371 | 7429 | 7487 | 7544 | 7602 | 7659 | 7717 | 7774 | 7832 | 7889 | 58 |
| 755 | 7947 | 8004 | 8062 | 8119 | 8177 | 8234 | 8292 | 8349 | 8407 | 8464 | 57 |
| 756 | 8522 | 8579 | 8637 | 8694 | 8752 | 8809 | 8866 | 8924 | 8981 | 9039 | 57 |
| 757 | 9096 | 9153 | 9211 | 9268 | 9325 | 9383 | 9440 | 9497 | 9555 | 9612 | 57 |
| 758 | *9669 | 9726 | 9784 | 9841 | 9898 | 9956 | *013 | 0070 | 0127 | 0185 | 57 |
| 759 | 88 0242 | 0299 | 0356 | 0413 | 0471 | 0528 | 0585 | 0642 | 0699 | 0756 | 57 |

N. 0 1 2 3 4 5 6 7 8 9 D.

| N. | 0 | 1 | 2 | 3 | 4 | 5 | 6 | 7 | 8 | 9 | D. |
|---|---|---|---|---|---|---|---|---|---|---|---|
| 760 | 88 0814 | 0871 | 0928 | 0985 | 1042 | 1099 | 1156 | 1213 | 1271 | 1328 | 57 |
| 761 | 1385 | 1442 | 1499 | 1556 | 1613 | 1670 | 1727 | 1784 | 1841 | 1898 | 57 |
| 762 | 1955 | 2012 | 2069 | 2126 | 2183 | 2240 | 2297 | 2354 | 2411 | 2468 | 57 |
| 763 | 2525 | 2581 | 2638 | 2695 | 2752 | 2809 | 2866 | 2923 | 2980 | 3037 | 57 |
| 764 | 3093 | 3150 | 3207 | 3264 | 3321 | 3377 | 3434 | 3491 | 3548 | 3605 | 57 |
| 765 | 3661 | 3718 | 3775 | 3832 | 3888 | 3945 | 4002 | 4059 | 4115 | 4172 | 57 |
| 766 | 4229 | 4285 | 4342 | 4399 | 4455 | 4512 | 4569 | 4625 | 4682 | 4739 | 57 |
| 767 | 4795 | 4852 | 4909 | 4965 | 5022 | 5078 | 5135 | 5192 | 5248 | 5305 | 57 |
| 768 | 5361 | 5418 | 5474 | 5531 | 5587 | 5644 | 5700 | 5757 | 5813 | 5870 | 57 |
| 769 | 5926 | 5983 | 6039 | 6096 | 6152 | 6209 | 6265 | 6321 | 6378 | 6434 | 56 |
| 770 | 6491 | 6547 | 6604 | 6660 | 6716 | 6773 | 6829 | 6885 | 6942 | 6998 | 56 |
| 771 | 7054 | 7111 | 7167 | 7223 | 7280 | 7336 | 7392 | 7449 | 7505 | 7561 | 56 |
| 772 | 7617 | 7674 | 7730 | 7786 | 7842 | 7898 | 7955 | 8011 | 8067 | 8123 | 56 |
| 773 | 8179 | 8236 | 8292 | 8348 | 8404 | 8460 | 8516 | 8573 | 8629 | 8685 | 56 |
| 774 | 8741 | 8797 | 8853 | 8909 | 8965 | 9021 | 9077 | 9134 | 9190 | 9246 | 56 |
| 775 | 9302 | 9358 | 9414 | 9470 | 9526 | 9582 | 9638 | 9694 | 9750 | 9806 | 56 |
| 776 | *9862 | 9918 | 9974 | •030 | 0086 | 0141 | 0197 | 0253 | 0309 | 0365 | 56 |
| 777 | 89 0421 | 0477 | 0533 | 0589 | 0645 | 0700 | 0756 | 0812 | 0868 | 0924 | 56 |
| 778 | 0980 | 1035 | 1091 | 1147 | 1203 | 1259 | 1314 | 1370 | 1426 | 1482 | 56 |
| 779 | 1537 | 1593 | 1649 | 1705 | 1760 | 1816 | 1872 | 1928 | 1983 | 2039 | 56 |
| 780 | 2095 | 2150 | 2206 | 2262 | 2317 | 2373 | 2429 | 2484 | 2540 | 2595 | 56 |
| 781 | 2651 | 2707 | 2762 | 2818 | 2873 | 2929 | 2985 | 3040 | 3096 | 3151 | 56 |
| 782 | 3207 | 3262 | 3318 | 3373 | 3429 | 3484 | 3540 | 3595 | 3651 | 3706 | 56 |
| 783 | 3762 | 3817 | 3873 | 3928 | 3984 | 4039 | 4094 | 4150 | 4205 | 4261 | 55 |
| 784 | 4316 | 4371 | 4427 | 4482 | 4538 | 4593 | 4648 | 4704 | 4759 | 4814 | 55 |
| 785 | 4870 | 4925 | 4980 | 5036 | 5091 | 5146 | 5201 | 5257 | 5312 | 5367 | 55 |
| 786 | 5423 | 5478 | 5533 | 5588 | 5644 | 5699 | 5754 | 5809 | 5864 | 5920 | 55 |
| 787 | 5975 | 6030 | 6085 | 6140 | 6195 | 6251 | 6306 | 6361 | 6416 | 6471 | 55 |
| 788 | 6526 | 6581 | 6636 | 6692 | 6747 | 6802 | 6857 | 6912 | 6967 | 7022 | 55 |
| 789 | 7077 | 7132 | 7187 | 7242 | 7297 | 7352 | 7407 | 7462 | 7517 | 7572 | 55 |
| 790 | 7627 | 7682 | 7737 | 7792 | 7847 | 7902 | 7957 | 8012 | 8067 | 8122 | 55 |
| 791 | 8176 | 8231 | 8286 | 8341 | 8396 | 8451 | 8506 | 8561 | 8615 | 8670 | 55 |
| 792 | 8725 | 8780 | 8835 | 8890 | 8944 | 8999 | 9054 | 9109 | 9164 | 9218 | 55 |
| 793 | 9273 | 9328 | 9383 | 9437 | 9492 | 9547 | 9602 | 9656 | 9711 | 9766 | 55 |
| 794 | *9821 | 9875 | 9930 | 9985 | •039 | 0094 | 0149 | 0203 | 0258 | 0312 | 55 |
| 795 | 90 0367 | 0422 | 0476 | 0531 | 0586 | 0640 | 0695 | 0749 | 0804 | 0859 | 55 |
| 796 | 0913 | 0968 | 1022 | 1077 | 1131 | 1186 | 1240 | 1295 | 1349 | 1404 | 55 |
| 797 | 1458 | 1513 | 1567 | 1622 | 1676 | 1731 | 1785 | 1840 | 1894 | 1948 | 54 |
| 798 | 2003 | 2057 | 2112 | 2166 | 2221 | 2275 | 2329 | 2384 | 2438 | 2492 | 54 |
| 799 | 2547 | 2601 | 2655 | 2710 | 2764 | 2818 | 2873 | 2927 | 2981 | 3036 | 54 |
| 800 | 3090 | 3144 | 3199 | 3253 | 3307 | 3361 | 3416 | 3470 | 3524 | 3578 | 54 |
| 801 | 3633 | 3687 | 3741 | 3795 | 3849 | 3904 | 3958 | 4012 | 4066 | 4120 | 54 |
| 802 | 4174 | 4229 | 4283 | 4337 | 4391 | 4445 | 4499 | 4553 | 4607 | 4661 | 54 |
| 803 | 4716 | 4770 | 4824 | 4878 | 4932 | 4986 | 5040 | 5094 | 5148 | 5202 | 54 |
| 804 | 5256 | 5310 | 5364 | 5418 | 5472 | 5526 | 5580 | 5634 | 5688 | 5742 | 54 |
| 805 | 5796 | 5850 | 5904 | 5958 | 6012 | 6066 | 6119 | 6173 | 6227 | 6281 | 54 |
| 806 | 6335 | 6389 | 6443 | 6497 | 6551 | 6604 | 6658 | 6712 | 6766 | 6820 | 54 |
| 807 | 6874 | 6927 | 6981 | 7035 | 7089 | 7143 | 7196 | 7250 | 7304 | 7358 | 54 |
| 808 | 7411 | 7465 | 7519 | 7573 | 7626 | 7680 | 7734 | 7787 | 7841 | 7895 | 54 |
| 809 | 7949 | 8002 | 8056 | 8110 | 8163 | 8217 | 8270 | 8324 | 8378 | 8431 | 54 |
| 810 | 8485 | 8539 | [illegible] | 8646 | [illegible] | [illegible] | [illegible] | [illegible] | [illegible] | [illegible] | 54 |
| 811 | 9021 | 9074 | 9128 | 9181 | 9235 | 9289 | 9342 | 9396 | 9449 | 9503 | 54 |
| 812 | *9556 | 9610 | 9663 | 9716 | 9770 | 9823 | 9877 | 9930 | 9984 | •037 | 53 |
| 813 | 91 0091 | 0144 | 0197 | 0251 | 0304 | 0358 | 0411 | 0464 | 0518 | 0571 | 53 |
| 814 | 0624 | 0678 | 0731 | 0784 | 0838 | 0891 | 0944 | 0998 | 1051 | 1104 | 53 |
| 815 | 1158 | 1211 | 1264 | 1317 | 1371 | 1424 | 1477 | 1530 | 1584 | 1637 | 53 |
| 816 | 1690 | 1743 | 1797 | 1850 | 1903 | 1956 | 2009 | 2063 | 2116 | 2169 | 53 |
| 817 | 2222 | 2275 | 2328 | 2381 | 2435 | 2488 | 2541 | 2594 | 2647 | 2700 | 53 |
| 818 | 2753 | 2806 | 2859 | 2913 | 2966 | 3019 | 3072 | 3125 | 3178 | 3231 | 53 |
| 819 | 3284 | 3337 | 3390 | 3443 | 3496 | 3549 | 3602 | 3655 | 3708 | 3761 | 53 |
| N. | 0 | 1 | 2 | 3 | 4 | 5 | 6 | 7 | 8 | 9 | D. |

| N. | 0 | 1 | 2 | 3 | 4 | 5 | 6 | 7 | 8 | 9 | D. |
|---|---|---|---|---|---|---|---|---|---|---|---|
| 820 | 91 3814 | 3867 | 3920 | 3973 | 4026 | 4079 | 4132 | 4184 | 4237 | 4290 | 53 |
| 821 | 4343 | 4396 | 4449 | 4502 | 4555 | 4608 | 4660 | 4713 | 4766 | 4819 | 53 |
| 822 | 4872 | 4925 | 4977 | 5030 | 5083 | 5136 | 5189 | 5241 | 5294 | 5347 | 53 |
| 823 | 5400 | 5453 | 5505 | 5558 | 5611 | 5664 | 5716 | 5769 | 5822 | 5875 | 53 |
| 824 | 5927 | 5980 | 6033 | 6085 | 6138 | 6191 | 6243 | 6296 | 6349 | 6401 | 53 |
| 825 | 6454 | 6507 | 6559 | 6612 | 6664 | 6717 | 6770 | 6822 | 6875 | 6927 | 53 |
| 826 | 6980 | 7033 | 7085 | 7138 | 7190 | 7243 | 7295 | 7348 | 7400 | 7453 | 53 |
| 827 | 7506 | 7558 | 7611 | 7663 | 7716 | 7768 | 7820 | 7873 | 7925 | 7978 | 52 |
| 828 | 8030 | 8083 | 8135 | 8188 | 8240 | 8293 | 8345 | 8397 | 8450 | 8502 | 52 |
| 829 | 8555 | 8607 | 8659 | 8712 | 8764 | 8816 | 8869 | 8921 | 8973 | 9026 | 52 |
| 830 | 9078 | 9130 | 9183 | 9235 | 9287 | 9340 | 9392 | 9444 | 9496 | 9549 | 52 |
| 831 | *9601 | 9653 | 9706 | 9758 | 9810 | 9862 | 9914 | 9967 | *019 | 0071 | 52 |
| 832 | 92 0123 | 0176 | 0228 | 0280 | 0332 | 0384 | 0436 | 0489 | 0541 | 0593 | 52 |
| 833 | 0645 | 0697 | 0749 | 0801 | 0853 | 0906 | 0958 | 1010 | 1062 | 1114 | 52 |
| 834 | 1166 | 1218 | 1270 | 1322 | 1374 | 1426 | 1478 | 1530 | 1582 | 1634 | 52 |
| 835 | 1686 | 1738 | 1790 | 1842 | 1894 | 1946 | 1998 | 2050 | 2102 | 2154 | 52 |
| 836 | 2206 | 2258 | 2310 | 2362 | 2414 | 2466 | 2518 | 2570 | 2622 | 2674 | 52 |
| 837 | 2725 | 2777 | 2829 | 2881 | 2933 | 2985 | 3037 | 3089 | 3140 | 3192 | 52 |
| 838 | 3244 | 3296 | 3348 | 3399 | 3451 | 3503 | 3555 | 3607 | 3658 | 3710 | 52 |
| 839 | 3762 | 3814 | 3865 | 3917 | 3969 | 4021 | 4072 | 4124 | 4176 | 4228 | 52 |
| 840 | 4279 | 4331 | 4383 | 4434 | 4486 | 4538 | 4589 | 4641 | 4693 | 4744 | 52 |
| 841 | 4796 | 4848 | 4899 | 4951 | 5003 | 5054 | 5106 | 5157 | 5209 | 5261 | 52 |
| 842 | 5312 | 5364 | 5415 | 5467 | 5518 | 5570 | 5621 | 5673 | 5725 | 5776 | 52 |
| 843 | 5828 | 5879 | 5931 | 5982 | 6034 | 6085 | 6137 | 6188 | 6240 | 6291 | 51 |
| 844 | 6342 | 6394 | 6445 | 6497 | 6548 | 6600 | 6651 | 6702 | 6754 | 6805 | 51 |
| 845 | 6857 | 6908 | 6959 | 7011 | 7062 | 7114 | 7165 | 7216 | 7268 | 7319 | 51 |
| 846 | 7370 | 7422 | 7473 | 7524 | 7576 | 7627 | 7678 | 7730 | 7781 | 7832 | 51 |
| 847 | 7883 | 7935 | 7986 | 8037 | 8088 | 8140 | 8191 | 8242 | 8293 | 8345 | 51 |
| 848 | 8396 | 8447 | 8498 | 8549 | 8601 | 8652 | 8703 | 8754 | 8805 | 8857 | 51 |
| 849 | 8908 | 8959 | 9010 | 9061 | 9112 | 9163 | 9215 | 9266 | 9317 | 9368 | 51 |
| 850 | 9419 | 9470 | 9521 | 9572 | 9623 | 9674 | 9725 | 9776 | 9827 | 9879 | 51 |
| 851 | *9930 | 9981 | *032 | 0083 | 0134 | 0185 | 0236 | 0287 | 0338 | 0389 | 51 |
| 852 | 93 0440 | 0491 | 0542 | 0592 | 0643 | 0694 | 0745 | 0796 | 0847 | 0898 | 51 |
| 853 | 0949 | 1000 | 1051 | 1102 | 1153 | 1204 | 1254 | 1305 | 1356 | 1407 | 51 |
| 854 | 1458 | 1509 | 1560 | 1610 | 1661 | 1712 | 1763 | 1814 | 1865 | 1915 | 51 |
| 855 | 1966 | 2017 | 2068 | 2118 | 2169 | 2220 | 2271 | 2322 | 2372 | 2423 | 51 |
| 856 | 2474 | 2524 | 2575 | 2626 | 2677 | 2727 | 2778 | 2829 | 2879 | 2930 | 51 |
| 857 | 2981 | 3031 | 3082 | 3133 | 3183 | 3234 | 3285 | 3335 | 3386 | 3437 | 51 |
| 858 | 3487 | 3538 | 3589 | 3639 | 3690 | 3740 | 3791 | 3841 | 3892 | 3943 | 51 |
| 859 | 3993 | 4044 | 4094 | 4145 | 4195 | 4246 | 4296 | 4347 | 4397 | 4448 | 51 |
| 860 | 4498 | 4549 | 4599 | 4650 | 4700 | 4751 | 4801 | 4852 | 4902 | 4953 | 50 |
| 861 | 5003 | 5054 | 5104 | 5154 | 5205 | 5255 | 5306 | 5356 | 5406 | 5457 | 50 |
| 862 | 5507 | 5558 | 5608 | 5658 | 5709 | 5759 | 5809 | 5860 | 5910 | 5960 | 50 |
| 863 | 6011 | 6061 | 6111 | 6162 | 6212 | 6262 | 6313 | 6363 | 6413 | 6463 | 50 |
| 864 | 6514 | 6564 | 6614 | 6665 | 6715 | 6765 | 6815 | 6865 | 6916 | 6966 | 50 |
| 865 | 7016 | 7066 | 7117 | 7167 | 7217 | 7267 | 7317 | 7367 | 7418 | 7468 | 50 |
| 866 | 7518 | 7568 | 7618 | 7668 | 7718 | 7769 | 7819 | 7869 | 7919 | 7969 | 50 |
| 867 | 8019 | 8069 | 8119 | 8169 | 8219 | 8269 | 8320 | 8370 | 8420 | 8470 | 50 |
| 868 | 8520 | 8570 | 8620 | 8670 | 8720 | 8770 | 8820 | 8870 | 8920 | 8970 | 50 |
| 869 | 9020 | 9070 | 9120 | 9170 | 9220 | 9270 | 9320 | 9369 | 9419 | 9469 | 50 |
| 870 | 9519 | 9569 | 9619 | 9669 | 9719 | 9769 | 9819 | 9869 | 9918 | 9968 | 50 |
| 871 | 94 0018 | 0068 | 0118 | 0168 | 0218 | 0267 | 0317 | 0367 | 0417 | 0467 | 50 |
| 872 | 0516 | 0566 | 0616 | 0666 | 0716 | 0765 | 0815 | 0865 | 0915 | 0964 | 50 |
| 873 | 1014 | 1064 | 1114 | 1163 | 1213 | 1263 | 1313 | 1362 | 1412 | 1462 | 50 |
| 874 | 1511 | 1561 | 1611 | 1660 | 1710 | 1760 | 1809 | 1859 | 1909 | 1958 | 50 |
| 875 | 2008 | 2058 | 2107 | 2157 | 2207 | 2256 | 2306 | 2355 | 2405 | 2455 | 50 |
| 876 | 2504 | 2554 | 2603 | 2653 | 2702 | 2752 | 2801 | 2851 | 2901 | 2950 | 50 |
| 877 | 3000 | 3049 | 3099 | 3148 | 3198 | 3247 | 3297 | 3346 | 3396 | 3445 | 49 |
| 878 | 3495 | 3544 | 3593 | 3643 | 3692 | 3742 | 3791 | 3841 | 3890 | 3939 | 49 |
| 879 | 3989 | 4038 | 4088 | 4137 | 4186 | 4236 | 4285 | 4335 | 4384 | 4433 | 49 |
| N. | 0 | 1 | 2 | 3 | 4 | 5 | 6 | 7 | 8 | 9 | D. |

| N. | 0 | 1 | 2 | 3 | 4 | 5 | 6 | 7 | 8 | 9 | D. |
|---|---|---|---|---|---|---|---|---|---|---|---|
| 880 | 94 4483 | 4532 | 4581 | 4631 | 4680 | 4729 | 4779 | 4828 | 4877 | 4927 | 49 |
| 881 | 4976 | 5025 | 5074 | 5124 | 5173 | 5222 | 5272 | 5321 | 5370 | 5419 | 49 |
| 882 | 5469 | 5518 | 5567 | 5616 | 5665 | 5715 | 5764 | 5813 | 5862 | 5912 | 49 |
| 883 | 5961 | 6010 | 6059 | 6108 | 6157 | 6207 | 6256 | 6305 | 6354 | 6403 | 49 |
| 884 | 6452 | 6501 | 6551 | 6600 | 6649 | 6698 | 6747 | 6796 | 6845 | 6894 | 49 |
| 885 | 6943 | 6992 | 7041 | 7090 | 7140 | 7189 | 7238 | 7287 | 7336 | 7385 | 49 |
| 886 | 7434 | 7483 | 7532 | 7581 | 7630 | 7679 | 7728 | 7777 | 7826 | 7875 | 49 |
| 887 | 7924 | 7973 | 8022 | 8070 | 8119 | 8168 | 8217 | 8266 | 8315 | 8364 | 49 |
| 888 | 8413 | 8462 | 8511 | 8560 | 8609 | 8657 | 8706 | 8755 | 8804 | 8853 | 49 |
| 889 | 8902 | 8951 | 8999 | 9048 | 9097 | 9146 | 9195 | 9244 | 9292 | 9341 | 49 |
| 890 | 9390 | 9439 | 9488 | 9536 | 9585 | 9634 | 9683 | 9731 | 9780 | 9829 | 49 |
| 891 | *9878 | 9926 | 9975 | ✦024 | 0073 | 0121 | 0170 | 0219 | 0267 | 0316 | 49 |
| 892 | 95 0365 | 0414 | 0462 | 0511 | 0560 | 0608 | 0657 | 0706 | 0754 | 0803 | 49 |
| 893 | 0851 | 0900 | 0949 | 0997 | 1046 | 1095 | 1143 | 1192 | 1240 | 1289 | 49 |
| 894 | 1338 | 1386 | 1435 | 1483 | 1532 | 1580 | 1629 | 1677 | 1726 | 1775 | 49 |
| 895 | 1823 | 1872 | 1920 | 1969 | 2017 | 2066 | 2114 | 2163 | 2211 | 2260 | 48 |
| 896 | 2308 | 2356 | 2405 | 2453 | 2502 | 2550 | 2599 | 2647 | 2696 | 2744 | 48 |
| 897 | 2792 | 2841 | 2889 | 2938 | 2986 | 3034 | 3083 | 3131 | 3180 | 3228 | 48 |
| 898 | 3276 | 3325 | 3373 | 3421 | 3470 | 3518 | 3566 | 3615 | 3663 | 3711 | 48 |
| 899 | 3760 | 3808 | 3856 | 3905 | 3953 | 4001 | 4049 | 4098 | 4146 | 4194 | 48 |
| 900 | 4243 | 4291 | 4339 | 4387 | 4435 | 4484 | 4532 | 4580 | 4628 | 4677 | 48 |
| 901 | 4725 | 4773 | 4821 | 4869 | 4918 | 4966 | 5014 | 5062 | 5110 | 5158 | 48 |
| 902 | 5207 | 5255 | 5303 | 5351 | 5399 | 5447 | 5495 | 5543 | 5592 | 5640 | 48 |
| 903 | 5688 | 5736 | 5784 | 5832 | 5880 | 5928 | 5976 | 6024 | 6072 | 6120 | 48 |
| 904 | 6168 | 6216 | 6265 | 6313 | 6361 | 6409 | 6457 | 6505 | 6553 | 6601 | 48 |
| 905 | 6649 | 6697 | 6745 | 6793 | 6840 | 6888 | 6936 | 6984 | 7032 | 7080 | 48 |
| 906 | 7128 | 7176 | 7224 | 7272 | 7320 | 7368 | 7416 | 7464 | 7512 | 7559 | 48 |
| 907 | 7607 | 7655 | 7703 | 7751 | 7799 | 7847 | 7894 | 7942 | 7990 | 8038 | 48 |
| 908 | 8086 | 8134 | 8181 | 8229 | 8277 | 8325 | 8373 | 8421 | 8468 | 8516 | 48 |
| 909 | 8564 | 8612 | 8659 | 8707 | 8755 | 8803 | 8850 | 8898 | 8946 | 8994 | 48 |
| 910 | 9041 | 9089 | 9137 | 9185 | 9232 | 9280 | 9328 | 9375 | 9423 | 9471 | 48 |
| 911 | 9518 | 9566 | 9614 | 9661 | 9709 | 9757 | 9804 | 9852 | 9900 | 9947 | 48 |
| 912 | *9995 | ✦042 | 0090 | 0138 | 0185 | 0233 | 0280 | 0328 | 0376 | 0423 | 48 |
| 913 | 96 0471 | 0518 | 0566 | 0613 | 0661 | 0709 | 0756 | 0804 | 0851 | 0899 | 48 |
| 914 | 0946 | 0994 | 1041 | 1089 | 1136 | 1184 | 1231 | 1279 | 1326 | 1374 | 47 |
| 915 | 1421 | 1469 | 1516 | 1563 | 1611 | 1658 | 1706 | 1753 | 1801 | 1848 | 47 |
| 916 | 1895 | 1943 | 1990 | 2038 | 2085 | 2132 | 2180 | 2227 | 2275 | 2322 | 47 |
| 917 | 2369 | 2417 | 2464 | 2511 | 2559 | 2606 | 2653 | 2701 | 2748 | 2795 | 47 |
| 918 | 2843 | 2890 | 2937 | 2985 | 3032 | 3079 | 3126 | 3174 | 3221 | 3268 | 47 |
| 919 | 3316 | 3363 | 3410 | 3457 | 3504 | 3552 | 3599 | 3646 | 3693 | 3741 | 47 |
| 920 | 3788 | 3835 | 3882 | 3929 | 3977 | 4024 | 4071 | 4118 | 4165 | 4212 | 47 |
| 921 | 4260 | 4307 | 4354 | 4401 | 4448 | 4495 | 4542 | 4590 | 4637 | 4684 | 47 |
| 922 | 4731 | 4778 | 4825 | 4872 | 4919 | 4966 | 5013 | 5061 | 5108 | 5155 | 47 |
| 923 | 5202 | 5249 | 5296 | 5343 | 5390 | 5437 | 5484 | 5531 | 5578 | 5625 | 47 |
| 924 | 5672 | 5719 | 5766 | 5813 | 5860 | 5907 | 5954 | 6001 | 6048 | 6095 | 47 |
| 925 | 6142 | 6189 | 6236 | 6283 | 6329 | 6376 | 6423 | 6470 | 6517 | 6564 | 47 |
| 926 | 6611 | 6658 | 6705 | 6752 | 6799 | 6845 | 6892 | 6939 | 6986 | 7033 | 47 |
| 927 | 7080 | 7127 | 7173 | 7220 | 7267 | 7314 | 7361 | 7408 | 7454 | 7501 | 47 |
| 928 | 7548 | 7595 | 7642 | 7688 | 7735 | 7782 | 7829 | 7875 | 7922 | 7969 | 47 |
| 929 | 8016 | 8062 | 8109 | 8156 | 8203 | 8249 | 8296 | 8343 | 8390 | 8436 | 47 |
| 930 | 8483 | 8530 | 8576 | 8623 | 8670 | 8716 | 8763 | 8810 | 8856 | 8903 | 47 |
| 931 | 8950 | 8996 | 9043 | 9090 | 9136 | 9183 | 9229 | 9276 | 9323 | 9369 | 47 |
| 932 | 9416 | 9463 | 9509 | 9556 | 9602 | 9649 | 9695 | 9742 | 9789 | 9835 | 47 |
| 933 | *9882 | 9928 | 9975 | ✦021 | 0068 | 0114 | 0161 | 0207 | 0254 | 0300 | 47 |
| 934 | 97 0347 | 0393 | 0440 | 0486 | 0533 | 0579 | 0626 | 0672 | 0719 | 0765 | 46 |
| 935 | 0812 | 0858 | 0904 | 0951 | 0997 | 1044 | 1090 | 1137 | 1183 | 1229 | 46 |
| 936 | 1276 | 1322 | 1369 | 1415 | 1461 | 1508 | 1554 | 1601 | 1647 | 1693 | 46 |
| 937 | 1740 | 1786 | 1832 | 1879 | 1925 | 1971 | 2018 | 2064 | 2110 | 2157 | 46 |
| 938 | 2203 | 2249 | 2295 | 2342 | 2388 | 2434 | 2481 | 2527 | 2573 | 2619 | 46 |
| 939 | 2666 | 2712 | 2758 | 2804 | 2851 | 2897 | 2943 | 2989 | 3035 | 3082 | 46 |
| N. | 0 | 1 | 2 | 3 | 4 | 5 | 6 | 7 | 8 | 9 | D. |

| N. | 0 | 1 | 2 | 3 | 4 | 5 | 6 | 7 | 8 | 9 | D. |
|---|---|---|---|---|---|---|---|---|---|---|---|
| 940 | 97 3128 | 3174 | 3220 | 3266 | 3313 | 3359 | 3405 | 3451 | 3497 | 3543 | 46 |
| 941 | 3590 | 3636 | 3682 | 3728 | 3774 | 3820 | 3866 | 3913 | 3959 | 4005 | 46 |
| 942 | 4051 | 4097 | 4143 | 4189 | 4235 | 4281 | 4327 | 4374 | 4420 | 4466 | 46 |
| 943 | 4512 | 4558 | 4604 | 4650 | 4696 | 4742 | 4788 | 4834 | 4880 | 4926 | 46 |
| 944 | 4972 | 5018 | 5064 | 5110 | 5156 | 5202 | 5248 | 5294 | 5340 | 5386 | 46 |
| 945 | 5432 | 5478 | 5524 | 5570 | 5616 | 5662 | 5707 | 5753 | 5799 | 5845 | 46 |
| 946 | 5891 | 5937 | 5983 | 6029 | 6075 | 6121 | 6167 | 6212 | 6258 | 6304 | 46 |
| 947 | 6350 | 6396 | 6442 | 6488 | 6533 | 6579 | 6625 | 6671 | 6717 | 6763 | 46 |
| 948 | 6808 | 6854 | 6900 | 6946 | 6992 | 7037 | 7083 | 7129 | 7175 | 7220 | 46 |
| 949 | 7266 | 7312 | 7358 | 7403 | 7449 | 7495 | 7541 | 7586 | 7632 | 7678 | 46 |
| 950 | 7724 | 7769 | 7815 | 7861 | 7906 | 7952 | 7998 | 8043 | 8089 | 8135 | 46 |
| 951 | 8181 | 8226 | 8272 | 8317 | 8363 | 8409 | 8454 | 8500 | 8546 | 8591 | 46 |
| 952 | 8637 | 8683 | 8728 | 8774 | 8819 | 8865 | 8911 | 8956 | 9002 | 9047 | 46 |
| 953 | 9093 | 9138 | 9184 | 9230 | 9275 | 9321 | 9366 | 9412 | 9457 | 9503 | 46 |
| 954 | 9548 | 9594 | 9639 | 9685 | 9730 | 9776 | 9821 | 9867 | 9912 | 9958 | 46 |
| 955 | 98 0003 | 0049 | 0094 | 0140 | 0185 | 0231 | 0276 | 0322 | 0367 | 0412 | 45 |
| 956 | 0458 | 0503 | 0549 | 0594 | 0640 | 0685 | 0730 | 0776 | 0821 | 0867 | 45 |
| 957 | 0912 | 0957 | 1003 | 1048 | 1093 | 1139 | 1184 | 1229 | 1275 | 1320 | 45 |
| 958 | 1366 | 1411 | 1456 | 1501 | 1547 | 1592 | 1637 | 1683 | 1728 | 1773 | 45 |
| 959 | 1819 | 1864 | 1909 | 1954 | 2000 | 2045 | 2090 | 2135 | 2181 | 2226 | 45 |
| 960 | 2271 | 2316 | 2362 | 2407 | 2452 | 2497 | 2543 | 2588 | 2633 | 2678 | 45 |
| 961 | 2723 | 2769 | 2814 | 2859 | 2904 | 2949 | 2994 | 3040 | 3085 | 3130 | 45 |
| 962 | 3175 | 3220 | 3265 | 3310 | 3356 | 3401 | 3446 | 3491 | 3536 | 3581 | 45 |
| 963 | 3626 | 3671 | 3716 | 3762 | 3807 | 3852 | 3897 | 3942 | 3987 | 4032 | 45 |
| 964 | 4077 | 4122 | 4167 | 4212 | 4257 | 4302 | 4347 | 4392 | 4437 | 4482 | 45 |
| 965 | 4527 | 4572 | 4617 | 4662 | 4707 | 4752 | 4797 | 4842 | 4887 | 4932 | 45 |
| 966 | 4977 | 5022 | 5067 | 5112 | 5157 | 5202 | 5247 | 5292 | 5337 | 5382 | 45 |
| 967 | 5426 | 5471 | 5516 | 5561 | 5606 | 5651 | 5696 | 5741 | 5786 | 5830 | 45 |
| 968 | 5875 | 5920 | 5965 | 6010 | 6055 | 6100 | 6144 | 6189 | 6234 | 6279 | 45 |
| 969 | 6324 | 6369 | 6413 | 6458 | 6503 | 6548 | 6593 | 6637 | 6682 | 6727 | 45 |
| 970 | 6772 | 6817 | 6861 | 6906 | 6951 | 6996 | 7040 | 7085 | 7130 | 7175 | 45 |
| 971 | 7219 | 7264 | 7309 | 7353 | 7398 | 7443 | 7488 | 7532 | 7577 | 7622 | 45 |
| 972 | 7666 | 7711 | 7756 | 7800 | 7845 | 7890 | 7934 | 7979 | 8024 | 8068 | 45 |
| 973 | 8113 | 8157 | 8202 | 8247 | 8291 | 8336 | 8381 | 8425 | 8470 | 8514 | 45 |
| 974 | 8559 | 8604 | 8648 | 8693 | 8737 | 8782 | 8826 | 8871 | 8916 | 8960 | 45 |
| 975 | 9005 | 9049 | 9094 | 9138 | 9183 | 9227 | 9272 | 9316 | 9361 | 9405 | 45 |
| 976 | 9450 | 9494 | 9539 | 9583 | 9628 | 9672 | 9717 | 9761 | 9806 | 9850 | 44 |
| 977 | * 9895 | 9939 | 9983 | *028 | 0072 | 0117 | 0161 | 0206 | 0250 | 0294 | 44 |
| 978 | 99 0339 | 0383 | 0428 | 0472 | 0516 | 0561 | 0605 | 0650 | 0694 | 0738 | 44 |
| 979 | 0783 | 0827 | 0871 | 0916 | 0960 | 1004 | 1049 | 1093 | 1137 | 1182 | 44 |
| 980 | 1226 | 1270 | 1315 | 1359 | 1403 | 1448 | 1492 | 1536 | 1580 | 1625 | 44 |
| 981 | 1669 | 1713 | 1758 | 1802 | 1846 | 1890 | 1935 | 1979 | 2023 | 2067 | 44 |
| 982 | 2111 | 2156 | 2200 | 2244 | 2288 | 2333 | 2377 | 2421 | 2465 | 2509 | 44 |
| 983 | 2554 | 2598 | 2642 | 2686 | 2730 | 2774 | 2819 | 2863 | 2907 | 2951 | 44 |
| 984 | 2995 | 3039 | 3083 | 3127 | 3172 | 3216 | 3260 | 3304 | 3348 | 3392 | 44 |
| 985 | 3436 | 3480 | 3524 | 3568 | 3613 | 3657 | 3701 | 3745 | 3789 | 3833 | 44 |
| 986 | 3877 | 3921 | 3965 | 4009 | 4053 | 4097 | 4141 | 4185 | 4229 | 4273 | 44 |
| 987 | 4317 | 4361 | 4405 | 4449 | 4493 | 4537 | 4581 | 4625 | 4669 | 4713 | 44 |
| 988 | 4757 | 4801 | 4845 | 4889 | 4933 | 4977 | 5021 | 5065 | 5108 | 5152 | 44 |
| 989 | 5196 | 5240 | 5284 | 5328 | 5372 | 5416 | 5460 | 5504 | 5547 | 5591 | 44 |
| 990 | 5635 | 5679 | 5723 | 5767 | 5811 | 5854 | 5898 | 5942 | 5986 | 6030 | 44 |
| 991 | 6074 | 6117 | 6161 | 6205 | 6249 | 6293 | 6337 | 6380 | 6424 | 6468 | 44 |
| 992 | 6512 | 6555 | 6599 | 6643 | 6687 | 6731 | 6774 | 6818 | 6862 | 6906 | 44 |
| 993 | 6949 | 6993 | 7037 | 7080 | 7124 | 7168 | 7212 | 7255 | 7299 | 7343 | 44 |
| 994 | 7386 | 7430 | 7474 | 7517 | 7561 | 7605 | 7648 | 7692 | 7736 | 7779 | 44 |
| 995 | 7823 | 7867 | 7910 | 7954 | 7998 | 8041 | 8085 | 8129 | 8172 | 8216 | 44 |
| 996 | 8259 | 8303 | 8347 | 8390 | 8434 | 8477 | 8521 | 8564 | 8608 | 8652 | 44 |
| 997 | 8695 | 8739 | 8782 | 8826 | 8869 | 8913 | 8956 | 9000 | 9043 | 9087 | 44 |
| 998 | 9131 | 9174 | 9218 | 9261 | 9305 | 9348 | 9392 | 9435 | 9479 | 9522 | 44 |
| 999 | 9565 | 9609 | 9652 | 9696 | 9739 | 9783 | 9826 | 9870 | 9913 | 9957 | 43 |
| N. | 0 | 1 | 2 | 3 | 4 | 5 | 6 | 7 | 8 | 9 | D. |

# TABLE II.

## LOGARITHMIC SINES AND TANGENTS,

FOR

EVERY DEGREE AND MINUTE OF THE QUADRANT.

If the logarithms of the values in Table III. be each increased by 10, the results will be the values of this table.

The logarithmic Secants and Cosecants are not given. They may be readily obtained, as follows:—Subtract the logarithmic Cosine from 20, and the remainder will be the logarithmic Secant; subtract the logarithmic Sine from 20, and the remainder will be the logarithmic Cosecant.

| 0° | | | | | | | | 179° |
|---|---|---|---|---|---|---|---|---|
| ′ | Sine. | D. | Cosine. | D. | Tang. | D. | Cotang. | ′ |
| 0 | Inf. Neg. | | 10·000000 | | Inf. Neg. | | Infinite. | 60 |
| 1 | 6·463726 | 501717 | 000000 | 00 | 6·463726 | 501717 | 13·536274 | 59 |
| 2 | 764756 | 293485 | 000000 | 00 | 764756 | 293483 | 235244 | 58 |
| 3 | 940847 | 208231 | 000000 | 00 | 940847 | 208231 | 059153 | 57 |
| 4 | 7·065786 | 161517 | 000000 | 00 | 7·065786 | 161517 | 12·934214 | 56 |
| 5 | 162696 | 131968 | 000000 | 00 | 162696 | 131969 | 837304 | 55 |
| 6 | 241877 | 111575 | 9·999999 | 01 | 241878 | 111578 | 758122 | 54 |
| 7 | 308824 | 96653 | 999999 | 01 | 308825 | 99653 | 691175 | 53 |
| 8 | 366816 | 85254 | 999999 | 01 | 366817 | 85254 | 633183 | 52 |
| 9 | 417968 | 76263 | 999999 | 01 | 417970 | 76263 | 582030 | 51 |
| 10 | 463726 | 68988 | 999998 | 01 | 463727 | 68988 | 536273 | 50 |
| 11 | 7·505118 | 62981 | 9·999998 | 01 | 7·505120 | 62981 | 12·494880 | 49 |
| 12 | 542906 | 57936 | 999997 | 01 | 542909 | 57933 | 457091 | 48 |
| 13 | 577668 | 53641 | 999997 | 01 | 577672 | 53642 | 422328 | 47 |
| 14 | 609853 | 49938 | 999996 | 01 | 609857 | 49939 | 390143 | 46 |
| 15 | 639816 | 46714 | 999996 | 01 | 639820 | 46715 | 360180 | 45 |
| 16 | 667845 | 43881 | 999995 | 01 | 667849 | 43882 | 332151 | 44 |
| 17 | 694173 | 41372 | 999995 | 01 | 694179 | 41373 | 305821 | 43 |
| 18 | 718997 | 39135 | 999994 | 01 | 719003 | 39136 | 280997 | 42 |
| 19 | 742478 | 37127 | 999993 | 01 | 742484 | 37128 | 257516 | 41 |
| 20 | 764754 | 35315 | 999993 | 01 | 764761 | 35136 | 235239 | 40 |
| 21 | 7·785943 | 33672 | 9·999992 | 01 | 7·785951 | 33673 | 12·214049 | 39 |
| 22 | 806146 | 32175 | 999991 | 01 | 806155 | 32176 | 193845 | 38 |
| 23 | 825451 | 30805 | 999990 | 01 | 825460 | 30806 | 174540 | 37 |
| 24 | 843934 | 29547 | 999989 | 02 | 843944 | 29549 | 156056 | 36 |
| 25 | 861662 | 28388 | 999989 | 02 | 861674 | 28390 | 138326 | 35 |
| 26 | 878695 | 27317 | 999988 | 02 | 878708 | 27318 | 121292 | 34 |
| 27 | 895085 | 26323 | 999987 | 02 | 895099 | 26325 | 104901 | 33 |
| 28 | 910879 | 25399 | 999986 | 02 | 910894 | 25401 | 089106 | 32 |
| 29 | 926119 | 24538 | 999985 | 02 | 926134 | 24540 | 073866 | 31 |
| 30 | 940842 | 23733 | 999983 | 02 | 940858 | 23735 | 059142 | 30 |
| 31 | 7·955082 | 22980 | 9·999982 | 02 | 7·955100 | 22981 | 12·044900 | 29 |
| 32 | 968870 | 22273 | 999981 | 02 | 968889 | 22275 | 031111 | 28 |
| 33 | 982233 | 21608 | 999980 | 02 | 982253 | 21610 | 017747 | 27 |
| 34 | 995198 | 20981 | 999979 | 02 | 995219 | 20983 | 004781 | 26 |
| 35 | 8·007787 | 20390 | 999977 | 02 | 8·007809 | 20392 | 11·992191 | 25 |
| 36 | 020021 | 19831 | 999976 | 02 | 020044 | 19833 | 979956 | 24 |
| 37 | 031919 | 19302 | 999975 | 02 | 031945 | 19305 | 968055 | 23 |
| 38 | 043501 | 18801 | 999973 | 02 | 043527 | 18803 | 956473 | 22 |
| 39 | 054781 | 18325 | 999972 | 02 | 054809 | 18327 | 945191 | 21 |
| 40 | 065776 | 17872 | 999971 | 02 | 065806 | 17874 | 934194 | 20 |
| 41 | 8·076500 | 17441 | 9·999969 | 02 | 8·076531 | 17444 | 11·923469 | 19 |
| 42 | 086965 | 17031 | 999968 | 02 | 086997 | 17034 | 913003 | 18 |
| 43 | 097183 | 16639 | 999966 | 02 | 097217 | 16642 | 902783 | 17 |
| 44 | 107167 | 16265 | 999964 | 03 | 107203 | 16268 | 892797 | 16 |
| 45 | 116926 | 15908 | 999963 | 03 | 116963 | 15910 | 883037 | 15 |
| 46 | 126471 | 15566 | 999961 | 03 | 126510 | 15568 | 873490 | 14 |
| 47 | 135810 | 15238 | 999959 | 03 | 135851 | 15241 | 864149 | 13 |
| 48 | 144953 | 14924 | 999958 | 03 | 144996 | 14927 | 855004 | 12 |
| 49 | 153907 | 14622 | 999956 | 03 | 153952 | 14627 | 846048 | 11 |
| 50 | 162681 | 14333 | 999954 | 03 | 162727 | 14336 | 837273 | 10 |
| 51 | 8·171280 | 14054 | 9·999952 | 03 | 8·171328 | 14057 | 11·828672 | 9 |
| 52 | 179713 | 13786 | 999950 | 03 | 179763 | 13790 | 820237 | 8 |
| 53 | 187985 | 13529 | 999948 | 03 | 188036 | 13532 | 811964 | 7 |
| 54 | 196102 | 13280 | 999946 | 03 | 196156 | 13284 | 803844 | 6 |
| 55 | 204070 | 13041 | 999944 | 03 | 204126 | 13044 | 795874 | 5 |
| 56 | 211895 | 12810 | 999942 | 04 | 211953 | 12814 | 788047 | 4 |
| 57 | 219581 | 12587 | 999940 | 04 | 219641 | 12590 | 780359 | 3 |
| 58 | 227134 | 12372 | 999938 | 04 | 227195 | 12376 | 772805 | 2 |
| 59 | 234557 | 12164 | 999936 | 04 | 234621 | 12168 | 765379 | 1 |
| 60 | 241855 | 11963 | 999934 | 04 | 241921 | 11967 | 758079 | 0 |
| ′ | Cosine. | D. | Sine. | D. | Cotang. | D. | Tang. | ′ |
| 90° | | | | | | | | 89° |

1° 178°

| ′ | Sine. | D. | Cosine. | D. | Tang. | D. | Cotang. | ′ |
|---|---|---|---|---|---|---|---|---|
| 0 | 8·241855 | 11963 | 9·999934 | 04 | 8·241921 | 11967 | 11·758079 | 60 |
| 1 | 249033 | 11768 | 999932 | 04 | 249102 | 11772 | 750898 | 59 |
| 2 | 256094 | 11580 | 999929 | 04 | 256165 | 11584 | 743835 | 58 |
| 3 | 263042 | 11398 | 999927 | 04 | 263115 | 11402 | 736885 | 57 |
| 4 | 269881 | 11221 | 999925 | 04 | 269956 | 11225 | 730044 | 56 |
| 5 | 276614 | 11050 | 999922 | 04 | 276691 | 11054 | 723309 | 55 |
| 6 | 283243 | 10883 | 999920 | 04 | 283323 | 10887 | 716677 | 54 |
| 7 | 289773 | 10721 | 999918 | 04 | 289856 | 10726 | 710144 | 53 |
| 8 | 296207 | 10565 | 999915 | 04 | 296292 | 10570 | 703708 | 52 |
| 9 | 302546 | 10413 | 999913 | 04 | 302634 | 10418 | 697366 | 51 |
| 10 | 308794 | 10266 | 999910 | 04 | 308884 | 10270 | 691116 | 50 |
| 11 | 8·314954 | 10122 | 9·999907 | 04 | 8·315046 | 10126 | 11·684954 | 49 |
| 12 | 321027 | 9982 | 999905 | 04 | 321122 | 9987 | 678878 | 48 |
| 13 | 327016 | 9847 | 999902 | 04 | 327114 | 9851 | 672886 | 47 |
| 14 | 332924 | 9714 | 999899 | 05 | 333025 | 9719 | 666975 | 46 |
| 15 | 338753 | 9586 | 999897 | 05 | 338856 | 9590 | 661144 | 45 |
| 16 | 344504 | 9460 | 999894 | 05 | 344610 | 9465 | 655390 | 44 |
| 17 | 350181 | 9338 | 999891 | 05 | 350289 | 9343 | 649711 | 43 |
| 18 | 355783 | 9219 | 999888 | 05 | 355895 | 9224 | 644105 | 42 |
| 19 | 361315 | 9103 | 999885 | 05 | 361430 | 9108 | 638570 | 41 |
| 20 | 366777 | 8990 | 999882 | 05 | 366895 | 8995 | 633105 | 40 |
| 21 | 8·372171 | 8880 | 9·999879 | 05 | 8·372292 | 8885 | 11·627708 | 39 |
| 22 | 377499 | 8772 | 999876 | 05 | 377622 | 8777 | 622378 | 38 |
| 23 | 382762 | 8667 | 999873 | 05 | 382889 | 8672 | 617111 | 37 |
| 24 | 387962 | 8564 | 999870 | 05 | 388092 | 8570 | 611908 | 36 |
| 25 | 393101 | 8464 | 999867 | 05 | 393234 | 8470 | 606766 | 35 |
| 26 | 398179 | 8366 | 999864 | 05 | 398315 | 8371 | 601685 | 34 |
| 27 | 403199 | 8271 | 999861 | 05 | 403338 | 8276 | 596662 | 33 |
| 28 | 408161 | 8177 | 999858 | 05 | 408304 | 8182 | 591696 | 32 |
| 29 | 413068 | 8086 | 999854 | 05 | 413213 | 8091 | 586787 | 31 |
| 30 | 417919 | 7996 | 999851 | 06 | 418068 | 8002 | 581932 | 30 |
| 31 | 8·422717 | 7909 | 9·999848 | 06 | 8·422869 | 7914 | 11·577131 | 29 |
| 32 | 427462 | 7823 | 999844 | 06 | 427618 | 7830 | 572382 | 28 |
| 33 | 432156 | 7740 | 999841 | 06 | 432315 | 7745 | 567685 | 27 |
| 34 | 436800 | 7657 | 999838 | 06 | 436962 | 7663 | 563038 | 26 |
| 35 | 441394 | 7577 | 999834 | 06 | 441560 | 7583 | 558440 | 25 |
| 36 | 445941 | 7499 | 999831 | 06 | 446110 | 7505 | 553890 | 24 |
| 37 | 450440 | 7422 | 999827 | 06 | 450613 | 7428 | 549387 | 23 |
| 38 | 454893 | 7346 | 999824 | 06 | 455070 | 7352 | 544930 | 22 |
| 39 | 459301 | 7273 | 999820 | 06 | 459481 | 7279 | 540519 | 21 |
| 40 | 463665 | 7200 | 999816 | 06 | 463849 | 7206 | 536151 | 20 |
| 41 | 8·467985 | 7129 | 9·999813 | 06 | 8·468172 | 7135 | 11·531828 | 19 |
| 42 | 472263 | 7060 | 999809 | 06 | 472454 | 7066 | 527546 | 18 |
| 43 | 476498 | 6991 | 999805 | 06 | 476693 | 6998 | 523307 | 17 |
| 44 | 480693 | 6924 | 999801 | 06 | 480892 | 6931 | 519108 | 16 |
| 45 | 484848 | 6859 | 999797 | 07 | 485050 | 6865 | 514950 | 15 |
| 46 | 488963 | 6794 | 999794 | 07 | 489170 | 6801 | 510830 | 14 |
| 47 | 493040 | 6731 | 999790 | 07 | 493250 | 6738 | 506750 | 13 |
| 48 | 497078 | 6669 | 999786 | 07 | 497293 | 6676 | 502707 | 12 |
| 49 | 501080 | 6608 | 999782 | 07 | 501298 | 6615 | 498702 | 11 |
| 50 | 505045 | 6548 | 999778 | 07 | 505267 | 6555 | 494733 | 10 |
| 51 | 8·508974 | 6489 | 9·999774 | 07 | 8·509200 | 6496 | 11·490800 | 9 |
| 52 | 512867 | 6431 | 999769 | 07 | 513098 | 6439 | 486902 | 8 |
| 53 | 516726 | 6375 | 999765 | 07 | 516961 | 6382 | 483039 | 7 |
| 54 | 520551 | 6319 | 999761 | 07 | 520790 | 6326 | 479210 | 6 |
| 55 | 524343 | 6264 | 999757 | 07 | 524586 | 6272 | 475414 | 5 |
| 56 | 528102 | 6211 | 999753 | 07 | 528349 | 6218 | 471651 | 4 |
| 57 | 531828 | 6158 | 999748 | 07 | 532080 | 6165 | 467920 | 3 |
| 58 | 535523 | 6106 | 999744 | 07 | 535779 | 6113 | 464221 | 2 |
| 59 | 539186 | 6055 | 999740 | 07 | 539447 | 6062 | 460553 | 1 |
| 60 | 542819 | 6004 | 999735 | 07 | 543084 | 6012 | 456916 | 0 |
| ′ | Cosine. | D. | Sine. | D. | Cotang. | D. | Tang. | ′ |

91° 88°

2° 177°

| ′ | Sine. | D. | Cosine. | D. | Tang. | D. | Cotang. | ′ |
|---|---|---|---|---|---|---|---|---|
| 0 | 8·542819 | 6004 | 9·999735 | 07 | 8·543084 | 6012 | 11·456916 | 60 |
| 1 | 546422 | 5955 | 999731 | 07 | 546691 | 5962 | 453309 | 59 |
| 2 | 549995 | 5906 | 999726 | 07 | 550268 | 5914 | 449732 | 58 |
| 3 | 553539 | 5858 | 999722 | 08 | 553817 | 5866 | 446183 | 57 |
| 4 | 557054 | 5811 | 999717 | 08 | 557336 | 5819 | 442664 | 56 |
| 5 | 560540 | 5765 | 999713 | 08 | 560828 | 5773 | 439172 | 55 |
| 6 | 563999 | 5719 | 999708 | 08 | 564291 | 5727 | 435709 | 54 |
| 7 | 567431 | 5674 | 999704 | 08 | 567727 | 5682 | 432273 | 53 |
| 8 | 570836 | 5630 | 999699 | 08 | 571137 | 5638 | 428863 | 52 |
| 9 | 574214 | 5587 | 999694 | 08 | 574520 | 5595 | 425480 | 51 |
| 10 | 577566 | 5544 | 999689 | 08 | 577877 | 5552 | 422123 | 50 |
| 11 | 8·580892 | 5502 | 9·999685 | 08 | 8·581208 | 5510 | 11·418792 | 49 |
| 12 | 584193 | 5460 | 999680 | 08 | 584514 | 5468 | 415486 | 48 |
| 13 | 587469* | 5419 | 999675 | 08 | 587795 | 5427 | 412205 | 47 |
| 14 | 590721 | 5379 | 999670 | 08 | 591051 | 5387 | 408949 | 46 |
| 15 | 593948 | 5339 | 999665 | 08 | 594283 | 5347 | 405717 | 45 |
| 16 | 597152 | 5300 | 999660 | 08 | 597492 | 5308 | 402508 | 44 |
| 17 | 600332 | 5261 | 999655 | 08 | 600677 | 5270 | 399323 | 43 |
| 18 | 603489 | 5223 | 999650 | 08 | 603839 | 5232 | 396161 | 42 |
| 19 | 606623 | 5186 | 999645 | 09 | 606978 | 5194 | 393022 | 41 |
| 20 | 609734 | 5149 | 999640 | 09 | 610094 | 5158 | 389906 | 40 |
| 21 | 8·612823 | 5112 | 9·999635 | 09 | 8·613189 | 5121 | 11·386811 | 39 |
| 22 | 615891 | 5076 | 999629 | 09 | 616262 | 5085 | 383738 | 38 |
| 23 | 618937 | 5041 | 999624 | 09 | 619313 | 5050 | 380687 | 37 |
| 24 | 621962 | 5006 | 999619 | 09 | 622343 | 5015 | 377657 | 36 |
| 25 | 624965 | 4972 | 999614 | 09 | 625352 | 4981 | 374648 | 35 |
| 26 | 627948 | 4938 | 999608 | 09 | 628340 | 4947 | 371660 | 34 |
| 27 | 630911 | 4904 | 999603 | 09 | 631308 | 4913 | 368692 | 33 |
| 28 | 633854 | 4871 | 999597 | 09 | 634256 | 4880 | 365744 | 32 |
| 29 | 636776 | 4839 | 999592 | 09 | 637184 | 4848 | 362816 | 31 |
| 30 | 639680 | 4806 | 999586 | 09 | 640093 | 4816 | 359907 | 30 |
| 31 | 8·642563 | 4775 | 9·999581 | 09 | 8·642982 | 4784 | 11·357018 | 29 |
| 32 | 645428 | 4743 | 999575 | 09 | 645853 | 4753 | 354147 | 28 |
| 33 | 648274 | 4712 | 999570 | 09 | 648704 | 4722 | 351296 | 27 |
| 34 | 651102 | 4682 | 999564 | 09 | 651537 | 4691 | 348463 | 26 |
| 35 | 653911 | 4652 | 999558 | 10 | 654352 | 4661 | 345648 | 25 |
| 36 | 656702 | 4622 | 999553 | 10 | 657149 | 4631 | 342851 | 24 |
| 37 | 659475 | 4592 | 999547 | 10 | 659928 | 4602 | 340072 | 23 |
| 38 | 662230 | 4563 | 999541 | 10 | 662689 | 4573 | 337311 | 22 |
| 39 | 664968 | 4535 | 999535 | 10 | 665433 | 4544 | 334567 | 21 |
| 40 | 667689 | 4506 | 999529 | 10 | 668160 | 4526 | 331840 | 20 |
| 41 | 8·670393 | 4479 | 9·999524 | 10 | 8·670870 | 4488 | 11·329130 | 19 |
| 42 | 673080 | 4451 | 999518 | 10 | 673563 | 4461 | 326437 | 18 |
| 43 | 675751 | 4424 | 999512 | 10 | 676239 | 4434 | 323761 | 17 |
| 44 | 678405 | 4397 | 999506 | 10 | 678900 | 4417 | 321100 | 16 |
| 45 | 681043 | 4370 | 999500 | 10 | 681544 | 4380 | 318456 | 15 |
| 46 | 683665 | 4344 | 999493 | 10 | 684172 | 4354 | 315828 | 14 |
| 47 | 686272 | 4318 | 999487 | 10 | 686784 | 4328 | 313216 | 13 |
| 48 | 688863 | 4292 | 999481 | 10 | 689381 | 4303 | 310619 | 12 |
| 49 | 691438 | 4267 | 999475 | 10 | 691963 | 4277 | 308037 | 11 |
| 50 | 693998 | 4242 | 999469 | 10 | 694529 | 4252 | 305471 | 10 |
| 51 | 8·696543 | 4217 | 9·999463 | 11 | 8·697081 | 4228 | 11·302919 | 9 |
| 52 | 699073 | 4192 | 999456 | 11 | 699617 | 4203 | 300383 | 8 |
| 53 | 701589 | 4168 | 999450 | 11 | 702139 | 4179 | 297861 | 7 |
| 54 | 704090 | 4144 | 999443 | 11 | 704646 | 4155 | 295354 | 6 |
| 55 | 706577 | 4121 | 999437 | 11 | 707140 | 4132 | 292860 | 5 |
| 56 | 709049 | 4097 | 999431 | 11 | 709618 | 4108 | 290382 | 4 |
| 57 | 711507 | 4074 | 999424 | 11 | 712083 | 4085 | 287917 | 3 |
| 58 | 713952 | 4051 | 999418 | 11 | 714534 | 4062 | 285466 | 2 |
| 59 | 716383 | 4029 | 999411 | 11 | 716972 | 4040 | 283028 | 1 |
| 60 | 718800 | 4006 | 999404 | 11 | 719396 | 4017 | 280604 | 0 |
| ′ | Cosine. | D. | Sine. | D. | Cotang. | D. | Tang. | ′ |

92° 87°

3° | 176°

| ′ | Sine. | D. | Cosine. | D. | Tang. | D. | Cotang. | ′ |
|---|---|---|---|---|---|---|---|---|
| 0 | 8·718800 | 4006 | 9·999404 | 11 | 8·719396 | 4017 | 11·280604 | 60 |
| 1 | 721204 | 3984 | 999398 | 11 | 721806 | 3995 | 278194 | 59 |
| 2 | 723595 | 3962 | 999391 | 11 | 724204 | 3974 | 275796 | 58 |
| 3 | 725972 | 3941 | 999384 | 11 | 726588 | 3952 | 273412 | 57 |
| 4 | 728337 | 3919 | 999378 | 11 | 728959 | 3930 | 271041 | 56 |
| 5 | 730688 | 3898 | 999371 | 11 | 731317 | 3909 | 268683 | 55 |
| 6 | 733027 | 3877 | 999364 | 12 | 733663 | 3889 | 266337 | 54 |
| 7 | 735354 | 3857 | 999357 | 12 | 735996 | 3868 | 264004 | 53 |
| 8 | 737667 | 3836 | 999350 | 12 | 738317 | 3848 | 261683 | 52 |
| 9 | 739969 | 3816 | 999343 | 12 | 740626 | 3827 | 259374 | 51 |
| 10 | 742259 | 3796 | 999336 | 12 | 742922 | 3807 | 257078 | 50 |
| 11 | 8·744536 | 3776 | 9·999329 | 12 | 8·745207 | 3787 | 11·254793 | 49 |
| 12 | 746802 | 3756 | 999322 | 12 | 747479 | 3768 | 252521 | 48 |
| 13 | 749055 | 3737 | 999315 | 12 | 749740 | 3749 | 250260 | 47 |
| 14 | 751297 | 3717 | 999308 | 12 | 751989 | 3729 | 248011 | 46 |
| 15 | 753528 | 3698 | 999301 | 12 | 754227 | 3710 | 245773 | 45 |
| 16 | 755747 | 3679 | 999294 | 12 | 756453 | 3692 | 243547 | 44 |
| 17 | 757955 | 3661 | 999287 | 12 | 758668 | 3673 | 241332 | 43 |
| 18 | 760151 | 3642 | 999279 | 12 | 760872 | 3655 | 239128 | 42 |
| 19 | 762337 | 3624 | 999272 | 12 | 763065 | 3636 | 236935 | 41 |
| 20 | 764511 | 3606 | 999265 | 12 | 765246 | 3618 | 234754 | 40 |
| 21 | 8·766675 | 3588 | 9·999257 | 12 | 8·767417 | 3600 | 11·232583 | 39 |
| 22 | 768828 | 3570 | 999250 | 13 | 769578 | 3583 | 230422 | 38 |
| 23 | 770970 | 3553 | 999242 | 13 | 771727 | 3565 | 228273 | 37 |
| 24 | 773101 | 3535 | 999235 | 13 | 773866 | 3548 | 226134 | 36 |
| 25 | 775223 | 3518 | 999227 | 13 | 775995 | 3531 | 224005 | 35 |
| 26 | 777333 | 3501 | 999220 | 13 | 778114 | 3514 | 221886 | 34 |
| 27 | 779434 | 3484 | 999212 | 13 | 780222 | 3497 | 219778 | 33 |
| 28 | 781524 | 3467 | 999205 | 13 | 782320 | 3480 | 217680 | 32 |
| 29 | 783605 | 3451 | 999197 | 13 | 784408 | 3464 | 215592 | 31 |
| 30 | 785675 | 3431 | 999189 | 13 | 786486 | 3447 | 213514 | 30 |
| 31 | 8·787736 | 3418 | 9·999181 | 13 | 8·788554 | 3431 | 11·211446 | 29 |
| 32 | 789787 | 3402 | 999174 | 13 | 790613 | 3414 | 209387 | 28 |
| 33 | 791828 | 3386 | 999166 | 13 | 792662 | 3399 | 207338 | 27 |
| 34 | 793859 | 3370 | 999158 | 13 | 794701 | 3383 | 205299 | 26 |
| 35 | 795881 | 3354 | 999150 | 13 | 796731 | 3368 | 203269 | 25 |
| 36 | 797894 | 3339 | 999142 | 13 | 798752 | 3352 | 201248 | 24 |
| 37 | 799897 | 3323 | 999134 | 13 | 800763 | 3337 | 199237 | 23 |
| 38 | 801892 | 3308 | 999126 | 13 | 802765 | 3322 | 197235 | 22 |
| 39 | 803876 | 3293 | 999118 | 13 | 804758 | 3307 | 195242 | 21 |
| 40 | 805852 | 3278 | 999110 | 13 | 806742 | 3292 | 193258 | 20 |
| 41 | 8·807819 | 3263 | 9·999102 | 13 | 8·808717 | 3278 | 11·191283 | 19 |
| 42 | 809777 | 3249 | 999094 | 14 | 810683 | 3262 | 189317 | 18 |
| 43 | 811726 | 3234 | 999086 | 14 | 812641 | 3248 | 187359 | 17 |
| 44 | 813667 | 3219 | 999077 | 14 | 814589 | 3233 | 185411 | 16 |
| 45 | 815599 | 3205 | 999069 | 14 | 816529 | 3219 | 183471 | 15 |
| 46 | 817522 | 3191 | 999061 | 14 | 818461 | 3205 | 181539 | 14 |
| 47 | 819436 | 3177 | 999053 | 14 | 820384 | 3191 | 179616 | 13 |
| 48 | 821343 | 3163 | 999044 | 14 | 822298 | 3177 | 177702 | 12 |
| 49 | 823240 | 3149 | 999036 | 14 | 824205 | 3163 | 175795 | 11 |
| 50 | 825130 | 3135 | 999027 | 14 | 826103 | 3150 | 173897 | 10 |
| 51 | 8·827011 | 3122 | 9·999019 | 14 | 8·827992 | 3136 | 11·172008 | 9 |
| 52 | 828884 | 3108 | 999010 | 14 | 829874 | 3123 | 170126 | 8 |
| 53 | 830749 | 3095 | 999002 | 14 | 831748 | 3110 | 168252 | 7 |
| 54 | 832607 | 3082 | 998993 | 14 | 833613 | 3096 | 166387 | 6 |
| 55 | 834456 | 3069 | 998984 | 14 | 835471 | 3083 | 164529 | 5 |
| 56 | 836297 | 3056 | 998976 | 14 | 837321 | 3070 | 162679 | 4 |
| 57 | 838130 | 3043 | 998967 | 15 | 839163 | 3057 | 160837 | 3 |
| 58 | 839956 | 3030 | 998958 | 15 | 840998 | 3045 | 159002 | 2 |
| 59 | 841774 | 3017 | 998950 | 15 | 842825 | 3032 | 157175 | 1 |
| 60 | 843585 | 3000 | 998941 | 15 | 844644 | 3019 | 155356 | 0 |
| ′ | Cosine. | D. | Sine. | D. | Cotang. | D. | Tang. | ′ |

93° | 86°

4° 175°

| ′ | Sine. | D. | Cosine. | D. | Tang. | D. | Cotang. | ′ |
|---|---|---|---|---|---|---|---|---|
| 0 | 8·843585 | 3005 | 9·998941 | 15 | 8·844644 | 3019 | 11·155356 | 60 |
| 1 | 845387 | 2992 | 998932 | 15 | 846455 | 3007 | 153545 | 59 |
| 2 | 847183 | 2980 | 998923 | 15 | 848260 | 2995 | 151740 | 58 |
| 3 | 848971 | 2967 | 998914 | 15 | 850057 | 2982 | 149943 | 57 |
| 4 | 850751 | 2955 | 998905 | 15 | 851846 | 2970 | 148154 | 56 |
| 5 | 852525 | 2943 | 998896 | 15 | 853628 | 2958 | 146372 | 55 |
| 6 | 854291 | 2931 | 998887 | 15 | 855403 | 2946 | 144597 | 54 |
| 7 | 856049 | 2919 | 998878 | 15 | 857171 | 2935 | 142829 | 53 |
| 8 | 857801 | 2907 | 998869 | 15 | 858932 | 2923 | 141068 | 52 |
| 9 | 859546 | 2896 | 998860 | 15 | 860686 | 2911 | 139314 | 51 |
| 10 | 861283 | 2884 | 998851 | 15 | 862433 | 2900 | 137567 | 50 |
| 11 | 8·863014 | 2873 | 9·998841 | 15 | 8·864173 | 2888 | 11·135827 | 49 |
| 12 | 864738 | 2861 | 998832 | 15 | 865906 | 2877 | 134094 | 48 |
| 13 | 866455 | 2850 | 998823 | 16 | 867632 | 2866 | 132368 | 47 |
| 14 | 868165 | 2839 | 998813 | 16 | 869351 | 2854 | 130649 | 46 |
| 15 | 869868 | 2828 | 998804 | 16 | 871064 | 2843 | 128936 | 45 |
| 16 | 871565 | 2817 | 998795 | 16 | 872770 | 2832 | 127230 | 44 |
| 17 | 873255 | 2806 | 998785 | 16 | 874469 | 2821 | 125531 | 43 |
| 18 | 874938 | 2795 | 998776 | 16 | 876162 | 2811 | 123838 | 42 |
| 19 | 876615 | 2786 | 998766 | 16 | 877849 | 2800 | 122151 | 41 |
| 20 | 878285 | 2773 | 998757 | 16 | 879529 | 2789 | 120471 | 40 |
| 21 | 8·879949 | 2763 | 9·998747 | 16 | 8·881202 | 2779 | 11·118798 | 39 |
| 22 | 881607 | 2752 | 998738 | 16 | 882869 | 2768 | 117131 | 38 |
| 23 | 883258 | 2742 | 998728 | 16 | 884530 | 2758 | 115470 | 37 |
| 24 | 884903 | 2731 | 998718 | 16 | 886185 | 2747 | 113815 | 36 |
| 25 | 886542 | 2721 | 998708 | 16 | 887833 | 2737 | 112167 | 35 |
| 26 | 888174 | 2711 | 998699 | 16 | 889476 | 2727 | 110524 | 34 |
| 27 | 889801 | 2700 | 998689 | 16 | 891112 | 2717 | 108888 | 33 |
| 28 | 891421 | 2690 | 998679 | 16 | 892742 | 2707 | 107258 | 32 |
| 29 | 893035 | 2680 | 998669 | 17 | 894366 | 2697 | 105634 | 31 |
| 30 | 894643 | 2670 | 998659 | 17 | 895984 | 2687 | 104016 | 30 |
| 31 | 8·896246 | 2660 | 9·998649 | 17 | 8·897596 | 2677 | 11 102404 | 29 |
| 32 | 897842 | 2651 | 998639 | 17 | 899203 | 2667 | 100797 | 28 |
| 33 | 899432 | 2641 | 998629 | 17 | 900803 | 2658 | 099197 | 27 |
| 34 | 901017 | 2631 | 998619 | 17 | 902398 | 2648 | 097602 | 26 |
| 35 | 902596 | 2622 | 998609 | 17 | 903987 | 2638 | 096013 | 25 |
| 36 | 904169 | 2612 | 998599 | 17 | 905570 | 2629 | 094430 | 24 |
| 37 | 905736 | 2603 | 998589 | 17 | 907147 | 2620 | 092853 | 23 |
| 38 | 907297 | 2593 | 998578 | 17 | 908719 | 2610 | 091281 | 22 |
| 39 | 908853 | 2584 | 998568 | 17 | 910285 | 2601 | 089715 | 21 |
| 40 | 910404 | 2575 | 998558 | 17 | 911846 | 2592 | 088154 | 20 |
| 41 | 8·911949 | 2566 | 9·998548 | 17 | 8·913401 | 2583 | 11·086599 | 19 |
| 42 | 913488 | 2556 | 998537 | 17 | 914951 | 2574 | 085049 | 18 |
| 43 | 915022 | 2547 | 998527 | 17 | 916495 | 2565 | 083505 | 17 |
| 44 | 916550 | 2538 | 998516 | 18 | 918034 | 2556 | 081966 | 16 |
| 45 | 918073 | 2529 | 998506 | 18 | 919568 | 2547 | 080432 | 15 |
| 46 | 919591 | 2520 | 998495 | 18 | 921096 | 2538 | 078904 | 14 |
| 47 | 921103 | 2512 | 998485 | 18 | 922619 | 2530 | 077381 | 13 |
| 48 | 922610 | 2503 | 998474 | 18 | 924136 | 2521 | 075864 | 12 |
| 49 | 924112 | 2494 | 998464 | 18 | 925649 | 2512 | 074351 | 11 |
| 50 | 925609 | 2486 | 998453 | 18 | 927156 | 2503 | 072844 | 10 |
| 51 | 8·927100 | 2477 | 9·998442 | 18 | 8·928658 | 2495 | 11·071342 | 9 |
| 52 | 928587 | 2469 | 998431 | 18 | 930155 | 2486 | 069845 | 8 |
| 53 | 930068 | 2460 | 998421 | 18 | 931647 | 2478 | 068353 | 7 |
| 54 | 931544 | 2452 | 998410 | 18 | 933134 | 2470 | 066866 | 6 |
| 55 | 933015 | 2443 | 998399 | 18 | 934616 | 2461 | 065384 | 5 |
| 56 | 934481 | 2435 | 998388 | 18 | 936093 | 2453 | 063907 | 4 |
| 57 | 935942 | 2427 | 998377 | 18 | 937565 | 2445 | 062435 | 3 |
| 58 | 937398 | 2419 | 998366 | 18 | 939032 | 2437 | 060968 | 2 |
| 59 | 938850 | 2411 | 998355 | 18 | 940494 | 2430 | 059506 | 1 |
| 60 | 940296 | 2403 | 998344 | 18 | 941952 | 2421 | 058048 | 0 |
| ′ | Cosine. | D. | Sine. | D. | Cotang. | D. | Tang. | ′ |

94° 85°

5° 174°

| ′ | Sine. | D. | Cosine. | D. | Tang. | D. | Cotang. | ′ |
|---|---|---|---|---|---|---|---|---|
| 0 | 8·940296 | 2403 | 9·998344 | 19 | 8·941952 | 2421 | 11·058048 | 60 |
| 1 | 941738 | 2394 | 998333 | 19 | 943404 | 2413 | 056596 | 59 |
| 2 | 943174 | 2387 | 998322 | 19 | 944852 | 2405 | 055148 | 58 |
| 3 | 944606 | 2379 | 998311 | 19 | 946295 | 2397 | 053705 | 57 |
| 4 | 946034 | 2371 | 998300 | 19 | 947734 | 2390 | 052266 | 56 |
| 5 | 947456 | 2363 | 998289 | 19 | 949168 | 2382 | 050832 | 55 |
| 6 | 948874 | 2355 | 998277 | 19 | 950597 | 2374 | 049403 | 54 |
| 7 | 950287 | 2348 | 998266 | 19 | 952021 | 2366 | 047979 | 53 |
| 8 | 951696 | 2340 | 998255 | 19 | 953441 | 2360 | 046559 | 52 |
| 9 | 953100 | 2332 | 998243 | 19 | 954856 | 2351 | 045144 | 51 |
| 10 | 954499 | 2325 | 998232 | 19 | 956267 | 2344 | 043733 | 50 |
| 11 | 8·955894 | 2317 | 9·998220 | 19 | 8·957674 | 2337 | 11·042326 | 49 |
| 12 | 957284 | 2310 | 998209 | 19 | 959075 | 2329 | 040925 | 48 |
| 13 | 958670 | 2302 | 998197 | 19 | 960473 | 2323 | 039527 | 47 |
| 14 | 960052 | 2295 | 998186 | 19 | 961866 | 2314 | 038134 | 46 |
| 15 | 961429 | 2288 | 998174 | 19 | 963255 | 2307 | 036745 | 45 |
| 16 | 962801 | 2280 | 998163 | 19 | 964639 | 2300 | 035361 | 44 |
| 17 | 964170 | 2273 | 998151 | 19 | 966019 | 2293 | 033981 | 43 |
| 18 | 965534 | 2266 | 998139 | 20 | 967394 | 2286 | 032606 | 42 |
| 19 | 966893 | 2259 | 998128 | 20 | 968766 | 2279 | 031234 | 41 |
| 20 | 968249 | 2252 | 998116 | 20 | 970133 | 2271 | 029867 | 40 |
| 21 | 8·969600 | 2244 | 9·998104 | 20 | 8·971496 | 2265 | 11·028504 | 39 |
| 22 | 970947 | 2238 | 998092 | 20 | 972855 | 2257 | 027145 | 38 |
| 23 | 972289 | 2231 | 998080 | 20 | 974209 | 2251 | 025791 | 37 |
| 24 | 973628 | 2224 | 998068 | 20 | 975560 | 2244 | 024440 | 36 |
| 25 | 974962 | 2217 | 998056 | 20 | 976906 | 2237 | 023094 | 35 |
| 26 | 976293 | 2210 | 998044 | 20 | 978248 | 2230 | 021752 | 34 |
| 27 | 977619 | 2203 | 998032 | 20 | 979586 | 2223 | 020414 | 33 |
| 28 | 978941 | 2197 | 998020 | 20 | 980921 | 2217 | 019079 | 32 |
| 29 | 980259 | 2190 | 998008 | 20 | 982251 | 2210 | 017749 | 31 |
| 30 | 981573 | 2183 | 997996 | 20 | 983577 | 2204 | 016423 | 30 |
| 31 | 8·982883 | 2177 | 9·997984 | 20 | 8·984899 | 2197 | 11·015101 | 29 |
| 32 | 984189 | 2170 | 997972 | 20 | 986217 | 2191 | 013783 | 28 |
| 33 | 985491 | 2163 | 997959 | 20 | 987532 | 2184 | 012468 | 27 |
| 34 | 986789 | 2157 | 997947 | 20 | 988842 | 2178 | 011158 | 26 |
| 35 | 988083 | 2150 | 997935 | 21 | 990149 | 2171 | 009851 | 25 |
| 36 | 989374 | 2144 | 997922 | 21 | 991451 | 2165 | 008549 | 24 |
| 37 | 990660 | 2138 | 997910 | 21 | 992750 | 2158 | 007250 | 23 |
| 38 | 991943 | 2131 | 997897 | 21 | 994045 | 2152 | 005955 | 22 |
| 39 | 993222 | 2125 | 997885 | 21 | 995337 | 2146 | 004663 | 21 |
| 40 | 994497 | 2119 | 997872 | 21 | 996624 | 2140 | 003376 | 20 |
| 41 | 8·995768 | 2112 | 9·997860 | 21 | 8·997908 | 2134 | 11·002092 | 19 |
| 42 | 997036 | 2106 | 997847 | 21 | 999188 | 2127 | 000812 | 18 |
| 43 | 998299 | 2100 | 997835 | 21 | 9·000465 | 2121 | 10·999535 | 17 |
| 44 | 999560 | 2094 | 997822 | 21 | 001738 | 2115 | 998262 | 16 |
| 45 | 9·000816 | 2087 | 997809 | 21 | 003007 | 2109 | 996993 | 15 |
| 46 | 002069 | 2082 | 997797 | 21 | 004272 | 2103 | 995728 | 14 |
| 47 | 003318 | 2076 | 997784 | 21 | 005534 | 2097 | 994466 | 13 |
| 48 | 004563 | 2070 | 997771 | 21 | 006792 | 2091 | 993208 | 12 |
| 49 | 005805 | 2064 | 997758 | 21 | 008047 | 2085 | 991953 | 11 |
| 50 | 007044 | 2058 | 997745 | 21 | 009298 | 2080 | 990702 | 10 |
| 51 | 9·008278 | 2052 | 9·997732 | 21 | 9·010546 | 2074 | 10·989454 | 9 |
| 52 | 009510 | 2046 | 997719 | 21 | 011790 | 2068 | 988210 | 8 |
| 53 | 010737 | 2040 | 997706 | 21 | 013031 | 2062 | 986969 | 7 |
| 54 | 011962 | 2034 | 997693 | 22 | 014268 | 2056 | 985732 | 6 |
| 55 | 013182 | 2029 | 997680 | 22 | 015502 | 2051 | 984498 | 5 |
| 56 | 014400 | 2023 | 997667 | 22 | 016732 | 2045 | 983268 | 4 |
| 57 | 015613 | 2017 | 997654 | 22 | 017959 | 2040 | 982041 | 3 |
| 58 | 016824 | 2012 | 997641 | 22 | 019183 | 2033 | 980817 | 2 |
| 59 | 018031 | 2006 | 997628 | 22 | 020403 | 2028 | 979597 | 1 |
| 60 | 019235 | 2000 | 997614 | 22 | 021620 | 2023 | 978380 | 0 |
| ′ | Cosine. | D. | Sine. | D. | Cotang. | D. | Tang. | ′ |

95° 84°

| 6° | | | | | | | | 173° |
|---|---|---|---|---|---|---|---|---|
| ′ | Sine. | D. | Cosine. | D. | Tang. | D. | Cotang. | ′ |
| 0 | 9·019235 | 2000 | 9·997614 | 22 | 9·021620 | 2023 | 10·978380 | 60 |
| 1 | 020435 | 1995 | 997601 | 22 | 022834 | 2017 | 977166 | 59 |
| 2 | 021632 | 1989 | 997588 | 22 | 024044 | 2011 | 975956 | 58 |
| 3 | 022825 | 1984 | 997574 | 22 | 025251 | 2006 | 974749 | 57 |
| 4 | 024016 | 1978 | 997561 | 22 | 026455 | 2000 | 973545 | 56 |
| 5 | 025203 | 1973 | 997547 | 22 | 027655 | 1995 | 972345 | 55 |
| 6 | 026386 | 1967 | 997534 | 23 | 028852 | 1990 | 971148 | 54 |
| 7 | 027567 | 1962 | 997520 | 23 | 030046 | 1985 | 969954 | 53 |
| 8 | 028744 | 1957 | 997507 | 23 | 031237 | 1979 | 968763 | 52 |
| 9 | 029918 | 1951 | 997493 | 23 | 032425 | 1974 | 967575 | 51 |
| 10 | 031089 | 1947 | 997480 | 23 | 033609 | 1969 | 966391 | 50 |
| 11 | 9·032257 | 1941 | 9·997466 | 23 | 9·034791 | 1964 | 10·965209 | 49 |
| 12 | 033421 | 1936 | 997452 | 23 | 035969 | 1958 | 964031 | 48 |
| 13 | 034582 | 1930 | 997439 | 23 | 037144 | 1953 | 962856 | 47 |
| 14 | 035741 | 1925 | 997425 | 23 | 038316 | 1948 | 961684 | 46 |
| 15 | 036896 | 1920 | 997411 | 23 | 039485 | 1943 | 960515 | 45 |
| 16 | 038048 | 1915 | 997397 | 23 | 040651 | 1938 | 959349 | 44 |
| 17 | 039197 | 1910 | 997383 | 23 | 041813 | 1933 | 958187 | 43 |
| 18 | 040342 | 1905 | 997369 | 23 | 042973 | 1928 | 957027 | 42 |
| 19 | 041485 | 1899 | 997355 | 23 | 044130 | 1923 | 955870 | 41 |
| 20 | 042625 | 1894 | 997341 | 23 | 045284 | 1918 | 954716 | 40 |
| 21 | 9·043762 | 1889 | 9·997327 | 24 | 9·046434 | 1913 | 10·953566 | 39 |
| 22 | 044895 | 1884 | 997313 | 24 | 047582 | 1908 | 952418 | 38 |
| 23 | 046026 | 1879 | 997299 | 24 | 048727 | 1903 | 951273 | 37 |
| 24 | 047154 | 1875 | 997285 | 24 | 049869 | 1898 | 950131 | 36 |
| 25 | 048279 | 1870 | 997271 | 24 | 051008 | 1893 | 948992 | 35 |
| 26 | 049400 | 1865 | 997257 | 24 | 052144 | 1889 | 947856 | 34 |
| 27 | 050519 | 1860 | 997242 | 24 | 053277 | 1884 | 946723 | 33 |
| 28 | 051635 | 1855 | 997228 | 24 | 054407 | 1879 | 945593 | 32 |
| 29 | 952749 | 1850 | 997214 | 24 | 055535 | 1874 | 944465 | 31 |
| 30 | 053859 | 1845 | 997199 | 24 | 056659 | 1870 | 943341 | 30 |
| 31 | 9·054966 | 1841 | 9·997185 | 24 | 9·057781 | 1865 | 10·942219 | 29 |
| 32 | 056071 | 1836 | 997170 | 24 | 058900 | 1869 | 941100 | 28 |
| 33 | 057172 | 1831 | 997156 | 24 | 060016 | 1855 | 939984 | 27 |
| 34 | 058271 | 1827 | 997141 | 24 | 061130 | 1851 | 938870 | 26 |
| 35 | 059367 | 1822 | 997127 | 24 | 062240 | 1846 | 937760 | 25 |
| 36 | 060460 | 1817 | 997112 | 24 | 063348 | 1842 | 936652 | 24 |
| 37 | 061551 | 1813 | 997098 | 24 | 064453 | 1837 | 935547 | 23 |
| 38 | 062639 | 1808 | 997083 | 25 | 065556 | 1833 | 934444 | 22 |
| 39 | 063724 | 1804 | 997068 | 25 | 066655 | 1828 | 933345 | 21 |
| 40 | 064806 | 1799 | 997053 | 25 | 067752 | 1824 | 932248 | 20 |
| 41 | 9·065885 | 1794 | 9·997039 | 25 | 9·068846 | 1819 | 10·931154 | 19 |
| 42 | 066962 | 1790 | 997024 | 25 | 069938 | 1815 | 930062 | 18 |
| 43 | 068036 | 1786 | 997009 | 25 | 071027 | 1810 | 928973 | 17 |
| 44 | 069107 | 1781 | 996994 | 25 | 072113 | 1806 | 927887 | 16 |
| 45 | 070176 | 1777 | 996979 | 25 | 073197 | 1802 | 926803 | 15 |
| 46 | 071242 | 1772 | 996964 | 25 | 074278 | 1797 | 925722 | 14 |
| 47 | 072306 | 1768 | 996949 | 25 | 075356 | 1793 | 924644 | 13 |
| 48 | 073366 | 1763 | 996934 | 25 | 076432 | 1789 | 923568 | 12 |
| 49 | 074424 | 1759 | 996919 | 25 | 077505 | 1784 | 922495 | 11 |
| 50 | 075480 | 1755 | 996904 | 25 | 078576 | 1780 | 921424 | 10 |
| 51 | 9·076533 | 1750 | 9·996889 | 25 | 9·079644 | 1776 | 10·920356 | 9 |
| 52 | 077583 | 1746 | 996874 | 25 | 080710 | 1772 | 919290 | 8 |
| 53 | 078631 | 1742 | 996858 | 25 | 081773 | 1767 | 918227 | 7 |
| 54 | 079676 | 1738 | 996843 | 25 | 082833 | 1763 | 917167 | 6 |
| 55 | 080719 | 1733 | 996828 | 25 | 083891 | 1759 | 916109 | 5 |
| 56 | 081759 | 1729 | 996812 | 26 | 084947 | 1755 | 915053 | 4 |
| 57 | 082797 | 1725 | 996797 | 26 | 086000 | 1751 | 914000 | 3 |
| 58 | 083832 | 1721 | 996782 | 26 | 087050 | 1747 | 912950 | 2 |
| 59 | 084864 | 1717 | 996766 | 26 | 088098 | 1743 | 911902 | 1 |
| 60 | 085894 | 1713 | 996751 | 26 | 089144 | 1738 | 910856 | 0 |
| ′ | Cosine. | D. | Sine. | D. | Cotang. | D. | Tang. | ′ |
| 96° | | | | | | | | 83° |

7° | 172°

| ′ | Sine. | D. | Cosine. | D. | Tang. | D. | Cotang. | ′ |
|---|---|---|---|---|---|---|---|---|
| 0 | 9·085894 | 1713 | 9·996751 | 26 | 9·089144 | 1738 | 10·910856 | 60 |
| 1 | 086922 | 1709 | 996735 | 26 | 090187 | 1734 | 909813 | 59 |
| 2 | 087947 | 1704 | 996720 | 26 | 091228 | 1730 | 908772 | 58 |
| 3 | 088970 | 1700 | 996704 | 26 | 092266 | 1727 | 907734 | 57 |
| 4 | 089990 | 1696 | 996688 | 26 | 093302 | 1722 | 906698 | 56 |
| 5 | 091008 | 1692 | 996673 | 26 | 094336 | 1719 | 905664 | 55 |
| 6 | 092024 | 1688 | 996657 | 26 | 095367 | 1715 | 904633 | 54 |
| 7 | 093037 | 1684 | 996641 | 26 | 096395 | 1711 | 903605 | 53 |
| 8 | 094047 | 1680 | 996625 | 26 | 097422 | 1707 | 902578 | 52 |
| 9 | 095056 | 1676 | 996610 | 26 | 098446 | 1703 | 901554 | 51 |
| 10 | 096062 | 1673 | 996594 | 26 | 099468 | 1699 | 900532 | 50 |
| 11 | 9·097065 | 1668 | 9·996578 | 27 | 9·100487 | 1695 | 10·899513 | 49 |
| 12 | 098066 | 1665 | 996562 | 27 | 101504 | 1691 | 898496 | 48 |
| 13 | 099065 | 1661 | 996546 | 27 | 102519 | 1687 | 897481 | 47 |
| 14 | 100062 | 1657 | 996530 | 27 | 103532 | 1684 | 896468 | 46 |
| 15 | 101056 | 1653 | 996514 | 27 | 104542 | 1680 | 895458 | 45 |
| 16 | 102048 | 1649 | 996498 | 27 | 105550 | 1676 | 894450 | 44 |
| 17 | 103037 | 1645 | 996482 | 27 | 106556 | 1672 | 893444 | 43 |
| 18 | 104025 | 1641 | 996465 | 27 | 107559 | 1669 | 892441 | 42 |
| 19 | 105010 | 1638 | 996449 | 27 | 108560 | 1665 | 891440 | 41 |
| 20 | 105992 | 1634 | 996433 | 27 | 109559 | 1661 | 890441 | 40 |
| 21 | 9·106973 | 1630 | 9·996417 | 27 | 9·110556 | 1658 | 10·889444 | 39 |
| 22 | 107951 | 1627 | 996400 | 27 | 111551 | 1654 | 888449 | 38 |
| 23 | 108927 | 1623 | 996384 | 27 | 112543 | 1650 | 887457 | 37 |
| 24 | 109901 | 1619 | 996368 | 27 | 113533 | 1646 | 886467 | 36 |
| 25 | 110873 | 1616 | 996351 | 27 | 114521 | 1643 | 885479 | 35 |
| 26 | 111842 | 1612 | 996335 | 27 | 115507 | 1639 | 884493 | 34 |
| 27 | 112809 | 1608 | 996318 | 27 | 116491 | 1636 | 883509 | 33 |
| 28 | 113774 | 1605 | 996302 | 28 | 117472 | 1632 | 882528 | 32 |
| 29 | 114737 | 1601 | 996285 | 28 | 118452 | 1629 | 881548 | 31 |
| 30 | 115698 | 1597 | 996269 | 28 | 119429 | 1625 | 880571 | 30 |
| 31 | 9·116656 | 1594 | 9·996252 | 28 | 9·120404 | 1622 | 10·879596 | 29 |
| 32 | 117613 | 1590 | 996235 | 28 | 121377 | 1618 | 878623 | 28 |
| 33 | 118567 | 1587 | 996219 | 28 | 122348 | 1615 | 877652 | 27 |
| 34 | 119515 | 1583 | 996202 | 28 | 123317 | 1611 | 876683 | 26 |
| 35 | 120469 | 1580 | 996185 | 28 | 124284 | 1607 | 875716 | 25 |
| 36 | 121417 | 1576 | 996168 | 28 | 125249 | 1604 | 874751 | 24 |
| 37 | 122362 | 1573 | 996151 | 28 | 126211 | 1601 | 873789 | 23 |
| 38 | 123306 | 1569 | 996134 | 28 | 127172 | 1597 | 872828 | 22 |
| 39 | 124248 | 1566 | 996117 | 28 | 128130 | 1594 | 871870 | 21 |
| 40 | 125187 | 1562 | 996100 | 28 | 129087 | 1591 | 870913 | 20 |
| 41 | 9·126125 | 1559 | 9·996083 | 29 | 9·130041 | 1587 | 10·869959 | 19 |
| 42 | 127060 | 1556 | 996066 | 29 | 130994 | 1584 | 869006 | 18 |
| 43 | 127993 | 1552 | 996049 | 29 | 131944 | 1581 | 868056 | 17 |
| 44 | 128925 | 1549 | 996032 | 29 | 132893 | 1577 | 867107 | 16 |
| 45 | 129854 | 1545 | 996015 | 29 | 133839 | 1574 | 866161 | 15 |
| 46 | 130781 | 1542 | 995998 | 29 | 134784 | 1571 | 865216 | 14 |
| 47 | 131706 | 1539 | 995980 | 29 | 135726 | 1567 | 864274 | 13 |
| 48 | 132630 | 1535 | 995963 | 29 | 136667 | 1564 | 863333 | 12 |
| 49 | 133551 | 1532 | 995946 | 29 | 137605 | 1561 | 862395 | 11 |
| 50 | 134470 | 1529 | 995928 | 29 | 138542 | 1558 | 861458 | 10 |
| 51 | 9·135387 | 1525 | 9·995911 | 29 | 9·139476 | 1555 | 10·860524 | 9 |
| 52 | 136303 | 1522 | 995894 | 29 | 140409 | 1551 | 859591 | 8 |
| 53 | 137216 | 1519 | 995876 | 29 | 141340 | 1548 | 858660 | 7 |
| 54 | 138128 | 1516 | 995859 | 29 | 142269 | 1545 | 857731 | 6 |
| 55 | 139037 | 1512 | 995841 | 29 | 143196 | 1542 | 856804 | 5 |
| 56 | 139944 | 1509 | 995823 | 29 | 144121 | 1539 | 855879 | 4 |
| 57 | 140850 | 1506 | 995806 | 29 | 145044 | 1535 | 854956 | 3 |
| 58 | 141754 | 1503 | 995788 | 29 | 145966 | 1532 | 854034 | 2 |
| 59 | 142655 | 1500 | 995771 | 29 | 146885 | 1529 | 853115 | 1 |
| 60 | 143555 | 1496 | 995753 | 29 | 147803 | 1526 | 852197 | 0 |
| ′ | Cosine. | D. | Sine. | D. | Cotang. | D. | Tang. | ′ |

97° | 82°

8° 171°

| ′ | Sine. | D. | Cosine. | D. | Tang. | D. | Cotang. | ′ |
|---|---|---|---|---|---|---|---|---|
| 0 | 9·143555 | 1496 | 9·995753 | 30 | 9·147803 | 1526 | 10·852197 | 60 |
| 1 | 144453 | 1493 | 995735 | 30 | 148718 | 1523 | 851282 | 59 |
| 2 | 145349 | 1490 | 995717 | 30 | 149632 | 1520 | 850368 | 58 |
| 3 | 146243 | 1487 | 995699 | 30 | 150544 | 1517 | 849456 | 57 |
| 4 | 147136 | 1484 | 995681 | 30 | 151454 | 1514 | 848546 | 56 |
| 5 | 148026 | 1481 | 995664 | 30 | 152363 | 1511 | 847637 | 55 |
| 6 | 148915 | 1478 | 995646 | 30 | 153269 | 1508 | 846731 | 54 |
| 7 | 149802 | 1475 | 995628 | 30 | 154174 | 1505 | 845826 | 53 |
| 8 | 150686 | 1472 | 995610 | 30 | 155077 | 1502 | 844923 | 52 |
| 9 | 151569 | 1469 | 995591 | 30 | 155978 | 1499 | 844022 | 51 |
| 10 | 152451 | 1466 | 995573 | 30 | 156877 | 1496 | 843123 | 50 |
| 11 | 9·153330 | 1463 | 9·995555 | 30 | 9·157775 | 1493 | 10·842225 | 49 |
| 12 | 154208 | 1460 | 995537 | 30 | 158671 | 1490 | 841329 | 48 |
| 13 | 155083 | 1457 | 995519 | 30 | 159565 | 1487 | 840435 | 47 |
| 14 | 155957 | 1454 | 995501 | 31 | 160457 | 1484 | 839543 | 46 |
| 15 | 156830 | 1451 | 995482 | 31 | 161347 | 1481 | 838653 | 45 |
| 16 | 157700 | 1448 | 995464 | 31 | 162236 | 1479 | 837764 | 44 |
| 17 | 158569 | 1445 | 995446 | 31 | 163123 | 1476 | 836877 | 43 |
| 18 | 159435 | 1442 | 995427 | 31 | 164008 | 1473 | 835992 | 42 |
| 19 | 160301 | 1439 | 995409 | 31 | 164892 | 1470 | 835108 | 41 |
| 20 | 161164 | 1436 | 995390 | 31 | 165774 | 1467 | 834226 | 40 |
| 21 | 9·162025 | 1433 | 9·995372 | 31 | 9·166654 | 1464 | 10·833346 | 39 |
| 22 | 162885 | 1430 | 995353 | 31 | 167532 | 1461 | 832468 | 38 |
| 23 | 163743 | 1427 | 995334 | 31 | 168409 | 1458 | 831591 | 37 |
| 24 | 164600 | 1424 | 995316 | 31 | 169284 | 1455 | 830716 | 36 |
| 25 | 165454 | 1422 | 995297 | 31 | 170157 | 1453 | 829843 | 35 |
| 26 | 166307 | 1419 | 995278 | 31 | 171029 | 1450 | 828971 | 34 |
| 27 | 167159 | 1416 | 995260 | 31 | 171899 | 1447 | 828101 | 33 |
| 28 | 168008 | 1413 | 995241 | 32 | 172767 | 1444 | 827233 | 32 |
| 29 | 168856 | 1410 | 995222 | 32 | 173634 | 1442 | 826366 | 31 |
| 30 | 169702 | 1407 | 995203 | 32 | 174499 | 1439 | 825501 | 30 |
| 31 | 9·170547 | 1405 | 9·995184 | 32 | 9·175362 | 1436 | 10·824638 | 29 |
| 32 | 171389 | 1402 | 995165 | 32 | 176224 | 1433 | 823776 | 28 |
| 33 | 172230 | 1399 | 995146 | 32 | 177084 | 1431 | 822916 | 27 |
| 34 | 173070 | 1396 | 995127 | 32 | 177942 | 1428 | 822058 | 26 |
| 35 | 173908 | 1394 | 995108 | 32 | 178799 | 1425 | 821201 | 25 |
| 36 | 174744 | 1391 | 995089 | 32 | 179655 | 1423 | 820345 | 24 |
| 37 | 175578 | 1388 | 995070 | 32 | 180508 | 1420 | 819492 | 23 |
| 38 | 176411 | 1386 | 995051 | 32 | 181360 | 1417 | 818640 | 22 |
| 39 | 177242 | 1383 | 995032 | 32 | 182211 | 1415 | 817789 | 21 |
| 40 | 178072 | 1380 | 995013 | 32 | 183059 | 1412 | 816941 | 20 |
| 41 | 9·178900 | 1377 | 9·994993 | 32 | 9·183907 | 1409 | 10·816093 | 19 |
| 42 | 179726 | 1374 | 994974 | 32 | 184752 | 1407 | 815248 | 18 |
| 43 | 180551 | 1372 | 994955 | 32 | 185597 | 1404 | 814403 | 17 |
| 44 | 181374 | 1369 | 994935 | 32 | 186439 | 1402 | 813561 | 16 |
| 45 | 182196 | 1366 | 994916 | 33 | 187280 | 1399 | 812720 | 15 |
| 46 | 183016 | 1364 | 994896 | 33 | 188120 | 1396 | 811880 | 14 |
| 47 | 183834 | 1361 | 994877 | 33 | 188958 | 1393 | 811042 | 13 |
| 48 | 184651 | 1359 | 994857 | 33 | 189794 | 1391 | 810206 | 12 |
| 49 | 185466 | 1356 | 994838 | 33 | 190629 | 1389 | 809371 | 11 |
| 50 | 186280 | 1353 | 994818 | 33 | 191462 | 1386 | 808538 | 10 |
| 51 | 9·187092 | 1351 | 9·994798 | 33 | 9·192294 | 1384 | 10·807706 | 9 |
| 52 | 187903 | 1348 | 994779 | 33 | 193124 | 1381 | 806876 | 8 |
| 53 | 188712 | 1346 | 994759 | 33 | 193953 | 1379 | 806047 | 7 |
| 54 | 189519 | 1343 | 994739 | 33 | 194780 | 1376 | 805220 | 6 |
| 55 | 190325 | 1341 | 994720 | 33 | 195606 | 1374 | 804394 | 5 |
| 56 | 191130 | 1338 | 994700 | 33 | 196430 | 1371 | 803570 | 4 |
| 57 | 191933 | 1336 | 994680 | 33 | 197253 | 1369 | 802747 | 3 |
| 58 | 192734 | 1333 | 994660 | 33 | 198074 | 1366 | 801926 | 2 |
| 59 | 193534 | 1330 | 994640 | 33 | 198894 | 1364 | 801106 | 1 |
| 60 | 194332 | 1328 | 994620 | 33 | 199713 | 1361 | 800287 | 0 |
| ′ | Cosine. | D. | Sine. | D. | Cotang. | D. | Tang. | ′ |

98° 81°

| 9° | | | | | | | | 170° |
|---|---|---|---|---|---|---|---|---|
| ′ | Sine. | D. | Cosine. | D. | Tang. | D. | Cotang. | ′ |
| 0 | 9·194332 | 1328 | 9·994620 | 33 | 9·199713 | 1361 | 10·800287 | 60 |
| 1 | 195129 | 1326 | 994600 | 33 | 200529 | 1359 | 799471 | 59 |
| 2 | 195925 | 1323 | 994580 | 33 | 201345 | 1356 | 798655 | 58 |
| 3 | 196719 | 1321 | 994560 | 34 | 202159 | 1354 | 797841 | 57 |
| 4 | 197511 | 1318 | 994540 | 34 | 202971 | 1352 | 797029 | 56 |
| 5 | 198302 | 1316 | 994519 | 34 | 203782 | 1349 | 796218 | 55 |
| 6 | 199091 | 1313 | 994499 | 34 | 204592 | 1347 | 795408 | 54 |
| 7 | 199879 | 1311 | 994479 | 34 | 205400 | 1345 | 794600 | 53 |
| 8 | 200666 | 1308 | 994459 | 34 | 206207 | 1342 | 793793 | 52 |
| 9 | 201451 | 1306 | 994438 | 34 | 207013 | 1340 | 792987 | 51 |
| 10 | 202234 | 1304 | 994418 | 34 | 207817 | 1338 | 792183 | 50 |
| 11 | 9·203017 | 1301 | 9·994398 | 34 | 9·208619 | 1335 | 10·791381 | 49 |
| 12 | 203797 | 1299 | 994377 | 34 | 209420 | 1333 | 790580 | 48 |
| 13 | 204577 | 1296 | 994357 | 34 | 210220 | 1331 | 789780 | 47 |
| 14 | 205354 | 1294 | 994336 | 34 | 211018 | 1328 | 788982 | 46 |
| 15 | 206131 | 1292 | 994316 | 34 | 211815 | 1326 | 788185 | 45 |
| 16 | 206906 | 1289 | 994295 | 34 | 212611 | 1324 | 787389 | 44 |
| 17 | 207679 | 1287 | 994274 | 35 | 213405 | 1321 | 786595 | 43 |
| 18 | 208452 | 1285 | 994254 | 35 | 214198 | 1319 | 785802 | 42 |
| 19 | 209222 | 1282 | 994233 | 35 | 214989 | 1317 | 785011 | 41 |
| 20 | 209992 | 1280 | 994212 | 35 | 215780 | 1315 | 784220 | 40 |
| 21 | 9·210760 | 1278 | 9·994191 | 35 | 9·216568 | 1312 | 10·783432 | 39 |
| 22 | 211526 | 1275 | 994171 | 35 | 217356 | 1310 | 782644 | 38 |
| 23 | 212291 | 1273 | 994150 | 35 | 218142 | 1308 | 781858 | 37 |
| 24 | 213055 | 1271 | 994129 | 35 | 218926 | 1305 | 781074 | 36 |
| 25 | 213818 | 1268 | 994108 | 35 | 219710 | 1303 | 780290 | 35 |
| 26 | 214579 | 1266 | 994087 | 35 | 220492 | 1301 | 779508 | 34 |
| 27 | 215338 | 1264 | 994066 | 35 | 221272 | 1299 | 778728 | 33 |
| 28 | 216097 | 1261 | 994045 | 35 | 222052 | 1297 | 777948 | 32 |
| 29 | 216854 | 1259 | 994024 | 35 | 222830 | 1294 | 777170 | 31 |
| 30 | 217609 | 1257 | 994003 | 35 | 223607 | 1292 | 776393 | 30 |
| 31 | 9·218363 | 1255 | 9·993982 | 35 | 9·224382 | 1290 | 10·775618 | 29 |
| 32 | 219116 | 1253 | 993960 | 35 | 225156 | 1288 | 774844 | 28 |
| 33 | 219868 | 1250 | 993939 | 35 | 225929 | 1286 | 774071 | 27 |
| 34 | 220618 | 1248 | 993918 | 35 | 226700 | 1284 | 773300 | 26 |
| 35 | 221367 | 1246 | 993897 | 36 | 227471 | 1281 | 772529 | 25 |
| 36 | 222115 | 1244 | 993875 | 36 | 228239 | 1279 | 771761 | 24 |
| 37 | 222861 | 1242 | 993854 | 36 | 229007 | 1277 | 770993 | 23 |
| 38 | 223606 | 1239 | 993832 | 36 | 229773 | 1275 | 770227 | 22 |
| 39 | 224349 | 1237 | 993811 | 36 | 230539 | 1273 | 769461 | 21 |
| 40 | 225092 | 1235 | 993789 | 36 | 231302 | 1271 | 768698 | 20 |
| 41 | 9·225833 | 1233 | 9·993768 | 36 | 9·232065 | 1269 | 10·767935 | 19 |
| 42 | 226573 | 1231 | 993746 | 36 | 232826 | 1267 | 767174 | 18 |
| 43 | 227311 | 1228 | 993725 | 36 | 233586 | 1265 | 766414 | 17 |
| 44 | 228048 | 1226 | 993703 | 36 | 234345 | 1262 | 765655 | 16 |
| 45 | 228784 | 1224 | 993681 | 36 | 235103 | 1260 | 764897 | 15 |
| 46 | 229518 | 1222 | 993660 | 36 | 235859 | 1258 | 764141 | 14 |
| 47 | 230252 | 1220 | 993638 | 36 | 236614 | 1256 | 763386 | 13 |
| 48 | 230984 | 1218 | 993616 | 36 | 237368 | 1254 | 762632 | 12 |
| 49 | 231715 | 1216 | 993594 | 37 | 238120 | 1252 | 761880 | 11 |
| 50 | 232444 | 1214 | 993572 | 37 | 238872 | 1250 | 761128 | 10 |
| 51 | 9·233172 | 1212 | 9·993550 | 37 | 9·239622 | 1248 | 10·760378 | 9 |
| 52 | 233899 | 1209 | 993528 | 37 | [illegible] | 1246 | 759629 | 8 |
| 53 | [illegible] | 1207 | 993506 | 37 | 241118 | 1244 | 758882 | 7 |
| 54 | 235349 | 1205 | 993484 | 37 | 241865 | 1242 | 758135 | 6 |
| 55 | 236073 | 1203 | 993462 | 37 | 242610 | 1240 | 757390 | 5 |
| 56 | 236795 | 1201 | 993440 | 37 | 243354 | 1238 | 756646 | 4 |
| 57 | 237515 | 1199 | 993418 | 37 | 244097 | 1236 | 755903 | 3 |
| 58 | 238235 | 1197 | 993396 | 37 | 244839 | 1234 | 755161 | 2 |
| 59 | 238953 | 1195 | 993374 | 37 | 245579 | 1232 | 754421 | 1 |
| 60 | 239670 | 1193 | 993351 | 37 | 246319 | 1230 | 753681 | 0 |
| ′ | Cosine. | D. | Sine. | D. | Cotang. | D. | Tang. | ′ |
| 99° | | | | | | | | 80° |

10° 169°

| ′ | Sine. | D. | Cosine. | D. | Tang. | D. | Cotang. | ′ |
|---|---|---|---|---|---|---|---|---|
| 0 | 9·239670 | 1193 | 9·993351 | 37 | 9·246319 | 1230 | 10·753681 | 60 |
| 1 | 240386 | 1191 | 993329 | 37 | 247057 | 1228 | 752943 | 59 |
| 2 | 241101 | 1189 | 993307 | 37 | 247794 | 1226 | 752206 | 58 |
| 3 | 241814 | 1187 | 993284 | 37 | 248530 | 1224 | 751470 | 57 |
| 4 | 242526 | 1185 | 993262 | 37 | 249264 | 1222 | 750736 | 56 |
| 5 | 243237 | 1183 | 993240 | 37 | 249998 | 1220 | 750002 | 55 |
| 6 | 243947 | 1181 | 993217 | 38 | 250730 | 1218 | 749270 | 54 |
| 7 | 244656 | 1179 | 993195 | 38 | 251461 | 1217 | 748539 | 53 |
| 8 | 245363 | 1177 | 993172 | 38 | 252191 | 1215 | 747809 | 52 |
| 9 | 246069 | 1175 | 993149 | 38 | 252920 | 1213 | 747080 | 51 |
| 10 | 246775 | 1173 | 993127 | 38 | 253648 | 1211 | 746352 | 50 |
| 11 | 9·247478 | 1171 | 9·993104 | 38 | 9·254374 | 1209 | 10·745626 | 49 |
| 12 | 248181 | 1169 | 993081 | 38 | 255100 | 1207 | 744900 | 48 |
| 13 | 248883 | 1167 | 993059 | 38 | 255824 | 1205 | 744176 | 47 |
| 14 | 249583 | 1165 | 993036 | 38 | 256547 | 1203 | 743453 | 46 |
| 15 | 250282 | 1163 | 993013 | 38 | 257269 | 1201 | 742731 | 45 |
| 16 | 250980 | 1161 | 992990 | 38 | 257990 | 1200 | 742010 | 44 |
| 17 | 251677 | 1159 | 992967 | 38 | 258710 | 1198 | 741290 | 43 |
| 18 | 252373 | 1158 | 992944 | 38 | 259429 | 1196 | 740571 | 42 |
| 19 | 253067 | 1156 | 992921 | 38 | 260146 | 1194 | 739854 | 41 |
| 20 | 253761 | 1154 | 992898 | 38 | 260863 | 1192 | 739137 | 40 |
| 21 | 9·254453 | 1152 | 9·992875 | 38 | 9·261578 | 1190 | 10·738422 | 39 |
| 22 | 255144 | 1150 | 992852 | 38 | 262292 | 1189 | 737708 | 38 |
| 23 | 255834 | 1148 | 992829 | 39 | 263005 | 1187 | 736995 | 37 |
| 24 | 256523 | 1146 | 992806 | 39 | 263717 | 1185 | 736283 | 36 |
| 25 | 257211 | 1144 | 992783 | 39 | 264428 | 1183 | 735572 | 35 |
| 26 | 257898 | 1142 | 992759 | 39 | 265138 | 1181 | 734862 | 34 |
| 27 | 258583 | 1141 | 992736 | 39 | 265847 | 1179 | 734153 | 33 |
| 28 | 259268 | 1139 | 992713 | 39 | 266555 | 1178 | 733445 | 32 |
| 29 | 259951 | 1137 | 992690 | 39 | 267261 | 1176 | 732739 | 31 |
| 30 | 260633 | 1135 | 992666 | 39 | 267967 | 1174 | 732033 | 30 |
| 31 | 9·261314 | 1133 | 9·992643 | 39 | 9·268671 | 1172 | 10·731329 | 29 |
| 32 | 261994 | 1131 | 992619 | 39 | 269375 | 1170 | 730625 | 28 |
| 33 | 262673 | 1130 | 992596 | 39 | 270077 | 1169 | 729923 | 27 |
| 34 | 263351 | 1128 | 992572 | 39 | 270779 | 1167 | 729221 | 26 |
| 35 | 264027 | 1126 | 992549 | 39 | 271479 | 1165 | 728521 | 25 |
| 36 | 264703 | 1124 | 992525 | 39 | 272178 | 1164 | 727822 | 24 |
| 37 | 265377 | 1122 | 992501 | 39 | 272876 | 1162 | 727124 | 23 |
| 38 | 266051 | 1120 | 992478 | 40 | 273573 | 1160 | 726427 | 22 |
| 39 | 266723 | 1119 | 992454 | 40 | 274269 | 1158 | 725731 | 21 |
| 40 | 267395 | 1117 | 992430 | 40 | 274964 | 1157 | 725036 | 20 |
| 41 | 9·268065 | 1115 | 9·992406 | 40 | 9·275658 | 1155 | 10·724342 | 19 |
| 42 | 268734 | 1113 | 992382 | 40 | 276351 | 1153 | 723649 | 18 |
| 43 | 269402 | 1111 | 992359 | 40 | 277043 | 1151 | 722957 | 17 |
| 44 | 270069 | 1110 | 992335 | 40 | 277734 | 1150 | 722266 | 16 |
| 45 | 270735 | 1108 | 992311 | 40 | 278424 | 1148 | 721576 | 15 |
| 46 | 271400 | 1106 | 992287 | 40 | 279113 | 1147 | 720887 | 14 |
| 47 | 272064 | 1105 | 992263 | 40 | 279801 | 1145 | 720199 | 13 |
| 48 | 272726 | 1103 | 992239 | 40 | 280488 | 1143 | 719512 | 12 |
| 49 | 273388 | 1101 | 992214 | 40 | 281174 | 1141 | 718826 | 11 |
| 50 | 274049 | 1099 | 992190 | 40 | 281858 | 1140 | 718142 | 10 |
| 51 | 9·274708 | 1098 | 9·992166 | 40 | 9·282542 | 1138 | 10·717458 | 9 |
| 52 | 275367 | 1096 | 992142 | 40 | 283225 | 1136 | 716775 | 8 |
| 53 | 276025 | 1094 | 992118 | 41 | 283907 | 1135 | 716093 | 7 |
| 54 | 276681 | 1092 | 992093 | 41 | 284588 | 1133 | 715412 | 6 |
| 55 | 277337 | 1091 | 992069 | 41 | 285268 | 1131 | 714732 | 5 |
| 56 | 277991 | 1089 | 992044 | 41 | 285947 | 1130 | 714053 | 4 |
| 57 | 278645 | 1087 | 992020 | 41 | 286624 | 1128 | 713376 | 3 |
| 58 | 279297 | 1086 | 991996 | 41 | 287301 | 1126 | 712699 | 2 |
| 59 | 279948 | 1084 | 991971 | 41 | 287977 | 1125 | 712023 | 1 |
| 60 | 280599 | 1082 | 991947 | 41 | 288652 | 1123 | 711348 | 0 |
| ′ | Cosine. | D. | Sine. | D. | Cotang. | D. | Tang. | ′ |

100° 79°

| 11° | | | | | | | | 168° |
|---|---|---|---|---|---|---|---|---|
| ′ | Sine. | D. | Cosine. | D. | Tang. | D. | Cotang. | ′ |
| 0 | 9·280599 | 1082 | 9·991947 | 41 | 9·288652 | 1123 | 10·711348 | 60 |
| 1 | 281248 | 1081 | 991922 | 41 | 289326 | 1122 | 710674 | 59 |
| 2 | 281897 | 1079 | 991897 | 41 | 289999 | 1120 | 710001 | 58 |
| 3 | 282544 | 1077 | 991873 | 41 | 290671 | 1118 | 709329 | 57 |
| 4 | 283190 | 1076 | 991848 | 41 | 291342 | 1117 | 708658 | 56 |
| 5 | 283836 | 1074 | 991823 | 41 | 292013 | 1115 | 707987 | 55 |
| 6 | 284480 | 1072 | 991799 | 41 | 292682 | 1114 | 707318 | 54 |
| 7 | 285124 | 1071 | 991774 | 42 | 293350 | 1112 | 706650 | 53 |
| 8 | 285766 | 1069 | 991749 | 42 | 294017 | 1111 | 705983 | 52 |
| 9 | 286408 | 1067 | 991724 | 42 | 294684 | 1109 | 705316 | 51 |
| 10 | 287048 | 1066 | 991699 | 42 | 295349 | 1107 | 704651 | 50 |
| 11 | 9·287688 | 1064 | 9·991674 | 42 | 9·296013 | 1106 | 10·703987 | 49 |
| 12 | 288326 | 1063 | 991649 | 42 | 296677 | 1104 | 703323 | 48 |
| 13 | 288964 | 1061 | 991624 | 42 | 297339 | 1103 | 702661 | 47 |
| 14 | 289600 | 1059 | 991599 | 42 | 298001 | 1101 | 701999 | 46 |
| 15 | 290236 | 1058 | 991574 | 42 | 298662 | 1100 | 701338 | 45 |
| 16 | 290870 | 1056 | 991549 | 42 | 299322 | 1098 | 700678 | 44 |
| 17 | 291504 | 1054 | 991524 | 42 | 299980 | 1096 | 700020 | 43 |
| 18 | 292137 | 1053 | 991498 | 42 | 300638 | 1095 | 699362 | 42 |
| 19 | 292768 | 1051 | 991473 | 42 | 301295 | 1093 | 698705 | 41 |
| 20 | 293399 | 1050 | 991448 | 42 | 301951 | 1092 | 698049 | 40 |
| 21 | 9·294029 | 1048 | 9·991422 | 42 | 9·302607 | 1090 | 10·697393 | 39 |
| 22 | 294658 | 1046 | 991397 | 42 | 303261 | 1089 | 696739 | 38 |
| 23 | 295286 | 1045 | 991372 | 43 | 303914 | 1087 | 696086 | 37 |
| 24 | 295913 | 1043 | 991346 | 43 | 304567 | 1086 | 695433 | 36 |
| 25 | 296539 | 1042 | 991321 | 43 | 305218 | 1084 | 694782 | 35 |
| 26 | 297164 | 1040 | 991295 | 43 | 305869 | 1083 | 694131 | 34 |
| 27 | 297788 | 1039 | 991270 | 43 | 306519 | 1081 | 693481 | 33 |
| 28 | 298412 | 1037 | 991244 | 43 | 307168 | 1080 | 692832 | 32 |
| 29 | 299034 | 1036 | 991218 | 43 | 307816 | 1078 | 692184 | 31 |
| 30 | 299655 | 1034 | 991193 | 43 | 308463 | 1077 | 691537 | 30 |
| 31 | 9·300276 | 1032 | 9·991167 | 43 | 9·309109 | 1075 | 10·690891 | 29 |
| 32 | 300895 | 1031 | 991141 | 43 | 309754 | 1074 | 690246 | 28 |
| 33 | 301514 | 1029 | 991115 | 43 | 310399 | 1073 | 689601 | 27 |
| 34 | 302132 | 1028 | 991090 | 43 | 311042 | 1071 | 688958 | 26 |
| 35 | 302748 | 1026 | 991064 | 43 | 311685 | 1070 | 688315 | 25 |
| 36 | 303364 | 1025 | 991038 | 43 | 312327 | 1068 | 687673 | 24 |
| 37 | 303979 | 1023 | 991012 | 43 | 312968 | 1067 | 687032 | 23 |
| 38 | 304593 | 1022 | 990986 | 43 | 313608 | 1065 | 686392 | 22 |
| 39 | 305207 | 1020 | 990960 | 43 | 314247 | 1064 | 685753 | 21 |
| 40 | 305819 | 1019 | 990934 | 44 | 314885 | 1062 | 685115 | 20 |
| 41 | 9·306430 | 1017 | 9·990908 | 44 | 9·315523 | 1061 | 10·684477 | 19 |
| 42 | 307041 | 1016 | 990882 | 44 | 316159 | 1060 | 683841 | 18 |
| 43 | 307650 | 1014 | 990855 | 44 | 316795 | 1058 | 683205 | 17 |
| 44 | 308259 | 1013 | 990829 | 44 | 317430 | 1057 | 682570 | 16 |
| 45 | 308867 | 1011 | 990803 | 44 | 318064 | 1055 | 681936 | 15 |
| 46 | 309474 | 1010 | 990777 | 44 | 318697 | 1054 | 681303 | 14 |
| 47 | 310080 | 1008 | 990750 | 44 | 319330 | 1053 | 680670 | 13 |
| 48 | 310685 | 1007 | 990724 | 44 | 319961 | 1051 | 680039 | 12 |
| 49 | 311289 | 1005 | 990697 | 44 | 320592 | 1050 | 679408 | 11 |
| 50 | 311893 | 1004 | 990671 | 44 | 321222 | 1048 | 678778 | 10 |
| 51 | 9·312495 | 1003 | 9·990645 | 44 | 9·321851 | 1047 | 10·678149 | 9 |
| 52 | 313097 | 1001 | 990618 | 44 | 322470 | 1045 | 677521 | 8 |
| 53 | 313698 | 1000 | 990591 | 44 | 323106 | 1044 | 676894 | 7 |
| 54 | 314297 | 998 | 990565 | 44 | 323733 | 1043 | 676267 | 6 |
| 55 | 314897 | 997 | 990538 | 44 | 324358 | 1041 | 675642 | 5 |
| 56 | 315495 | 996 | 990511 | 45 | 324983 | 1040 | 675017 | 4 |
| 57 | 316092 | 994 | 990485 | 45 | 325607 | 1039 | 674393 | 3 |
| 58 | 316689 | 993 | 990458 | 45 | 326231 | 1037 | 673769 | 2 |
| 59 | 317284 | 991 | 990431 | 45 | 326853 | 1036 | 673147 | 1 |
| 60 | 317879 | 990 | 990404 | 45 | 327475 | 1035 | 672525 | 0 |
| ′ | Cosine. | D. | Sine. | D. | Cotang. | D. | Tang. | ′ |
| 101° | | | | | | | | 78° |

12° 167°

| ′ | Sine. | D. | Cosine. | D. | Tang. | D. | Cotang. | ′ |
|---|---|---|---|---|---|---|---|---|
| 0 | 9·317879 | 990 | 9·990404 | 45 | 9·327475 | 1035 | 10·672525 | 60 |
| 1 | 318473 | 988 | 990378 | 45 | 328095 | 1033 | 671905 | 59 |
| 2 | 319066 | 987 | 990351 | 45 | 328715 | 1032 | 671285 | 58 |
| 3 | 319658 | 986 | 990324 | 45 | 329334 | 1030 | 670666 | 57 |
| 4 | 320249 | 984 | 990297 | 45 | 329953 | 1029 | 670047 | 56 |
| 5 | 320840 | 983 | 990270 | 45 | 330570 | 1028 | 669430 | 55 |
| 6 | 321430 | 982 | 990243 | 45 | 331187 | 1026 | 668813 | 54 |
| 7 | 322019 | 980 | 990215 | 45 | 331803 | 1025 | 668197 | 53 |
| 8 | 322607 | 979 | 990188 | 45 | 332418 | 1024 | 667582 | 52 |
| 9 | 323194 | 977 | 990161 | 45 | 333033 | 1023 | 666967 | 51 |
| 10 | 323780 | 976 | 990134 | 45 | 333646 | 1021 | 666354 | 50 |
| 11 | 9·324366 | 975 | 9·990107 | 46 | 9·334259 | 1020 | 10·665741 | 49 |
| 12 | 324950 | 973 | 990079 | 46 | 334871 | 1019 | 665129 | 48 |
| 13 | 325534 | 972 | 990052 | 46 | 335482 | 1017 | 664518 | 47 |
| 14 | 326117 | 970 | 990025 | 46 | 336093 | 1016 | 663907 | 46 |
| 15 | 326700 | 969 | 989997 | 46 | 336702 | 1015 | 663298 | 45 |
| 16 | 327281 | 968 | 989970 | 46 | 337311 | 1013 | 662689 | 44 |
| 17 | 327862 | 966 | 989942 | 46 | 337919 | 1012 | 662081 | 43 |
| 18 | 328442 | 965 | 989915 | 46 | 338527 | 1011 | 661473 | 42 |
| 19 | 329021 | 964 | 989887 | 46 | 339133 | 1010 | 660867 | 41 |
| 20 | 329599 | 962 | 989860 | 46 | 339739 | 1008 | 660261 | 40 |
| 21 | 9·330176 | 961 | 9·989832 | 46 | 9·340344 | 1007 | 10·659656 | 39 |
| 22 | 330753 | 960 | 989804 | 46 | 340948 | 1006 | 659052 | 38 |
| 23 | 331329 | 958 | 989777 | 46 | 341552 | 1004 | 658448 | 37 |
| 24 | 331903 | 957 | 989749 | 47 | 342155 | 1003 | 657845 | 36 |
| 25 | 332478 | 956 | 989721 | 47 | 342757 | 1002 | 657243 | 35 |
| 26 | 333051 | 954 | 989693 | 47 | 343358 | 1000 | 656642 | 34 |
| 27 | 333624 | 953 | 989665 | 47 | 343958 | 999 | 656042 | 33 |
| 28 | 334195 | 952 | 989637 | 47 | 344558 | 998 | 655442 | 32 |
| 29 | 334767 | 950 | 989610 | 47 | 345157 | 997 | 654843 | 31 |
| 30 | 335337 | 949 | 989582 | 47 | 345755 | 996 | 654245 | 30 |
| 31 | 9·335906 | 948 | 9·989553 | 47 | 9·346353 | 994 | 10·653647 | 29 |
| 32 | 336475 | 946 | 989525 | 47 | 346949 | 993 | 653051 | 28 |
| 33 | 337043 | 945 | 989497 | 47 | 347545 | 992 | 652455 | 27 |
| 34 | 337610 | 944 | 989469 | 47 | 348141 | 991 | 651859 | 26 |
| 35 | 338176 | 943 | 989441 | 47 | 348735 | 990 | 651265 | 25 |
| 36 | 338742 | 941 | 989413 | 47 | 349329 | 988 | 650671 | 24 |
| 37 | 339307 | 940 | 989385 | 47 | 349922 | 987 | 650078 | 23 |
| 38 | 339871 | 939 | 989356 | 47 | 350514 | 986 | 649486 | 22 |
| 39 | 340434 | 937 | 989328 | 47 | 351106 | 985 | 648894 | 21 |
| 40 | 340996 | 936 | 989300 | 47 | 351697 | 983 | 648303 | 20 |
| 41 | 9·341558 | 935 | 9·989271 | 47 | 9·352287 | 982 | 10·647713 | 19 |
| 42 | 342119 | 934 | 989243 | 47 | 352876 | 981 | 647124 | 18 |
| 43 | 342679 | 932 | 989214 | 47 | 353465 | 980 | 646535 | 17 |
| 44 | 343239 | 931 | 989186 | 47 | 354053 | 979 | 645947 | 16 |
| 45 | 343797 | 930 | 989157 | 47 | 354640 | 977 | 645360 | 15 |
| 46 | 344355 | 929 | 989128 | 48 | 355227 | 976 | 644773 | 14 |
| 47 | 344912 | 927 | 989100 | 48 | 355813 | 975 | 644187 | 13 |
| 48 | 345469 | 926 | 989071 | 48 | 356398 | 974 | 643602 | 12 |
| 49 | 346024 | 925 | 989042 | 48 | 356982 | 973 | 643018 | 11 |
| 50 | 346579 | 924 | 989014 | 48 | 357566 | 971 | 642434 | 10 |
| 51 | 9·347134 | 922 | 9·988985 | 48 | 9·358149 | 970 | 10·641851 | 9 |
| 52 | 347687 | 921 | 988956 | 48 | 358731 | 969 | 641269 | 8 |
| 53 | 348240 | 920 | 988927 | 48 | 359313 | 968 | 640687 | 7 |
| 54 | 348792 | 919 | 988898 | 48 | 359893 | 967 | 640107 | 6 |
| 55 | 349343 | 917 | 988869 | 48 | 360474 | 966 | 639526 | 5 |
| 56 | 349893 | 916 | 988840 | 48 | 361053 | 965 | 638947 | 4 |
| 57 | 350443 | 915 | 988811 | 49 | 361632 | 963 | 638368 | 3 |
| 58 | 350992 | 914 | 988782 | 49 | 362210 | 962 | 637790 | 2 |
| 59 | 351540 | 913 | 988753 | 49 | 362787 | 961 | 637213 | 1 |
| 60 | 352088 | 911 | 988724 | 49 | 363364 | 960 | 636636 | 0 |
| ′ | Cosine. | D. | Sine. | D. | Cotang. | D. | Tang. | ′ |

102° 77°

| 13° | | | | | | | | 166° |
|---|---|---|---|---|---|---|---|---|
| ′ | Sine. | D. | Cosine. | D. | Tang. | D. | Cotang. | ′ |
| 0 | 9·352088 | 911 | 9·988724 | 49 | 9·363364 | 960 | 10·636636 | 60 |
| 1 | 352635 | 910 | 988695 | 49 | 363940 | 959 | 636060 | 59 |
| 2 | 353181 | 909 | 988666 | 49 | 364515 | 958 | 635485 | 58 |
| 3 | 353726 | 908 | 988636 | 49 | 365090 | 957 | 634910 | 57 |
| 4 | 354271 | 907 | 988607 | 49 | 365664 | 955 | 634336 | 56 |
| 5 | 354815 | 905 | 988578 | 49 | 366237 | 954 | 633763 | 55 |
| 6 | 355358 | 904 | 988548 | 49 | 366810 | 953 | 633190 | 54 |
| 7 | 355901 | 903 | 988519 | 49 | 367382 | 952 | 632618 | 53 |
| 8 | 356443 | 902 | 988489 | 49 | 367953 | 951 | 632047 | 52 |
| 9 | 356984 | 901 | 988460 | 49 | 368524 | 950 | 631476 | 51 |
| 10 | 357524 | 899 | 988430 | 49 | 369094 | 949 | 630906 | 50 |
| 11 | 9·358064 | 898 | 9·988401 | 49 | 9·369663 | 948 | 10·630337 | 49 |
| 12 | 358603 | 897 | 988371 | 49 | 370232 | 946 | 629768 | 48 |
| 13 | 359141 | 896 | 988342 | 49 | 370799 | 945 | 629201 | 47 |
| 14 | 359678 | 895 | 988312 | 50 | 371367 | 944 | 628633 | 46 |
| 15 | 360215 | 893 | 988282 | 50 | 371933 | 943 | 628067 | 45 |
| 16 | 360752 | 892 | 988252 | 50 | 372499 | 942 | 627501 | 44 |
| 17 | 361287 | 891 | 988223 | 50 | 373064 | 941 | 626936 | 43 |
| 18 | 361822 | 890 | 988193 | 50 | 373629 | 940 | 626371 | 42 |
| 19 | 362356 | 889 | 988163 | 50 | 374193 | 939 | 625807 | 41 |
| 20 | 362889 | 888 | 988133 | 50 | 374756 | 938 | 625244 | 40 |
| 21 | 9·363422 | 887 | 9·988103 | 50 | 9·375319 | 937 | 10·624681 | 39 |
| 22 | 363954 | 885 | 988073 | 50 | 375881 | 935 | 624119 | 38 |
| 23 | 364485 | 884 | 988043 | 50 | 376442 | 934 | 623558 | 37 |
| 24 | 365016 | 883 | 988013 | 50 | 377003 | 933 | 622997 | 36 |
| 25 | 365546 | 882 | 987983 | 50 | 377563 | 932 | 622437 | 35 |
| 26 | 366075 | 881 | 987953 | 50 | 378122 | 931 | 621878 | 34 |
| 27 | 366604 | 880 | 987922 | 50 | 378681 | 930 | 621319 | 33 |
| 28 | 367131 | 879 | 987892 | 50 | 379239 | 929 | 620761 | 32 |
| 29 | 367659 | 877 | 987862 | 50 | 379797 | 928 | 620203 | 31 |
| 30 | 368185 | 876 | 987832 | 51 | 380354 | 927 | 619646 | 30 |
| 31 | 9·368711 | 875 | 9·987801 | 51 | 9·380910 | 926 | 10·619090 | 29 |
| 32 | 369236 | 874 | 987771 | 51 | 381466 | 925 | 618534 | 28 |
| 33 | 369761 | 873 | 987740 | 51 | 382020 | 924 | 617980 | 27 |
| 34 | 370285 | 872 | 987710 | 51 | 382575 | 923 | 617425 | 26 |
| 35 | 370808 | 871 | 987679 | 51 | 383129 | 922 | 616871 | 25 |
| 36 | 371330 | 870 | 987649 | 51 | 383682 | 921 | 616318 | 24 |
| 37 | 371852 | 869 | 987618 | 51 | 384234 | 920 | 615766 | 23 |
| 38 | 372373 | 867 | 987588 | 51 | 384786 | 919 | 615214 | 22 |
| 39 | 372894 | 866 | 987557 | 51 | 385337 | 918 | 614663 | 21 |
| 40 | 373414 | 865 | 987526 | 51 | 385888 | 917 | 614112 | 20 |
| 41 | 9·373933 | 864 | 9·987496 | 51 | 9·386438 | 915 | 10·613562 | 19 |
| 42 | 374452 | 863 | 987465 | 51 | 386987 | 914 | 613013 | 18 |
| 43 | 374970 | 862 | 987434 | 51 | 387536 | 913 | 612464 | 17 |
| 44 | 375487 | 861 | 987403 | 52 | 388084 | 912 | 611916 | 16 |
| 45 | 376003 | 860 | 987372 | 52 | 388631 | 911 | 611369 | 15 |
| 46 | 376519 | 859 | 987341 | 52 | 389178 | 910 | 610822 | 14 |
| 47 | 377035 | 858 | 987310 | 52 | 389724 | 909 | 610276 | 13 |
| 48 | 377549 | 857 | 987279 | 52 | 390270 | 908 | 609730 | 12 |
| 49 | 378063 | 856 | 987248 | 52 | 390815 | 907 | 609185 | 11 |
| 50 | 378577 | 854 | 987217 | 52 | 391360 | 906 | 608640 | 10 |
| 51 | 9·379089 | 853 | 9·987186 | 52 | 9·391903 | 905 | 10·608097 | 9 |
| 52 | 379601 | 852 | 987155 | 52 | 392447 | 904 | 607553 | 8 |
| 53 | 380113 | 851 | 987124 | 52 | 392989 | 903 | 607011 | 7 |
| 54 | 380624 | 850 | 987092 | 52 | 393531 | 902 | 606469 | 6 |
| 55 | 381134 | 849 | 987061 | 52 | 394073 | 901 | 605927 | 5 |
| 56 | 381643 | 848 | 987030 | 52 | 394614 | 900 | 605386 | 4 |
| 57 | 382152 | 847 | 986998 | 52 | 395154 | 899 | 604846 | 3 |
| 58 | 382661 | 846 | 986967 | 52 | 395694 | 898 | 604306 | 2 |
| 59 | 383168 | 845 | 986936 | 52 | 396233 | 897 | 603767 | 1 |
| 60 | 383675 | 844 | 986904 | 52 | 396771 | 896 | 603229 | 0 |
| ′ | Cosine. | D. | Sine. | D. | Cotang. | D. | Tang. | ′ |
| 103° | | | | | | | | 76° |

14° 165°

| ′ | Sine. | D. | Cosine. | D. | Tang. | D. | Cotang. | ′ |
|---|---|---|---|---|---|---|---|---|
| 0 | 9·383675 | 844 | 9·986904 | 52 | 9·396771 | 896 | 10·603229 | 60 |
| 1 | 384182 | 843 | 986873 | 53 | 397309 | 896 | 602691 | 59 |
| 2 | 384687 | 842 | 986841 | 53 | 397846 | 895 | 602154 | 58 |
| 3 | 385192 | 841 | 986809 | 53 | 398383 | 894 | 601617 | 57 |
| 4 | 385697 | 840 | 986778 | 53 | 398919 | 893 | 601081 | 56 |
| 5 | 386201 | 839 | 986746 | 53 | 399455 | 892 | 600545 | 55 |
| 6 | 386704 | 838 | 986714 | 53 | 399990 | 891 | 600010 | 54 |
| 7 | 387207 | 837 | 986683 | 53 | 400524 | 890 | 599476 | 53 |
| 8 | 387709 | 836 | 986651 | 53 | 401058 | 889 | 598942 | 52 |
| 9 | 388210 | 835 | 986619 | 53 | 401591 | 888 | 598409 | 51 |
| 10 | 388711 | 834 | 986587 | 53 | 402124 | 887 | 597876 | 50 |
| 11 | 9·389211 | 833 | 9·986555 | 53 | 9·402656 | 886 | 10·597344 | 49 |
| 12 | 389711 | 832 | 986523 | 53 | 403187 | 885 | 596813 | 48 |
| 13 | 390210 | 831 | 986491 | 53 | 403718 | 884 | 596282 | 47 |
| 14 | 390708 | 830 | 986459 | 53 | 404249 | 883 | 595751 | 46 |
| 15 | 391206 | 828 | 986427 | 53 | 404778 | 882 | 595222 | 45 |
| 16 | 391703 | 827 | 986395 | 53 | 405308 | 881 | 594692 | 44 |
| 17 | 392199 | 826 | 986363 | 54 | 405836 | 880 | 594164 | 43 |
| 18 | 392695 | 825 | 986331 | 54 | 406364 | 879 | 593636 | 42 |
| 19 | 393191 | 824 | 986299 | 54 | 406892 | 878 | 593108 | 41 |
| 20 | 393685 | 823 | 986266 | 54 | 407419 | 877 | 592581 | 40 |
| 21 | 9·394179 | 822 | 9·986234 | 54 | 9·407945 | 876 | 10·592055 | 39 |
| 22 | 394673 | 821 | 986202 | 54 | 408471 | 875 | 591529 | 38 |
| 23 | 395166 | 820 | 986169 | 54 | 408996 | 874 | 591004 | 37 |
| 24 | 395658 | 819 | 986137 | 54 | 409521 | 874 | 590479 | 36 |
| 25 | 396150 | 818 | 986104 | 54 | 410045 | 873 | 589955 | 35 |
| 26 | 396641 | 817 | 986072 | 54 | 410569 | 872 | 589431 | 34 |
| 27 | 397132 | 817 | 986039 | 54 | 411092 | 871 | 588908 | 33 |
| 28 | 397621 | 816 | 986007 | 54 | 411615 | 870 | 588385 | 32 |
| 29 | 398111 | 815 | 985974 | 54 | 412137 | 869 | 587863 | 31 |
| 30 | 398600 | 814 | 985942 | 54 | 412658 | 868 | 587342 | 30 |
| 31 | 9·399088 | 813 | 9·985909 | 55 | 9·413179 | 867 | 10·586821 | 29 |
| 32 | 399575 | 812 | 985876 | 55 | 413699 | 866 | 586301 | 28 |
| 33 | 400062 | 811 | 985843 | 55 | 414219 | 865 | 585781 | 27 |
| 34 | 400549 | 810 | 985811 | 55 | 414738 | 864 | 585262 | 26 |
| 35 | 401035 | 809 | 985778 | 55 | 415257 | 864 | 584743 | 25 |
| 36 | 401520 | 808 | 985745 | 55 | 415775 | 863 | 584225 | 24 |
| 37 | 402005 | 807 | 985712 | 55 | 416293 | 862 | 583707 | 23 |
| 38 | 402489 | 806 | 985679 | 55 | 416810 | 861 | 583190 | 22 |
| 39 | 402972 | 805 | 985646 | 55 | 417326 | 860 | 582674 | 21 |
| 40 | 403455 | 804 | 985613 | 55 | 417842 | 859 | 582158 | 20 |
| 41 | 9·403938 | 803 | 9·985580 | 55 | 9·418358 | 858 | 10·581642 | 19 |
| 42 | 404420 | 802 | 985547 | 55 | 418873 | 857 | 581127 | 18 |
| 43 | 404901 | 801 | 985514 | 55 | 419387 | 856 | 580613 | 17 |
| 44 | 405382 | 800 | 985480 | 55 | 419901 | 855 | 580099 | 16 |
| 45 | 405862 | 799 | 985447 | 55 | 420415 | 855 | 579585 | 15 |
| 46 | 406341 | 798 | 985414 | 56 | 420927 | 854 | 579073 | 14 |
| 47 | 406820 | 797 | 985381 | 56 | 421440 | 853 | 578560 | 13 |
| 48 | 407299 | 796 | 985347 | 56 | 421952 | 852 | 578048 | 12 |
| 49 | 407777 | 795 | 985314 | 56 | 422463 | 851 | 577537 | 11 |
| 50 | 408254 | 794 | 985280 | 56 | 422974 | 850 | 577026 | 10 |
| 51 | 9·408731 | 794 | 9·985247 | 56 | 9·423484 | 849 | 10·576516 | 9 |
| 52 | 409207 | 793 | 985213 | 56 | 423993 | 848 | 576007 | 8 |
| 53 | 409682 | 792 | 985180 | 56 | 424503 | 848 | 575497 | 7 |
| 54 | 410157 | 791 | 985146 | 56 | 425011 | 847 | 574989 | 6 |
| 55 | 410632 | 790 | 985113 | 56 | 425519 | 846 | 574481 | 5 |
| 56 | 411106 | 789 | 985079 | 56 | 426027 | 845 | 573973 | 4 |
| 57 | 411579 | 788 | 985045 | 56 | 426534 | 844 | 573466 | 3 |
| 58 | 412052 | 787 | 985011 | 56 | 427041 | 843 | 572959 | 2 |
| 59 | 412524 | 786 | 984978 | 56 | 427547 | 843 | 572453 | 1 |
| 60 | 412996 | 785 | 984944 | 56 | 428052 | 842 | 571948 | 0 |
| ′ | Cosine. | D. | Sine. | D. | Cotang. | D. | Tang. | ′ |

104° 75°

15° 164°

| ′ | Sine. | D. | Cosine. | D. | Tang. | D. | Cotang. | ′ |
|---|---|---|---|---|---|---|---|---|
| 0 | 9·412996 | 785 | 9·984944 | 57 | 9·428052 | 842 | 10·571948 | 60 |
| 1 | 413467 | 784 | 984910 | 57 | 428558 | 841 | 571442 | 59 |
| 2 | 413938 | 783 | 984876 | 57 | 429062 | 840 | 570938 | 58 |
| 3 | 414408 | 783 | 984842 | 57 | 429566 | 839 | 570434 | 57 |
| 4 | 414878 | 782 | 984808 | 57 | 430070 | 838 | 569930 | 56 |
| 5 | 415347 | 781 | 984774 | 57 | 430573 | 838 | 569427 | 55 |
| 6 | 415815 | 780 | 984740 | 57 | 431075 | 837 | 568925 | 54 |
| 7 | 416283 | 779 | 984706 | 57 | 431577 | 836 | 568423 | 53 |
| 8 | 416751 | 778 | 984672 | 57 | 432079 | 835 | 567921 | 52 |
| 9 | 417217 | 777 | 984638 | 57 | 432580 | 834 | 567420 | 51 |
| 10 | 417684 | 776 | 984603 | 57 | 433080 | 833 | 566920 | 50 |
| 11 | 9·418150 | 775 | 9·984569 | 57 | 9·433580 | 832 | 10·566420 | 49 |
| 12 | 418615 | 774 | 984535 | 57 | 434080 | 832 | 565920 | 48 |
| 13 | 419079 | 773 | 984500 | 57 | 434579 | 831 | 565421 | 47 |
| 14 | 419544 | 773 | 984466 | 57 | 435078 | 830 | 564922 | 46 |
| 15 | 420007 | 772 | 984432 | 58 | 435576 | 829 | 564424 | 45 |
| 16 | 420470 | 771 | 984397 | 58 | 436073 | 828 | 563927 | 44 |
| 17 | 420933 | 770 | 984363 | 58 | 436570 | 828 | 563430 | 43 |
| 18 | 421395 | 769 | 984328 | 58 | 437067 | 827 | 562933 | 42 |
| 19 | 421857 | 768 | 984294 | 58 | 437563 | 826 | 562437 | 41 |
| 20 | 422318 | 767 | 984259 | 58 | 438059 | 825 | 561941 | 40 |
| 21 | 9·422778 | 767 | 9·984224 | 58 | 9·438554 | 824 | 10·561446 | 39 |
| 22 | 423238 | 766 | 984190 | 58 | 439048 | 823 | 560952 | 38 |
| 23 | 423697 | 765 | 984155 | 58 | 439543 | 823 | 560457 | 37 |
| 24 | 424156 | 764 | 984120 | 58 | 440036 | 822 | 559964 | 36 |
| 25 | 424615 | 763 | 984085 | 58 | 440529 | 821 | 559471 | 35 |
| 26 | 425073 | 762 | 984050 | 58 | 441022 | 820 | 558978 | 34 |
| 27 | 425530 | 761 | 984015 | 58 | 441514 | 819 | 558486 | 33 |
| 28 | 425987 | 760 | 983981 | 58 | 442006 | 819 | 557994 | 32 |
| 29 | 426443 | 760 | 983946 | 58 | 442497 | 818 | 557503 | 31 |
| 30 | 426899 | 759 | 983911 | 58 | 442988 | 817 | 557012 | 30 |
| 31 | 9·427354 | 758 | 9·983875 | 58 | 9·443479 | 816 | 10·556521 | 29 |
| 32 | 427809 | 757 | 983840 | 59 | 443968 | 816 | 556032 | 28 |
| 33 | 428263 | 756 | 983805 | 59 | 444458 | 815 | 555542 | 27 |
| 34 | 428717 | 755 | 983770 | 59 | 444947 | 814 | 555053 | 26 |
| 35 | 429170 | 754 | 983735 | 59 | 445435 | 813 | 554565 | 25 |
| 36 | 429623 | 753 | 983700 | 59 | 445923 | 812 | 554077 | 24 |
| 37 | 430075 | 752 | 983664 | 59 | 446411 | 812 | 553589 | 23 |
| 38 | 430527 | 752 | 983629 | 59 | 446898 | 811 | 553102 | 22 |
| 39 | 430978 | 751 | 983594 | 59 | 447384 | 810 | 552616 | 21 |
| 40 | 431429 | 750 | 983558 | 59 | 447870 | 809 | 552130 | 20 |
| 41 | 9·431879 | 749 | 9·983523 | 59 | 9·448356 | 809 | 10·551644 | 19 |
| 42 | 432329 | 749 | 983487 | 59 | 448841 | 808 | 551159 | 18 |
| 43 | 432778 | 748 | 983452 | 59 | 449326 | 807 | 550674 | 17 |
| 44 | 433226 | 747 | 983416 | 59 | 449810 | 806 | 550190 | 16 |
| 45 | 433675 | 746 | 983381 | 59 | 450294 | 806 | 549706 | 15 |
| 46 | 434122 | 745 | 983345 | 59 | 450777 | 805 | 549223 | 14 |
| 47 | 434569 | 744 | 983309 | 59 | 451260 | 804 | 548740 | 13 |
| 48 | 435016 | 744 | 983273 | 60 | 451743 | 803 | 548257 | 12 |
| 49 | 435462 | 743 | 983238 | 60 | 452225 | 802 | 547775 | 11 |
| 50 | 435908 | 742 | 983202 | 60 | 452706 | 802 | 547294 | 10 |
| 51 | 9·436353 | 741 | 9·983166 | 60 | 9·453187 | 801 | 10·546813 | 9 |
| 52 | 436798 | 740 | 983130 | 60 | 453668 | 800 | 546330 | 8 |
| 53 | 437242 | 740 | 983094 | 60 | 454148 | 799 | 545852 | 7 |
| 54 | 437686 | 739 | 983058 | 60 | 454628 | 799 | 545372 | 6 |
| 55 | 438129 | 738 | 983022 | 60 | 455107 | 798 | 544893 | 5 |
| 56 | 438572 | 737 | 982986 | 60 | 455586 | 797 | 544414 | 4 |
| 57 | 439014 | 736 | 982950 | 60 | 456064 | 796 | 543936 | 3 |
| 58 | 439456 | 736 | 982914 | 60 | 456542 | 796 | 543458 | 2 |
| 59 | 439897 | 735 | 982878 | 60 | 457019 | 795 | 542981 | 1 |
| 60 | 440338 | 734 | 982842 | 60 | 457496 | 794 | 542504 | 0 |
| ′ | Cosine. | D. | Sine. | D. | Cotang. | D. | Tang. | ′ |

105° 74°

16° 163°

| ′ | Sine. | D. | Cosine. | D. | Tang. | D. | Cotang. | ′ |
|---|---|---|---|---|---|---|---|---|
| 0 | 9·440338 | 734 | 9·982842 | 60 | 9·457496 | 794 | 10·542504 | 60 |
| 1 | 440778 | 733 | 982805 | 60 | 457973 | 793 | 542027 | 59 |
| 2 | 441218 | 732 | 982769 | 61 | 458449 | 793 | 541551 | 58 |
| 3 | 441658 | 731 | 982733 | 61 | 458925 | 792 | 541075 | 57 |
| 4 | 442096 | 731 | 982696 | 61 | 459400 | 791 | 540600 | 56 |
| 5 | 442535 | 730 | 982660 | 61 | 459875 | 790 | 540125 | 55 |
| 6 | 442973 | 729 | 982624 | 61 | 460349 | 790 | 539651 | 54 |
| 7 | 443410 | 728 | 982587 | 61 | 460823 | 789 | 539177 | 53 |
| 8 | 443847 | 727 | 982551 | 61 | 461297 | 788 | 538703 | 52 |
| 9 | 444284 | 727 | 982514 | 61 | 461770 | 788 | 538230 | 51 |
| 10 | 444720 | 726 | 982477 | 61 | 462242 | 787 | 537758 | 50 |
| 11 | 9·445155 | 725 | 9·982441 | 61 | 9·462715 | 786 | 10·537285 | 49 |
| 12 | 445590 | 724 | 982404 | 61 | 463186 | 785 | 536814 | 48 |
| 13 | 446025 | 723 | 982367 | 61 | 463658 | 785 | 536342 | 47 |
| 14 | 446459 | 723 | 982331 | 61 | 464128 | 784 | 535872 | 46 |
| 15 | 446893 | 722 | 982294 | 61 | 464599 | 783 | 535401 | 45 |
| 16 | 447326 | 721 | 982257 | 61 | 465069 | 783 | 534931 | 44 |
| 17 | 447759 | 720 | 982220 | 62 | 465539 | 782 | 534461 | 43 |
| 18 | 448191 | 720 | 982183 | 62 | 466008 | 781 | 533992 | 42 |
| 19 | 448623 | 719 | 982146 | 62 | 466477 | 780 | 533523 | 41 |
| 20 | 449054 | 718 | 982109 | 62 | 466945 | 780 | 533055 | 40 |
| 21 | 9·449485 | 717 | 9·982072 | 62 | 9·467413 | 779 | 10·532587 | 39 |
| 22 | 449915 | 716 | 982035 | 62 | 467880 | 778 | 532120 | 38 |
| 23 | 450345 | 716 | 981998 | 62 | 468347 | 778 | 531653 | 37 |
| 24 | 450775 | 715 | 981961 | 62 | 468814 | 777 | 531186 | 36 |
| 25 | 451204 | 714 | 981924 | 62 | 469280 | 776 | 530720 | 35 |
| 26 | 451632 | 713 | 981886 | 62 | 469746 | 775 | 530254 | 34 |
| 27 | 452060 | 713 | 981849 | 62 | 470211 | 775 | 529789 | 33 |
| 28 | 452488 | 712 | 981812 | 62 | 470676 | 774 | 529324 | 32 |
| 29 | 452915 | 711 | 981774 | 62 | 471141 | 773 | 528859 | 31 |
| 30 | 453342 | 710 | 981737 | 62 | 471605 | 773 | 528395 | 30 |
| 31 | 9·453768 | 710 | 9·981700 | 63 | 9·472069 | 772 | 10·527931 | 29 |
| 32 | 454194 | 709 | 981662 | 63 | 472532 | 771 | 527468 | 28 |
| 33 | 454619 | 708 | 981625 | 63 | 472995 | 771 | 527005 | 27 |
| 34 | 455044 | 707 | 981587 | 63 | 473457 | 770 | 526543 | 26 |
| 35 | 455469 | 707 | 981549 | 63 | 473919 | 769 | 526081 | 25 |
| 36 | 455893 | 706 | 981512 | 63 | 474381 | 769 | 525619 | 24 |
| 37 | 456316 | 705 | 981474 | 63 | 474842 | 768 | 525158 | 23 |
| 38 | 456739 | 704 | 981436 | 63 | 475303 | 767 | 524697 | 22 |
| 39 | 457162 | 704 | 981399 | 63 | 475763 | 767 | 524237 | 21 |
| 40 | 457584 | 703 | 981361 | 63 | 476223 | 766 | 523777 | 20 |
| 41 | 9·458006 | 702 | 9·981323 | 63 | 9·476683 | 765 | 10·523317 | 19 |
| 42 | 458427 | 701 | 981285 | 63 | 477142 | 765 | 522858 | 18 |
| 43 | 458848 | 701 | 981247 | 63 | 477601 | 764 | 522399 | 17 |
| 44 | 459268 | 700 | 981209 | 63 | 478059 | 763 | 521941 | 16 |
| 45 | 459688 | 699 | 981171 | 63 | 478517 | 763 | 521483 | 15 |
| 46 | 460108 | 698 | 981133 | 64 | 478975 | 762 | 521025 | 14 |
| 47 | 460527 | 698 | 981095 | 64 | 479432 | 761 | 520568 | 13 |
| 48 | 460946 | 697 | 981057 | 64 | 479889 | 761 | 520111 | 12 |
| 49 | 461364 | 696 | 981019 | 64 | 480345 | 760 | 519655 | 11 |
| 50 | 461782 | 695 | 980981 | 64 | 480801 | 759 | 519199 | 10 |
| 51 | 9·462199 | 695 | 9·980942 | 64 | 9·481257 | 759 | 10·518743 | 9 |
| 52 | 462616 | 694 | 980904 | 64 | 481712 | 758 | 518288 | 8 |
| 53 | 463032 | 693 | 980866 | 64 | 482167 | 757 | 517833 | 7 |
| 54 | 463448 | 693 | 980827 | 64 | 482621 | 757 | 517379 | 6 |
| 55 | 463864 | 692 | 980789 | 64 | 483075 | 756 | 516925 | 5 |
| 56 | 464279 | 691 | 980750 | 64 | 483529 | 755 | 516471 | 4 |
| 57 | 464694 | 690 | 980712 | 64 | 483982 | 755 | 516018 | 3 |
| 58 | 465108 | 690 | 980673 | 64 | 484435 | 754 | 515565 | 2 |
| 59 | 465522 | 689 | 980635 | 64 | 484887 | 753 | 515113 | 1 |
| 60 | 465935 | 688 | 980596 | 64 | 485339 | 753 | 514661 | 0 |
| ′ | Cosine. | D. | Sine. | D. | Cotang. | D. | Tang. | ′ |

106° 73°

17° 162°

| ′ | Sine. | D. | Cosine. | D. | Tang. | D. | Cotang. | ′ |
|---|---|---|---|---|---|---|---|---|
| 0 | 9·465935 | 688 | 9·980596 | 64 | 9·485339 | 755 | 10·514661 | 60 |
| 1 | 466348 | 688 | 980558 | 64 | 485791 | 752 | 514209 | 59 |
| 2 | 466761 | 687 | 980519 | 65 | 486242 | 751 | 513758 | 58 |
| 3 | 467173 | 686 | 980480 | 65 | 486693 | 751 | 513307 | 57 |
| 4 | 467585 | 685 | 980442 | 65 | 487143 | 750 | 512857 | 56 |
| 5 | 467996 | 685 | 980403 | 65 | 487593 | 749 | 512407 | 55 |
| 6 | 468407 | 684 | 980364 | 65 | 488043 | 749 | 511957 | 54 |
| 7 | 468817 | 683 | 980325 | 65 | 488492 | 748 | 511508 | 53 |
| 8 | 469227 | 683 | 980286 | 65 | 488941 | 747 | 511059 | 52 |
| 9 | 469637 | 682 | 980247 | 65 | 489390 | 747 | 510610 | 51 |
| 10 | 470046 | 681 | 980208 | 65 | 489838 | 746 | 510162 | 50 |
| 11 | 9·470455 | 680 | 9·980169 | 65 | 9·490286 | 746 | 10·509714 | 49 |
| 12 | 470863 | 680 | 980130 | 65 | 490733 | 745 | 509267 | 48 |
| 13 | 471271 | 679 | 980091 | 65 | 491180 | 744 | 508820 | 47 |
| 14 | 471679 | 678 | 980052 | 65 | 491627 | 744 | 508373 | 46 |
| 15 | 472086 | 678 | 980012 | 65 | 492073 | 743 | 507927 | 45 |
| 16 | 472492 | 677 | 979973 | 65 | 492519 | 743 | 507481 | 44 |
| 17 | 472898 | 676 | 979934 | 66 | 492965 | 742 | 507035 | 43 |
| 18 | 473304 | 676 | 979895 | 66 | 493410 | 741 | 506590 | 42 |
| 19 | 473710 | 675 | 979855 | 66 | 493854 | 740 | 506146 | 41 |
| 20 | 474115 | 674 | 979816 | 66 | 494299 | 740 | 505701 | 40 |
| 21 | 9·474519 | 674 | 9·979776 | 66 | 9·494743 | 740 | 10·505257 | 39 |
| 22 | 474923 | 673 | 979737 | 66 | 495186 | 739 | 504814 | 38 |
| 23 | 475327 | 672 | 979697 | 66 | 495630 | 738 | 504370 | 37 |
| 24 | 475730 | 672 | 979658 | 66 | 496073 | 737 | 503927 | 36 |
| 25 | 476133 | 671 | 979618 | 66 | 496515 | 737 | 503485 | 35 |
| 26 | 476536 | 670 | 979579 | 66 | 496957 | 736 | 503043 | 34 |
| 27 | 476938 | 669 | 979539 | 66 | 497399 | 736 | 502601 | 33 |
| 28 | 477340 | 669 | 979499 | 66 | 497841 | 735 | 502159 | 32 |
| 29 | 477741 | 668 | 979459 | 66 | 498282 | 734 | 501718 | 31 |
| 30 | 478142 | 667 | 979420 | 66 | 498722 | 734 | 501278 | 30 |
| 31 | 9·478542 | 667 | 9·979380 | 66 | 9·499163 | 733 | 10·500837 | 29 |
| 32 | 478942 | 666 | 979340 | 66 | 499603 | 733 | 500397 | 28 |
| 33 | 479342 | 665 | 979300 | 67 | 500042 | 732 | 499958 | 27 |
| 34 | 479741 | 665 | 979260 | 67 | 500481 | 731 | 499519 | 26 |
| 35 | 480140 | 664 | 979220 | 67 | 500920 | 731 | 499080 | 25 |
| 36 | 480539 | 663 | 979180 | 67 | 501359 | 730 | 498641 | 24 |
| 37 | 480937 | 663 | 979140 | 67 | 501797 | 730 | 498203 | 23 |
| 38 | 481334 | 662 | 979100 | 67 | 502235 | 729 | 497765 | 22 |
| 39 | 481731 | 661 | 979059 | 67 | 502672 | 728 | 497328 | 21 |
| 40 | 482128 | 661 | 979019 | 67 | 503109 | 728 | 496891 | 20 |
| 41 | 9·482525 | 660 | 9·978979 | 67 | 9·503546 | 727 | 10·496454 | 19 |
| 42 | 482921 | 659 | 978939 | 67 | 503982 | 727 | 496018 | 18 |
| 43 | 483316 | 659 | 978898 | 67 | 504418 | 726 | 495582 | 17 |
| 44 | 483712 | 658 | 978858 | 67 | 504854 | 725 | 495146 | 16 |
| 45 | 484107 | 657 | 978817 | 67 | 505289 | 725 | 494711 | 15 |
| 46 | 484501 | 657 | 978777 | 67 | 505724 | 724 | 494276 | 14 |
| 47 | 484895 | 656 | 978737 | 67 | 506159 | 724 | 493841 | 13 |
| 48 | 485289 | 655 | 978696 | 68 | 506593 | 723 | 493407 | 12 |
| 49 | 485682 | 655 | 978655 | 68 | 507027 | 722 | 492973 | 11 |
| 50 | 486075 | 654 | 978615 | 68 | 507460 | 722 | 492540 | 10 |
| 51 | 9·486467 | 653 | 9·978574 | 68 | 9·507893 | 721 | 10·492107 | 9 |
| 52 | 486860 | 653 | 978533 | 68 | 508326 | 721 | 491674 | 8 |
| 53 | 487251 | 652 | 978493 | 68 | 508759 | 720 | 491241 | 7 |
| 54 | 487643 | 651 | 978452 | 68 | 509191 | 719 | 490809 | 6 |
| 55 | 488034 | 651 | 978411 | 68 | 509622 | 719 | 490378 | 5 |
| 56 | 488424 | 650 | 978370 | 68 | 510054 | 718 | 489946 | 4 |
| 57 | 488814 | 650 | 978329 | 68 | 510485 | 718 | 489515 | 3 |
| 58 | 489204 | 649 | 978288 | 68 | 510916 | 717 | 489084 | 2 |
| 59 | 489593 | 648 | 978247 | 68 | 511346 | 716 | 488654 | 1 |
| 60 | 489982 | 648 | 978206 | 68 | 511776 | 716 | 488224 | 0 |
| ′ | Cosine. | D. | Sine. | D. | Cotang. | D. | Tang. | ′ |

107° 72°

18° 161°

| ′ | Sine. | D. | Cosine. | D. | Tang. | D. | Cotang. | ′ |
|---|---|---|---|---|---|---|---|---|
| 0 | 9·489982 | 648 | 9·978206 | 68 | 9·511776 | 716 | 10·488224 | 60 |
| 1 | 490371 | 648 | 978165 | 68 | 512206 | 716 | 487794 | 59 |
| 2 | 490759 | 647 | 978124 | 68 | 512635 | 715 | 487365 | 58 |
| 3 | 491147 | 646 | 978083 | 69 | 513064 | 714 | 486936 | 57 |
| 4 | 491535 | 646 | 978042 | 69 | 513493 | 714 | 486507 | 56 |
| 5 | 491922 | 645 | 978001 | 69 | 513921 | 713 | 486079 | 55 |
| 6 | 492308 | 644 | 977959 | 69 | 514349 | 713 | 485651 | 54 |
| 7 | 492695 | 644 | 977918 | 69 | 514777 | 712 | 485223 | 53 |
| 8 | 493081 | 643 | 977877 | 69 | 515204 | 712 | 484796 | 52 |
| 9 | 493466 | 642 | 977835 | 69 | 515631 | 711 | 484369 | 51 |
| 10 | 493851 | 642 | 977794 | 69 | 516057 | 710 | 483943 | 50 |
| 11 | 9·494236 | 641 | 9·977752 | 69 | 9·516484 | 710 | 10·483516 | 49 |
| 12 | 494621 | 641 | 977711 | 69 | 516910 | 709 | 483090 | 48 |
| 13 | 495005 | 640 | 977669 | 69 | 517335 | 709 | 482665 | 47 |
| 14 | 495388 | 639 | 977628 | 69 | 517761 | 708 | 482239 | 46 |
| 15 | 495772 | 639 | 977586 | 69 | 518186 | 708 | 481814 | 45 |
| 16 | 496154 | 638 | 977544 | 70 | 518610 | 707 | 481390 | 44 |
| 17 | 496537 | 637 | 977503 | 70 | 519034 | 706 | 480966 | 43 |
| 18 | 496919 | 637 | 977461 | 70 | 519458 | 706 | 480542 | 42 |
| 19 | 497301 | 636 | 977419 | 70 | 519882 | 705 | 480118 | 41 |
| 20 | 497682 | 636 | 977377 | 70 | 520305 | 705 | 479695 | 40 |
| 21 | 9·498064 | 635 | 9·977335 | 70 | 9·520728 | 704 | 10·479272 | 39 |
| 22 | 498444 | 634 | 977293 | 70 | 521151 | 703 | 478849 | 38 |
| 23 | 498825 | 634 | 977251 | 70 | 521573 | 703 | 478427 | 37 |
| 24 | 499204 | 633 | 977209 | 70 | 521995 | 703 | 478005 | 36 |
| 25 | 499584 | 632 | 977167 | 70 | 522417 | 702 | 477583 | 35 |
| 26 | 499963 | 632 | 977125 | 70 | 522838 | 702 | 477162 | 34 |
| 27 | 500342 | 631 | 977083 | 70 | 523259 | 701 | 476741 | 33 |
| 28 | 500721 | 631 | 977041 | 70 | 523680 | 701 | 476320 | 32 |
| 29 | 501099 | 630 | 976999 | 70 | 524100 | 700 | 475900 | 31 |
| 30 | 501476 | 629 | 976957 | 70 | 524520 | 699 | 475480 | 30 |
| 31 | 9·501854 | 629 | 9·976914 | 70 | 9·524940 | 699 | 10·475060 | 29 |
| 32 | 502231 | 628 | 976872 | 71 | 525359 | 698 | 474641 | 28 |
| 33 | 502607 | 628 | 976830 | 71 | 525778 | 698 | 474222 | 27 |
| 34 | 502984 | 627 | 976787 | 71 | 526197 | 697 | 473803 | 26 |
| 35 | 503360 | 626 | 976745 | 71 | 526615 | 697 | 473385 | 25 |
| 36 | 503735 | 626 | 976702 | 71 | 527033 | 696 | 472967 | 24 |
| 37 | 504110 | 625 | 976660 | 71 | 527451 | 696 | 472549 | 23 |
| 38 | 504485 | 625 | 976617 | 71 | 527868 | 695 | 472132 | 22 |
| 39 | 504860 | 624 | 976574 | 71 | 528285 | 695 | 471715 | 21 |
| 40 | 505234 | 623 | 976532 | 71 | 528702 | 694 | 471298 | 20 |
| 41 | 9·505608 | 623 | 9·976489 | 71 | 9·529119 | 693 | 10·470881 | 19 |
| 42 | 505981 | 622 | 976446 | 71 | 529535 | 693 | 470465 | 18 |
| 43 | 506354 | 622 | 976404 | 71 | 529951 | 693 | 470049 | 17 |
| 44 | 506727 | 621 | 976361 | 71 | 530366 | 692 | 469634 | 16 |
| 45 | 507099 | 620 | 976318 | 71 | 530781 | 691 | 469219 | 15 |
| 46 | 507471 | 620 | 976275 | 71 | 531196 | 691 | 468804 | 14 |
| 47 | 507843 | 619 | 976232 | 72 | 531611 | 690 | 468389 | 13 |
| 48 | 508214 | 619 | 976189 | 72 | 532025 | 690 | 467975 | 12 |
| 49 | 508585 | 618 | 976146 | 72 | 532439 | 689 | 467561 | 11 |
| 50 | 508956 | 618 | 976103 | 72 | 532853 | 689 | 467147 | 10 |
| 51 | 9·509326 | 617 | 9·976060 | 72 | 9·533266 | 688 | 10·466734 | 9 |
| 52 | 509696 | 616 | 976017 | 72 | 533679 | 688 | 466321 | 8 |
| 53 | 510065 | 616 | 975974 | 72 | 534092 | 687 | 465908 | 7 |
| 54 | 510434 | 615 | 975930 | 72 | 534504 | 687 | 465496 | 6 |
| 55 | 510803 | 615 | 975887 | 72 | 534916 | 686 | 465084 | 5 |
| 56 | 511172 | 614 | 975844 | 72 | 535328 | 686 | 464672 | 4 |
| 57 | 511540 | 613 | 975800 | 72 | 535739 | 685 | 464261 | 3 |
| 58 | 511907 | 613 | 975757 | 72 | 536150 | 685 | 463850 | 2 |
| 59 | 512275 | 612 | 975714 | 72 | 536561 | 684 | 463439 | 1 |
| 60 | 512642 | 612 | 975670 | 72 | 536972 | 684 | 463028 | 0 |
| ′ | Cosine. | D. | Sine. | D. | Cotang. | D. | Tang. | ′ |

108° 71°

| 19° | | | | | | | | 160° |
|---|---|---|---|---|---|---|---|---|
| ′ | Sine. | D. | Cosine. | D. | Tang. | D. | Cotang. | ′ |
| 0 | 9·512642 | 612 | 9·975670 | 73 | 9·536972 | 684 | 10·463028 | 60 |
| 1 | 513009 | 611 | 975627 | 73 | 537382 | 683 | 462618 | 59 |
| 2 | 513375 | 611 | 975583 | 73 | 537792 | 683 | 462208 | 58 |
| 3 | 513741 | 610 | 975539 | 73 | 538202 | 682 | 461798 | 57 |
| 4 | 514107 | 609 | 975496 | 73 | 538611 | 682 | 461389 | 56 |
| 5 | 514472 | 609 | 975452 | 73 | 539020 | 681 | 460980 | 55 |
| 6 | 514837 | 608 | 975408 | 73 | 539429 | 681 | 460571 | 54 |
| 7 | 515202 | 608 | 975365 | 73 | 539837 | 680 | 460163 | 53 |
| 8 | 515566 | 607 | 975321 | 73 | 540245 | 680 | 459755 | 52 |
| 9 | 515930 | 607 | 975277 | 73 | 540653 | 679 | 459347 | 51 |
| 10 | 516294 | 606 | 975233 | 73 | 541061 | 679 | 458939 | 50 |
| 11 | 9·516657 | 605 | 9·975189 | 73 | 9·541468 | 678 | 10·458532 | 49 |
| 12 | 517020 | 605 | 975145 | 73 | 541875 | 678 | 458125 | 48 |
| 13 | 517382 | 604 | 975101 | 73 | 542281 | 677 | 457719 | 47 |
| 14 | 517745 | 604 | 975057 | 73 | 542688 | 677 | 457312 | 46 |
| 15 | 518107 | 603 | 975013 | 73 | 543094 | 676 | 456906 | 45 |
| 16 | 518468 | 603 | 974969 | 74 | 543499 | 676 | 456501 | 44 |
| 17 | 518829 | 602 | 974925 | 74 | 543905 | 675 | 456095 | 43 |
| 18 | 519190 | 601 | 974880 | 74 | 544310 | 675 | 455690 | 42 |
| 19 | 519551 | 601 | 974836 | 74 | 544715 | 674 | 455285 | 41 |
| 20 | 519911 | 600 | 974792 | 74 | 545119 | 674 | 454881 | 40 |
| 21 | 9·520271 | 600 | 9·974748 | 74 | 9·545524 | 673 | 10·454476 | 39 |
| 22 | 520631 | 599 | 974703 | 74 | 545928 | 673 | 454072 | 38 |
| 23 | 520990 | 599 | 974659 | 74 | 546331 | 672 | 453669 | 37 |
| 24 | 521349 | 598 | 974614 | 74 | 546735 | 672 | 453265 | 36 |
| 25 | 521707 | 598 | 974570 | 74 | 547138 | 671 | 452862 | 35 |
| 26 | 522066 | 597 | 974525 | 74 | 547540 | 671 | 452460 | 34 |
| 27 | 522424 | 596 | 974481 | 74 | 547943 | 670 | 452057 | 33 |
| 28 | 522781 | 596 | 974436 | 74 | 548345 | 670 | 451655 | 32 |
| 29 | 523138 | 595 | 974391 | 74 | 548747 | 669 | 451253 | 31 |
| 30 | 523495 | 595 | 974347 | 75 | 549149 | 669 | 450851 | 30 |
| 31 | 9·523852 | 594 | 9·974302 | 75 | 9·549550 | 668 | 10·450450 | 29 |
| 32 | 524208 | 594 | 974257 | 75 | 549951 | 668 | 450049 | 28 |
| 33 | 524564 | 593 | 974212 | 75 | 550352 | 667 | 449648 | 27 |
| 34 | 524920 | 593 | 974167 | 75 | 550752 | 667 | 449248 | 26 |
| 35 | 525275 | 592 | 974122 | 75 | 551153 | 666 | 448847 | 25 |
| 36 | 525630 | 591 | 974077 | 75 | 551552 | 666 | 448448 | 24 |
| 37 | 525984 | 591 | 974032 | 75 | 551952 | 665 | 448048 | 23 |
| 38 | 526339 | 590 | 973987 | 75 | 552351 | 665 | 447649 | 22 |
| 39 | 526693 | 590 | 973942 | 75 | 552750 | 665 | 447250 | 21 |
| 40 | 527046 | 589 | 973897 | 75 | 553149 | 664 | 446851 | 20 |
| 41 | 9·527400 | 589 | 9·973852 | 75 | 9·553548 | 664 | 10·446452 | 19 |
| 42 | 527753 | 588 | 973807 | 75 | 553946 | 663 | 446054 | 18 |
| 43 | 528105 | 588 | 973761 | 75 | 554344 | 663 | 445656 | 17 |
| 44 | 528458 | 587 | 973716 | 76 | 554741 | 662 | 445259 | 16 |
| 45 | 528810 | 587 | 973671 | 76 | 555139 | 662 | 444861 | 15 |
| 46 | 529161 | 586 | 973625 | 76 | 555536 | 661 | 444464 | 14 |
| 47 | 529513 | 586 | 973580 | 76 | 555933 | 661 | 444067 | 13 |
| 48 | 529864 | 585 | 973535 | 76 | 556329 | 660 | 443671 | 12 |
| 49 | 530215 | 585 | 973489 | 76 | 556725 | 660 | 443275 | 11 |
| 50 | 530565 | 584 | 973444 | 76 | 557121 | 659 | 442879 | 10 |
| 51 | 9·530915 | 584 | 9·973398 | 76 | 9·557517 | 659 | 10·442483 | 9 |
| 52 | 531265 | 583 | 973352 | 76 | 557913 | 659 | 442087 | 8 |
| 53 | 531614 | 582 | 973307 | 76 | 558308 | 658 | 441692 | 7 |
| 54 | 531963 | 582 | 973261 | 76 | 558703 | 658 | 441297 | 6 |
| 55 | 532312 | 581 | 973215 | 76 | 559097 | 657 | 440903 | 5 |
| 56 | 532661 | 581 | 973169 | 76 | 559491 | 657 | 440509 | 4 |
| 57 | 533009 | 580 | 973124 | 76 | 559885 | 656 | 440115 | 3 |
| 58 | 533357 | 580 | 973078 | 76 | 560279 | 656 | 439721 | 2 |
| 59 | 533704 | 579 | 973032 | 77 | 560673 | 655 | 439327 | 1 |
| 60 | 534052 | 578 | 972986 | 77 | 561066 | 655 | 438934 | 0 |
| ′ | Cosine. | D. | Sine. | D. | Cotang. | D. | Tang. | ′ |
| 109° | | | | | | | | 70° |

20° 159°

| ′ | Sine. | D. | Cosine. | D. | Tang. | D. | Cotang. | ′ |
|---|---|---|---|---|---|---|---|---|
| 0 | 9·534052 | 578 | 9·972986 | 77 | 9·561066 | 655 | 10·438934 | 60 |
| 1 | 534399 | 577 | 972940 | 77 | 561459 | 654 | 438541 | 59 |
| 2 | 534745 | 577 | 972894 | 77 | 561851 | 654 | 438149 | 58 |
| 3 | 535092 | 577 | 972848 | 77 | 562244 | 653 | 437756 | 57 |
| 4 | 535438 | 576 | 972802 | 77 | 562636 | 653 | 437364 | 56 |
| 5 | 535783 | 576 | 972755 | 77 | 563028 | 653 | 436972 | 55 |
| 6 | 536129 | 575 | 972709 | 77 | 563419 | 652 | 436581 | 54 |
| 7 | 536474 | 574 | 972663 | 77 | 563811 | 652 | 436189 | 53 |
| 8 | 536818 | 574 | 972617 | 77 | 564202 | 651 | 435798 | 52 |
| 9 | 537163 | 573 | 972570 | 77 | 564593 | 651 | 435407 | 51 |
| 10 | 537507 | 573 | 972524 | 77 | 564983 | 650 | 435017 | 50 |
| 11 | 9·537851 | 572 | 9·972478 | 77 | 9·565373 | 650 | 10·434627 | 49 |
| 12 | 538194 | 572 | 972431 | 78 | 565763 | 649 | 434237 | 48 |
| 13 | 538538 | 571 | 972385 | 78 | 566153 | 649 | 433847 | 47 |
| 14 | 538880 | 571 | 972338 | 78 | 566542 | 649 | 433458 | 46 |
| 15 | 539223 | 570 | 972291 | 78 | 566932 | 648 | 433068 | 45 |
| 16 | 539565 | 570 | 972245 | 78 | 567320 | 648 | 432680 | 44 |
| 17 | 539907 | 569 | 972198 | 78 | 567709 | 647 | 432291 | 43 |
| 18 | 540249 | 569 | 972151 | 78 | 568098 | 647 | 431902 | 42 |
| 19 | 540590 | 568 | 972105 | 78 | 568486 | 646 | 431514 | 41 |
| 20 | 540931 | 568 | 972058 | 78 | 568873 | 646 | 431127 | 40 |
| 21 | 9·541272 | 567 | 9·972011 | 78 | 9·569261 | 645 | 10·430739 | 39 |
| 22 | 541613 | 567 | 971964 | 78 | 569648 | 645 | 430352 | 38 |
| 23 | 541953 | 566 | 971917 | 78 | 570035 | 645 | 429965 | 37 |
| 24 | 542293 | 566 | 971870 | 78 | 570422 | 644 | 429578 | 36 |
| 25 | 542632 | 565 | 971823 | 78 | 570809 | 644 | 429191 | 35 |
| 26 | 542971 | 565 | 971776 | 78 | 571195 | 643 | 428805 | 34 |
| 27 | 543310 | 564 | 971729 | 79 | 571581 | 643 | 428419 | 33 |
| 28 | 543649 | 564 | 971682 | 79 | 571967 | 642 | 428033 | 32 |
| 29 | 543987 | 563 | 971635 | 79 | 572352 | 642 | 427648 | 31 |
| 30 | 544325 | 563 | 971588 | 79 | 572738 | 642 | 427262 | 30 |
| 31 | 9·544663 | 562 | 9·971540 | 79 | 9·573123 | 641 | 10·426877 | 29 |
| 32 | 545000 | 562 | 971493 | 79 | 573507 | 641 | 426493 | 28 |
| 33 | 545338 | 561 | 971446 | 79 | 573892 | 640 | 426108 | 27 |
| 34 | 545674 | 561 | 971398 | 79 | 574276 | 640 | 425724 | 26 |
| 35 | 546011 | 560 | 971351 | 79 | 574660 | 639 | 425340 | 25 |
| 36 | 546347 | 560 | 971303 | 79 | 575044 | 639 | 424956 | 24 |
| 37 | 546683 | 559 | 971256 | 79 | 575427 | 639 | 424573 | 23 |
| 38 | 547019 | 559 | 971208 | 79 | 575810 | 638 | 424190 | 22 |
| 39 | 547354 | 558 | 971161 | 79 | 576193 | 638 | 423807 | 21 |
| 40 | 547689 | 558 | 971113 | 79 | 576576 | 637 | 423424 | 20 |
| 41 | 9·548024 | 557 | 9·971066 | 80 | 9·576959 | 637 | 10·423041 | 19 |
| 42 | 548359 | 557 | 971018 | 80 | 577341 | 636 | 422659 | 18 |
| 43 | 548693 | 556 | 970970 | 80 | 577723 | 636 | 422277 | 17 |
| 44 | 549027 | 556 | 970922 | 80 | 578104 | 636 | 421896 | 16 |
| 45 | 549360 | 555 | 970874 | 80 | 578486 | 635 | 421514 | 15 |
| 46 | 549693 | 555 | 970827 | 80 | 578867 | 635 | 421133 | 14 |
| 47 | 550026 | 554 | 970779 | 80 | 579248 | 634 | 420752 | 13 |
| 48 | 550359 | 554 | 970731 | 80 | 579629 | 634 | 420371 | 12 |
| 49 | 550692 | 553 | 970683 | 80 | 580009 | 634 | 419991 | 11 |
| 50 | 551024 | 553 | 970635 | 80 | 580389 | 633 | 419611 | 10 |
| 51 | 9·551356 | 552 | 9·970586 | 80 | 9·580769 | 633 | 10·419231 | 9 |
| 52 | 551687 | 552 | 970538 | 80 | 581149 | 632 | 418851 | 8 |
| 53 | 552018 | 552 | 970490 | 80 | 581528 | 632 | 418472 | 7 |
| 54 | 552349 | 551 | 970442 | 80 | 581907 | 632 | 418093 | 6 |
| 55 | 552680 | 551 | 970394 | 80 | 582286 | 631 | 417714 | 5 |
| 56 | 553010 | 550 | 970345 | 81 | 582665 | 631 | 417335 | 4 |
| 57 | 553341 | 550 | 970297 | 81 | 583044 | 630 | 416956 | 3 |
| 58 | 553670 | 549 | 970249 | 81 | 583422 | 630 | 416578 | 2 |
| 59 | 554000 | 549 | 970200 | 81 | 583800 | 629 | 416200 | 1 |
| 60 | 554329 | 548 | 970152 | 81 | 584177 | 629 | 415823 | 0 |
| ′ | Cosine. | D. | Sine. | D. | Cotang. | D. | Tang. | ′ |

110° 69°

21° 158°

| ′ | Sine. | D. | Cosine. | D. | Tang. | D. | Cotang. | ′ |
|---|---|---|---|---|---|---|---|---|
| 0 | 9·554329 | 548 | 9·970152 | 81 | 9·584177 | 629 | 10·415823 | 60 |
| 1 | 554658 | 548 | 970103 | 81 | 584555 | 629 | 415445 | 59 |
| 2 | 554987 | 547 | 970055 | 81 | 584932 | 628 | 415068 | 58 |
| 3 | 555315 | 547 | 970006 | 81 | 585309 | 628 | 414691 | 57 |
| 4 | 555643 | 546 | 969957 | 81 | 585686 | 627 | 414314 | 56 |
| 5 | 555971 | 546 | 969909 | 81 | 586062 | 627 | 413938 | 55 |
| 6 | 556299 | 545 | 969860 | 81 | 586439 | 627 | 413561 | 54 |
| 7 | 556626 | 545 | 969811 | 81 | 586815 | 626 | 413185 | 53 |
| 8 | 556953 | 544 | 969762 | 81 | 587190 | 626 | 412810 | 52 |
| 9 | 557280 | 544 | 969714 | 81 | 587566 | 625 | 412434 | 51 |
| 10 | 557606 | 543 | 969665 | 81 | 587941 | 625 | 412059 | 50 |
| 11 | 9·557932 | 543 | 9·969616 | 82 | 9·588316 | 625 | 10·411684 | 49 |
| 12 | 558258 | 543 | 969567 | 82 | 588691 | 624 | 411309 | 48 |
| 13 | 558583 | 542 | 969518 | 82 | 589066 | 624 | 410934 | 47 |
| 14 | 558909 | 542 | 969469 | 82 | 589440 | 623 | 410560 | 46 |
| 15 | 559234 | 541 | 969420 | 82 | 589814 | 623 | 410186 | 45 |
| 16 | 559558 | 541 | 969370 | 82 | 590188 | 623 | 409812 | 44 |
| 17 | 559883 | 540 | 969321 | 82 | 590562 | 622 | 409438 | 43 |
| 18 | 560207 | 540 | 969272 | 82 | 590935 | 622 | 409065 | 42 |
| 19 | 560531 | 539 | 969223 | 82 | 591308 | 622 | 408692 | 41 |
| 20 | 560855 | 539 | 969173 | 82 | 591681 | 621 | 408319 | 40 |
| 21 | 9·561178 | 538 | 9·969124 | 82 | 9·592054 | 621 | 10·407946 | 39 |
| 22 | 561501 | 538 | 969075 | 82 | 592426 | 620 | 407574 | 38 |
| 23 | 561824 | 537 | 969025 | 82 | 592799 | 620 | 407201 | 37 |
| 24 | 562146 | 537 | 968976 | 82 | 593171 | 619 | 406829 | 36 |
| 25 | 562468 | 536 | 968926 | 83 | 593542 | 619 | 406458 | 35 |
| 26 | 562790 | 536 | 968877 | 83 | 593914 | 618 | 406086 | 34 |
| 27 | 563112 | 536 | 968827 | 83 | 594285 | 618 | 405715 | 33 |
| 28 | 563433 | 535 | 968777 | 83 | 594656 | 618 | 405344 | 32 |
| 29 | 563755 | 535 | 968728 | 83 | 595027 | 617 | 404973 | 31 |
| 30 | 564075 | 534 | 968678 | 83 | 595398 | 617 | 404602 | 30 |
| 31 | 9·564396 | 534 | 9·968628 | 83 | 9·595768 | 617 | 10·404232 | 29 |
| 32 | 564716 | 533 | 968578 | 83 | 596138 | 616 | 403862 | 28 |
| 33 | 565036 | 533 | 968528 | 83 | 596508 | 616 | 403492 | 27 |
| 34 | 565356 | 532 | 968479 | 83 | 596878 | 616 | 403122 | 26 |
| 35 | 565676 | 532 | 968429 | 83 | 597247 | 615 | 402753 | 25 |
| 36 | 565995 | 531 | 968379 | 83 | 597616 | 615 | 402384 | 24 |
| 37 | 566314 | 531 | 968329 | 83 | 597985 | 615 | 402015 | 23 |
| 38 | 566632 | 531 | 968278 | 83 | 598354 | 614 | 401646 | 22 |
| 39 | 566951 | 530 | 968228 | 84 | 598722 | 614 | 401278 | 21 |
| 40 | 567269 | 530 | 968178 | 84 | 599091 | 613 | 400909 | 20 |
| 41 | 9·567587 | 529 | 9·968128 | 84 | 9·599459 | 613 | 10·400541 | 19 |
| 42 | 567904 | 529 | 968078 | 84 | 599827 | 613 | 400173 | 18 |
| 43 | 568222 | 528 | 968027 | 84 | 600194 | 612 | 399806 | 17 |
| 44 | 568539 | 528 | 967977 | 84 | 600562 | 612 | 399438 | 16 |
| 45 | 568856 | 528 | 967927 | 84 | 600929 | 611 | 399071 | 15 |
| 46 | 569172 | 527 | 967876 | 84 | 601296 | 611 | 398704 | 14 |
| 47 | 569488 | 527 | 967826 | 84 | 601663 | 611 | 398337 | 13 |
| 48 | 569804 | 526 | 967775 | 84 | 602029 | 610 | 397971 | 12 |
| 49 | 570120 | 526 | 967725 | 84 | 602395 | 610 | 397605 | 11 |
| 50 | 570435 | 525 | 967674 | 84 | 602761 | 610 | 397239 | 10 |
| 51 | 9·570751 | 525 | 9·967624 | 84 | 9·603127 | 609 | 10·396873 | 9 |
| 52 | 571066 | 524 | 967573 | 84 | 603493 | 609 | 396507 | 8 |
| 53 | 571380 | 524 | 967522 | 85 | 603858 | 609 | 396142 | 7 |
| 54 | 571695 | 523 | 967471 | 85 | 604223 | 608 | 395777 | 6 |
| 55 | 572009 | 523 | 967421 | 85 | 604588 | 608 | 395412 | 5 |
| 56 | 572323 | 523 | 967370 | 85 | 604953 | 607 | 395047 | 4 |
| 57 | 572636 | 522 | 967319 | 85 | 605317 | 607 | 394683 | 3 |
| 58 | 572950 | 522 | 967268 | 85 | 605682 | 607 | 394318 | 2 |
| 59 | 573263 | 521 | 967217 | 85 | 606046 | 606 | 393954 | 1 |
| 60 | 573575 | 521 | 967166 | 85 | 606410 | 606 | 393590 | 0 |
| ′ | Cosine. | D. | Sine. | D. | Cotang. | D. | Tang. | ′ |

111° 68°

| 22° | | | | | | | | 157° |
|---|---|---|---|---|---|---|---|---|
| ′ | Sine. | D. | Cosine. | D. | Tang. | D. | Cotang. | ′ |
| 0 | 9·573575 | 521 | 9·967166 | 85 | 9·606410 | 606 | 10·393590 | 60 |
| 1 | 573888 | 520 | 967115 | 85 | 606773 | 606 | 393227 | 59 |
| 2 | 574200 | 520 | 967064 | 85 | 607137 | 605 | 392863 | 58 |
| 3 | 574512 | 519 | 967013 | 85 | 607500 | 605 | 392500 | 57 |
| 4 | 574824 | 519 | 966961 | 85 | 607863 | 604 | 392137 | 56 |
| 5 | 575136 | 519 | 966910 | 85 | 608225 | 604 | 391775 | 55 |
| 6 | 575447 | 518 | 966859 | 85 | 608588 | 604 | 391412 | 54 |
| 7 | 575758 | 518 | 966808 | 85 | 608950 | 603 | 391050 | 53 |
| 8 | 576069 | 517 | 966756 | 86 | 609312 | 603 | 390688 | 52 |
| 9 | 576379 | 517 | 966705 | 86 | 609674 | 603 | 390326 | 51 |
| 10 | 576689 | 516 | 966653 | 86 | 610036 | 602 | 389964 | 50 |
| 11 | 9·576999 | 516 | 9·966602 | 86 | 9·610397 | 602 | 10·389603 | 49 |
| 12 | 577309 | 516 | 966550 | 86 | 610759 | 602 | 389241 | 48 |
| 13 | 577618 | 515 | 966499 | 86 | 611120 | 601 | 388880 | 47 |
| 14 | 577927 | 515 | 966447 | 86 | 611480 | 601 | 388520 | 46 |
| 15 | 578236 | 514 | 966395 | 86 | 611841 | 601 | 388159 | 45 |
| 16 | 578545 | 514 | 966344 | 86 | 612201 | 600 | 387799 | 44 |
| 17 | 578853 | 513 | 966292 | 86 | 612561 | 600 | 387439 | 43 |
| 18 | 579162 | 513 | 966240 | 86 | 612921 | 600 | 387079 | 42 |
| 19 | 579470 | 513 | 966188 | 86 | 613281 | 599 | 386719 | 41 |
| 20 | 579777 | 512 | 966136 | 86 | 613641 | 599 | 386359 | 40 |
| 21 | 9·580085 | 512 | 9·966085 | 87 | 9·614000 | 598 | 10·386000 | 39 |
| 22 | 580392 | 511 | 966033 | 87 | 614359 | 598 | 385641 | 38 |
| 23 | 580699 | 511 | 965981 | 87 | 614718 | 598 | 385282 | 37 |
| 24 | 581005 | 511 | 965928 | 87 | 615077 | 597 | 384923 | 36 |
| 25 | 581312 | 510 | 965876 | 87 | 615435 | 597 | 384565 | 35 |
| 26 | 581618 | 510 | 965824 | 87 | 615793 | 597 | 384207 | 34 |
| 27 | 581924 | 509 | 965772 | 87 | 616151 | 596 | 383849 | 33 |
| 28 | 582229 | 509 | 965720 | 87 | 616509 | 596 | 383491 | 32 |
| 29 | 582535 | 509 | 965668 | 87 | 616867 | 596 | 383133 | 31 |
| 30 | 582840 | 508 | 965615 | 87 | 617224 | 595 | 382776 | 30 |
| 31 | 9·583145 | 508 | 9·965563 | 87 | 9·617582 | 595 | 10·382418 | 29 |
| 32 | 583449 | 507 | 965511 | 87 | 617939 | 595 | 382061 | 28 |
| 33 | 583754 | 507 | 965458 | 87 | 618295 | 594 | 381705 | 27 |
| 34 | 584058 | 506 | 965406 | 87 | 618652 | 594 | 381348 | 26 |
| 35 | 584361 | 506 | 965353 | 88 | 619008 | 594 | 380992 | 25 |
| 36 | 584665 | 506 | 965301 | 88 | 619364 | 593 | 380636 | 24 |
| 37 | 584968 | 505 | 965248 | 88 | 619720 | 593 | 380280 | 23 |
| 38 | 585272 | 505 | 965195 | 88 | 620076 | 593 | 379924 | 22 |
| 39 | 585574 | 504 | 965143 | 88 | 620432 | 592 | 379568 | 21 |
| 40 | 585877 | 504 | 965090 | 88 | 620787 | 592 | 379213 | 20 |
| 41 | 9·586179 | 503 | 9·965037 | 88 | 9·621142 | 592 | 10·378858 | 19 |
| 42 | 586482 | 503 | 964984 | 88 | 621497 | 591 | 378503 | 18 |
| 43 | 586783 | 503 | 964931 | 88 | 621852 | 591 | 378148 | 17 |
| 44 | 587085 | 502 | 964879 | 88 | 622207 | 590 | 377793 | 16 |
| 45 | 587386 | 502 | 964826 | 88 | 622561 | 590 | 377439 | 15 |
| 46 | 587688 | 501 | 964773 | 88 | 622915 | 590 | 377085 | 14 |
| 47 | 587989 | 501 | 964720 | 88 | 623269 | 589 | 376731 | 13 |
| 48 | 588289 | 501 | 964666 | 89 | 623623 | 589 | 376377 | 12 |
| 49 | 588590 | 500 | 964613 | 89 | 623976 | 589 | 376024 | 11 |
| 50 | 588890 | 500 | 964560 | 89 | 624330 | 588 | 375670 | 10 |
| 51 | 9·589190 | 499 | 9·964507 | 89 | 9·624683 | 588 | 10·375317 | 9 |
| 52 | 589489 | 499 | 964454 | 89 | 625036 | 588 | 374964 | 8 |
| 53 | 589789 | 499 | 964400 | 89 | 625388 | 587 | 374612 | 7 |
| 54 | 590088 | 498 | 964347 | 89 | 625741 | 587 | 374259 | 6 |
| 55 | 590387 | 498 | 964294 | 89 | 626093 | 587 | 373907 | 5 |
| 56 | 590686 | 497 | 964240 | 89 | 626445 | 586 | 373555 | 4 |
| 57 | 590984 | 497 | 964187 | 89 | 626797 | 586 | 373203 | 3 |
| 58 | 591282 | 497 | 964133 | 89 | 627149 | 586 | 372851 | 2 |
| 59 | 591580 | 496 | 964080 | 89 | 627501 | 585 | 372499 | 1 |
| 60 | 591878 | 496 | 964026 | 89 | 627852 | 585 | 372148 | 0 |
| ′ | Cosine. | D. | Sine. | D. | Cotang. | D. | Tang. | ′ |
| 112° | | | | | | | | 67° |

| 23° | | | | | | | | 156° |
|---|---|---|---|---|---|---|---|---|
| ′ | Sine. | D. | Cosine. | D. | Tang. | D. | Cotang. | ′ |
| 0 | 9·591878 | 496 | 9·964026 | 89 | 9·627852 | 585 | 10·372148 | 60 |
| 1 | 592176 | 495 | 963972 | 89 | 628203 | 585 | 371797 | 59 |
| 2 | 592473 | 495 | 963919 | 89 | 628554 | 585 | 371446 | 58 |
| 3 | 592770 | 495 | 963865 | 90 | 628905 | 584 | 371095 | 57 |
| 4 | 593067 | 494 | 963811 | 90 | 629255 | 584 | 370745 | 56 |
| 5 | 593363 | 494 | 963757 | 90 | 629606 | 583 | 370394 | 55 |
| 6 | 593659 | 493 | 963704 | 90 | 629956 | 583 | 370044 | 54 |
| 7 | 593955 | 493 | 963650 | 90 | 630306 | 583 | 369694 | 53 |
| 8 | 594251 | 493 | 963596 | 90 | 630656 | 583 | 369344 | 52 |
| 9 | 594547 | 492 | 963542 | 90 | 631005 | 582 | 368995 | 51 |
| 10 | 594842 | 492 | 963488 | 90 | 631355 | 582 | 368645 | 50 |
| 11 | 9·595137 | 491 | 9·963434 | 90 | 9·631704 | 582 | 10·368296 | 49 |
| 12 | 595432 | 491 | 963379 | 90 | 632053 | 581 | 367947 | 48 |
| 13 | 595727 | 491 | 963325 | 90 | 632402 | 581 | 367598 | 47 |
| 14 | 596021 | 490 | 963271 | 90 | 632750 | 581 | 367250 | 46 |
| 15 | 596315 | 490 | 963217 | 90 | 633099 | 580 | 366901 | 45 |
| 16 | 596609 | 489 | 963163 | 90 | 633447 | 580 | 366553 | 44 |
| 17 | 596903 | 489 | 963108 | 91 | 633795 | 580 | 366205 | 43 |
| 18 | 597196 | 489 | 963054 | 91 | 634143 | 579 | 365857 | 42 |
| 19 | 597490 | 488 | 962999 | 91 | 634490 | 579 | 365510 | 41 |
| 20 | 597783 | 488 | 962945 | 91 | 634838 | 579 | 365162 | 40 |
| 21 | 9·598075 | 487 | 9·962890 | 91 | 9·635185 | 578 | 10·364815 | 39 |
| 22 | 598368 | 487 | 962836 | 91 | 635532 | 578 | 364468 | 38 |
| 23 | 598660 | 487 | 962781 | 91 | 635879 | 578 | 364121 | 37 |
| 24 | 598952 | 486 | 962727 | 91 | 636226 | 577 | 363774 | 36 |
| 25 | 599244 | 486 | 962672 | 91 | 636572 | 577 | 363428 | 35 |
| 26 | 599536 | 485 | 962617 | 91 | 636919 | 577 | 363081 | 34 |
| 27 | 599827 | 485 | 962562 | 91 | 637265 | 577 | 362735 | 33 |
| 28 | 600118 | 485 | 962508 | 91 | 637611 | 576 | 362389 | 32 |
| 29 | 600409 | 484 | 962453 | 91 | 637956 | 576 | 362044 | 31 |
| 30 | 600700 | 484 | 962398 | 92 | 638302 | 576 | 361698 | 30 |
| 31 | 9·600990 | 484 | 9·962343 | 92 | 9·638647 | 575 | 10·361353 | 29 |
| 32 | 601280 | 483 | 962288 | 92 | 638992 | 575 | 361008 | 28 |
| 33 | 601570 | 483 | 962233 | 92 | 639337 | 575 | 360663 | 27 |
| 34 | 601860 | 482 | 962178 | 92 | 639682 | 574 | 360318 | 26 |
| 35 | 602150 | 482 | 962123 | 92 | 640027 | 574 | 359973 | 25 |
| 36 | 602439 | 482 | 962067 | 92 | 640371 | 574 | 359629 | 24 |
| 37 | 602728 | 481 | 962012 | 92 | 640716 | 573 | 359284 | 23 |
| 38 | 603017 | 481 | 961957 | 92 | 641060 | 573 | 358940 | 22 |
| 39 | 603305 | 481 | 961902 | 92 | 641404 | 573 | 358596 | 21 |
| 40 | 603594 | 480 | 961846 | 92 | 641747 | 572 | 358253 | 20 |
| 41 | 9·603882 | 480 | 9·961791 | 92 | 9·642091 | 572 | 10·357909 | 19 |
| 42 | 604170 | 479 | 961735 | 92 | 642434 | 572 | 357566 | 18 |
| 43 | 604457 | 479 | 961680 | 92 | 642777 | 572 | 357223 | 17 |
| 44 | 604745 | 479 | 961624 | 93 | 643120 | 571 | 356880 | 16 |
| 45 | 605032 | 478 | 961569 | 93 | 643463 | 571 | 356537 | 15 |
| 46 | 605319 | 478 | 961513 | 93 | 643806 | 571 | 356194 | 14 |
| 47 | 605606 | 478 | 961458 | 93 | 644148 | 570 | 355852 | 13 |
| 48 | 605892 | 477 | 961402 | 93 | 644490 | 570 | 355510 | 12 |
| 49 | 606179 | 477 | 961346 | 93 | 644832 | 570 | 355168 | 11 |
| 50 | 606465 | 476 | 961290 | 93 | 645174 | 569 | 354826 | 10 |
| 51 | 9·606751 | 476 | 9·961235 | 93 | 9·645516 | 569 | 10·354484 | 9 |
| 52 | 607036 | 476 | 961179 | 93 | 645857 | 569 | 354143 | 8 |
| 53 | 607322 | 475 | 961123 | 93 | 646199 | 569 | 353801 | 7 |
| 54 | 607607 | 475 | 961067 | 93 | 646540 | 568 | 353460 | 6 |
| 55 | 607892 | 474 | 961011 | 93 | 646881 | 568 | 353119 | 5 |
| 56 | 608177 | 474 | 960955 | 93 | 647222 | 568 | 352778 | 4 |
| 57 | 608461 | 474 | 960899 | 93 | 647562 | 567 | 352438 | 3 |
| 58 | 608745 | 473 | 960843 | 94 | 647903 | 567 | 352097 | 2 |
| 59 | 609029 | 473 | 960786 | 94 | 648243 | 567 | 351757 | 1 |
| 60 | 609313 | 473 | 960730 | 94 | 648583 | 566 | 351417 | 0 |
| ′ | Cosine. | D. | Sine. | D. | Cotang. | D. | Tang. | ′ |
| 113° | | | | | | | | 66° |

| 24° | | | | | | | | | 155° |
|---|---|---|---|---|---|---|---|---|---|
| ′ | Sine. | D. | Cosine. | D. | Tang. | D. | Cotang. | | ′ |
| 0 | 9·609313 | 473 | 9·960730 | 94 | 9·648583 | 566 | 10·351417 | | 60 |
| 1 | 609597 | 472 | 960674 | 94 | 648923 | 566 | 351077 | | 59 |
| 2 | 609880 | 472 | 960618 | 94 | 649263 | 566 | 350737 | | 58 |
| 3 | 610164 | 472 | 960561 | 94 | 649602 | 566 | 350398 | | 57 |
| 4 | 610447 | 471 | 960505 | 94 | 649942 | 565 | 350058 | | 56 |
| 5 | 610729 | 471 | 960448 | 94 | 650281 | 565 | 349719 | | 55 |
| 6 | 611012 | 470 | 960392 | 94 | 650620 | 565 | 349380 | | 54 |
| 7 | 611294 | 470 | 960335 | 94 | 650959 | 564 | 349041 | | 53 |
| 8 | 611576 | 470 | 960279 | 94 | 651297 | 564 | 348703 | | 52 |
| 9 | 611858 | 469 | 960222 | 94 | 651636 | 564 | 348364 | | 51 |
| 10 | 612140 | 469 | 960165 | 94 | 651974 | 563 | 348026 | | 50 |
| 11 | 9·612421 | 469 | 9·960109 | 95 | 9·652312 | 563 | 10·347688 | | 49 |
| 12 | 612702 | 468 | 960052 | 95 | 652650 | 563 | 347350 | | 48 |
| 13 | 612983 | 468 | 959995 | 95 | 652988 | 563 | 347012 | | 47 |
| 14 | 613264 | 467 | 959938 | 95 | 653326 | 562 | 346674 | | 46 |
| 15 | 613545 | 467 | 959882 | 95 | 653663 | 562 | 346337 | | 45 |
| 16 | 613825 | 467 | 959825 | 95 | 654000 | 562 | 346000 | | 44 |
| 17 | 614105 | 466 | 959768 | 95 | 654337 | 561 | 345663 | | 43 |
| 18 | 614385 | 466 | 959711 | 95 | 654674 | 561 | 345326 | | 42 |
| 19 | 614665 | 466 | 959654 | 95 | 655011 | 561 | 344989 | | 41 |
| 20 | 614944 | 465 | 959596 | 95 | 655348 | 561 | 344652 | | 40 |
| 21 | 9·615223 | 465 | 9·959539 | 95 | 9·655684 | 560 | 10·344316 | | 39 |
| 22 | 615502 | 465 | 959482 | 95 | 656020 | 560 | 343980 | | 38 |
| 23 | 615781 | 464 | 959425 | 95 | 656356 | 560 | 343644 | | 37 |
| 24 | 616060 | 464 | 959368 | 95 | 656692 | 559 | 343308 | | 36 |
| 25 | 616338 | 464 | 959310 | 96 | 657028 | 559 | 342972 | | 35 |
| 26 | 616616 | 463 | 959253 | 96 | 657364 | 559 | 342636 | | 34 |
| 27 | 616894 | 463 | 959195 | 96 | 657699 | 559 | 342301 | | 33 |
| 28 | 617172 | 462 | 959138 | 96 | 658034 | 558 | 341966 | | 32 |
| 29 | 617450 | 462 | 959080 | 96 | 658369 | 558 | 341631 | | 31 |
| 30 | 617727 | 462 | 959023 | 96 | 658704 | 558 | 341296 | | 30 |
| 31 | 9·618004 | 461 | 9·958965 | 96 | 9·659039 | 558 | 10·340961 | | 29 |
| 32 | 618281 | 461 | 958908 | 96 | 659373 | 557 | 340627 | | 28 |
| 33 | 618558 | 461 | 958850 | 96 | 659708 | 557 | 340292 | | 27 |
| 34 | 618834 | 460 | 958792 | 96 | 660042 | 557 | 339958 | | 26 |
| 35 | 619110 | 460 | 958734 | 66 | 660376 | 557 | 339624 | | 25 |
| 36 | 619386 | 460 | 958677 | 96 | 660710 | 556 | 339290 | | 24 |
| 37 | 619662 | 459 | 958619 | 96 | 661043 | 556 | 338957 | | 23 |
| 38 | 619938 | 459 | 958561 | 96 | 661377 | 556 | 338623 | | 22 |
| 39 | 620213 | 459 | 958503 | 97 | 661710 | 555 | 338290 | | 21 |
| 40 | 620488 | 458 | 958445 | 97 | 662043 | 555 | 337957 | | 20 |
| 41 | 9·620763 | 458 | 9·958387 | 97 | 9·662376 | 555 | 10·337624 | | 19 |
| 42 | 621038 | 457 | 958329 | 97 | 662709 | 554 | 337291 | | 18 |
| 43 | 621313 | 457 | 958271 | 97 | 663042 | 554 | 336958 | | 17 |
| 44 | 621587 | 457 | 958213 | 97 | 663375 | 554 | 336625 | | 16 |
| 45 | 621861 | 456 | 958154 | 97 | 663707 | 554 | 336293 | | 15 |
| 46 | 622135 | 456 | 958096 | 97 | 664039 | 553 | 335961 | | 14 |
| 47 | 622409 | 456 | 958038 | 97 | 664371 | 553 | 335629 | | 13 |
| 48 | 622682 | 455 | 957979 | 97 | 664703 | 553 | 335297 | | 12 |
| 49 | 622956 | 455 | 957921 | 97 | 665035 | 553 | 334965 | | 11 |
| 50 | 623229 | 455 | 957863 | 97 | 665366 | 552 | 334634 | | 10 |
| 51 | 9·623502 | 454 | 9·957804 | 97 | 9·665698 | 552 | 10·334302 | | 9 |
| 52 | 623774 | 454 | 957746 | 98 | 666029 | 552 | 333971 | | 8 |
| 53 | 624047 | 454 | 957687 | 98 | 666360 | 551 | 333640 | | 7 |
| 54 | 624319 | 453 | 957628 | 98 | 666691 | 551 | 333309 | | 6 |
| 55 | 624591 | 453 | 957570 | 98 | 667021 | 551 | 332979 | | 5 |
| 56 | 624863 | 453 | 957511 | 98 | 667352 | 551 | 332648 | | 4 |
| 57 | 625135 | 452 | 957452 | 98 | 667682 | 550 | 332318 | | 3 |
| 58 | 625406 | 452 | 957393 | 98 | 668013 | 550 | 331987 | | 2 |
| 59 | 625677 | 452 | 957335 | 98 | 668343 | 550 | 331657 | | 1 |
| 60 | 625948 | 451 | 957276 | 98 | 668673 | 550 | 331327 | | 0 |
| ′ | Cosine. | D. | Sine. | D. | Cotang. | D. | Tang. | | ′ |
| 114° | | | | | | | | | 65° |

| 25° | | | | | | | | 154° |
|---|---|---|---|---|---|---|---|---|
| ′ | Sine. | D. | Cosine. | D. | Tang. | D. | Cotang. | ′ |
| 0 | 9·625948 | 451 | 9·957276 | 98 | 9·668673 | 550 | 10·331327 | 60 |
| 1 | 626219 | 451 | 957217 | 98 | 669002 | 549 | 330998 | 59 |
| 2 | 626490 | 451 | 957158 | 98 | 669332 | 549 | 330668 | 58 |
| 3 | 626760 | 450 | 957099 | 98 | 669661 | 549 | 330339 | 57 |
| 4 | 627030 | 450 | 957040 | 98 | 669991 | 548 | 330009 | 56 |
| 5 | 627300 | 450 | 956981 | 98 | 670320 | 548 | 329680 | 55 |
| 6 | 627570 | 449 | 956921 | 99 | 670649 | 548 | 329351 | 54 |
| 7 | 627840 | 449 | 956862 | 99 | 670977 | 548 | 329023 | 53 |
| 8 | 628109 | 449 | 956803 | 99 | 671306 | 547 | 328694 | 52 |
| 9 | 628378 | 448 | 956744 | 99 | 671635 | 547 | 328365 | 51 |
| 10 | 628647 | 448 | 956684 | 99 | 671963 | 547 | 328037 | 50 |
| 11 | 9·628916 | 457 | 9·956625 | 99 | 9·672291 | 547 | 10·327709 | 49 |
| 12 | 629185 | 447 | 956566 | 99 | 672619 | 546 | 327381 | 48 |
| 13 | 629453 | 447 | 956506 | 99 | 672947 | 546 | 327053 | 47 |
| 14 | 629721 | 446 | 956447 | 99 | 673274 | 546 | 326726 | 46 |
| 15 | 629989 | 446 | 956387 | 99 | 673602 | 546 | 326398 | 45 |
| 16 | 630257 | 446 | 956327 | 99 | 673929 | 545 | 326071 | 44 |
| 17 | 630524 | 446 | 956268 | 99 | 674257 | 545 | 325743 | 43 |
| 18 | 630792 | 445 | 956208 | 100 | 674584 | 545 | 325416 | 42 |
| 19 | 631059 | 445 | 956148 | 100 | 674911 | 544 | 325089 | 41 |
| 20 | 631326 | 445 | 956089 | 100 | 675237 | 544 | 324763 | 40 |
| 21 | 9·631593 | 444 | 9·956029 | 100 | 9·675564 | 544 | 10·324436 | 39 |
| 22 | 631859 | 444 | 955969 | 100 | 675890 | 544 | 324110 | 38 |
| 23 | 632125 | 444 | 955909 | 100 | 676217 | 543 | 323783 | 37 |
| 24 | 632392 | 443 | 955849 | 100 | 676543 | 543 | 323457 | 36 |
| 25 | 632658 | 443 | 955789 | 100 | 676869 | 543 | 323131 | 35 |
| 26 | 632923 | 443 | 955729 | 100 | 677194 | 543 | 322806 | 34 |
| 27 | 633189 | 442 | 955669 | 100 | 677520 | 542 | 322480 | 33 |
| 28 | 633454 | 442 | 955609 | 100 | 677846 | 542 | 322154 | 32 |
| 29 | 633719 | 442 | 955548 | 100 | 678171 | 542 | 321829 | 31 |
| 30 | 633984 | 441 | 955488 | 100 | 678496 | 542 | 321504 | 30 |
| 31 | 9·634249 | 441 | 9·955428 | 101 | 9·678821 | 541 | 10·321179 | 29 |
| 32 | 634514 | 440 | 955368 | 101 | 679146 | 541 | 320854 | 28 |
| 33 | 634778 | 440 | 955307 | 101 | 679471 | 541 | 320529 | 27 |
| 34 | 635042 | 440 | 955247 | 101 | 679795 | 541 | 320205 | 26 |
| 35 | 635306 | 439 | 955186 | 101 | 680120 | 540 | 319880 | 25 |
| 36 | 635570 | 439 | 955126 | 101 | 680444 | 540 | 319556 | 24 |
| 37 | 635834 | 439 | 955065 | 101 | 680768 | 540 | 319232 | 23 |
| 38 | 636097 | 438 | 955005 | 101 | 681092 | 540 | 318908 | 22 |
| 39 | 636360 | 438 | 954944 | 101 | 681416 | 539 | 318584 | 21 |
| 40 | 636623 | 438 | 954883 | 101 | 681740 | 539 | 318260 | 20 |
| 41 | 9·636886 | 437 | 9·954823 | 101 | 9·682063 | 539 | 10·317937 | 19 |
| 42 | 637148 | 437 | 954762 | 101 | 682387 | 539 | 317613 | 18 |
| 43 | 637411 | 437 | 954701 | 101 | 682710 | 538 | 317290 | 17 |
| 44 | 637673 | 437 | 954640 | 101 | 683033 | 538 | 316967 | 16 |
| 45 | 637935 | 436 | 954579 | 101 | 683356 | 538 | 316644 | 15 |
| 46 | 638197 | 436 | 954518 | 102 | 683679 | 538 | 316321 | 14 |
| 47 | 638458 | 436 | 954457 | 102 | 684001 | 537 | 315999 | 13 |
| 48 | 638720 | 435 | 954396 | 102 | 684324 | 537 | 315676 | 12 |
| 49 | 638981 | 435 | 954335 | 102 | 684646 | 537 | 315354 | 11 |
| 50 | 639242 | 435 | 954274 | 102 | 684968 | 537 | 315032 | 10 |
| 51 | 9·639503 | 434 | 9·954213 | 102 | 9·685290 | 536 | 10·314710 | 9 |
| 52 | 639764 | 434 | 954152 | 102 | 685612 | 536 | 314388 | 8 |
| 53 | 640024 | 434 | 954090 | 102 | 685934 | 536 | 314066 | 7 |
| 54 | 640284 | 433 | 954029 | 102 | 686255 | 536 | 313745 | 6 |
| 55 | 640544 | 433 | 953968 | 102 | 686577 | 535 | 313423 | 5 |
| 56 | 640804 | 433 | 953906 | 102 | 686898 | 535 | 313102 | 4 |
| 57 | 641064 | 432 | 953845 | 102 | 687219 | 535 | 312781 | 3 |
| 58 | 641324 | 432 | 953783 | 102 | 687540 | 535 | 312460 | 2 |
| 59 | 641583 | 432 | 953722 | 103 | 687861 | 534 | 312139 | 1 |
| 60 | 641842 | 431 | 953660 | 103 | 688182 | 534 | 311818 | 0 |
| ′ | Cosine. | D. | Sine. | D. | Cotang. | D. | Tang. | ′ |
| 115° | | | | | | | | 64° |

| 26° | | | | | | | | 153° |
|---|---|---|---|---|---|---|---|---|
| ′ | Sine. | D. | Cosine. | D. | Tang. | D. | Cotang. | ′ |
| 0 | 9·641842 | 431 | 9·953660 | 103 | 9·688182 | 534 | 10·311818 | 60 |
| 1 | 642101 | 431 | 953599 | 103 | 688502 | 534 | 311498 | 59 |
| 2 | 642360 | 431 | 953537 | 103 | 688823 | 534 | 311177 | 58 |
| 3 | 642618 | 430 | 953475 | 103 | 689143 | 533 | 310857 | 57 |
| 4 | 642877 | 430 | 953413 | 103 | 689463 | 533 | 310537 | 56 |
| 5 | 643135 | 430 | 953352 | 103 | 689783 | 533 | 310217 | 55 |
| 6 | 643393 | 430 | 953290 | 103 | 690103 | 533 | 309897 | 54 |
| 7 | 643650 | 429 | 953228 | 103 | 690423 | 533 | 309577 | 53 |
| 8 | 643908 | 429 | 953166 | 103 | 690742 | 532 | 309258 | 52 |
| 9 | 644165 | 429 | 953104 | 103 | 691062 | 532 | 308938 | 51 |
| 10 | 644423 | 428 | 953042 | 103 | 691381 | 532 | 308619 | 50 |
| 11 | 9·644680 | 428 | 9·952980 | 104 | 9·691700 | 531 | 10·308300 | 49 |
| 12 | 644936 | 428 | 952918 | 104 | 692019 | 531 | 307981 | 48 |
| 13 | 645193 | 427 | 952855 | 104 | 692338 | 531 | 307662 | 47 |
| 14 | 645450 | 427 | 952793 | 104 | 692656 | 531 | 307344 | 46 |
| 15 | 645706 | 427 | 952731 | 104 | 692975 | 531 | 307025 | 45 |
| 16 | 645962 | 426 | 952669 | 104 | 693293 | 530 | 306707 | 44 |
| 17 | 646218 | 426 | 952606 | 104 | 693612 | 530 | 306388 | 43 |
| 18 | 646474 | 426 | 952544 | 104 | 693930 | 530 | 306070 | 42 |
| 19 | 646729 | 425 | 952481 | 104 | 694248 | 530 | 305752 | 41 |
| 20 | 646984 | 425 | 952419 | 104 | 694566 | 529 | 305434 | 40 |
| 21 | 9·647240 | 425 | 9·952356 | 104 | 9·694883 | 529 | 10·305117 | 39 |
| 22 | 647494 | 424 | 952294 | 104 | 695201 | 529 | 304799 | 38 |
| 23 | 647749 | 424 | 952231 | 104 | 695518 | 529 | 304482 | 37 |
| 24 | 648004 | 424 | 952168 | 105 | 695836 | 529 | 304164 | 36 |
| 25 | 648258 | 424 | 952106 | 105 | 696153 | 528 | 303847 | 35 |
| 26 | 648512 | 423 | 952043 | 105 | 696470 | 528 | 303530 | 34 |
| 27 | 648766 | 423 | 951980 | 105 | 696787 | 528 | 303213 | 33 |
| 28 | 649020 | 423 | 951917 | 105 | 697103 | 528 | 302897 | 32 |
| 29 | 649274 | 422 | 951854 | 105 | 697420 | 527 | 302580 | 31 |
| 30 | 649527 | 422 | 951791 | 105 | 697736 | 527 | 302264 | 30 |
| 31 | 9·649781 | 422 | 9·951728 | 105 | 9·698053 | 527 | 10·301947 | 29 |
| 32 | 650034 | 422 | 951665 | 105 | 698369 | 527 | 301631 | 28 |
| 33 | 650287 | 421 | 951602 | 105 | 698685 | 526 | 301315 | 27 |
| 34 | 650539 | 421 | 951539 | 105 | 699001 | 526 | 300999 | 26 |
| 35 | 650792 | 421 | 951476 | 105 | 699316 | 526 | 300684 | 25 |
| 36 | 651044 | 420 | 951412 | 105 | 699632 | 526 | 300368 | 24 |
| 37 | 651297 | 420 | 951349 | 106 | 699947 | 526 | 300053 | 23 |
| 38 | 651549 | 420 | 951286 | 106 | 700263 | 525 | 299737 | 22 |
| 39 | 651800 | 419 | 951222 | 106 | 700578 | 525 | 299422 | 21 |
| 40 | 652052 | 419 | 951159 | 106 | 700893 | 525 | 299107 | 20 |
| 41 | 9·652304 | 419 | 9·951096 | 106 | 9·701208 | 524 | 10·298792 | 19 |
| 42 | 652555 | 418 | 951032 | 106 | 701523 | 524 | 298477 | 18 |
| 43 | 652806 | 418 | 950968 | 106 | 701837 | 524 | 298163 | 17 |
| 44 | 653057 | 418 | 950905 | 106 | 702152 | 524 | 297848 | 16 |
| 45 | 653308 | 418 | 950841 | 106 | 702466 | 524 | 297534 | 15 |
| 46 | 653558 | 417 | 950778 | 106 | 702781 | 523 | 297219 | 14 |
| 47 | 653808 | 417 | 950714 | 106 | 703095 | 523 | 296905 | 13 |
| 48 | 654059 | 417 | 950650 | 106 | 703409 | 523 | 296591 | 12 |
| 49 | 654309 | 416 | 950586 | 106 | 703722 | 523 | 296278 | 11 |
| 50 | 654558 | 416 | 950522 | 107 | 704036 | 522 | 295964 | 10 |
| 51 | 9·654808 | 416 | 9·950458 | 107 | 9·704350 | 522 | 10·295650 | 9 |
| 52 | 655058 | 416 | 950394 | 107 | 704663 | 522 | 295337 | 8 |
| 53 | 655307 | 415 | 950330 | 107 | 704976 | 522 | 295024 | 7 |
| 54 | 655556 | 415 | 950266 | 107 | 705290 | 522 | 294710 | 6 |
| 55 | 655805 | 415 | 950202 | 107 | 705603 | 521 | 294397 | 5 |
| 56 | 656054 | 414 | 950138 | 107 | 705916 | 521 | 294084 | 4 |
| 57 | 656302 | 414 | 950074 | 107 | 706228 | 521 | 293772 | 3 |
| 58 | 656551 | 414 | 950010 | 107 | 706541 | 521 | 293459 | 2 |
| 59 | 656799 | 413 | 949945 | 107 | 706854 | 521 | 293146 | 1 |
| 60 | 657047 | 413 | 949881 | 107 | 707166 | 520 | 292834 | 0 |
| ′ | Cosine. | D. | Sine. | D. | Cotang. | D. | Tang. | ′ |
| 116° | | | | | | | | 63° |

| 27° | | | | | | | | 152° |
|---|---|---|---|---|---|---|---|---|
| ′ | Sine. | D. | Cosine. | D. | Tang. | D. | Cotang. | ′ |
| 0 | 9·657047 | 413 | 9·949881 | 107 | 9·707166 | 520 | 10·292834 | 60 |
| 1 | 657295 | 413 | 949816 | 107 | 707478 | 520 | 292522 | 59 |
| 2 | 657542 | 412 | 949752 | 107 | 707790 | 520 | 292210 | 58 |
| 3 | 657790 | 412 | 949688 | 108 | 708102 | 520 | 291898 | 57 |
| 4 | 658037 | 412 | 949623 | 108 | 708414 | 519 | 291586 | 56 |
| 5 | 658284 | 412 | 949558 | 108 | 708726 | 519 | 291274 | 55 |
| 6 | 658531 | 411 | 949494 | 108 | 709037 | 519 | 290963 | 54 |
| 7 | 658778 | 411 | 949429 | 108 | 709349 | 519 | 290651 | 53 |
| 8 | 659025 | 411 | 949364 | 108 | 709660 | 519 | 290340 | 52 |
| 9 | 659271 | 410 | 949300 | 108 | 709971 | 518 | 290029 | 51 |
| 10 | 659517 | 410 | 949235 | 108 | 710282 | 518 | 289718 | 50 |
| 11 | 9·659763 | 410 | 9·949170 | 108 | 9·710593 | 518 | 10·289407 | 49 |
| 12 | 660009 | 409 | 949105 | 108 | 710904 | 518 | 289096 | 48 |
| 13 | 660255 | 409 | 949040 | 108 | 711215 | 518 | 288785 | 47 |
| 14 | 660501 | 409 | 948975 | 108 | 711525 | 517 | 288475 | 46 |
| 15 | 660746 | 409 | 948910 | 108 | 711836 | 517 | 288164 | 45 |
| 16 | 660991 | 408 | 948845 | 108 | 712146 | 517 | 287854 | 44 |
| 17 | 661236 | 408 | 948780 | 109 | 712456 | 517 | 287544 | 43 |
| 18 | 661481 | 408 | 948715 | 109 | 712766 | 516 | 287234 | 42 |
| 19 | 661726 | 407 | 948650 | 109 | 713076 | 516 | 286924 | 41 |
| 20 | 661970 | 407 | 948584 | 109 | 713386 | 516 | 286614 | 40 |
| 21 | 9·662214 | 407 | 9·948519 | 109 | 9·713696 | 516 | 10·286304 | 39 |
| 22 | 662459 | 407 | 948454 | 109 | 714005 | 516 | 285995 | 38 |
| 23 | 662703 | 406 | 948388 | 109 | 714314 | 515 | 285686 | 37 |
| 24 | 662946 | 406 | 948323 | 109 | 714624 | 515 | 285376 | 36 |
| 25 | 663190 | 406 | 948257 | 109 | 714933 | 515 | 285067 | 35 |
| 26 | 663433 | 405 | 948192 | 109 | 715242 | 515 | 284758 | 34 |
| 27 | 663677 | 405 | 948126 | 109 | 715551 | 514 | 284449 | 33 |
| 28 | 663920 | 405 | 948060 | 109 | 715860 | 514 | 284140 | 32 |
| 29 | 664163 | 405 | 947995 | 110 | 716168 | 514 | 283832 | 31 |
| 30 | 664406 | 404 | 947929 | 110 | 716477 | 514 | 283523 | 30 |
| 31 | 9·664648 | 404 | 9·947863 | 110 | 9·716785 | 514 | 10·283215 | 29 |
| 32 | 664891 | 404 | 947797 | 110 | 717093 | 513 | 282907 | 28 |
| 33 | 665133 | 403 | 947731 | 110 | 717401 | 513 | 282599 | 27 |
| 34 | 665375 | 403 | 947665 | 110 | 717709 | 513 | 282291 | 26 |
| 35 | 665617 | 403 | 947600 | 110 | 718017 | 513 | 281983 | 25 |
| 36 | 665859 | 402 | 947533 | 110 | 718325 | 513 | 281675 | 24 |
| 37 | 666100 | 402 | 947467 | 110 | 718633 | 512 | 281367 | 23 |
| 38 | 666342 | 402 | 947401 | 110 | 718940 | 512 | 281060 | 22 |
| 39 | 666583 | 402 | 947335 | 110 | 719248 | 512 | 280752 | 21 |
| 40 | 666824 | 401 | 947269 | 110 | 719555 | 512 | 280445 | 20 |
| 41 | 9·667065 | 401 | 9·947203 | 110 | 9·719862 | 512 | 10·280138 | 19 |
| 42 | 667305 | 401 | 947136 | 111 | 720169 | 511 | 279831 | 18 |
| 43 | 667546 | 401 | 947070 | 111 | 720476 | 511 | 279524 | 17 |
| 44 | 667786 | 400 | 947004 | 111 | 720783 | 511 | 279217 | 16 |
| 45 | 668027 | 400 | 946937 | 111 | 721089 | 511 | 278911 | 15 |
| 46 | 668267 | 400 | 946871 | 111 | 721396 | 511 | 278604 | 14 |
| 47 | 668506 | 399 | 946804 | 111 | 721702 | 510 | 278298 | 13 |
| 48 | 668746 | 399 | 946738 | 111 | 722009 | 510 | 277991 | 12 |
| 49 | 668986 | 399 | 946671 | 111 | 722315 | 510 | 277685 | 11 |
| 50 | 669225 | 399 | 946604 | 111 | 722621 | 510 | 277379 | 10 |
| 51 | 9·669464 | 398 | 9·946538 | 111 | 9·722927 | 510 | 10·277073 | 9 |
| 52 | 669703 | 398 | 946471 | 111 | 723232 | 509 | 276768 | 8 |
| 53 | 669942 | 398 | 946404 | 111 | 723538 | 509 | 276462 | 7 |
| 54 | 670181 | 397 | 946337 | 111 | 723844 | 509 | 276156 | 6 |
| 55 | 670419 | 397 | 946270 | 112 | 724149 | 509 | 275851 | 5 |
| 56 | 670658 | 397 | 946203 | 112 | 724454 | 509 | 275546 | 4 |
| 57 | 670896 | 397 | 946136 | 112 | 724760 | 508 | 275240 | 3 |
| 58 | 671134 | 396 | 946069 | 112 | 725065 | 508 | 274935 | 2 |
| 59 | 671372 | 396 | 946002 | 112 | 725370 | 508 | 274630 | 1 |
| 60 | 671609 | 396 | 945935 | 112 | 725674 | 508 | 274326 | 0 |
| ′ | Cosine. | D. | Sine. | D. | Cotang. | D. | Tang. | ′ |
| 117° | | | | | | | | 62° |

28° | 151°

| ′ | Sine. | D. | Cosine. | D. | Tang. | D. | Cotang. | ′ |
|---|---|---|---|---|---|---|---|---|
| 0 | 9·671609 | 396 | 9·945935 | 112 | 9·725674 | 508 | 10·274326 | 60 |
| 1 | 671847 | 395 | 945868 | 112 | 725979 | 508 | 274021 | 59 |
| 2 | 672084 | 395 | 945800 | 112 | 726284 | 507 | 273716 | 58 |
| 3 | 672321 | 395 | 945733 | 112 | 726588 | 507 | 273412 | 57 |
| 4 | 672558 | 395 | 945666 | 112 | 726892 | 507 | 273108 | 56 |
| 5 | 672795 | 394 | 945598 | 112 | 727197 | 507 | 272803 | 55 |
| 6 | 673032 | 394 | 945531 | 112 | 727501 | 507 | 272499 | 54 |
| 7 | 673268 | 394 | 945464 | 113 | 727805 | 506 | 272195 | 53 |
| 8 | 673505 | 394 | 945396 | 113 | 728109 | 506 | 271891 | 52 |
| 9 | 673741 | 393 | 945328 | 113 | 728412 | 506 | 271588 | 51 |
| 10 | 673977 | 393 | 945261 | 113 | 728716 | 506 | 271284 | 50 |
| 11 | 9·674213 | 393 | 9·945193 | 113 | 9·729020 | 506 | 10·270980 | 49 |
| 12 | 674448 | 392 | 945125 | 113 | 729323 | 505 | 270677 | 48 |
| 13 | 674684 | 392 | 945058 | 113 | 729626 | 505 | 270374 | 47 |
| 14 | 674919 | 392 | 944990 | 113 | 729929 | 505 | 270071 | 46 |
| 15 | 675155 | 392 | 944922 | 113 | 730233 | 505 | 269767 | 45 |
| 16 | 675390 | 391 | 944854 | 113 | 730535 | 505 | 269465 | 44 |
| 17 | 675624 | 391 | 944786 | 113 | 730838 | 504 | 269162 | 43 |
| 18 | 675859 | 391 | 944718 | 113 | 731141 | 504 | 268859 | 42 |
| 19 | 676094 | 391 | 944650 | 113 | 731444 | 504 | 268556 | 41 |
| 20 | 676328 | 390 | 944582 | 114 | 731746 | 504 | 268254 | 40 |
| 21 | 9·676562 | 390 | 9·944514 | 114 | 9·732048 | 504 | 10·267952 | 39 |
| 22 | 676796 | 390 | 944446 | 114 | 732351 | 503 | 267649 | 38 |
| 23 | 677030 | 390 | 944377 | 114 | 732653 | 503 | 267347 | 37 |
| 24 | 677264 | 389 | 944309 | 114 | 732955 | 503 | 267045 | 36 |
| 25 | 677498 | 389 | 944241 | 114 | 733257 | 503 | 266743 | 35 |
| 26 | 677731 | 389 | 944172 | 114 | 733558 | 503 | 266442 | 34 |
| 27 | 677964 | 388 | 944104 | 114 | 733860 | 502 | 266140 | 33 |
| 28 | 678197 | 388 | 944036 | 114 | 734162 | 502 | 265838 | 32 |
| 29 | 678430 | 388 | 943967 | 114 | 734463 | 502 | 265537 | 31 |
| 30 | 678663 | 388 | 943899 | 114 | 734764 | 502 | 265236 | 30 |
| 31 | 9·678895 | 387 | 9·943830 | 114 | 9·735066 | 502 | 10·264934 | 29 |
| 32 | 679128 | 387 | 943761 | 114 | 735367 | 502 | 264633 | 28 |
| 33 | 679360 | 387 | 943693 | 115 | 735668 | 501 | 264332 | 27 |
| 34 | 679592 | 387 | 943624 | 115 | 735969 | 501 | 264031 | 26 |
| 35 | 679824 | 386 | 943555 | 115 | 736269 | 501 | 263731 | 25 |
| 36 | 680056 | 386 | 943486 | 115 | 736570 | 501 | 263430 | 24 |
| 37 | 680288 | 386 | 943417 | 115 | 736870 | 501 | 263130 | 23 |
| 38 | 680519 | 385 | 943348 | 115 | 737171 | 500 | 262829 | 22 |
| 39 | 680750 | 385 | 943279 | 115 | 737471 | 500 | 262529 | 21 |
| 40 | 680982 | 385 | 943210 | 115 | 737771 | 500 | 262229 | 20 |
| 41 | 9·681213 | 385 | 9·943141 | 115 | 9·738071 | 500 | 10·261929 | 19 |
| 42 | 681443 | 384 | 943072 | 115 | 738371 | 500 | 261629 | 18 |
| 43 | 681674 | 384 | 943003 | 115 | 738671 | 499 | 261329 | 17 |
| 44 | 681905 | 384 | 942934 | 115 | 738971 | 499 | 261029 | 16 |
| 45 | 682135 | 384 | 942864 | 115 | 739271 | 499 | 260729 | 15 |
| 46 | 682365 | 383 | 942795 | 116 | 739570 | 499 | 260430 | 14 |
| 47 | 682595 | 383 | 942726 | 116 | 739870 | 499 | 260130 | 13 |
| 48 | 682825 | 383 | 942656 | 116 | 740169 | 499 | 259831 | 12 |
| 49 | 683055 | 383 | 942587 | 116 | 740468 | 498 | 259532 | 11 |
| 50 | 683284 | 382 | 942517 | 116 | 740767 | 498 | 259233 | 10 |
| 51 | 9·683514 | 382 | 9·942448 | 116 | 9·741066 | 498 | 10·258934 | 9 |
| 52 | 683743 | 382 | 942378 | 116 | 741365 | 498 | 258635 | 8 |
| 53 | 683972 | 382 | 942308 | 116 | 741664 | 498 | 258336 | 7 |
| 54 | 684201 | 381 | 942239 | 116 | 741962 | 497 | 258038 | 6 |
| 55 | 684430 | 381 | 942169 | 116 | 742261 | 497 | 257739 | 5 |
| 56 | 684658 | 381 | 942099 | 116 | 742559 | 497 | 257441 | 4 |
| 57 | 684887 | 380 | 942029 | 116 | 742858 | 497 | 257142 | 3 |
| 58 | 685115 | 380 | 941959 | 116 | 743156 | 497 | 256844 | 2 |
| 59 | 685343 | 380 | 941889 | 117 | 743454 | 497 | 256546 | 1 |
| 60 | 685571 | 380 | 941819 | 117 | 743752 | 496 | 256248 | 0 |
| ′ | Cosine. | D. | Sine. | D. | Cotang. | D. | Tang. | ′ |

118° | 61°

29° 150°

| ′ | Sine. | D. | Cosine. | D. | Tang. | D. | Cotang. | ′ |
|---|---|---|---|---|---|---|---|---|
| 0 | 9·685571 | 380 | 9·941819 | 117 | 9·743752 | 496 | 10·256248 | 60 |
| 1 | 685799 | 379 | 941749 | 117 | 744050 | 496 | 255950 | 59 |
| 2 | 686027 | 379 | 941679 | 117 | 744348 | 496 | 255652 | 58 |
| 3 | 686254 | 379 | 941609 | 117 | 744645 | 496 | 255355 | 57 |
| 4 | 686482 | 379 | 941539 | 117 | 744943 | 496 | 255057 | 56 |
| 5 | 686709 | 378 | 941469 | 117 | 745240 | 496 | 254760 | 55 |
| 6 | 686936 | 378 | 941398 | 117 | 745538 | 495 | 254462 | 54 |
| 7 | 687163 | 378 | 941328 | 117 | 745835 | 495 | 254165 | 53 |
| 8 | 687389 | 378 | 941258 | 117 | 746132 | 495 | 253868 | 52 |
| 9 | 687616 | 377 | 941187 | 117 | 746429 | 495 | 253571 | 51 |
| 10 | 687843 | 377 | 941117 | 117 | 746726 | 495 | 253274 | 50 |
| 11 | 9·688069 | 377 | 9·941046 | 118 | 9·747023 | 494 | 10·252977 | 49 |
| 12 | 688295 | 377 | 940975 | 118 | 747319 | 494 | 252681 | 48 |
| 13 | 688521 | 376 | 940905 | 118 | 747616 | 494 | 252384 | 47 |
| 14 | 688747 | 376 | 940834 | 118 | 747913 | 494 | 252087 | 46 |
| 15 | 688972 | 376 | 940763 | 118 | 748209 | 494 | 251791 | 45 |
| 16 | 689198 | 376 | 940693 | 118 | 748505 | 493 | 251495 | 44 |
| 17 | 689423 | 375 | 940622 | 118 | 748801 | 493 | 251199 | 43 |
| 18 | 689648 | 375 | 940551 | 118 | 749097 | 493 | 250903 | 42 |
| 19 | 689873 | 375 | 940480 | 118 | 749393 | 493 | 250607 | 41 |
| 20 | 690098 | 375 | 940409 | 118 | 749689 | 493 | 250311 | 40 |
| 21 | 9·690323 | 374 | 9·940338 | 118 | 9·749985 | 493 | 10·250015 | 39 |
| 22 | 690548 | 374 | 940267 | 118 | 750281 | 492 | 249719 | 38 |
| 23 | 690772 | 374 | 940196 | 118 | 750576 | 492 | 249424 | 37 |
| 24 | 690996 | 374 | 940125 | 119 | 750872 | 492 | 249128 | 36 |
| 25 | 691220 | 373 | 940054 | 119 | 751167 | 492 | 248833 | 35 |
| 26 | 691444 | 373 | 939982 | 119 | 751462 | 492 | 248538 | 34 |
| 27 | 691668 | 373 | 939911 | 119 | 751757 | 492 | 248243 | 33 |
| 28 | 691892 | 373 | 939840 | 119 | 752052 | 491 | 247948 | 32 |
| 29 | 692115 | 372 | 939768 | 119 | 752347 | 491 | 247653 | 31 |
| 30 | 692339 | 372 | 939697 | 119 | 752642 | 491 | 247358 | 30 |
| 31 | 9·692562 | 372 | 9·939625 | 119 | 9·752937 | 491 | 10·247063 | 29 |
| 32 | 692785 | 371 | 939554 | 119 | 753231 | 491 | 246769 | 28 |
| 33 | 693008 | 371 | 939482 | 119 | 753526 | 491 | 246474 | 27 |
| 34 | 693231 | 371 | 939410 | 119 | 753820 | 490 | 246180 | 26 |
| 35 | 693453 | 371 | 939339 | 119 | 754115 | 490 | 245885 | 25 |
| 36 | 693676 | 370 | 939267 | 120 | 754409 | 490 | 245591 | 24 |
| 37 | 693898 | 370 | 939195 | 120 | 754703 | 490 | 245297 | 23 |
| 38 | 694120 | 370 | 939123 | 120 | 754997 | 490 | 245003 | 22 |
| 39 | 694342 | 370 | 939052 | 120 | 755291 | 490 | 244709 | 21 |
| 40 | 694564 | 369 | 938980 | 120 | 755585 | 489 | 244415 | 20 |
| 41 | 9·694786 | 369 | 9·938908 | 120 | 9·755878 | 489 | 10·244122 | 19 |
| 42 | 695007 | 369 | 938836 | 120 | 756172 | 489 | 243828 | 18 |
| 43 | 695229 | 369 | 938763 | 120 | 756465 | 489 | 243535 | 17 |
| 44 | 695450 | 368 | 938691 | 120 | 756759 | 489 | 243241 | 16 |
| 45 | 695671 | 368 | 938619 | 120 | 757052 | 489 | 242948 | 15 |
| 46 | 695892 | 368 | 938547 | 120 | 757345 | 488 | 242655 | 14 |
| 47 | 696113 | 368 | 938475 | 120 | 757638 | 488 | 242362 | 13 |
| 48 | 696334 | 367 | 938402 | 121 | 757931 | 488 | 242069 | 12 |
| 49 | 696554 | 367 | 938330 | 121 | 758224 | 488 | 241776 | 11 |
| 50 | 696775 | 367 | 938258 | 121 | 758517 | 488 | 241483 | 10 |
| 51 | 9·696995 | 367 | 9·938185 | 121 | 9·758810 | 488 | 10·241190 | 9 |
| 52 | 697215 | 366 | 938113 | 121 | 759102 | 487 | 240898 | 8 |
| 53 | 697435 | 366 | 938040 | 121 | 759395 | 487 | 240605 | 7 |
| 54 | 697654 | 366 | 937967 | 121 | 759687 | 487 | 240313 | 6 |
| 55 | 697874 | 366 | 937895 | 121 | 759979 | 487 | 240021 | 5 |
| 56 | 698094 | 365 | 937822 | 121 | 760272 | 487 | 239728 | 4 |
| 57 | 698313 | 365 | 937749 | 121 | 760564 | 487 | 239436 | 3 |
| 58 | 698532 | 365 | 937676 | 121 | 760856 | 486 | 239144 | 2 |
| 59 | 698751 | 365 | 937604 | 121 | 761148 | 486 | 238852 | 1 |
| 60 | 698970 | 364 | 937531 | 121 | 761439 | 486 | 238561 | 0 |
| ′ | Cosine. | D. | Sine. | D. | Cotang. | D. | Tang. | ′ |

119° 60°

30° | 149°

| ′ | Sine. | D. | Cosine. | D. | Tang. | D. | Cotang. | ′ |
|---|---|---|---|---|---|---|---|---|
| 0 | 9·698970 | 364 | 9·937531 | 121 | 9·761439 | 486 | 10·238561 | 60 |
| 1 | 699189 | 364 | 937458 | 122 | 761731 | 486 | 238269 | 59 |
| 2 | 699407 | 364 | 937385 | 122 | 762023 | 486 | 237977 | 58 |
| 3 | 699626 | 364 | 937312 | 122 | 762314 | 486 | 237686 | 57 |
| 4 | 699844 | 363 | 937238 | 122 | 762606 | 485 | 237394 | 56 |
| 5 | 700062 | 363 | 937165 | 122 | 762897 | 485 | 237103 | 55 |
| 6 | 700280 | 363 | 937092 | 122 | 763188 | 485 | 236812 | 54 |
| 7 | 700498 | 363 | 937019 | 122 | 763479 | 485 | 236521 | 53 |
| 8 | 700716 | 363 | 936946 | 122 | 763770 | 485 | 236230 | 52 |
| 9 | 700933 | 362 | 936872 | 122 | 764061 | 485 | 235939 | 51 |
| 10 | 701151 | 362 | 936799 | 122 | 764352 | 484 | 235648 | 50 |
| 11 | 9·701368 | 362 | 9·936725 | 122 | 9·764643 | 484 | 10·235357 | 49 |
| 12 | 701585 | 362 | 936652 | 123 | 764933 | 484 | 235067 | 48 |
| 13 | 701802 | 361 | 936578 | 123 | 765224 | 484 | 234776 | 47 |
| 14 | 702019 | 361 | 936505 | 123 | 765514 | 484 | 234486 | 46 |
| 15 | 702236 | 361 | 936431 | 123 | 765805 | 484 | 234195 | 45 |
| 16 | 702452 | 361 | 936357 | 123 | 766095 | 484 | 233905 | 44 |
| 17 | 702669 | 360 | 936284 | 123 | 766385 | 483 | 233615 | 43 |
| 18 | 702885 | 360 | 936210 | 123 | 766675 | 483 | 233325 | 42 |
| 19 | 703101 | 360 | 936136 | 123 | 766965 | 483 | 233035 | 41 |
| 20 | 703317 | 360 | 936062 | 123 | 767255 | 483 | 232745 | 40 |
| 21 | 9·703533 | 359 | 9·935988 | 123 | 9·767545 | 483 | 10·232455 | 39 |
| 22 | 703749 | 359 | 935914 | 123 | 767834 | 483 | 232166 | 38 |
| 23 | 703964 | 359 | 935840 | 123 | 768124 | 482 | 231876 | 37 |
| 24 | 704179 | 359 | 935766 | 124 | 768414 | 482 | 231586 | 36 |
| 25 | 704395 | 359 | 935692 | 124 | 768703 | 482 | 231297 | 35 |
| 26 | 704610 | 358 | 935618 | 124 | 768992 | 482 | 231008 | 34 |
| 27 | 704825 | 358 | 935543 | 124 | 769281 | 482 | 230719 | 33 |
| 28 | 705040 | 358 | 935469 | 124 | 769571 | 482 | 230429 | 32 |
| 29 | 705254 | 358 | 935395 | 124 | 769860 | 481 | 230140 | 31 |
| 30 | 705469 | 357 | 935320 | 124 | 770148 | 481 | 229852 | 30 |
| 31 | 9·705683 | 357 | 9·935246 | 124 | 9·770437 | 481 | 10·229563 | 29 |
| 32 | 705898 | 357 | 935171 | 124 | 770726 | 481 | 229274 | 28 |
| 33 | 706112 | 357 | 935097 | 124 | 771015 | 481 | 228985 | 27 |
| 34 | 706326 | 356 | 935022 | 124 | 771303 | 481 | 228697 | 26 |
| 35 | 706539 | 356 | 934948 | 124 | 771592 | 481 | 228408 | 25 |
| 36 | 706753 | 356 | 934873 | 124 | 771880 | 480 | 228120 | 24 |
| 37 | 706967 | 356 | 934798 | 125 | 772168 | 480 | 227832 | 23 |
| 38 | 707180 | 355 | 934723 | 125 | 772457 | 480 | 227543 | 22 |
| 39 | 707393 | 355 | 934649 | 125 | 772745 | 480 | 227255 | 21 |
| 40 | 707606 | 355 | 934574 | 125 | 773033 | 480 | 226967 | 20 |
| 41 | 9·707819 | 355 | 9·934499 | 125 | 9·773321 | 480 | 10·226679 | 19 |
| 42 | 708032 | 354 | 934424 | 125 | 773608 | 479 | 226392 | 18 |
| 43 | 708245 | 354 | 934349 | 125 | 773896 | 479 | 226104 | 17 |
| 44 | 708458 | 354 | 934274 | 125 | 774184 | 479 | 225816 | 16 |
| 45 | 708670 | 354 | 934199 | 125 | 774471 | 479 | 225529 | 15 |
| 46 | 708882 | 353 | 934123 | 125 | 774759 | 479 | 225241 | 14 |
| 47 | 709094 | 353 | 934048 | 125 | 775046 | 479 | 224954 | 13 |
| 48 | 709306 | 353 | 933973 | 125 | 775333 | 479 | 224667 | 12 |
| 49 | 709518 | 353 | 933898 | 126 | 775621 | 478 | 224379 | 11 |
| 50 | 709730 | 353 | 933822 | 126 | 775908 | 478 | 224092 | 10 |
| 51 | 9·709941 | 352 | 9·933747 | 126 | 9·776195 | 478 | 10·223805 | 9 |
| 52 | 710153 | 352 | 933671 | 126 | 776482 | 478 | 223518 | 8 |
| 53 | 710364 | 352 | 933596 | 126 | 776768 | 478 | 223232 | 7 |
| 54 | 710575 | 352 | 933520 | 126 | 777055 | 478 | 222945 | 6 |
| 55 | 710786 | 351 | 933445 | 126 | 777342 | 478 | 222658 | 5 |
| 56 | 710997 | 351 | 933369 | 126 | 777628 | 477 | 222372 | 4 |
| 57 | 711208 | 351 | 933293 | 126 | 777915 | 477 | 222085 | 3 |
| 58 | 711419 | 351 | 933217 | 126 | 778201 | 477 | 221799 | 2 |
| 59 | 711629 | 350 | 933141 | 126 | 778488 | 477 | 221512 | 1 |
| 60 | 711839 | 350 | 933066 | 126 | 778774 | 477 | 221226 | 0 |
| ′ | Cosine. | D. | Sine. | D. | Cotang. | D. | Tang. | ′ |

120° | 59°

31° | | | | | | | | 148°

| ′ | Sine. | D. | Cosine. | D. | Tang. | D. | Cotang. | ′ |
|---|---|---|---|---|---|---|---|---|
| 0 | 9·711839 | 350 | 9·933066 | 126 | 9·778774 | 477 | 10·221226 | 60 |
| 1 | 712050 | 350 | 932990 | 127 | 779060 | 477 | 220940 | 59 |
| 2 | 712260 | 350 | 932914 | 127 | 779346 | 476 | 220654 | 58 |
| 3 | 712469 | 349 | 932838 | 127 | 779632 | 476 | 220368 | 57 |
| 4 | 712679 | 349 | 932762 | 127 | 779918 | 476 | 220082 | 56 |
| 5 | 712889 | 349 | 932685 | 127 | 780203 | 476 | 219797 | 55 |
| 6 | 713098 | 349 | 932609 | 127 | 780489 | 476 | 219511 | 54 |
| 7 | 713308 | 349 | 932533 | 127 | 780775 | 476 | 219225 | 53 |
| 8 | 713517 | 348 | 932457 | 127 | 781060 | 476 | 218940 | 52 |
| 9 | 713726 | 348 | 932380 | 127 | 781346 | 475 | 218654 | 51 |
| 10 | 713935 | 348 | 932304 | 127 | 781631 | 475 | 218369 | 50 |
| 11 | 9·714144 | 348 | 9·932228 | 127 | 9·781916 | 475 | 10·218084 | 49 |
| 12 | 714352 | 347 | 932151 | 127 | 782201 | 475 | 217799 | 48 |
| 13 | 714561 | 347 | 932075 | 128 | 782486 | 475 | 217514 | 47 |
| 14 | 714769 | 347 | 931998 | 128 | 782771 | 475 | 217229 | 46 |
| 15 | 714978 | 347 | 931921 | 128 | 783056 | 475 | 216944 | 45 |
| 16 | 715186 | 347 | 931845 | 128 | 783341 | 475 | 216659 | 44 |
| 17 | 715394 | 346 | 931768 | 128 | 783626 | 474 | 216374 | 43 |
| 18 | 715602 | 346 | 931691 | 128 | 783910 | 474 | 216090 | 42 |
| 19 | 715809 | 346 | 931614 | 128 | 784195 | 474 | 215805 | 41 |
| 20 | 716017 | 346 | 931537 | 128 | 784479 | 474 | 215521 | 40 |
| 21 | 9·716224 | 345 | 9·931460 | 128 | 9·784764 | 474 | 10·215236 | 39 |
| 22 | 716432 | 345 | 931383 | 128 | 785048 | 474 | 214952 | 38 |
| 23 | 716639 | 345 | 931306 | 128 | 785332 | 473 | 214668 | 37 |
| 24 | 716846 | 345 | 931229 | 129 | 785616 | 473 | 214384 | 36 |
| 25 | 717053 | 345 | 931152 | 129 | 785900 | 473 | 214100 | 35 |
| 26 | 717259 | 344 | 931075 | 129 | 786184 | 473 | 213816 | 34 |
| 27 | 717466 | 344 | 930998 | 129 | 786468 | 473 | 213532 | 33 |
| 28 | 717673 | 344 | 930921 | 129 | 786752 | 473 | 213248 | 32 |
| 29 | 717879 | 344 | 930843 | 129 | 787036 | 473 | 212964 | 31 |
| 30 | 718085 | 343 | 930766 | 129 | 787319 | 472 | 212681 | 30 |
| 31 | 9·718291 | 343 | 9·930688 | 129 | 9·787603 | 472 | 10·212397 | 29 |
| 32 | 718497 | 343 | 930611 | 129 | 787886 | 472 | 212114 | 28 |
| 33 | 718703 | 343 | 930533 | 129 | 788170 | 472 | 211830 | 27 |
| 34 | 718909 | 343 | 930456 | 129 | 788453 | 472 | 211547 | 26 |
| 35 | 719114 | 342 | 930378 | 129 | 788736 | 472 | 211264 | 25 |
| 36 | 719320 | 342 | 930300 | 130 | 789019 | 472 | 210981 | 24 |
| 37 | 719525 | 342 | 930223 | 130 | 789302 | 471 | 210698 | 23 |
| 38 | 719730 | 342 | 930145 | 130 | 789585 | 471 | 210415 | 22 |
| 39 | 719935 | 341 | 930067 | 130 | 789868 | 471 | 210132 | 21 |
| 40 | 720140 | 341 | 929989 | 130 | 790151 | 471 | 209849 | 20 |
| 41 | 9·720345 | 341 | 9·929911 | 130 | 9·790434 | 471 | 10·209566 | 19 |
| 42 | 720549 | 341 | 929833 | 130 | 790716 | 471 | 209284 | 18 |
| 43 | 720754 | 340 | 929755 | 130 | 790999 | 471 | 209001 | 17 |
| 44 | 720958 | 340 | 929677 | 130 | 791281 | 471 | 208719 | 16 |
| 45 | 721162 | 340 | 929599 | 130 | 791563 | 470 | 208437 | 15 |
| 46 | 721366 | 340 | 929521 | 130 | 791846 | 470 | 208154 | 14 |
| 47 | 721570 | 340 | 929442 | 130 | 792128 | 470 | 207872 | 13 |
| 48 | 721774 | 339 | 929364 | 131 | 792410 | 470 | 207590 | 12 |
| 49 | 721978 | 339 | 929286 | 131 | 792692 | 470 | 207308 | 11 |
| 50 | 722181 | 339 | 929207 | 131 | 792974 | 470 | 207026 | 10 |
| 51 | 9·722385 | 339 | 9·929129 | 131 | 9·793256 | 470 | 10·206744 | 9 |
| 52 | 722588 | 339 | 929050 | 131 | 793538 | 469 | 206462 | 8 |
| 53 | 722791 | 338 | 928972 | 131 | 793819 | 469 | 206181 | 7 |
| 54 | 722994 | 338 | 928893 | 131 | 794101 | 469 | 205899 | 6 |
| 55 | 723197 | 338 | 928815 | 131 | 794383 | 469 | 205617 | 5 |
| 56 | 723400 | 338 | 928736 | 131 | 794664 | 469 | 205336 | 4 |
| 57 | 723603 | 337 | 928657 | 131 | 794946 | 469 | 205054 | 3 |
| 58 | 723805 | 337 | 928578 | 131 | 795227 | 469 | 204773 | 2 |
| 59 | 724007 | 337 | 928499 | 131 | 795508 | 468 | 204492 | 1 |
| 60 | 724210 | 337 | 928420 | 131 | 795789 | 468 | 204211 | 0 |
| ′ | Cosine. | D. | Sine. | D. | Cotang. | D. | Tang. | ′ |

121° | | | | | | | | 58°

32° 147°

| ′ | Sine. | D. | Cosine. | D. | Tang. | D. | Cotang. | ′ |
|---|---|---|---|---|---|---|---|---|
| 0 | 9·724210 | 337 | 9·928420 | 132 | 9·795789 | 468 | 10·204211 | 60 |
| 1 | 724412 | 337 | 928342 | 132 | 796070 | 468 | 203930 | 59 |
| 2 | 724614 | 336 | 928263 | 132 | 796351 | 468 | 203649 | 58 |
| 3 | 724816 | 336 | 928183 | 132 | 796632 | 468 | 203368 | 57 |
| 4 | 725017 | 336 | 928104 | 132 | 796913 | 468 | 203087 | 56 |
| 5 | 725219 | 336 | 928025 | 132 | 797194 | 468 | 202806 | 55 |
| 6 | 725420 | 335 | 927946 | 132 | 797474 | 468 | 202526 | 54 |
| 7 | 725622 | 335 | 927867 | 132 | 797755 | 468 | 202245 | 53 |
| 8 | 725823 | 335 | 927787 | 132 | 798036 | 467 | 201964 | 52 |
| 9 | 726024 | 335 | 927708 | 132 | 798316 | 467 | 201684 | 51 |
| 10 | 726225 | 335 | 927629 | 132 | 798596 | 467 | 201404 | 50 |
| 11 | 9·726426 | 334 | 9·927549 | 132 | 9·798877 | 467 | 10·201123 | 49 |
| 12 | 726626 | 334 | 927470 | 133 | 799157 | 467 | 200843 | 48 |
| 13 | 726827 | 334 | 927390 | 133 | 799437 | 467 | 200563 | 47 |
| 14 | 727027 | 334 | 927310 | 133 | 799717 | 467 | 200283 | 46 |
| 15 | 727228 | 334 | 927231 | 133 | 799997 | 466 | 200003 | 45 |
| 16 | 727428 | 333 | 927151 | 133 | 800277 | 466 | 199723 | 44 |
| 17 | 727628 | 333 | 927071 | 133 | 800557 | 466 | 199443 | 43 |
| 18 | 727828 | 333 | 926991 | 133 | 800836 | 466 | 199164 | 42 |
| 19 | 728027 | 333 | 926911 | 133 | 801116 | 466 | 198884 | 41 |
| 20 | 728227 | 333 | 926831 | 133 | 801396 | 466 | 198604 | 40 |
| 21 | 9·728427 | 332 | 9·926751 | 133 | 9·801675 | 466 | 10·198325 | 39 |
| 22 | 728626 | 332 | 926671 | 133 | 801955 | 466 | 198045 | 38 |
| 23 | 728825 | 332 | 926591 | 133 | 802234 | 465 | 197766 | 37 |
| 24 | 729024 | 332 | 926511 | 134 | 802513 | 465 | 197487 | 36 |
| 25 | 729223 | 331 | 926431 | 134 | 802792 | 465 | 197208 | 35 |
| 26 | 729422 | 331 | 926351 | 134 | 803072 | 465 | 196928 | 34 |
| 27 | 729621 | 331 | 926270 | 134 | 803351 | 465 | 196649 | 33 |
| 28 | 729820 | 331 | 926190 | 134 | 803630 | 465 | 196370 | 32 |
| 29 | 730018 | 330 | 926110 | 134 | 803909 | 465 | 196091 | 31 |
| 30 | 730217 | 330 | 926029 | 134 | 804187 | 465 | 195813 | 30 |
| 31 | 9·730415 | 330 | 9·925949 | 134 | 9·804466 | 464 | 10·195534 | 29 |
| 32 | 730613 | 330 | 925868 | 134 | 804745 | 464 | 195255 | 28 |
| 33 | 730811 | 330 | 925788 | 134 | 805023 | 464 | 194977 | 27 |
| 34 | 731009 | 329 | 925707 | 134 | 805302 | 464 | 194698 | 26 |
| 35 | 731206 | 329 | 925626 | 134 | 805580 | 464 | 194420 | 25 |
| 36 | 731404 | 329 | 925545 | 135 | 805859 | 464 | 194141 | 24 |
| 37 | 731602 | 329 | 925465 | 135 | 806137 | 464 | 193863 | 23 |
| 38 | 731799 | 329 | 925384 | 135 | 806415 | 463 | 193585 | 22 |
| 39 | 731996 | 328 | 925303 | 135 | 806693 | 463 | 193307 | 21 |
| 40 | 732193 | 328 | 925222 | 135 | 806971 | 463 | 193029 | 20 |
| 41 | 9·732390 | 328 | 9·925141 | 135 | 9·807249 | 463 | 10·192751 | 19 |
| 42 | 732587 | 328 | 925060 | 135 | 807527 | 463 | 192473 | 18 |
| 43 | 732784 | 328 | 924979 | 135 | 807805 | 463 | 192195 | 17 |
| 44 | 732980 | 327 | 924897 | 135 | 808083 | 463 | 191917 | 16 |
| 45 | 733177 | 327 | 924816 | 135 | 808361 | 463 | 191639 | 15 |
| 46 | 733373 | 327 | 924735 | 136 | 808638 | 462 | 191362 | 14 |
| 47 | 733569 | 327 | 924654 | 136 | 808916 | 462 | 191084 | 13 |
| 48 | 733765 | 327 | 924572 | 136 | 809193 | 462 | 190807 | 12 |
| 49 | 733961 | 326 | 924491 | 136 | 809471 | 462 | 190529 | 11 |
| 50 | 734157 | 326 | 924409 | 136 | 809748 | 462 | 190252 | 10 |
| 51 | 9·734353 | 326 | 9·924328 | 136 | 9·810025 | 462 | 10·189975 | 9 |
| 52 | 734549 | 326 | 924246 | 136 | 810302 | 462 | 189698 | 8 |
| 53 | 734744 | 325 | 924164 | 136 | 810580 | 462 | 189420 | 7 |
| 54 | 734939 | 325 | 924083 | 136 | 810857 | 462 | 189143 | 6 |
| 55 | 735135 | 325 | 924001 | 136 | 811134 | 461 | 188866 | 5 |
| 56 | 735330 | 325 | 923919 | 136 | 811410 | 461 | 188590 | 4 |
| 57 | 735525 | 325 | 923837 | 136 | 811687 | 461 | 188313 | 3 |
| 58 | 735719 | 324 | 923755 | 137 | 811964 | 461 | 188036 | 2 |
| 59 | 735914 | 324 | 923673 | 137 | 812241 | 461 | 187759 | 1 |
| 60 | 736109 | 324 | 923591 | 137 | 812517 | 461 | 187483 | 0 |
| ′ | Cosine. | D. | Sine. | D. | Cotang. | D. | Tang. | ′ |

122° 57°

33° 146°

| ′ | Sine. | D. | Cosine. | D. | Tang. | D. | Cotang. | ′ |
|---|---|---|---|---|---|---|---|---|
| 0 | 9·736109 | 324 | 9·923591 | 137 | 9·812517 | 461 | 10·187483 | 60 |
| 1 | 736303 | 324 | 923509 | 137 | 812794 | 461 | 187206 | 59 |
| 2 | 736498 | 324 | 923427 | 137 | 813070 | 461 | 186930 | 58 |
| 3 | 736692 | 323 | 923345 | 137 | 813347 | 460 | 186653 | 57 |
| 4 | 736886 | 323 | 923263 | 137 | 813623 | 460 | 186377 | 56 |
| 5 | 737080 | 323 | 923181 | 137 | 813899 | 460 | 186101 | 55 |
| 6 | 737274 | 323 | 923098 | 137 | 814176 | 460 | 185824 | 54 |
| 7 | 737467 | 323 | 923016 | 137 | 814452 | 460 | 185548 | 53 |
| 8 | 737661 | 322 | 922933 | 137 | 814728 | 460 | 185272 | 52 |
| 9 | 737855 | 322 | 922851 | 137 | 815004 | 460 | 184996 | 51 |
| 10 | 738048 | 322 | 922768 | 138 | 815280 | 460 | 184720 | 50 |
| 11 | 9·738241 | 322 | 9·922686 | 138 | 9·815555 | 459 | 10·184445 | 49 |
| 12 | 738434 | 322 | 922603 | 138 | 815831 | 459 | 184169 | 48 |
| 13 | 738627 | 321 | 922520 | 138 | 816107 | 459 | 183893 | 47 |
| 14 | 738820 | 321 | 922438 | 138 | 816382 | 459 | 183618 | 46 |
| 15 | 739013 | 321 | 922355 | 138 | 816658 | 459 | 183342 | 45 |
| 16 | 739206 | 321 | 922272 | 138 | 816933 | 459 | 183067 | 44 |
| 17 | 739398 | 321 | 922189 | 138 | 817209 | 459 | 182791 | 43 |
| 18 | 739590 | 320 | 922106 | 138 | 817484 | 459 | 182516 | 42 |
| 19 | 739783 | 320 | 922023 | 138 | 817759 | 459 | 182241 | 41 |
| 20 | 739975 | 320 | 921940 | 138 | 818035 | 458 | 181965 | 40 |
| 21 | 9·740167 | 320 | 9·921857 | 139 | 9·818310 | 458 | 10·181690 | 39 |
| 22 | 740359 | 320 | 921774 | 139 | 818585 | 458 | 181415 | 38 |
| 23 | 740550 | 319 | 921691 | 139 | 818860 | 458 | 181140 | 37 |
| 24 | 740742 | 319 | 921607 | 139 | 819135 | 458 | 180865 | 36 |
| 25 | 740934 | 319 | 921524 | 139 | 819410 | 458 | 180590 | 35 |
| 26 | 741125 | 319 | 921441 | 139 | 819684 | 458 | 180316 | 34 |
| 27 | 741316 | 319 | 921357 | 139 | 819959 | 458 | 180041 | 33 |
| 28 | 741508 | 318 | 921274 | 139 | 820234 | 458 | 179766 | 32 |
| 29 | 741699 | 318 | 921190 | 139 | 820508 | 457 | 179492 | 31 |
| 30 | 741889 | 318 | 921107 | 139 | 820783 | 457 | 179217 | 30 |
| 31 | 9·742080 | 318 | 9·921023 | 139 | 9·821057 | 457 | 10·178943 | 29 |
| 32 | 742271 | 318 | 920939 | 140 | 821332 | 457 | 178668 | 28 |
| 33 | 742462 | 317 | 920856 | 140 | 821606 | 457 | 178394 | 27 |
| 34 | 742652 | 317 | 920772 | 140 | 821880 | 457 | 178120 | 26 |
| 35 | 742842 | 317 | 920688 | 140 | 822154 | 457 | 177846 | 25 |
| 36 | 743033 | 317 | 920604 | 140 | 822429 | 457 | 177571 | 24 |
| 37 | 743223 | 317 | 920520 | 140 | 822703 | 457 | 177297 | 23 |
| 38 | 743413 | 316 | 920436 | 140 | 822977 | 456 | 177023 | 22 |
| 39 | 743602 | 316 | 920352 | 140 | 823251 | 456 | 176749 | 21 |
| 40 | 743792 | 316 | 920268 | 140 | 823524 | 456 | 176476 | 20 |
| 41 | 9·743982 | 316 | 9·920184 | 140 | 9·823798 | 456 | 10·176202 | 19 |
| 42 | 744171 | 316 | 920099 | 140 | 824072 | 456 | 175928 | 18 |
| 43 | 744361 | 315 | 920015 | 140 | 824345 | 456 | 175655 | 17 |
| 44 | 744550 | 315 | 919931 | 141 | 824619 | 456 | 175381 | 16 |
| 45 | 744739 | 315 | 919846 | 141 | 824893 | 456 | 175107 | 15 |
| 46 | 744928 | 315 | 919762 | 141 | 825166 | 456 | 174834 | 14 |
| 47 | 745117 | 315 | 919677 | 141 | 825439 | 455 | 174561 | 13 |
| 48 | 745306 | 314 | 919593 | 141 | 825713 | 455 | 174287 | 12 |
| 49 | 745494 | 314 | 919508 | 141 | 825986 | 455 | 174014 | 11 |
| 50 | 745683 | 314 | 919424 | 141 | 826259 | 455 | 173741 | 10 |
| 51 | 9·745871 | 314 | 9·919339 | 141 | 9·826532 | 455 | 10·173468 | 9 |
| 52 | 746060 | 314 | 919254 | 141 | 826805 | 455 | 173195 | 8 |
| 53 | 746248 | 313 | 919169 | 141 | 827078 | 455 | 172922 | 7 |
| 54 | 746436 | 313 | 919085 | 141 | 827351 | 455 | 172649 | 6 |
| 55 | 746624 | 313 | 919000 | 141 | 827624 | 455 | 172376 | 5 |
| 56 | 746812 | 313 | 918915 | 142 | 827897 | 454 | 172103 | 4 |
| 57 | 746999 | 313 | 918830 | 142 | 828170 | 454 | 171830 | 3 |
| 58 | 747187 | 312 | 918745 | 142 | 828442 | 454 | 171558 | 2 |
| 59 | 747374 | 312 | 918659 | 142 | 828715 | 454 | 171285 | 1 |
| 60 | 747562 | 312 | 918574 | 142 | 828987 | 454 | 171013 | 0 |
| ′ | Cosine. | D. | Sine. | D. | Cotang. | D. | Tang. | ′ |

123° 56°

| 34° | | | | | | | | 145° |
|---|---|---|---|---|---|---|---|---|
| ′ | Sine. | D. | Cosine. | D. | Tang. | D. | Cotang. | ′ |
| 0 | 9·747562 | 312 | 9·918574 | 142 | 9·828987 | 454 | 10·171013 | 60 |
| 1 | 747749 | 312 | 918489 | 142 | 829260 | 454 | 170740 | 59 |
| 2 | 747936 | 312 | 918404 | 142 | 829532 | 454 | 170468 | 58 |
| 3 | 748123 | 311 | 918318 | 142 | 829805 | 454 | 170195 | 57 |
| 4 | 748310 | 311 | 918233 | 142 | 830077 | 454 | 169923 | 56 |
| 5 | 748497 | 311 | 918147 | 142 | 830349 | 453 | 169651 | 55 |
| 6 | 748683 | 311 | 918062 | 142 | 830621 | 453 | 169379 | 54 |
| 7 | 748870 | 311 | 917976 | 143 | 830893 | 453 | 169107 | 53 |
| 8 | 749056 | 310 | 917891 | 143 | 831165 | 453 | 168835 | 52 |
| 9 | 749243 | 310 | 917805 | 143 | 831437 | 453 | 168563 | 51 |
| 10 | 749429 | 310 | 917719 | 143 | 831709 | 453 | 168291 | 50 |
| 11 | 9·749615 | 310 | 9·917634 | 143 | 9·831981 | 453 | 10·168019 | 49 |
| 12 | 749801 | 310 | 917548 | 143 | 832253 | 453 | 167747 | 48 |
| 13 | 749987 | 309 | 917462 | 143 | 832525 | 453 | 167475 | 47 |
| 14 | 750172 | 309 | 917376 | 143 | 832796 | 453 | 167204 | 46 |
| 15 | 750358 | 309 | 917290 | 143 | 833068 | 452 | 166932 | 45 |
| 16 | 750543 | 309 | 917204 | 143 | 833339 | 452 | 166661 | 44 |
| 17 | 750729 | 309 | 917118 | 144 | 833611 | 452 | 166389 | 43 |
| 18 | 750914 | 308 | 917032 | 144 | 833882 | 452 | 166118 | 42 |
| 19 | 751099 | 308 | 916946 | 144 | 834154 | 452 | 165846 | 41 |
| 20 | 751284 | 308 | 916859 | 144 | 834425 | 452 | 165575 | 40 |
| 21 | 9·751469 | 308 | 9·916773 | 144 | 9·834696 | 452 | 10·165304 | 39 |
| 22 | 751654 | 308 | 916687 | 144 | 834967 | 452 | 165033 | 38 |
| 23 | 751839 | 308 | 916600 | 144 | 835238 | 452 | 164762 | 37 |
| 24 | 752023 | 307 | 916514 | 144 | 835509 | 452 | 164491 | 36 |
| 25 | 752208 | 307 | 916427 | 144 | 835780 | 451 | 164220 | 35 |
| 26 | 752392 | 307 | 916341 | 144 | 836051 | 451 | 163949 | 34 |
| 27 | 752576 | 307 | 916254 | 144 | 836322 | 451 | 163678 | 33 |
| 28 | 752760 | 307 | 916167 | 145 | 836593 | 451 | 163407 | 32 |
| 29 | 752944 | 306 | 916081 | 145 | 836864 | 451 | 163136 | 31 |
| 30 | 753128 | 306 | 915994 | 145 | 837134 | 451 | 162866 | 30 |
| 31 | 9·753312 | 306 | 9·915907 | 145 | 9·837405 | 451 | 10·162595 | 29 |
| 32 | 753495 | 306 | 915820 | 145 | 837675 | 451 | 162325 | 28 |
| 33 | 753679 | 306 | 915733 | 145 | 837946 | 451 | 162054 | 27 |
| 34 | 753862 | 305 | 915646 | 145 | 838216 | 451 | 161784 | 26 |
| 35 | 754046 | 305 | 915559 | 145 | 838487 | 450 | 161513 | 25 |
| 36 | 754229 | 305 | 915472 | 145 | 838757 | 450 | 161243 | 24 |
| 37 | 754412 | 305 | 915385 | 145 | 839027 | 450 | 160973 | 23 |
| 38 | 754595 | 305 | 915297 | 145 | 839297 | 450 | 160703 | 22 |
| 39 | 754778 | 304 | 915210 | 145 | 839568 | 450 | 160432 | 21 |
| 40 | 754960 | 304 | 915123 | 146 | 839838 | 450 | 160162 | 20 |
| 41 | 9·755143 | 304 | 9·915035 | 146 | 9·840108 | 450 | 10·159892 | 19 |
| 42 | 755326 | 304 | 914948 | 146 | 840378 | 450 | 159622 | 18 |
| 43 | 755508 | 304 | 914860 | 146 | 840648 | 450 | 159352 | 17 |
| 44 | 755690 | 304 | 914773 | 146 | 840917 | 449 | 159083 | 16 |
| 45 | 755872 | 303 | 914685 | 146 | 841187 | 449 | 158813 | 15 |
| 46 | 756054 | 303 | 914598 | 146 | 841457 | 449 | 158543 | 14 |
| 47 | 756236 | 303 | 914510 | 146 | 841727 | 449 | 158273 | 13 |
| 48 | 756418 | 303 | 914422 | 146 | 841996 | 449 | 158004 | 12 |
| 49 | 756600 | 303 | 914334 | 146 | 842266 | 449 | 157734 | 11 |
| 50 | 756782 | 302 | 914246 | 147 | 842535 | 449 | 157465 | 10 |
| 51 | 9·756963 | 302 | 9·914158 | 147 | 9·842805 | 449 | 10·157195 | 9 |
| 52 | 757144 | 302 | 914070 | 147 | 843074 | 449 | 156926 | 8 |
| 53 | 757326 | 302 | 913982 | 147 | 843343 | 449 | 156657 | 7 |
| 54 | 757507 | 302 | 913894 | 147 | 843612 | 449 | 156388 | 6 |
| 55 | 757688 | 301 | 913806 | 147 | 843882 | 448 | 156118 | 5 |
| 56 | 757869 | 301 | 913718 | 147 | 844151 | 448 | 155849 | 4 |
| 57 | 758050 | 301 | 913630 | 147 | 844420 | 448 | 155580 | 3 |
| 58 | 758230 | 301 | 913541 | 147 | 844689 | 448 | 155311 | 2 |
| 59 | 758411 | 301 | 913453 | 147 | 844958 | 448 | 155042 | 1 |
| 60 | 758591 | 301 | 913365 | 147 | 845227 | 448 | 154773 | 0 |
| ′ | Cosine. | D. | Sine. | D. | Cotang. | D. | Tang. | ′ |
| 124° | | | | | | | | 55° |

| 35° | | | | | | | | 144° |
|---|---|---|---|---|---|---|---|---|
| ′ | Sine. | D. | Cosine. | D. | Tang. | D. | Cotang. | ′ |
| 0 | 9·758591 | 301 | 9·913365 | 147 | 9·845227 | 448 | 10·154773 | 60 |
| 1 | 758772 | 300 | 913276 | 147 | 845496 | 448 | 154504 | 59 |
| 2 | 758952 | 300 | 913187 | 148 | 845764 | 448 | 154236 | 58 |
| 3 | 759132 | 300 | 913099 | 148 | 846033 | 448 | 153967 | 57 |
| 4 | 759312 | 300 | 913010 | 148 | 846302 | 448 | 153698 | 56 |
| 5 | 759492 | 300 | 912922 | 148 | 846570 | 447 | 153430 | 55 |
| 6 | 759672 | 299 | 912833 | 148 | 846839 | 447 | 153161 | 54 |
| 7 | 759852 | 299 | 912744 | 148 | 847108 | 447 | 152892 | 53 |
| 8 | 760031 | 299 | 912655 | 148 | 847376 | 447 | 152624 | 52 |
| 9 | 760211 | 299 | 912566 | 148 | 847644 | 447 | 152356 | 51 |
| 10 | 760390 | 299 | 912477 | 148 | 847913 | 447 | 152087 | 50 |
| 11 | 9·760569 | 298 | 9·912388 | 148 | 9·848181 | 447 | 10·151819 | 49 |
| 12 | 760748 | 298 | 912299 | 149 | 848449 | 447 | 151551 | 48 |
| 13 | 760927 | 298 | 912210 | 149 | 848717 | 447 | 151283 | 47 |
| 14 | 761106 | 298 | 912121 | 149 | 848986 | 447 | 151014 | 46 |
| 15 | 761285 | 298 | 912031 | 149 | 849254 | 447 | 150746 | 45 |
| 16 | 761464 | 298 | 911942 | 149 | 849522 | 447 | 150478 | 44 |
| 17 | 761642 | 297 | 911853 | 149 | 849790 | 446 | 150210 | 43 |
| 18 | 761821 | 297 | 911763 | 149 | 850057 | 446 | 149943 | 42 |
| 19 | 761999 | 297 | 911674 | 149 | 850325 | 446 | 149675 | 41 |
| 20 | 762177 | 297 | 911584 | 149 | 850593 | 446 | 149407 | 40 |
| 21 | 9·762356 | 297 | 9·911495 | 149 | 9·850861 | 446 | 10·149139 | 39 |
| 22 | 762534 | 296 | 911405 | 149 | 851129 | 446 | 148871 | 38 |
| 23 | 762712 | 296 | 911315 | 150 | 851396 | 446 | 148604 | 37 |
| 24 | 762889 | 296 | 911226 | 150 | 851664 | 446 | 148336 | 36 |
| 25 | 763067 | 296 | 911136 | 150 | 851931 | 446 | 148069 | 35 |
| 26 | 763245 | 296 | 911046 | 150 | 852199 | 446 | 147801 | 34 |
| 27 | 763422 | 296 | 910956 | 150 | 852466 | 446 | 147534 | 33 |
| 28 | 763600 | 295 | 910866 | 150 | 852733 | 445 | 147267 | 32 |
| 29 | 763777 | 295 | 910776 | 150 | 853001 | 445 | 146999 | 31 |
| 30 | 763954 | 295 | 910686 | 150 | 853268 | 445 | 146732 | 30 |
| 31 | 9·764131 | 295 | 9·910596 | 150 | 9·853535 | 445 | 10·146465 | 29 |
| 32 | 764308 | 295 | 910506 | 150 | 853802 | 445 | 146198 | 28 |
| 33 | 764485 | 294 | 910415 | 150 | 854069 | 445 | 145931 | 27 |
| 34 | 764662 | 294 | 910325 | 151 | 854336 | 445 | 145664 | 26 |
| 35 | 764838 | 294 | 910235 | 151 | 854603 | 445 | 145397 | 25 |
| 36 | 765015 | 294 | 910144 | 151 | 854870 | 445 | 145130 | 24 |
| 37 | 765191 | 294 | 910054 | 151 | 855137 | 445 | 144863 | 23 |
| 38 | 765367 | 294 | 909963 | 151 | 855404 | 445 | 144596 | 22 |
| 39 | 765544 | 293 | 909873 | 151 | 855671 | 444 | 144329 | 21 |
| 40 | 765720 | 293 | 909782 | 151 | 855938 | 444 | 144062 | 20 |
| 41 | 9·765896 | 293 | 9·909691 | 151 | 9·856204 | 444 | 10·143796 | 19 |
| 42 | 766072 | 293 | 909601 | 151 | 856471 | 444 | 143529 | 18 |
| 43 | 766247 | 293 | 909510 | 151 | 856737 | 444 | 143263 | 17 |
| 44 | 766423 | 293 | 909419 | 151 | 857004 | 444 | 142996 | 16 |
| 45 | 766598 | 292 | 909328 | 152 | 857270 | 444 | 142730 | 15 |
| 46 | 766774 | 292 | 909237 | 152 | 857537 | 444 | 142463 | 14 |
| 47 | 766949 | 292 | 909146 | 152 | 857803 | 444 | 142197 | 13 |
| 48 | 767124 | 292 | 909055 | 152 | 858069 | 444 | 141931 | 12 |
| 49 | 767300 | 292 | 908964 | 152 | 858336 | 444 | 141664 | 11 |
| 50 | 767475 | 291 | 908873 | 152 | 858602 | 443 | 141398 | 10 |
| 51 | 9·767649 | 291 | 9·908781 | 152 | 9·858868 | 443 | 10·141132 | 9 |
| 52 | 767824 | 291 | 908690 | 152 | 859134 | 443 | 140866 | 8 |
| 53 | 767999 | 291 | 908599 | 152 | 859400 | 443 | 140600 | 7 |
| 54 | 768173 | 291 | 908507 | 152 | 859666 | 443 | 140334 | 6 |
| 55 | 768348 | 290 | 908416 | 153 | 859932 | 443 | 140068 | 5 |
| 56 | 768522 | 290 | 908324 | 153 | 860198 | 443 | 139802 | 4 |
| 57 | 768697 | 290 | 908233 | 153 | 860464 | 443 | 139536 | 3 |
| 58 | 768871 | 290 | 908141 | 153 | 860730 | 443 | 139270 | 2 |
| 59 | 769045 | 290 | 908049 | 153 | 860995 | 443 | 139005 | 1 |
| 60 | 769219 | 290 | 907958 | 153 | 861261 | 443 | 138739 | 0 |
| ′ | Cosine. | D. | Sine. | D. | Cotang. | D. | Tang. | ′ |
| 125° | | | | | | | | 54° |

36° 143°

| ′ | Sine. | D. | Cosine. | D. | Tang. | D. | Cotang. | ′ |
|---|---|---|---|---|---|---|---|---|
| 0 | 9·769219 | 290 | 9·907958 | 153 | 9·861261 | 443 | 10·138739 | 60 |
| 1 | 769393 | 289 | 907866 | 153 | 861527 | 443 | 138473 | 59 |
| 2 | 769566 | 289 | 907774 | 153 | 861792 | 442 | 138208 | 58 |
| 3 | 769740 | 289 | 907682 | 153 | 862058 | 442 | 137942 | 57 |
| 4 | 769913 | 289 | 907590 | 153 | 862323 | 442 | 137677 | 56 |
| 5 | 770087 | 289 | 907498 | 153 | 862589 | 442 | 137411 | 55 |
| 6 | 770260 | 288 | 907406 | 153 | 862854 | 442 | 137146 | 54 |
| 7 | 770433 | 288 | 907314 | 154 | 863119 | 442 | 136881 | 53 |
| 8 | 770606 | 288 | 907222 | 154 | 863385 | 442 | 136615 | 52 |
| 9 | 770779 | 288 | 907129 | 154 | 863650 | 442 | 136350 | 51 |
| 10 | 770952 | 288 | 907037 | 144 | 863915 | 442 | 136085 | 50 |
| 11 | 9·771125 | 288 | 9·906945 | 154 | 9·864180 | 442 | 10·135820 | 49 |
| 12 | 771298 | 287 | 906852 | 154 | 864445 | 442 | 135555 | 48 |
| 13 | 771470 | 287 | 906760 | 154 | 864710 | 442 | 135290 | 47 |
| 14 | 771643 | 287 | 906667 | 154 | 864975 | 441 | 135025 | 46 |
| 15 | 771815 | 287 | 906575 | 154 | 865240 | 441 | 134760 | 45 |
| 16 | 771987 | 287 | 906482 | 154 | 865505 | 441 | 134495 | 44 |
| 17 | 772159 | 287 | 906389 | 155 | 865770 | 441 | 134230 | 43 |
| 18 | 772331 | 286 | 906296 | 155 | 866035 | 441 | 133965 | 42 |
| 19 | 772503 | 286 | 906204 | 155 | 866300 | 441 | 133700 | 41 |
| 20 | 772675 | 286 | 906111 | 155 | 866564 | 441 | 133436 | 40 |
| 21 | 9·772847 | 286 | 9·906018 | 155 | 9·866829 | 441 | 10·133171 | 39 |
| 22 | 773018 | 286 | 905925 | 155 | 867094 | 441 | 132906 | 38 |
| 23 | 773190 | 286 | 905832 | 155 | 867358 | 441 | 132642 | 37 |
| 24 | 773361 | 285 | 905739 | 155 | 867623 | 441 | 132377 | 36 |
| 25 | 773533 | 285 | 905645 | 155 | 867887 | 441 | 132113 | 35 |
| 26 | 773704 | 285 | 905552 | 155 | 868152 | 440 | 131848 | 34 |
| 27 | 773875 | 285 | 905459 | 155 | 868416 | 440 | 131584 | 33 |
| 28 | 774046 | 285 | 905366 | 156 | 868680 | 440 | 131320 | 32 |
| 29 | 774217 | 285 | 905272 | 156 | 868945 | 440 | 131055 | 31 |
| 30 | 774388 | 284 | 905179 | 156 | 869209 | 440 | 130791 | 30 |
| 31 | 9·774558 | 284 | 9·905085 | 156 | 9·869473 | 440 | 10·130527 | 29 |
| 32 | 774729 | 284 | 904992 | 156 | 869737 | 440 | 130263 | 28 |
| 33 | 774899 | 284 | 904898 | 156 | 870001 | 440 | 129999 | 27 |
| 34 | 775070 | 284 | 904804 | 156 | 870265 | 440 | 129735 | 26 |
| 35 | 775240 | 284 | 904711 | 156 | 870529 | 440 | 129471 | 25 |
| 36 | 775410 | 283 | 904617 | 156 | 870793 | 440 | 129207 | 24 |
| 37 | 775580 | 283 | 904523 | 156 | 871057 | 440 | 128943 | 23 |
| 38 | 775750 | 283 | 904429 | 157 | 871321 | 440 | 128679 | 22 |
| 39 | 775920 | 283 | 904335 | 157 | 871585 | 440 | 128415 | 21 |
| 40 | 776090 | 283 | 904241 | 157 | 871849 | 439 | 128151 | 20 |
| 41 | 9·776259 | 283 | 9·904147 | 157 | 9·872112 | 439 | 10·127888 | 19 |
| 42 | 776429 | 282 | 904053 | 157 | 872376 | 439 | 127624 | 18 |
| 43 | 776598 | 282 | 903959 | 157 | 872640 | 439 | 127360 | 17 |
| 44 | 776768 | 282 | 903864 | 157 | 872903 | 439 | 127097 | 16 |
| 45 | 776937 | 282 | 903770 | 157 | 873167 | 439 | 126833 | 15 |
| 46 | 777106 | 282 | 903676 | 157 | 873430 | 439 | 126570 | 14 |
| 47 | 777275 | 281 | 903581 | 167 | 873694 | 439 | 126306 | 13 |
| 48 | 777444 | 281 | 903487 | 157 | 873957 | 439 | 126043 | 12 |
| 49 | 777613 | 281 | 903392 | 158 | 874220 | 439 | 125780 | 11 |
| 50 | 777781 | 281 | 903298 | 158 | 874484 | 439 | 125516 | 10 |
| 51 | 9·777950 | 281 | 9·903203 | 158 | 9·874747 | 439 | 10·125253 | 9 |
| 52 | 778119 | 281 | 903108 | 158 | 875010 | 439 | 124990 | 8 |
| 53 | 778287 | 280 | 903014 | 158 | 875273 | 438 | 124727 | 7 |
| 54 | 778455 | 280 | 902919 | 158 | 875537 | 438 | 124463 | 6 |
| 55 | 778624 | 280 | 902824 | 158 | 875800 | 438 | 124200 | 5 |
| 56 | 778792 | 280 | 902729 | 158 | 876063 | 438 | 123937 | 4 |
| 57 | 778960 | 280 | 902634 | 158 | 876326 | 438 | 123674 | 3 |
| 58 | 779128 | 280 | 902539 | 159 | 876589 | 438 | 123411 | 2 |
| 59 | 779295 | 279 | 902444 | 159 | 876852 | 438 | 123148 | 1 |
| 60 | 779463 | 279 | 902349 | 159 | 877114 | 438 | 122886 | 0 |
| ′ | Cosine. | D. | Sine. | D. | Cotang. | D. | Tang. | ′ |

126° 53°

| 37° | | | | | | | | 142° |
|---|---|---|---|---|---|---|---|---|
| ′ | Sine. | D. | Cosine. | D. | Tang. | D. | Cotang. | ′ |
| 0 | 9·779463 | 279 | 9·902349 | 159 | 9·877114 | 438 | 10·122886 | 60 |
| 1 | 779631 | 279 | 902253 | 159 | 877377 | 438 | 122623 | 59 |
| 2 | 779798 | 279 | 902158 | 159 | 877640 | 438 | 122360 | 58 |
| 3 | 779966 | 279 | 902063 | 159 | 877903 | 438 | 122097 | 57 |
| 4 | 780133 | 279 | 901967 | 159 | 878165 | 438 | 121835 | 56 |
| 5 | 780300 | 278 | 901872 | 159 | 878428 | 438 | 121572 | 55 |
| 6 | 780467 | 278 | 901776 | 159 | 878691 | 438 | 121309 | 54 |
| 7 | 780634 | 278 | 901681 | 159 | 878953 | 437 | 121047 | 53 |
| 8 | 780801 | 278 | 901585 | 159 | 879216 | 437 | 120784 | 52 |
| 9 | 780968 | 278 | 901490 | 159 | 879478 | 437 | 120522 | 51 |
| 10 | 781134 | 278 | 901394 | 160 | 879741 | 437 | 120259 | 50 |
| 11 | 9·781301 | 277 | 9·901298 | 160 | 9·880003 | 437 | 10·119997 | 49 |
| 12 | 781468 | 277 | 901202 | 160 | 880265 | 437 | 119735 | 48 |
| 13 | 781634 | 277 | 901106 | 160 | 880528 | 437 | 119472 | 47 |
| 14 | 781800 | 277 | 901010 | 160 | 880790 | 437 | 119210 | 46 |
| 15 | 781966 | 277 | 900914 | 160 | 881052 | 437 | 118948 | 45 |
| 16 | 782132 | 277 | 900818 | 160 | 881314 | 437 | 118686 | 44 |
| 17 | 782298 | 276 | 900722 | 160 | 881577 | 437 | 118423 | 43 |
| 18 | 782464 | 276 | 900626 | 160 | 881839 | 437 | 118161 | 42 |
| 19 | 782630 | 276 | 900529 | 160 | 882101 | 437 | 117899 | 41 |
| 20 | 782796 | 276 | 900433 | 161 | 882363 | 436 | 117637 | 40 |
| 21 | 9·782961 | 276 | 9·900337 | 161 | 9·882625 | 436 | 10·117375 | 39 |
| 22 | 783127 | 276 | 900240 | 161 | 882887 | 436 | 117113 | 38 |
| 23 | 783292 | 275 | 900144 | 161 | 883148 | 436 | 116852 | 37 |
| 24 | 783458 | 275 | 900047 | 161 | 883410 | 436 | 116590 | 36 |
| 25 | 783623 | 275 | 899951 | 161 | 883672 | 436 | 116328 | 35 |
| 26 | 783788 | 275 | 899854 | 161 | 883934 | 436 | 116066 | 34 |
| 27 | 783953 | 275 | 899757 | 161 | 884196 | 436 | 115804 | 33 |
| 28 | 784118 | 275 | 899660 | 161 | 884457 | 436 | 115543 | 32 |
| 29 | 784282 | 274 | 899564 | 161 | 884719 | 436 | 115281 | 31 |
| 30 | 784447 | 274 | 899467 | 162 | 884980 | 436 | 115020 | 30 |
| 31 | 9·784612 | 274 | 9·899370 | 162 | 9·885242 | 436 | 10·114758 | 29 |
| 32 | 784776 | 274 | 899273 | 162 | 885504 | 436 | 114496 | 28 |
| 33 | 784941 | 274 | 899176 | 162 | 885765 | 436 | 114235 | 27 |
| 34 | 785105 | 274 | 899078 | 162 | 886026 | 436 | 113974 | 26 |
| 35 | 785269 | 273 | 898981 | 162 | 886288 | 436 | 113712 | 25 |
| 36 | 785433 | 273 | 898884 | 162 | 886549 | 435 | 113451 | 24 |
| 37 | 785597 | 273 | 898787 | 162 | 886811 | 435 | 113189 | 23 |
| 38 | 785761 | 273 | 898689 | 162 | 887072 | 435 | 112928 | 22 |
| 39 | 785925 | 273 | 898592 | 162 | 887333 | 435 | 112667 | 21 |
| 40 | 786089 | 273 | 898494 | 163 | 887594 | 435 | 112406 | 20 |
| 41 | 9·786252 | 272 | 9·898397 | 163 | 9·887855 | 435 | 10·112145 | 19 |
| 42 | 786416 | 272 | 898299 | 163 | 888116 | 435 | 111884 | 18 |
| 43 | 786579 | 272 | 898202 | 163 | 888378 | 435 | 111622 | 17 |
| 44 | 786742 | 272 | 898104 | 163 | 888639 | 435 | 111361 | 16 |
| 45 | 786906 | 272 | 898006 | 163 | 888900 | 435 | 111100 | 15 |
| 46 | 787069 | 272 | 897908 | 163 | 889161 | 435 | 110839 | 14 |
| 47 | 787232 | 271 | 897810 | 163 | 889421 | 435 | 110579 | 13 |
| 48 | 787395 | 271 | 897712 | 163 | 889682 | 435 | 110318 | 12 |
| 49 | 787557 | 271 | 897614 | 163 | 889943 | 435 | 110057 | 11 |
| 50 | 787720 | 271 | 897516 | 163 | 890204 | 434 | 109796 | 10 |
| 51 | 9·787883 | 271 | 9·897418 | 164 | 9·890465 | 434 | 10·109535 | 9 |
| 52 | 788045 | 271 | 897320 | 164 | 890725 | 434 | 109275 | 8 |
| 53 | 788208 | 271 | 897222 | 164 | 890986 | 434 | 109014 | 7 |
| 54 | 788370 | 270 | 897123 | 164 | 891247 | 434 | 108753 | 6 |
| 55 | 788532 | 270 | 897025 | 164 | 891507 | 434 | 108493 | 5 |
| 56 | 788694 | 270 | 896926 | 164 | 891768 | 434 | 108232 | 4 |
| 57 | 788856 | 270 | 896828 | 164 | 892028 | 434 | 107972 | 3 |
| 58 | 789018 | 270 | 896729 | 164 | 892289 | 434 | 107711 | 2 |
| 59 | 789180 | 270 | 896631 | 164 | 892549 | 434 | 107451 | 1 |
| 60 | 789342 | 269 | 896532 | 164 | 892810 | 434 | 107190 | 0 |
| ′ | Cosine. | D. | Sine. | D. | Cotang. | D. | Tang. | ′ |
| 127° | | | | | | | | 52° |

| 38° | | | | | | | | 141° |
|---|---|---|---|---|---|---|---|---|
| ′ | Sine. | D. | Cosine. | D. | Tang. | D. | Cotang. | ′ |
| 0 | 9·789342 | 269 | 9·896532 | 164 | 9·892810 | 434 | 10·107190 | 60 |
| 1 | 789504 | 269 | 896433 | 165 | 893070 | 434 | 106930 | 59 |
| 2 | 789665 | 269 | 896335 | 165 | 893331 | 434 | 106669 | 58 |
| 3 | 789827 | 269 | 896236 | 165 | 893591 | 434 | 106409 | 57 |
| 4 | 789988 | 269 | 896137 | 165 | 893851 | 434 | 106149 | 56 |
| 5 | 790149 | 269 | 896038 | 165 | 894111 | 434 | 105889 | 55 |
| 6 | 790310 | 268 | 895939 | 165 | 894372 | 434 | 105628 | 54 |
| 7 | 790471 | 268 | 895840 | 165 | 894632 | 433 | 105368 | 53 |
| 8 | 790632 | 268 | 895741 | 165 | 894892 | 433 | 105108 | 52 |
| 9 | 790793 | 268 | 895641 | 165 | 895152 | 433 | 104848 | 51 |
| 10 | 790954 | 268 | 895542 | 165 | 895412 | 433 | 104588 | 50 |
| 11 | 9·791115 | 268 | 9·895443 | 166 | 9·895672 | 433 | 10·104328 | 49 |
| 12 | 791275 | 267 | 895343 | 166 | 895932 | 433 | 104068 | 48 |
| 13 | 791436 | 267 | 895244 | 166 | 896192 | 433 | 103808 | 47 |
| 14 | 791596 | 267 | 895145 | 166 | 896452 | 433 | 103548 | 46 |
| 15 | 791757 | 267 | 895045 | 166 | 896712 | 433 | 103288 | 45 |
| 16 | 791917 | 267 | 894945 | 166 | 896971 | 433 | 103029 | 44 |
| 17 | 792077 | 267 | 894846 | 166 | 897231 | 433 | 102769 | 43 |
| 18 | 792237 | 266 | 894746 | 166 | 897491 | 433 | 102509 | 42 |
| 19 | 792397 | 266 | 894646 | 166 | 897751 | 433 | 102249 | 41 |
| 20 | 792557 | 266 | 894546 | 166 | 898010 | 433 | 101990 | 40 |
| 21 | 9·792716 | 266 | 9·894446 | 167 | 9·898270 | 433 | 10·101730 | 39 |
| 22 | 792876 | 266 | 894346 | 167 | 898530 | 433 | 101470 | 38 |
| 23 | 793035 | 266 | 894246 | 167 | 898789 | 433 | 101211 | 37 |
| 24 | 793195 | 265 | 894146 | 167 | 899049 | 432 | 100951 | 36 |
| 25 | 793354 | 265 | 894046 | 167 | 899308 | 432 | 100692 | 35 |
| 26 | 793514 | 265 | 893946 | 167 | 899568 | 432 | 100432 | 34 |
| 27 | 793673 | 265 | 893846 | 167 | 899827 | 432 | 100173 | 33 |
| 28 | 793832 | 265 | 893745 | 167 | 900087 | 432 | 099913 | 32 |
| 29 | 793991 | 265 | 893645 | 167 | 900346 | 432 | 099654 | 31 |
| 30 | 794150 | 264 | 893544 | 167 | 900605 | 432 | 099395 | 30 |
| 31 | 9·794308 | 264 | 9·893444 | 168 | 9·900864 | 432 | 10·099136 | 29 |
| 32 | 794467 | 264 | 893343 | 168 | 901124 | 432 | 098876 | 28 |
| 33 | 794626 | 264 | 893243 | 168 | 901383 | 432 | 098617 | 27 |
| 34 | 794784 | 264 | 893142 | 168 | 901642 | 432 | 098358 | 26 |
| 35 | 794942 | 264 | 893041 | 168 | 901901 | 432 | 098099 | 25 |
| 36 | 795101 | 264 | 892940 | 168 | 902160 | 432 | 097840 | 24 |
| 37 | 795259 | 263 | 892839 | 168 | 902420 | 432 | 097580 | 23 |
| 38 | 795417 | 263 | 892739 | 168 | 902679 | 432 | 097321 | 22 |
| 39 | 795575 | 263 | 892638 | 168 | 902938 | 432 | 097062 | 21 |
| 40 | 795733 | 263 | 892536 | 168 | 903197 | 431 | 096803 | 20 |
| 41 | 9·795891 | 263 | 9·892435 | 169 | 9·903456 | 431 | 10·096544 | 19 |
| 42 | 796049 | 263 | 892334 | 169 | 903714 | 431 | 096286 | 18 |
| 43 | 796206 | 263 | 892233 | 169 | 903973 | 431 | 096027 | 17 |
| 44 | 796364 | 262 | 892132 | 169 | 904232 | 431 | 095768 | 16 |
| 45 | 796521 | 262 | 892030 | 169 | 904491 | 431 | 095509 | 15 |
| 46 | 796679 | 262 | 891929 | 169 | 904750 | 431 | 095250 | 14 |
| 47 | 796836 | 262 | 891827 | 169 | 905008 | 431 | 094992 | 13 |
| 48 | 796993 | 262 | 891726 | 169 | 905267 | 431 | 094733 | 12 |
| 49 | 797150 | 261 | 891624 | 169 | 905526 | 431 | 094474 | 11 |
| 50 | 797307 | 261 | 891523 | 170 | 905785 | 431 | 094215 | 10 |
| 51 | 9·797464 | 261 | 9·891421 | 170 | 9·906043 | 431 | 10·093957 | 9 |
| 52 | 797621 | 261 | 891319 | 170 | 906302 | 431 | 093698 | 8 |
| 53 | 797777 | 261 | 891217 | 170 | 906560 | 431 | 093440 | 7 |
| 54 | 797934 | 261 | 891115 | 170 | 906819 | 431 | 093181 | 6 |
| 55 | 798091 | 261 | 891013 | 170 | 907077 | 431 | 092923 | 5 |
| 56 | 798247 | 261 | 890911 | 170 | 907336 | 431 | 092664 | 4 |
| 57 | 798403 | 260 | 890809 | 170 | 907594 | 431 | 092406 | 3 |
| 58 | 798560 | 260 | 890707 | 170 | 907853 | 431 | 092147 | 2 |
| 59 | 798716 | 260 | 890605 | 170 | 908111 | 430 | 091889 | 1 |
| 60 | 798872 | 260 | 890503 | 170 | 908369 | 430 | 091631 | 0 |
| ′ | Cosine. | D. | Sine. | D. | Cotang. | D. | Tang. | ′ |
| 128° | | | | | | | | 51° |

39° 140°

| ′ | Sine. | D. | Cosine. | D. | Tang. | D. | Cotang. | ′ |
|---|---|---|---|---|---|---|---|---|
| 0 | 9·798872 | 260 | 9·890503 | 170 | 9·908369 | 430 | 10·091631 | 60 |
| 1 | 799028 | 260 | 890400 | 171 | 908628 | 430 | 091372 | 59 |
| 2 | 799184 | 260 | 890298 | 171 | 908886 | 430 | 091114 | 58 |
| 3 | 799339 | 259 | 890195 | 171 | 909144 | 430 | 090856 | 57 |
| 4 | 799495 | 259 | 890093 | 171 | 909402 | 430 | 090598 | 56 |
| 5 | 799651 | 259 | 889990 | 171 | 909660 | 430 | 090340 | 55 |
| 6 | 799806 | 259 | 889888 | 171 | 909918 | 430 | 090082 | 54 |
| 7 | 799962 | 259 | 889785 | 171 | 910177 | 430 | 089823 | 53 |
| 8 | 800117 | 259 | 889682 | 171 | 910435 | 430 | 089565 | 52 |
| 9 | 800272 | 258 | 889579 | 171 | 910693 | 430 | 089307 | 51 |
| 10 | 800427 | 258 | 889477 | 171 | 910951 | 430 | 089049 | 50 |
| 11 | 9·800582 | 258 | 9·889374 | 172 | 9·911209 | 430 | 10·088791 | 49 |
| 12 | 800737 | 258 | 889271 | 172 | 911467 | 430 | 088533 | 48 |
| 13 | 800892 | 258 | 889168 | 172 | 911725 | 430 | 088275 | 47 |
| 14 | 801047 | 258 | 889064 | 172 | 911982 | 430 | 088018 | 46 |
| 15 | 801201 | 258 | 888961 | 172 | 912240 | 430 | 087760 | 45 |
| 16 | 801356 | 257 | 888858 | 172 | 912498 | 430 | 087502 | 44 |
| 17 | 801511 | 257 | 888755 | 172 | 912756 | 430 | 087244 | 43 |
| 18 | 801665 | 257 | 888651 | 172 | 913014 | 429 | 086986 | 42 |
| 19 | 801819 | 257 | 888548 | 172 | 913271 | 429 | 086729 | 41 |
| 20 | 801973 | 257 | 888444 | 173 | 913529 | 429 | 086471 | 40 |
| 21 | 9·802128 | 257 | 9·888341 | 173 | 9·913787 | 429 | 10·086213 | 39 |
| 22 | 802282 | 256 | 888237 | 173 | 914044 | 429 | 085956 | 38 |
| 23 | 802436 | 256 | 888134 | 173 | 914302 | 429 | 085698 | 37 |
| 24 | 802589 | 256 | 888030 | 173 | 914560 | 429 | 085440 | 36 |
| 25 | 802743 | 256 | 887926 | 173 | 914817 | 429 | 085183 | 35 |
| 26 | 802897 | 256 | 887822 | 173 | 915075 | 429 | 084925 | 34 |
| 27 | 803050 | 256 | 887718 | 173 | 915332 | 429 | 084668 | 33 |
| 28 | 803204 | 256 | 887614 | 173 | 915590 | 429 | 084410 | 32 |
| 29 | 803357 | 255 | 887510 | 173 | 915847 | 429 | 084153 | 31 |
| 30 | 803511 | 255 | 887406 | 174 | 916104 | 429 | 083896 | 30 |
| 31 | 9·803664 | 255 | 9·887302 | 174 | 9·916362 | 429 | 10·083638 | 29 |
| 32 | 803817 | 255 | 887198 | 174 | 916619 | 429 | 083381 | 28 |
| 33 | 803970 | 255 | 887093 | 174 | 916877 | 429 | 083123 | 27 |
| 34 | 804123 | 255 | 886989 | 174 | 917134 | 429 | 082866 | 26 |
| 35 | 804276 | 254 | 886885 | 174 | 917391 | 429 | 082609 | 25 |
| 36 | 804428 | 254 | 886780 | 174 | 917648 | 429 | 082352 | 24 |
| 37 | 804581 | 254 | 886676 | 174 | 917906 | 429 | 082094 | 23 |
| 38 | 804734 | 254 | 886571 | 174 | 918163 | 428 | 081837 | 22 |
| 39 | 804886 | 254 | 886466 | 174 | 918420 | 428 | 081580 | 21 |
| 40 | 805039 | 254 | 886362 | 175 | 918677 | 428 | 081323 | 20 |
| 41 | 9·805191 | 254 | 9·886257 | 175 | 9·918934 | 428 | 10·081066 | 19 |
| 42 | 805343 | 253 | 886152 | 175 | 919191 | 428 | 080809 | 18 |
| 43 | 805495 | 253 | 886047 | 175 | 919448 | 428 | 080552 | 17 |
| 44 | 805647 | 253 | 885942 | 175 | 919705 | 428 | 080295 | 16 |
| 45 | 805799 | 253 | 885837 | 175 | 919962 | 428 | 080038 | 15 |
| 46 | 805951 | 253 | 885732 | 175 | 920219 | 428 | 079781 | 14 |
| 47 | 806103 | 253 | 885627 | 175 | 920476 | 428 | 079524 | 13 |
| 48 | 806254 | 253 | 885522 | 175 | 920733 | 428 | 079267 | 12 |
| 49 | 806406 | 252 | 885416 | 175 | 920990 | 428 | 079010 | 11 |
| 50 | 806557 | 252 | 885311 | 176 | 921247 | 428 | 078753 | 10 |
| 51 | 9·806709 | 252 | 9·885205 | 176 | 9·921503 | 428 | 10·078497 | 9 |
| 52 | 806860 | 252 | 885100 | 176 | 921760 | 428 | 078240 | 8 |
| 53 | 807011 | 252 | 884994 | 176 | 922017 | 428 | 077983 | 7 |
| 54 | 807163 | 252 | 884889 | 176 | 922274 | 428 | 077726 | 6 |
| 55 | 807314 | 252 | 884783 | 176 | 922530 | 428 | 077470 | 5 |
| 56 | 807465 | 251 | 884677 | 176 | 922787 | 428 | 077213 | 4 |
| 57 | 807615 | 251 | 884572 | 176 | 923044 | 428 | 076956 | 3 |
| 58 | 807766 | 251 | 884466 | 176 | 923300 | 428 | 076700 | 2 |
| 59 | 807917 | 251 | 884360 | 176 | 923557 | 427 | 076443 | 1 |
| 60 | 808067 | 251 | 884254 | 177 | 923814 | 427 | 076186 | 0 |
| ′ | Cosine. | D. | Sine. | D. | Cotang. | D. | Tang. | ′ |

129° 50°

| 40° | | | | | | | | 139° |
|---|---|---|---|---|---|---|---|---|
| ′ | Sine. | D. | Cosine. | D. | Tang. | D. | Cotang. | ′ |
| 0 | 9·808067 | 251 | 9·884254 | 177 | 9·923814 | 427 | 10·076186 | 60 |
| 1 | 808218 | 251 | 884148 | 177 | 924070 | 427 | 075930 | 59 |
| 2 | 808368 | 251 | 884042 | 177 | 924327 | 427 | 075673 | 58 |
| 3 | 808519 | 250 | 883936 | 177 | 924583 | 427 | 075417 | 57 |
| 4 | 808669 | 250 | 883829 | 177 | 924840 | 427 | 075160 | 56 |
| 5 | 808819 | 250 | 883723 | 177 | 925096 | 427 | 074904 | 55 |
| 6 | 808969 | 250 | 883617 | 177 | 925352 | 427 | 074648 | 54 |
| 7 | 809119 | 250 | 883510 | 177 | 925609 | 427 | 074391 | 53 |
| 8 | 809269 | 250 | 883404 | 177 | 925865 | 427 | 074135 | 52 |
| 9 | 809419 | 249 | 883297 | 178 | 926122 | 427 | 073878 | 51 |
| 10 | 809569 | 249 | 883191 | 178 | 926378 | 427 | 073622 | 50 |
| 11 | 9·809718 | 249 | 9·883084 | 178 | 9·926634 | 427 | 10·073366 | 49 |
| 12 | 809868 | 249 | 882977 | 178 | 926890 | 427 | 073110 | 48 |
| 13 | 810017 | 249 | 882871 | 178 | 927147 | 427 | 072853 | 47 |
| 14 | 810167 | 249 | 882764 | 178 | 927403 | 427 | 072597 | 46 |
| 15 | 810316 | 248 | 882657 | 178 | 927659 | 427 | 072341 | 45 |
| 16 | 810465 | 248 | 882550 | 178 | 927915 | 427 | 072085 | 44 |
| 17 | 810614 | 248 | 882443 | 178 | 928171 | 427 | 071829 | 43 |
| 18 | 810763 | 248 | 882336 | 179 | 928427 | 427 | 071573 | 42 |
| 19 | 810912 | 248 | 882229 | 179 | 928684 | 427 | 071316 | 41 |
| 20 | 811061 | 248 | 882121 | 179 | 928940 | 427 | 071060 | 40 |
| 21 | 9·811210 | 248 | 9·882014 | 179 | 9·929196 | 427 | 10·070804 | 39 |
| 22 | 811358 | 247 | 881907 | 179 | 929452 | 427 | 070548 | 38 |
| 23 | 811507 | 247 | 881799 | 179 | 929708 | 427 | 070292 | 37 |
| 24 | 811655 | 247 | 881692 | 179 | 029964 | 426 | 070036 | 36 |
| 25 | 811804 | 247 | 881584 | 179 | 930220 | 426 | 069780 | 35 |
| 26 | 811952 | 247 | 881477 | 179 | 930475 | 426 | 069525 | 34 |
| 27 | 812100 | 247 | 881369 | 179 | 930731 | 426 | 069269 | 33 |
| 28 | 812248 | 247 | 881261 | 180 | 930987 | 426 | 069013 | 32 |
| 29 | 812396 | 246 | 881153 | 180 | 931243 | 426 | 068757 | 31 |
| 30 | 812544 | 246 | 881046 | 180 | 931499 | 426 | 068501 | 30 |
| 31 | 9·812692 | 246 | 9·880938 | 180 | 9·931755 | 426 | 10·068245 | 29 |
| 32 | 812840 | 246 | 880830 | 180 | 932010 | 426 | 067990 | 28 |
| 33 | 812988 | 246 | 880722 | 180 | 932266 | 426 | 067734 | 27 |
| 34 | 813135 | 246 | 880613 | 180 | 932522 | 426 | 067478 | 26 |
| 35 | 813283 | 246 | 880505 | 180 | 932778 | 426 | 067222 | 25 |
| 36 | 813430 | 245 | 880397 | 180 | 933033 | 426 | 066967 | 24 |
| 37 | 813578 | 245 | 880289 | 181 | 933289 | 426 | 066711 | 23 |
| 38 | 813725 | 245 | 880180 | 181 | 933545 | 426 | 066455 | 22 |
| 39 | 813872 | 245 | 880072 | 181 | 933800 | 426 | 066200 | 21 |
| 40 | 814019 | 245 | 879963 | 181 | 934056 | 426 | 065944 | 20 |
| 41 | 9·814166 | 245 | 9·879855 | 181 | 9·934311 | 426 | 10·065689 | 19 |
| 42 | 814313 | 245 | 879746 | 181 | 934567 | 426 | 065433 | 18 |
| 43 | 814460 | 244 | 879637 | 181 | 934822 | 426 | 065178 | 17 |
| 44 | 814607 | 244 | 879529 | 181 | 935078 | 426 | 064922 | 16 |
| 45 | 814753 | 244 | 879420 | 181 | 935333 | 426 | 064667 | 15 |
| 46 | 814900 | 244 | 879311 | 181 | 935589 | 426 | 064411 | 14 |
| 47 | 815046 | 244 | 879202 | 182 | 935844 | 426 | 064156 | 13 |
| 48 | 815193 | 244 | 879093 | 182 | 936100 | 426 | 063900 | 12 |
| 49 | 815339 | 244 | 878984 | 182 | 936355 | 426 | 063645 | 11 |
| 50 | 815485 | 243 | 878875 | 182 | 936611 | 426 | 063389 | 10 |
| 51 | 9·815631 | 243 | 9·878766 | 182 | 9·936866 | 425 | 10·063134 | 9 |
| 52 | 815778 | 243 | 878656 | 182 | 937121 | 425 | 062879 | 8 |
| 53 | 815924 | 243 | 878547 | 182 | 937377 | 425 | 062623 | 7 |
| 54 | 816069 | 243 | 878438 | 182 | 937632 | 425 | 062368 | 6 |
| 55 | 816215 | 243 | 878328 | 182 | 937887 | 425 | 062113 | 5 |
| 56 | 816361 | 243 | 878219 | 183 | 938142 | 425 | 061858 | 4 |
| 57 | 816507 | 242 | 878109 | 183 | 938398 | 425 | 061602 | 3 |
| 58 | 816652 | 242 | 877999 | 183 | 938653 | 425 | 061347 | 2 |
| 59 | 816798 | 242 | 877890 | 183 | 938908 | 425 | 061092 | 1 |
| 60 | 816943 | 242 | 877780 | 183 | 939163 | 425 | 060837 | 0 |
| ′ | Cosine. | D. | Sine. | D. | Cotang. | D. | Tang. | ′ |
| 130° | | | | | | | | 49° |

41° 138°

| ′ | Sine. | D. | Cosine. | D. | Tang. | D. | Cotang. | ′ |
|---|---|---|---|---|---|---|---|---|
| 0 | 9·816943 | 242 | 9·877780 | 183 | 9·939163 | 425 | 10·060837 | 60 |
| 1 | 817088 | 242 | 877670 | 183 | 939418 | 425 | 060582 | 59 |
| 2 | 817233 | 242 | 877560 | 183 | 939673 | 425 | 060327 | 58 |
| 3 | 817379 | 242 | 877450 | 183 | 939928 | 425 | 060072 | 57 |
| 4 | 817524 | 241 | 877340 | 183 | 940183 | 425 | 059817 | 56 |
| 5 | 817668 | 241 | 877230 | 184 | 940439 | 425 | 059561 | 55 |
| 6 | 817813 | 241 | 877120 | 184 | 940694 | 425 | 059306 | 54 |
| 7 | 817958 | 241 | 877010 | 184 | 940949 | 425 | 059051 | 53 |
| 8 | 818103 | 241 | 876899 | 184 | 941204 | 425 | 058796 | 52 |
| 9 | 818247 | 241 | 876789 | 184 | 941459 | 425 | 058541 | 51 |
| 10 | 818392 | 241 | 876678 | 184 | 941713 | 425 | 058287 | 50 |
| 11 | 9·818536 | 240 | 9·876568 | 184 | 9·941968 | 425 | 10·058032 | 49 |
| 12 | 818681 | 240 | 876457 | 184 | 942223 | 425 | 057777 | 48 |
| 13 | 818825 | 240 | 876347 | 184 | 942478 | 425 | 057522 | 47 |
| 14 | 818969 | 240 | 876236 | 185 | 942733 | 425 | 057267 | 46 |
| 15 | 819113 | 240 | 876125 | 185 | 942988 | 425 | 057012 | 45 |
| 16 | 819257 | 240 | 876014 | 185 | 943243 | 425 | 056757 | 44 |
| 17 | 819401 | 240 | 875904 | 185 | 943498 | 425 | 056502 | 43 |
| 18 | 819545 | 239 | 875793 | 185 | 943752 | 425 | 056248 | 42 |
| 19 | 819689 | 239 | 875682 | 185 | 944007 | 425 | 055993 | 41 |
| 20 | 819832 | 239 | 875571 | 185 | 944262 | 425 | 055738 | 40 |
| 21 | 9·819976 | 239 | 9·875459 | 185 | 9·944517 | 425 | 10·055483 | 39 |
| 22 | 820120 | 239 | 875348 | 185 | 944771 | 424 | 055229 | 38 |
| 23 | 820263 | 239 | 875237 | 185 | 945026 | 424 | 054974 | 37 |
| 24 | 820406 | 239 | 875126 | 186 | 945281 | 424 | 054719 | 36 |
| 25 | 820550 | 238 | 875014 | 186 | 945535 | 424 | 054465 | 35 |
| 26 | 820693 | 238 | 874903 | 186 | 945790 | 424 | 054210 | 34 |
| 27 | 820836 | 238 | 874791 | 186 | 946045 | 424 | 053955 | 33 |
| 28 | 820979 | 238 | 874680 | 186 | 946299 | 424 | 053701 | 32 |
| 29 | 821122 | 238 | 874568 | 186 | 946554 | 424 | 053446 | 31 |
| 30 | 821265 | 238 | 874456 | 186 | 946808 | 424 | 053192 | 30 |
| 31 | 9·821407 | 238 | 9·874344 | 186 | 9·947063 | 424 | 10·052937 | 29 |
| 32 | 821550 | 238 | 874232 | 187 | 947318 | 424 | 052682 | 28 |
| 33 | 821693 | 237 | 874121 | 187 | 947572 | 424 | 052428 | 27 |
| 34 | 821835 | 237 | 874009 | 187 | 947827 | 424 | 052173 | 26 |
| 35 | 821977 | 237 | 873896 | 187 | 948081 | 424 | 051919 | 25 |
| 36 | 822120 | 237 | 873784 | 187 | 948335 | 424 | 051665 | 24 |
| 37 | 822262 | 237 | 873672 | 187 | 948590 | 424 | 051410 | 23 |
| 38 | 822404 | 237 | 873560 | 187 | 948844 | 424 | 051156 | 22 |
| 39 | 822546 | 237 | 873448 | 187 | 949099 | 424 | 050901 | 21 |
| 40 | 822688 | 236 | 873335 | 187 | 949353 | 424 | 050647 | 20 |
| 41 | 9·822830 | 236 | 9·873223 | 187 | 9·949608 | 424 | 10·050392 | 19 |
| 42 | 822972 | 236 | 873110 | 188 | 949862 | 424 | 050138 | 18 |
| 43 | 823114 | 236 | 872998 | 188 | 950116 | 424 | 049884 | 17 |
| 44 | 823255 | 236 | 872885 | 188 | 950371 | 424 | 049629 | 16 |
| 45 | 823397 | 236 | 872772 | 188 | 950625 | 424 | 049375 | 15 |
| 46 | 823539 | 236 | 872659 | 188 | 950879 | 424 | 049121 | 14 |
| 47 | 823680 | 235 | 872547 | 188 | 951133 | 424 | 048867 | 13 |
| 48 | 823821 | 235 | 872434 | 188 | 951388 | 424 | 048612 | 12 |
| 49 | 823963 | 235 | 872321 | 188 | 951642 | 424 | 048358 | 11 |
| 50 | 824104 | 235 | 872208 | 188 | 951896 | 424 | 048104 | 10 |
| 51 | 9·824245 | 235 | 9·872095 | 189 | 9·952150 | 424 | 10·047850 | 9 |
| 52 | 824386 | 235 | 871981 | 189 | 952405 | 424 | 047595 | 8 |
| 53 | 824527 | 235 | 871868 | 189 | 952659 | 424 | 047341 | 7 |
| 54 | 824668 | 234 | 871755 | 189 | 952913 | 424 | 047087 | 6 |
| 55 | 824808 | 234 | 871641 | 189 | 953167 | 423 | 046833 | 5 |
| 56 | 824949 | 234 | 871528 | 189 | 953421 | 423 | 046579 | 4 |
| 57 | 825090 | 234 | 871414 | 189 | 953675 | 423 | 046325 | 3 |
| 58 | 825230 | 234 | 871301 | 189 | 953929 | 423 | 046071 | 2 |
| 59 | 825371 | 234 | 871187 | 189 | 954183 | 423 | 045817 | 1 |
| 60 | 825511 | 234 | 871073 | 190 | 954437 | 423 | 045563 | 0 |
| ′ | Cosine. | D. | Sine. | D. | Cotang. | D. | Tang. | ′ |

131° 48°

| 42° | | | | | | | | 137° |
|---|---|---|---|---|---|---|---|---|
| ′ | Sine. | D. | Cosine. | D. | Tang. | D. | Cotang. | ′ |
| 0 | 9·825511 | 234 | 9·871073 | 190 | 9·954437 | 423 | 10·045563 | 60 |
| 1 | 825651 | 233 | 870960 | 190 | 954691 | 423 | 045309 | 59 |
| 2 | 825791 | 233 | 870846 | 190 | 954946 | 423 | 045054 | 58 |
| 3 | 825931 | 233 | 870732 | 190 | 955200 | 423 | 044800 | 57 |
| 4 | 826071 | 233 | 870618 | 190 | 955454 | 423 | 044546 | 56 |
| 5 | 826211 | 233 | 870504 | 190 | 955708 | 423 | 044292 | 55 |
| 6 | 826351 | 233 | 870390 | 190 | 955961 | 423 | 044039 | 54 |
| 7 | 826491 | 233 | 870276 | 190 | 956215 | 423 | 043785 | 53 |
| 8 | 826631 | 233 | 870161 | 190 | 956469 | 423 | 043531 | 52 |
| 9 | 826770 | 232 | 870047 | 191 | 956723 | 423 | 043277 | 51 |
| 10 | 826910 | 232 | 869933 | 191 | 956977 | 423 | 043023 | 50 |
| 11 | 9·827049 | 232 | 9·869818 | 191 | 9·957231 | 423 | 10·042769 | 49 |
| 12 | 827189 | 232 | 869704 | 191 | 957485 | 423 | 042515 | 48 |
| 13 | 827328 | 232 | 869589 | 191 | 957739 | 423 | 042261 | 47 |
| 14 | 827467 | 232 | 869474 | 191 | 957993 | 423 | 042007 | 46 |
| 15 | 827606 | 232 | 869360 | 191 | 958247 | 423 | 041753 | 45 |
| 16 | 827745 | 232 | 869245 | 191 | 958500 | 423 | 041500 | 44 |
| 17 | 827884 | 231 | 869130 | 191 | 958754 | 423 | 041246 | 43 |
| 18 | 828023 | 231 | 869015 | 192 | 959008 | 423 | 040992 | 42 |
| 19 | 828162 | 231 | 868900 | 192 | 959262 | 423 | 040738 | 41 |
| 20 | 828301 | 231 | 868785 | 192 | 959516 | 423 | 040484 | 40 |
| 21 | 9·828439 | 231 | 9·868670 | 192 | 9·959769 | 423 | 10·040231 | 39 |
| 22 | 828578 | 231 | 868555 | 192 | 960023 | 423 | 039977 | 38 |
| 23 | 828716 | 231 | 868440 | 192 | 960277 | 423 | 039723 | 37 |
| 24 | 828855 | 230 | 868324 | 192 | 960530 | 423 | 039470 | 36 |
| 25 | 828993 | 230 | 868209 | 192 | 960784 | 423 | 039216 | 35 |
| 26 | 829131 | 230 | 868093 | 192 | 961038 | 423 | 038962 | 34 |
| 27 | 829269 | 230 | 867978 | 193 | 961292 | 423 | 038708 | 33 |
| 28 | 829407 | 230 | 867862 | 193 | 961545 | 423 | 038455 | 32 |
| 29 | 829545 | 230 | 867747 | 193 | 961799 | 423 | 038201 | 31 |
| 30 | 829683 | 230 | 867631 | 193 | 962052 | 423 | 037948 | 30 |
| 31 | 9·829821 | 229 | 9·867515 | 193 | 9·962306 | 423 | 10·037694 | 29 |
| 32 | 829959 | 229 | 867399 | 193 | 962560 | 423 | 037440 | 28 |
| 33 | 830097 | 229 | 867283 | 193 | 962813 | 423 | 037187 | 27 |
| 34 | 830234 | 229 | 867167 | 193 | 963067 | 423 | 036933 | 26 |
| 35 | 830372 | 229 | 867051 | 193 | 963320 | 423 | 036680 | 25 |
| 36 | 830509 | 229 | 866935 | 194 | 963574 | 423 | 036426 | 24 |
| 37 | 830646 | 229 | 866819 | 194 | 963828 | 423 | 036172 | 23 |
| 38 | 830784 | 229 | 866703 | 194 | 964081 | 423 | 035919 | 22 |
| 39 | 830921 | 228 | 866586 | 194 | 964335 | 423 | 035665 | 21 |
| 40 | 831058 | 228 | 866470 | 194 | 964588 | 422 | 035412 | 20 |
| 41 | 9·831195 | 228 | 9·866353 | 194 | 9·964842 | 422 | 10·035158 | 19 |
| 42 | 831332 | 228 | 866237 | 194 | 965095 | 422 | 034905 | 18 |
| 43 | 831469 | 228 | 866120 | 194 | 965349 | 422 | 034651 | 17 |
| 44 | 831606 | 228 | 866004 | 195 | 965602 | 422 | 034398 | 16 |
| 45 | 831742 | 228 | 865887 | 195 | 965855 | 422 | 034145 | 15 |
| 46 | 831879 | 228 | 865770 | 195 | 966109 | 422 | 033891 | 14 |
| 47 | 832015 | 227 | 865653 | 195 | 966362 | 422 | 033638 | 13 |
| 48 | 832152 | 227 | 865536 | 195 | 966616 | 422 | 033384 | 12 |
| 49 | 832288 | 227 | 865419 | 195 | 966869 | 422 | 033131 | 11 |
| 50 | 832425 | 227 | 865302 | 195 | 967123 | 422 | 032877 | 10 |
| 51 | 9·832561 | 227 | 9·865185 | 195 | 9·967376 | 422 | 10·032624 | 9 |
| 52 | 832697 | 227 | 865068 | 195 | 967629 | 422 | 032371 | 8 |
| 53 | 832833 | 227 | 864950 | 195 | 967883 | 422 | 032117 | 7 |
| 54 | 832969 | 226 | 864833 | 196 | 968136 | 422 | 031864 | 6 |
| 55 | 833105 | 226 | 864716 | 196 | 968389 | 422 | 031611 | 5 |
| 56 | 833241 | 226 | 864598 | 196 | 968643 | 422 | 031357 | 4 |
| 57 | 833377 | 226 | 864481 | 196 | 968896 | 422 | 031104 | 3 |
| 58 | 833512 | 226 | 864363 | 196 | 969149 | 422 | 030851 | 2 |
| 59 | 833648 | 226 | 864245 | 196 | 969403 | 422 | 030597 | 1 |
| 60 | 833783 | 226 | 864127 | 196 | 969656 | 422 | 030344 | 0 |
| ′ | Cosine. | D. | Sine. | D. | Cotang. | D. | Tang. | ′ |
| 132° | | | | | | | | 47° |

| 43° | | | | | | | | 136° |
|---|---|---|---|---|---|---|---|---|
| ′ | Sine. | D. | Cosine. | D. | Tang. | D. | Cotang. | ′ |
| 0 | 9·833783 | 226 | 9·864127 | 196 | 9·969656 | 422 | 10·030344 | 60 |
| 1 | 833919 | 225 | 864010 | 196 | 969909 | 422 | 030091 | 59 |
| 2 | 834054 | 225 | 863892 | 197 | 970162 | 422 | 029838 | 58 |
| 3 | 834189 | 225 | 863774 | 197 | 970416 | 422 | 029584 | 57 |
| 4 | 834325 | 225 | 863656 | 197 | 970669 | 422 | 029331 | 56 |
| 5 | 834460 | 225 | 863538 | 197 | 970922 | 422 | 029078 | 55 |
| 6 | 834595 | 225 | 863419 | 197 | 971175 | 422 | 028825 | 54 |
| 7 | 834730 | 225 | 863301 | 197 | 971429 | 422 | 028571 | 53 |
| 8 | 834865 | 225 | 863183 | 197 | 971682 | 422 | 028318 | 52 |
| 9 | 834999 | 224 | 863064 | 197 | 971935 | 422 | 028065 | 51 |
| 10 | 835134 | 224 | 862946 | 198 | 972188 | 422 | 027812 | 50 |
| 11 | 9·835269 | 224 | 9·862827 | 198 | 9·972441 | 422 | 10·027559 | 49 |
| 12 | 835403 | 224 | 862709 | 198 | 972695 | 422 | 027305 | 48 |
| 13 | 835538 | 224 | 862590 | 198 | 972948 | 422 | 027052 | 47 |
| 14 | 835672 | 224 | 862471 | 198 | 973201 | 422 | 026799 | 46 |
| 15 | 835807 | 224 | 862353 | 198 | 973454 | 422 | 026546 | 45 |
| 16 | 835941 | 224 | 862234 | 198 | 973707 | 422 | 026293 | 44 |
| 17 | 836075 | 223 | 862115 | 198 | 973960 | 422 | 026040 | 43 |
| 18 | 836209 | 223 | 861996 | 198 | 974213 | 422 | 025787 | 42 |
| 19 | 836343 | 223 | 861877 | 198 | 974466 | 422 | 025534 | 41 |
| 20 | 836477 | 223 | 861758 | 199 | 974720 | 422 | 025280 | 40 |
| 21 | 9·836611 | 223 | 9·861638 | 199 | 9·974973 | 422 | 10·025027 | 39 |
| 22 | 836745 | 223 | 861519 | 199 | 975226 | 422 | 024774 | 38 |
| 23 | 836878 | 223 | 861400 | 199 | 975479 | 422 | 024521 | 37 |
| 24 | 837012 | 222 | 861280 | 199 | 975732 | 422 | 024268 | 36 |
| 25 | 837146 | 222 | 861161 | 199 | 975985 | 422 | 024015 | 35 |
| 26 | 837279 | 222 | 861041 | 199 | 976238 | 422 | 023762 | 34 |
| 27 | 837412 | 222 | 860922 | 199 | 976491 | 422 | 023509 | 33 |
| 28 | 837546 | 222 | 860802 | 199 | 976744 | 422 | 023256 | 32 |
| 29 | 837679 | 222 | 860682 | 200 | 976997 | 422 | 023003 | 31 |
| 30 | 837812 | 222 | 860562 | 200 | 977250 | 422 | 022750 | 30 |
| 31 | 9·837945 | 222 | 9·860442 | 200 | 9·977503 | 422 | 10·022497 | 29 |
| 32 | 838078 | 221 | 860322 | 200 | 977756 | 422 | 022244 | 28 |
| 33 | 838211 | 221 | 860202 | 200 | 978009 | 422 | 021991 | 27 |
| 34 | 838344 | 221 | 860082 | 200 | 978262 | 422 | 021738 | 26 |
| 35 | 838477 | 221 | 859962 | 200 | 978515 | 422 | 021485 | 25 |
| 36 | 838610 | 221 | 859842 | 200 | 978768 | 422 | 021232 | 24 |
| 37 | 838742 | 221 | 859721 | 201 | 979021 | 422 | 020979 | 23 |
| 38 | 838875 | 221 | 859601 | 201 | 979274 | 422 | 020726 | 22 |
| 39 | 839007 | 221 | 859480 | 201 | 979527 | 422 | 020473 | 21 |
| 40 | 839140 | 220 | 859360 | 201 | 979780 | 422 | 020220 | 20 |
| 41 | 9·839272 | 220 | 9·859239 | 201 | 9·980033 | 422 | 10·019967 | 19 |
| 42 | 839404 | 220 | 859119 | 201 | 980286 | 422 | 019714 | 18 |
| 43 | 839536 | 220 | 858998 | 201 | 980538 | 422 | 019462 | 17 |
| 44 | 839668 | 220 | 858877 | 201 | 980791 | 421 | 019209 | 16 |
| 45 | 839800 | 220 | 858756 | 202 | 981044 | 421 | 018956 | 15 |
| 46 | 839932 | 220 | 858635 | 202 | 981297 | 421 | 018703 | 14 |
| 47 | 840064 | 219 | 858514 | 202 | 981550 | 421 | 018450 | 13 |
| 48 | 840196 | 219 | 858393 | 202 | 981803 | 421 | 018197 | 12 |
| 49 | 840328 | 219 | 858272 | 202 | 982056 | 421 | 017944 | 11 |
| 50 | 840459 | 219 | 858151 | 202 | 982309 | 421 | 017691 | 10 |
| 51 | 9·840591 | 219 | 9·858029 | 202 | 9·982562 | 421 | 10·017438 | 9 |
| 52 | 840722 | 219 | 857908 | 202 | 982814 | 421 | 017186 | 8 |
| 53 | 840854 | 219 | 857786 | 202 | 983067 | 421 | 016933 | 7 |
| 54 | 840985 | 219 | 857665 | 203 | 983320 | 421 | 016680 | 6 |
| 55 | 841116 | 218 | 857543 | 203 | 983573 | 421 | 016427 | 5 |
| 56 | 841247 | 218 | 857422 | 203 | 983826 | 421 | 016174 | 4 |
| 57 | 841378 | 218 | 857300 | 203 | 984079 | 421 | 015921 | 3 |
| 58 | 841509 | 218 | 857178 | 203 | 984332 | 421 | 015668 | 2 |
| 59 | 841640 | 218 | 857056 | 203 | 984584 | 421 | 015416 | 1 |
| 60 | 841771 | 218 | 856934 | 203 | 984837 | 421 | 015163 | 0 |
| ′ | Cosine. | D. | Sine. | D. | Cotang. | D. | Tang. | ′ |
| 133° | | | | | | | | 46° |

44° 135°

| ′ | Sine. | D. | Cosine. | D. | Tang. | D. | Cotang. | ′ |
|---|---|---|---|---|---|---|---|---|
| 0 | 9·841771 | 218 | 9·856934 | 203 | 9·984837 | 421 | 10·015163 | 60 |
| 1 | 841902 | 218 | 856812 | 203 | 985090 | 421 | 014910 | 59 |
| 2 | 842033 | 218 | 856690 | 204 | 985343 | 421 | 014657 | 58 |
| 3 | 842163 | 217 | 856568 | 204 | 985596 | 421 | 014404 | 57 |
| 4 | 842294 | 217 | 856446 | 204 | 985848 | 421 | 014152 | 56 |
| 5 | 842424 | 217 | 856323 | 204 | 986101 | 421 | 013899 | 55 |
| 6 | 842555 | 217 | 856201 | 204 | 986354 | 421 | 013646 | 54 |
| 7 | 842685 | 217 | 856078 | 204 | 986607 | 421 | 013393 | 53 |
| 8 | 842815 | 217 | 855956 | 204 | 986860 | 421 | 013140 | 52 |
| 9 | 842946 | 217 | 855833 | 204 | 987112 | 421 | 012888 | 51 |
| 10 | 843076 | 217 | 855711 | 205 | 987365 | 421 | 012635 | 50 |
| 11 | 9·843206 | 216 | 9·855588 | 205 | 9·987618 | 421 | 10·012382 | 49 |
| 12 | 843336 | 216 | 855465 | 205 | 987871 | 421 | 012129 | 48 |
| 13 | 843466 | 216 | 855342 | 205 | 988123 | 421 | 011877 | 47 |
| 14 | 843595 | 216 | 855219 | 205 | 988376 | 421 | 011624 | 46 |
| 15 | 843725 | 216 | 855096 | 205 | 988629 | 421 | 011371 | 45 |
| 16 | 843855 | 216 | 854973 | 205 | 988882 | 421 | 011118 | 44 |
| 17 | 843984 | 216 | 854850 | 205 | 989134 | 421 | 010866 | 43 |
| 18 | 844114 | 215 | 854727 | 206 | 989387 | 421 | 010613 | 42 |
| 19 | 844243 | 215 | 854603 | 206 | 989640 | 421 | 010360 | 41 |
| 20 | 844372 | 215 | 854480 | 206 | 989893 | 421 | 010107 | 40 |
| 21 | 9·844502 | 215 | 9·854356 | 206 | 9·990145 | 421 | 10·009855 | 39 |
| 22 | 844631 | 215 | 854233 | 206 | 990398 | 421 | 009602 | 38 |
| 23 | 844760 | 215 | 854109 | 206 | 990651 | 421 | 009349 | 37 |
| 24 | 844889 | 215 | 853986 | 206 | 990903 | 421 | 009097 | 36 |
| 25 | 845018 | 215 | 853862 | 206 | 991156 | 421 | 008844 | 35 |
| 26 | 845147 | 215 | 853738 | 206 | 991409 | 421 | 008591 | 34 |
| 27 | 845276 | 214 | 853614 | 207 | 991662 | 421 | 008338 | 33 |
| 28 | 845405 | 214 | 853490 | 207 | 991914 | 421 | 008086 | 32 |
| 29 | 845533 | 214 | 853366 | 207 | 992167 | 421 | 007833 | 31 |
| 30 | 845662 | 214 | 853242 | 207 | 992420 | 421 | 007580 | 30 |
| 31 | 9·845790 | 214 | 9·853118 | 207 | 9·992672 | 421 | 10·007328 | 29 |
| 32 | 845919 | 214 | 852994 | 207 | 992925 | 421 | 007075 | 28 |
| 33 | 846047 | 214 | 852869 | 207 | 993178 | 421 | 006822 | 27 |
| 34 | 846175 | 214 | 852745 | 207 | 993431 | 421 | 006569 | 26 |
| 35 | 846304 | 214 | 852620 | 207 | 993683 | 421 | 006317 | 25 |
| 36 | 846432 | 213 | 852496 | 208 | 993936 | 421 | 006064 | 24 |
| 37 | 846560 | 213 | 852371 | 208 | 994189 | 421 | 005811 | 23 |
| 38 | 846688 | 213 | 852247 | 208 | 994441 | 421 | 005559 | 22 |
| 39 | 846816 | 213 | 852122 | 208 | 994694 | 421 | 005306 | 21 |
| 40 | 846944 | 213 | 851997 | 208 | 994947 | 421 | 005053 | 20 |
| 41 | 9·847071 | 213 | 9·851872 | 208 | 9·995199 | 421 | 10·004801 | 19 |
| 42 | 847199 | 213 | 851747 | 208 | 995452 | 421 | 004548 | 18 |
| 43 | 847327 | 213 | 851622 | 208 | 995705 | 421 | 004295 | 17 |
| 44 | 847454 | 212 | 851497 | 209 | 995957 | 421 | 004043 | 16 |
| 45 | 847582 | 212 | 851372 | 209 | 996210 | 421 | 003790 | 15 |
| 46 | 847709 | 212 | 851246 | 209 | 996463 | 421 | 003537 | 14 |
| 47 | 847836 | 212 | 851121 | 209 | 996715 | 421 | 003285 | 13 |
| 48 | 847964 | 212 | 850996 | 209 | 996968 | 421 | 003032 | 12 |
| 49 | 848091 | 212 | 850870 | 209 | 997221 | 421 | 002779 | 11 |
| 50 | 848218 | 212 | 850745 | 209 | 997473 | 421 | 002527 | 10 |
| 51 | 9·848345 | 212 | 9·850619 | 209 | 9·997726 | 421 | 10·002274 | 9 |
| 52 | 848472 | 211 | 850493 | 210 | 997979 | 421 | 002021 | 8 |
| 53 | 848599 | 211 | 850368 | 210 | 998231 | 421 | 001769 | 7 |
| 54 | 848726 | 211 | 850242 | 210 | 998484 | 421 | 001516 | 6 |
| 55 | 848852 | 211 | 850116 | 210 | 998737 | 421 | 001263 | 5 |
| 56 | 848979 | 211 | 849990 | 210 | 998989 | 421 | 001011 | 4 |
| 57 | 849106 | 211 | 849864 | 210 | 999242 | 421 | 000758 | 3 |
| 58 | 849232 | 211 | 849738 | 210 | 999495 | 421 | 000505 | 2 |
| 59 | 849359 | 211 | 849611 | 210 | 999747 | 421 | 000253 | 1 |
| 60 | 849485 | 211 | 849485 | 210 | 10·000000 | 421 | 10·000000 | 0 |
| ′ | Cosine. | D. | Sine. | D. | Cotang. | D. | Tang. | ′ |

134° 45°

# TABLE III.,

OF

# NATURAL SINES AND TANGENTS;

TO

## EVERY DEGREE AND MINUTE OF THE QUADRANT.

---

If the given angle is less than 45°, look for the degrees and the title of the column, at the *top* of the page; and for the minutes on the *left*. But if the angle is between 45° and 90°, look for the degrees and the title of the column, at the *bottom*; and for the minutes on the *right*.

The *Secants and Cosecants*, which are not inserted in this table, may be easily supplied. If 1 be divided by the cosine of an arc, the quotient will be the secant of that arc. And if 1 be divided by the sine, the quotient will be the cosecant.

The values of the Sines and Cosines are less than a unit, and are given in decimals, although the decimal point is not printed. So also, the tangents of arcs less than 45°, and cotangents of arcs greater than 45°, are less than a unit and are expressed in decimals with the decimal point omitted.

| ′ | 0° | | 1° | | 2° | | 3° | | 4° | | ′ |
|---|---|---|---|---|---|---|---|---|---|---|---|
| | Sine. | Cosine. | Sine. | Cosine. | Sine. | Cosine. | Sine. | Cosine. | Sine. | Cosine. | |
| 0 | 00000 | Unit. | 01745 | 99985 | 03490 | 99939 | 05234 | 99863 | 06976 | 99756 | 60 |
| 1 | 00029 | Unit. | 01774 | 99984 | 03519 | 99938 | 05263 | 99861 | 07005 | 99754 | 59 |
| 2 | 00058 | Unit. | 01803 | 99984 | 03548 | 99937 | 05292 | 99860 | 07034 | 99752 | 58 |
| 3 | 00087 | Unit. | 01832 | 99983 | 03577 | 99936 | 05321 | 99858 | 07063 | 99750 | 57 |
| 4 | 00116 | Unit. | 01862 | 99983 | 03606 | 99935 | 05350 | 99857 | 07092 | 99748 | 56 |
| 5 | 00145 | Unit. | 01891 | 99982 | 03635 | 99934 | 05379 | 99855 | 07121 | 99746 | 55 |
| 6 | 00175 | Unit. | 01920 | 99982 | 03664 | 99933 | 05408 | 99854 | 07150 | 99744 | 54 |
| 7 | 00204 | Unit. | 01949 | 99981 | 03693 | 99932 | 05437 | 99852 | 07179 | 99742 | 53 |
| 8 | 00233 | Unit. | 01978 | 99980 | 03723 | 99931 | 05466 | 99851 | 07208 | 99740 | 52 |
| 9 | 00262 | Unit. | 02007 | 99980 | 03752 | 99930 | 05495 | 99849 | 07237 | 99738 | 51 |
| 10 | 00291 | Unit. | 02036 | 99979 | 03781 | 99929 | 05524 | 99847 | 07266 | 99736 | 50 |
| 11 | 00320 | 99999 | 02065 | 99979 | 03810 | 99927 | 05553 | 99846 | 07295 | 99734 | 49 |
| 12 | 00349 | 99999 | 02094 | 99978 | 03839 | 99926 | 05582 | 99844 | 07324 | 99731 | 48 |
| 13 | 00378 | 99999 | 02123 | 99977 | 03868 | 99925 | 05611 | 99842 | 07353 | 99729 | 47 |
| 14 | 00407 | 99999 | 02152 | 99977 | 03897 | 99924 | 05640 | 99841 | 07382 | 99727 | 46 |
| 15 | 00436 | 99999 | 02181 | 99976 | 03926 | 99923 | 05669 | 99839 | 07411 | 99725 | 45 |
| 16 | 00465 | 99999 | 02211 | 99976 | 03955 | 99922 | 05698 | 99838 | 07440 | 99723 | 44 |
| 17 | 00495 | 99999 | 02240 | 99975 | 03984 | 99921 | 05727 | 99836 | 07469 | 99721 | 43 |
| 18 | 00524 | 99999 | 02269 | 99974 | 04013 | 99919 | 05756 | 99834 | 07498 | 99719 | 42 |
| 19 | 00553 | 99998 | 02298 | 99974 | 04042 | 99918 | 05785 | 99833 | 07527 | 99716 | 41 |
| 20 | 00582 | 99998 | 02327 | 99973 | 04071 | 99917 | 05814 | 99831 | 07556 | 99714 | 40 |
| 21 | 00611 | 99998 | 02356 | 99972 | 04100 | 99916 | 05844 | 99829 | 07585 | 99712 | 39 |
| 22 | 00640 | 99998 | 02385 | 99972 | 04129 | 99915 | 05873 | 99827 | 07614 | 99710 | 38 |
| 23 | 00669 | 99998 | 02414 | 99971 | 04159 | 99913 | 05902 | 99826 | 07643 | 99708 | 37 |
| 24 | 00698 | 99998 | 02443 | 99970 | 04188 | 99912 | 05931 | 99824 | 07672 | 99705 | 36 |
| 25 | 00727 | 99997 | 02472 | 99969 | 04217 | 99911 | 05960 | 99822 | 07701 | 99703 | 35 |
| 26 | 00756 | 99997 | 02501 | 99969 | 04246 | 99910 | 05989 | 99821 | 07730 | 99701 | 34 |
| 27 | 00785 | 99997 | 02530 | 99968 | 04275 | 99909 | 06018 | 99819 | 07759 | 99699 | 33 |
| 28 | 00814 | 99997 | 02560 | 99967 | 04304 | 99907 | 06047 | 99817 | 07788 | 99696 | 32 |
| 29 | 00844 | 99996 | 02589 | 99966 | 04333 | 99906 | 06076 | 99815 | 07817 | 99694 | 31 |
| 30 | 00873 | 99996 | 02618 | 99966 | 04362 | 99905 | 06105 | 99813 | 07846 | 99692 | 30 |
| 31 | 00902 | 99996 | 02647 | 99965 | 04391 | 99904 | 06134 | 99812 | 07875 | 99689 | 29 |
| 32 | 00931 | 99996 | 02676 | 99964 | 04420 | 99902 | 06163 | 99810 | 07904 | 99687 | 28 |
| 33 | 00960 | 99995 | 02705 | 99963 | 04449 | 99901 | 06192 | 99808 | 07933 | 99685 | 27 |
| 34 | 00989 | 99995 | 02734 | 99963 | 04478 | 99900 | 06221 | 99806 | 07962 | 99683 | 26 |
| 35 | 01018 | 99995 | 02763 | 99962 | 04507 | 99898 | 06250 | 99804 | 07991 | 99680 | 25 |
| 36 | 01047 | 99995 | 02792 | 99961 | 04536 | 99897 | 06279 | 99803 | 08020 | 99678 | 24 |
| 37 | 01076 | 99994 | 02821 | 99960 | 04565 | 99896 | 06308 | 99801 | 08049 | 99676 | 23 |
| 38 | 01105 | 99994 | 02850 | 99959 | 04594 | 99894 | 06337 | 99799 | 08078 | 99673 | 22 |
| 39 | 01134 | 99994 | 02879 | 99959 | 04623 | 99893 | 06366 | 99797 | 08107 | 99671 | 21 |
| 40 | 01164 | 99993 | 02908 | 99958 | 04653 | 99892 | 06395 | 99795 | 08136 | 99668 | 20 |
| 41 | 01193 | 99993 | 02938 | 99957 | 04682 | 99890 | 06424 | 99793 | 08165 | 99666 | 19 |
| 42 | 01222 | 99993 | 02967 | 99956 | 04711 | 99889 | 06453 | 99792 | 08194 | 99664 | 18 |
| 43 | 01251 | 99992 | 02996 | 99955 | 04740 | 99888 | 06482 | 99790 | 08223 | 99661 | 17 |
| 44 | 01280 | 99992 | 03025 | 99954 | 04769 | 99886 | 06511 | 99788 | 08252 | 99659 | 16 |
| 45 | 01309 | 99991 | 03054 | 99953 | 04798 | 99885 | 06540 | 99786 | 08281 | 99657 | 15 |
| 46 | 01338 | 99991 | 03083 | 99952 | 04827 | 99883 | 06569 | 99784 | 08310 | 99654 | 14 |
| 47 | 01367 | 99991 | 03112 | 99952 | 04856 | 99882 | 06598 | 99782 | 08339 | 99652 | 13 |
| 48 | 01396 | 99990 | 03141 | 99951 | 04885 | 99881 | 06627 | 99780 | 08368 | 99649 | 12 |
| 49 | 01425 | 99990 | 03170 | 99950 | 04914 | 99879 | 06656 | 99778 | 08397 | 99647 | 11 |
| 50 | 01454 | 99989 | 03199 | 99949 | 04943 | 99878 | 06685 | 99776 | 08426 | 99644 | 10 |
| 51 | 01483 | 99989 | 03228 | 99948 | 04972 | 99876 | 06714 | 99774 | 08455 | 99642 | 9 |
| 52 | 01513 | 99989 | 03257 | 99947 | 05001 | 99875 | 06743 | 99772 | 08484 | 99639 | 8 |
| 53 | 01542 | 99988 | 03286 | 99946 | 05030 | 99873 | 06773 | 99770 | 08513 | 99637 | 7 |
| 54 | 01571 | 99988 | 03316 | 99945 | 05059 | 99872 | 06802 | 99768 | 08542 | 99635 | 6 |
| 55 | 01600 | 99987 | 03345 | 99944 | 05088 | 99870 | 06831 | 99766 | 08571 | 99632 | 5 |
| 56 | 01629 | 99987 | 03374 | 99943 | 05117 | 99869 | 06860 | 99764 | 08600 | 99630 | 4 |
| 57 | 01658 | 99986 | 03403 | 99942 | 05146 | 99867 | 06889 | 99762 | 08629 | 99627 | 3 |
| 58 | 01687 | 99986 | 03432 | 99941 | 05175 | 99866 | 06918 | 99760 | 08658 | 99625 | 2 |
| 59 | 01716 | 99985 | 03461 | 99940 | 05205 | 99864 | 06947 | 99758 | 08687 | 99622 | 1 |
| 60 | 01745 | 99985 | 03490 | 99939 | 05234 | 99863 | 06976 | 99756 | 08716 | 99619 | 0 |
| | Cosine. | Sine. | Cosine. | Sine. | Cosine. | Sine. | Cosine. | Sine. | Cosine. | Sine. | |
| ′ | 89° | | 88° | | 87° | | 86° | | 85° | | ′ |

| ′ | 5° Sine | 5° Cosine. | 6° Sine. | 6° Cosine. | 7° Sine. | 7° Cosine. | 8° Sine. | 8° Cosine. | 9° Sine. | 9° Cosine. | ′ |
|---|---|---|---|---|---|---|---|---|---|---|---|
| 0 | 08716 | 99619 | 10453 | 99452 | 12187 | 99255 | 13917 | 99027 | 15643 | 98769 | 60 |
| 1 | 08745 | 99617 | 10482 | 99449 | 12216 | 99251 | 13946 | 99023 | 15672 | 98764 | 59 |
| 2 | 08774 | 99614 | 10511 | 99446 | 12245 | 99248 | 13975 | 99019 | 15701 | 98760 | 58 |
| 3 | 08803 | 99612 | 10540 | 99443 | 12274 | 99244 | 14004 | 99015 | 15730 | 98755 | 57 |
| 4 | 08831 | 99609 | 10569 | 99440 | 12302 | 99240 | 14033 | 99011 | 15758 | 98751 | 56 |
| 5 | 08860 | 99607 | 10597 | 99437 | 12331 | 99237 | 14061 | 99006 | 15787 | 98746 | 55 |
| 6 | 08889 | 99604 | 10626 | 99434 | 12360 | 99233 | 14090 | 99002 | 15816 | 98741 | 54 |
| 7 | 08918 | 99602 | 10655 | 99431 | 12389 | 99230 | 14119 | 98998 | 15845 | 98737 | 53 |
| 8 | 08947 | 99599 | 10684 | 99428 | 12418 | 99226 | 14148 | 98994 | 15873 | 98732 | 52 |
| 9 | 08976 | 99596 | 10713 | 99424 | 12447 | 99222 | 14177 | 98990 | 15902 | 98728 | 51 |
| 10 | 09005 | 99594 | 10742 | 99421 | 12476 | 99219 | 14205 | 98986 | 15931 | 98723 | 50 |
| 11 | 09034 | 99591 | 10771 | 99418 | 12504 | 99215 | 14234 | 98982 | 15959 | 98718 | 49 |
| 12 | 09063 | 99588 | 10800 | 99415 | 12533 | 99211 | 14263 | 98978 | 15988 | 98714 | 48 |
| 13 | 09092 | 99586 | 10829 | 99412 | 12562 | 99208 | 14292 | 98973 | 16017 | 98709 | 47 |
| 14 | 09121 | 99583 | 10858 | 99409 | 12591 | 99204 | 14320 | 98969 | 16046 | 98704 | 46 |
| 15 | 09150 | 99580 | 10887 | 99406 | 12620 | 99200 | 14349 | 98965 | 16074 | 98700 | 45 |
| 16 | 09179 | 99578 | 10916 | 99402 | 12649 | 99197 | 14378 | 98961 | 16103 | 98695 | 44 |
| 17 | 09208 | 99575 | 10945 | 99399 | 12678 | 99193 | 14407 | 98957 | 16132 | 98690 | 43 |
| 18 | 09237 | 99572 | 10973 | 99396 | 12706 | 99189 | 14436 | 98953 | 16160 | 98686 | 42 |
| 19 | 09266 | 99570 | 11002 | 99393 | 12735 | 99186 | 14464 | 98948 | 16189 | 98681 | 41 |
| 20 | 09295 | 99567 | 11031 | 99390 | 12764 | 99182 | 14493 | 98944 | 16218 | 98676 | 40 |
| 21 | 09324 | 99564 | 11060 | 99386 | 12793 | 99178 | 14522 | 98940 | 16246 | 98671 | 39 |
| 22 | 09353 | 99562 | 11089 | 99383 | 12822 | 99175 | 14551 | 98936 | 16275 | 98667 | 38 |
| 23 | 09382 | 99559 | 11118 | 99380 | 12851 | 99171 | 14580 | 98931 | 16304 | 98662 | 37 |
| 24 | 09411 | 99556 | 11147 | 99377 | 12880 | 99167 | 14608 | 98927 | 16333 | 98657 | 36 |
| 25 | 09440 | 99553 | 11176 | 99374 | 12908 | 99163 | 14637 | 98923 | 16361 | 98652 | 35 |
| 26 | 09469 | 99551 | 11205 | 99370 | 12937 | 99160 | 14666 | 98919 | 16390 | 98648 | 34 |
| 27 | 09498 | 99548 | 11234 | 99367 | 12966 | 99156 | 14695 | 98914 | 16419 | 98643 | 33 |
| 28 | 09527 | 99545 | 11263 | 99364 | 12995 | 99152 | 14723 | 98910 | 16447 | 98638 | 32 |
| 29 | 09556 | 99542 | 11291 | 99360 | 13024 | 99148 | 14752 | 98906 | 16476 | 98633 | 31 |
| 30 | 09585 | 99540 | 11320 | 99357 | 13053 | 99144 | 14781 | 98902 | 16505 | 98629 | 30 |
| 31 | 09614 | 99537 | 11349 | 99354 | 13081 | 99141 | 14810 | 98897 | 16533 | 98624 | 29 |
| 32 | 09642 | 99534 | 11378 | 99351 | 13110 | 99137 | 14838 | 98893 | 16562 | 98619 | 28 |
| 33 | 09671 | 99531 | 11407 | 99347 | 13139 | 99133 | 14867 | 98889 | 16591 | 98614 | 27 |
| 34 | 09700 | 99528 | 11436 | 99344 | 13168 | 99129 | 14896 | 98884 | 16620 | 98609 | 26 |
| 35 | 09729 | 99526 | 11465 | 99341 | 13197 | 99125 | 14925 | 98880 | 16648 | 98604 | 25 |
| 36 | 09758 | 99523 | 11494 | 99337 | 13226 | 99122 | 14954 | 98876 | 16677 | 98600 | 24 |
| 37 | 09787 | 99520 | 11523 | 99334 | 13254 | 99118 | 14982 | 98871 | 16706 | 98595 | 23 |
| 38 | 09816 | 99517 | 11552 | 99331 | 13283 | 99114 | 15011 | 98867 | 16734 | 98590 | 22 |
| 39 | 09845 | 99514 | 11580 | 99327 | 13312 | 99110 | 15040 | 98863 | 16763 | 98585 | 21 |
| 40 | 09874 | 99511 | 11609 | 69324 | 13341 | 99106 | 15069 | 98858 | 16792 | 98580 | 20 |
| 41 | 09903 | 99508 | 11638 | 99320 | 13370 | 99102 | 15097 | 98854 | 16820 | 98575 | 19 |
| 42 | 09932 | 99506 | 11667 | 99317 | 13399 | 99098 | 15126 | 98849 | 16849 | 98570 | 18 |
| 43 | 09961 | 99503 | 11696 | 99314 | 13427 | 99094 | 15155 | 98845 | 16878 | 98565 | 17 |
| 44 | 09990 | 99500 | 11725 | 99310 | 13456 | 99091 | 15184 | 98841 | 16906 | 98561 | 16 |
| 45 | 10019 | 99497 | 11754 | 99307 | 13485 | 99087 | 15212 | 98836 | 16935 | 98556 | 15 |
| 46 | 10048 | 99494 | 11783 | 99303 | 13514 | 99083 | 15241 | 98832 | 16964 | 98551 | 14 |
| 47 | 10077 | 99491 | 11812 | 99300 | 13543 | 99079 | 15270 | 98827 | 16992 | 98546 | 13 |
| 48 | 10106 | 99488 | 11840 | 99297 | 13572 | 99075 | 15299 | 98823 | 17021 | 98541 | 12 |
| 49 | 10135 | 99485 | 11869 | 99293 | 13600 | 99071 | 15327 | 98818 | 17050 | 98536 | 11 |
| 50 | 10164 | 99482 | 11898 | 99290 | 13629 | 99067 | 15356 | 98814 | 17078 | 98531 | 10 |
| 51 | 10192 | 99479 | 11927 | 99286 | 13658 | 99063 | 15385 | 98809 | 17107 | 98526 | 9 |
| 52 | 10221 | 99476 | 11956 | 99283 | 13687 | 99059 | 15414 | 98805 | 17136 | 98521 | 8 |
| 53 | 10250 | 99473 | 11985 | 99279 | 13716 | 99055 | 15442 | 98800 | 17164 | 98516 | 7 |
| 54 | 10279 | 99470 | 12014 | 99276 | 13744 | 99051 | 15471 | 98796 | 17193 | 98511 | 6 |
| 55 | 10308 | 99467 | 12043 | 99272 | 13773 | 99047 | 15500 | 98791 | 17222 | 98506 | 5 |
| 56 | 10337 | 99464 | 12071 | 99269 | 13802 | 99043 | 15529 | 98787 | 17250 | 98501 | 4 |
| 57 | 10366 | 99461 | 12100 | 99265 | 13831 | 99039 | 15557 | 98782 | 17279 | 98496 | 3 |
| 58 | 10395 | 99458 | 12129 | 99262 | 13860 | 99035 | 15586 | 98778 | 17308 | 98491 | 2 |
| 59 | 10424 | 99455 | 12158 | 99258 | 13889 | 99031 | 15615 | 98773 | 17336 | 98486 | 1 |
| 60 | 10453 | 99452 | 12187 | 99255 | 13917 | 99027 | 15643 | 98769 | 17365 | 98481 | 0 |
| ′ | Cosine. | Sine. | Cosine. | Sine. | Cosine. | Sine. | Cosine. | Sine. | Cosine. | Sine. | ′ |
|  | 84° |  | 83° |  | 82° |  | 81° |  | 80° |  |  |

| ′ | 10° | | 11° | | 12° | | 13° | | 14° | | ′ |
|---|---|---|---|---|---|---|---|---|---|---|---|
| | Sine. | Cosine. | Sine. | Cosine. | Sine. | Cosine. | Sine. | Cosine. | Sine. | Cosine. | |
| 0 | 17365 | 98481 | 19081 | 98163 | 20791 | 97815 | 22495 | 97437 | 24192 | 97030 | 60 |
| 1 | 17393 | 98476 | 19109 | 98157 | 20820 | 97809 | 22523 | 97430 | 24220 | 97023 | 59 |
| 2 | 17422 | 98471 | 19138 | 98152 | 20848 | 97803 | 22552 | 97424 | 24249 | 97015 | 58 |
| 3 | 17451 | 98466 | 19167 | 98146 | 20877 | 97797 | 22580 | 97417 | 24277 | 97008 | 57 |
| 4 | 17479 | 98461 | 19195 | 98140 | 20905 | 97791 | 22608 | 97411 | 24305 | 97001 | 56 |
| 5 | 17508 | 98455 | 19224 | 98135 | 20933 | 97784 | 22637 | 97404 | 24333 | 96994 | 55 |
| 6 | 17537 | 98450 | 19252 | 98129 | 20962 | 97778 | 22665 | 97398 | 24362 | 96987 | 54 |
| 7 | 17565 | 98445 | 19281 | 98124 | 20990 | 97772 | 22693 | 97391 | 24390 | 96980 | 53 |
| 8 | 17594 | 98440 | 19309 | 98118 | 21019 | 97766 | 22722 | 97384 | 24418 | 96973 | 52 |
| 9 | 17623 | 98435 | 19338 | 98112 | 21047 | 97760 | 22750 | 97378 | 24446 | 96966 | 51 |
| 10 | 17651 | 98430 | 19366 | 98107 | 21076 | 97754 | 22778 | 97371 | 24474 | 96959 | 50 |
| 11 | 17680 | 98425 | 19395 | 98101 | 21104 | 97748 | 22807 | 97365 | 24503 | 96952 | 49 |
| 12 | 17708 | 98420 | 19423 | 98096 | 21132 | 97742 | 22835 | 97358 | 24531 | 96945 | 48 |
| 13 | 17737 | 98414 | 19452 | 98090 | 21161 | 97735 | 22863 | 97351 | 24559 | 96937 | 47 |
| 14 | 17766 | 98409 | 19481 | 98084 | 21189 | 97729 | 22892 | 97345 | 24587 | 96930 | 46 |
| 15 | 17794 | 98404 | 19509 | 98079 | 21218 | 97723 | 22920 | 97338 | 24615 | 96923 | 45 |
| 16 | 17823 | 98399 | 19538 | 98073 | 21246 | 97717 | 22948 | 97331 | 24644 | 96916 | 44 |
| 17 | 17852 | 98394 | 19566 | 98067 | 21275 | 97711 | 22977 | 97325 | 24672 | 96909 | 43 |
| 18 | 17880 | 98389 | 19595 | 98061 | 21303 | 97705 | 23005 | 97318 | 24700 | 96902 | 42 |
| 19 | 17909 | 98383 | 19623 | 98056 | 21331 | 97698 | 23033 | 97311 | 24728 | 96894 | 41 |
| 20 | 17937 | 98378 | 19652 | 98050 | 21360 | 97692 | 23062 | 97304 | 24756 | 96887 | 40 |
| 21 | 17966 | 98373 | 19680 | 98044 | 21388 | 97686 | 23090 | 97298 | 24784 | 96880 | 39 |
| 22 | 17995 | 98368 | 19709 | 98039 | 21417 | 97680 | 23118 | 97291 | 24813 | 96873 | 38 |
| 23 | 18023 | 98362 | 19737 | 98033 | 21445 | 97673 | 23146 | 97284 | 24841 | 96866 | 37 |
| 24 | 18052 | 98357 | 19766 | 98027 | 21474 | 97667 | 23175 | 97278 | 24869 | 96858 | 36 |
| 25 | 18081 | 98352 | 19794 | 98021 | 21502 | 97661 | 23203 | 97271 | 24897 | 96851 | 35 |
| 26 | 18109 | 98347 | 19823 | 98016 | 21530 | 97655 | 23231 | 97264 | 24925 | 96844 | 34 |
| 27 | 18138 | 98341 | 19851 | 98010 | 21559 | 97648 | 23260 | 97257 | 24954 | 96837 | 33 |
| 28 | 18166 | 98336 | 19880 | 98004 | 21587 | 97642 | 23288 | 97251 | 24982 | 96829 | 32 |
| 29 | 18195 | 98331 | 19908 | 97998 | 21616 | 97636 | 23316 | 97244 | 25010 | 96822 | 31 |
| 30 | 18224 | 98325 | 19937 | 97992 | 21644 | 97630 | 23345 | 97237 | 25038 | 96815 | 30 |
| 31 | 18252 | 98320 | 19965 | 97987 | 21672 | 97623 | 23373 | 97230 | 25066 | 96807 | 29 |
| 32 | 18281 | 98315 | 19994 | 97981 | 21701 | 97617 | 23401 | 97223 | 25094 | 96800 | 28 |
| 33 | 18309 | 98310 | 20022 | 97975 | 21729 | 97611 | 23429 | 97217 | 25122 | 96793 | 27 |
| 34 | 18338 | 98304 | 20051 | 97969 | 21758 | 97604 | 23458 | 97210 | 25151 | 96786 | 26 |
| 35 | 18367 | 98299 | 20079 | 97963 | 21786 | 97598 | 23486 | 97203 | 25179 | 96778 | 25 |
| 36 | 18395 | 98294 | 20108 | 97958 | 21814 | 97592 | 23514 | 97196 | 25207 | 96771 | 24 |
| 37 | 18424 | 98288 | 20136 | 97952 | 21843 | 97585 | 23542 | 97189 | 25235 | 96764 | 23 |
| 38 | 18452 | 98283 | 20165 | 97946 | 21871 | 97579 | 23571 | 97182 | 25263 | 96756 | 22 |
| 39 | 18481 | 98277 | 20193 | 97940 | 21899 | 97573 | 23599 | 97176 | 25291 | 96749 | 21 |
| 40 | 18509 | 98272 | 20222 | 97934 | 21928 | 97566 | 23627 | 97169 | 25320 | 96742 | 20 |
| 41 | 18538 | 98267 | 20250 | 97928 | 21956 | 97560 | 23656 | 97162 | 25348 | 96734 | 19 |
| 42 | 18567 | 98261 | 20279 | 97922 | 21985 | 97553 | 23684 | 97155 | 25376 | 96727 | 18 |
| 43 | 18595 | 98256 | 20307 | 97916 | 22013 | 97547 | 23712 | 97148 | 25404 | 96719 | 17 |
| 44 | 18624 | 98250 | 20336 | 97910 | 22041 | 97541 | 23740 | 97141 | 25432 | 96712 | 16 |
| 45 | 18652 | 98245 | 20364 | 97905 | 22070 | 97534 | 23769 | 97134 | 25460 | 96705 | 15 |
| 46 | 18681 | 98240 | 20393 | 97899 | 22098 | 97528 | 23797 | 97127 | 25488 | 96697 | 14 |
| 47 | 18710 | 98234 | 20421 | 97893 | 22126 | 97521 | 23825 | 97120 | 25516 | 96690 | 13 |
| 48 | 18738 | 98229 | 20450 | 97887 | 22155 | 97515 | 23853 | 97113 | 25545 | 96682 | 12 |
| 49 | 18767 | 98223 | 20478 | 97881 | 22183 | 97508 | 23882 | 97106 | 25573 | 96675 | 11 |
| 50 | 18795 | 98218 | 20507 | 97875 | 22212 | 97502 | 23910 | 97100 | 25601 | 96667 | 10 |
| 51 | 18824 | 98212 | 20535 | 97869 | 22240 | 97496 | 23938 | 97093 | 25629 | 96660 | 9 |
| 52 | 18852 | 98207 | 20563 | 97863 | 22268 | 97489 | 23966 | 97086 | 25657 | 96653 | 8 |
| 53 | 18881 | 98201 | 20592 | 97857 | 22297 | 97483 | 23995 | 97079 | 25685 | 96645 | 7 |
| 54 | 18910 | 98196 | 20620 | 97851 | 22325 | 97476 | 24023 | 97072 | 25713 | 96638 | 6 |
| 55 | 18938 | 98190 | 20649 | 97845 | 22353 | 97470 | 24051 | 97065 | 25741 | 96630 | 5 |
| 56 | 18967 | 98185 | 20677 | 97839 | 22382 | 97463 | 24079 | 97058 | 25769 | 96623 | 4 |
| 57 | 18995 | 98179 | 20706 | 97833 | 22410 | 97457 | 24108 | 97051 | 25798 | 96615 | 3 |
| 58 | 19024 | 98174 | 20734 | 97827 | 22438 | 97450 | 24136 | 97044 | 25826 | 96608 | 2 |
| 59 | 19052 | 98168 | 20763 | 97821 | 22467 | 97444 | 24164 | 97037 | 25854 | 96600 | 1 |
| 60 | 19081 | 98163 | 20791 | 97815 | 22495 | 97437 | 24192 | 97030 | 25882 | 96593 | 0 |
| ′ | Cosine. | Sine. | Cosine. | Sine. | Cosine. | Sine. | Cosine. | Sine. | Cosine. | Sine. | ′ |
| | 79° | | 78° | | 77° | | 76° | | 75° | | |

| ′ | 15° Sine. | 15° Cosine. | 16° Sine. | 16° Cosine. | 17° Sine. | 17° Cosine. | 18° Sine. | 18° Cosine. | 19° Sine. | 19° Cosine. | ′ |
|---|---|---|---|---|---|---|---|---|---|---|---|
| 0 | 25882 | 96593 | 27564 | 96126 | 29237 | 95630 | 30902 | 95106 | 32557 | 94552 | 60 |
| 1 | 25910 | 96585 | 27592 | 96118 | 29265 | 95622 | 30929 | 95097 | 32584 | 94542 | 59 |
| 2 | 25938 | 96578 | 27620 | 96110 | 29293 | 95613 | 30957 | 95088 | 32612 | 94533 | 58 |
| 3 | 25966 | 96570 | 27648 | 96102 | 29321 | 95605 | 30985 | 95079 | 32639 | 94523 | 57 |
| 4 | 25994 | 96562 | 27676 | 96094 | 29348 | 95596 | 31012 | 95070 | 32667 | 94514 | 56 |
| 5 | 26022 | 96555 | 27704 | 96086 | 29376 | 95588 | 31040 | 95061 | 32694 | 94504 | 55 |
| 6 | 26050 | 96547 | 27731 | 96078 | 29404 | 95579 | 31068 | 95052 | 32722 | 94495 | 54 |
| 7 | 26079 | 96540 | 27759 | 96070 | 29432 | 95571 | 31095 | 95043 | 32749 | 94485 | 53 |
| 8 | 26107 | 96532 | 27787 | 96062 | 29460 | 95562 | 31123 | 95033 | 32777 | 94476 | 52 |
| 9 | 26135 | 96524 | 27815 | 96054 | 29487 | 95554 | 31151 | 95024 | 32804 | 94466 | 51 |
| 10 | 26163 | 96517 | 27843 | 96046 | 29515 | 95545 | 31178 | 95015 | 32832 | 94457 | 50 |
| 11 | 26191 | 96509 | 27871 | 96037 | 29543 | 95536 | 31206 | 95006 | 32859 | 94447 | 49 |
| 12 | 26219 | 96502 | 27899 | 96029 | 29571 | 95528 | 31233 | 94997 | 32887 | 94438 | 48 |
| 13 | 26247 | 96494 | 27927 | 96021 | 29599 | 95519 | 31261 | 94988 | 32914 | 94428 | 47 |
| 14 | 26275 | 96486 | 27955 | 96013 | 29626 | 95511 | 31289 | 94979 | 32942 | 94418 | 46 |
| 15 | 26303 | 96479 | 27983 | 96005 | 29654 | 95502 | 31316 | 94970 | 32969 | 94409 | 45 |
| 16 | 26331 | 96471 | 28011 | 95997 | 29682 | 95493 | 31344 | 94961 | 32997 | 94399 | 44 |
| 17 | 26359 | 96463 | 28039 | 95989 | 29710 | 95485 | 31372 | 94952 | 33024 | 94390 | 43 |
| 18 | 26387 | 96456 | 28067 | 95981 | 29737 | 95476 | 31399 | 94943 | 33051 | 94380 | 42 |
| 19 | 26415 | 96448 | 28095 | 95972 | 29765 | 95467 | 31427 | 94933 | 33079 | 94370 | 41 |
| 20 | 26443 | 96440 | 28123 | 95964 | 29793 | 95459 | 31454 | 94924 | 33106 | 94361 | 40 |
| 21 | 26471 | 96433 | 28150 | 95956 | 29821 | 95450 | 31482 | 94915 | 33134 | 94351 | 39 |
| 22 | 26500 | 96425 | 28178 | 95948 | 29849 | 95441 | 31510 | 94906 | 33161 | 94342 | 38 |
| 23 | 26528 | 96417 | 28206 | 95940 | 29876 | 95433 | 31537 | 94897 | 33189 | 94332 | 37 |
| 24 | 26556 | 96410 | 28234 | 95931 | 29904 | 95424 | 31565 | 94888 | 33216 | 94322 | 36 |
| 25 | 26584 | 96402 | 28262 | 95923 | 29932 | 95415 | 31593 | 94878 | 33244 | 94313 | 35 |
| 26 | 26612 | 96394 | 28290 | 95915 | 29960 | 95407 | 31620 | 94869 | 33271 | 94303 | 34 |
| 27 | 26640 | 96386 | 28318 | 95907 | 29987 | 95398 | 31648 | 94860 | 33298 | 94293 | 33 |
| 28 | 26668 | 96379 | 28346 | 95898 | 30015 | 95389 | 31675 | 94851 | 33326 | 94284 | 32 |
| 29 | 26696 | 96371 | 28374 | 95890 | 30043 | 95380 | 31703 | 94842 | 33353 | 94274 | 31 |
| 30 | 26724 | 96363 | 28402 | 95882 | 30071 | 95372 | 31730 | 94832 | 33381 | 94264 | 30 |
| 31 | 26752 | 96355 | 28429 | 95874 | 30098 | 95363 | 31758 | 94823 | 33408 | 94254 | 29 |
| 32 | 26780 | 96347 | 28457 | 95865 | 30126 | 95354 | 31786 | 94814 | 33436 | 94245 | 28 |
| 33 | 26808 | 96340 | 28485 | 95857 | 30154 | 95345 | 31813 | 94805 | 33463 | 94235 | 27 |
| 34 | 26836 | 96332 | 28513 | 95849 | 30182 | 95337 | 31841 | 94795 | 33490 | 94225 | 26 |
| 35 | 26864 | 96324 | 28541 | 95841 | 30209 | 95328 | 31868 | 94786 | 33518 | 94215 | 25 |
| 36 | 26892 | 96316 | 28569 | 95832 | 30237 | 95319 | 31896 | 94777 | 33545 | 94206 | 24 |
| 37 | 26920 | 96308 | 28597 | 95824 | 30265 | 95310 | 31923 | 94768 | 33573 | 94196 | 23 |
| 38 | 26948 | 96301 | 28625 | 95816 | 30292 | 95301 | 31951 | 94758 | 33600 | 94186 | 22 |
| 39 | 26976 | 96293 | 28652 | 95807 | 30320 | 95293 | 31979 | 94749 | 33627 | 94176 | 21 |
| 40 | 27004 | 96285 | 28680 | 95799 | 30348 | 95284 | 32006 | 94740 | 33655 | 94167 | 20 |
| 41 | 27032 | 96277 | 28708 | 95791 | 30376 | 95275 | 32034 | 94730 | 33682 | 94157 | 19 |
| 42 | 27060 | 96269 | 28736 | 95782 | 30403 | 95266 | 32061 | 94721 | 33710 | 94147 | 18 |
| 43 | 27088 | 96261 | 28764 | 95774 | 30431 | 95257 | 32089 | 94712 | 33737 | 94137 | 17 |
| 44 | 27116 | 96253 | 28792 | 95766 | 30459 | 95248 | 32116 | 94702 | 33764 | 94127 | 16 |
| 45 | 27144 | 96246 | 28820 | 95757 | 30486 | 95240 | 32144 | 94693 | 33792 | 94118 | 15 |
| 46 | 27172 | 96238 | 28847 | 95749 | 30514 | 95231 | 32171 | 94684 | 33819 | 94108 | 14 |
| 47 | 27200 | 96230 | 28875 | 95740 | 30542 | 95222 | 32199 | 94674 | 33846 | 94098 | 13 |
| 48 | 27228 | 96222 | 28903 | 95732 | 30570 | 95213 | 32227 | 94665 | 33874 | 94088 | 12 |
| 49 | 27256 | 96214 | 28931 | 95724 | 30597 | 95204 | 32254 | 94656 | 33901 | 94078 | 11 |
| 50 | 27284 | 96206 | 28959 | 95715 | 30625 | 95195 | 32282 | 94646 | 33929 | 94068 | 10 |
| 51 | 27312 | 96198 | 28987 | 95707 | 30653 | 95186 | 32309 | 94637 | 33956 | 94058 | 9 |
| 52 | 27340 | 96190 | 29015 | 95698 | 30680 | 95177 | 32337 | 94627 | 33983 | 94049 | 8 |
| 53 | 27368 | 96182 | 29042 | 95690 | 30708 | 95168 | 32364 | 94618 | 34011 | 94039 | 7 |
| 54 | 27396 | 96174 | 29070 | 95681 | 30736 | 95159 | 32392 | 94609 | 34038 | 94029 | 6 |
| 55 | 27424 | 96166 | 29098 | 95673 | 30763 | 95150 | 32419 | 94599 | 34065 | 94019 | 5 |
| 56 | 27452 | 96158 | 29126 | 95664 | 30791 | 95142 | 32447 | 94590 | 34093 | 94009 | 4 |
| 57 | 27480 | 96150 | 29154 | 95656 | 30819 | 95133 | 32474 | 94580 | 34120 | 93999 | 3 |
| 58 | 27508 | 96142 | 29182 | 95647 | 30846 | 95124 | 32502 | 94571 | 34147 | 93989 | 2 |
| 59 | 27536 | 96134 | 29209 | 95639 | 30874 | 95115 | 32529 | 94561 | 34175 | 93979 | 1 |
| 60 | 27564 | 96126 | 29237 | 95630 | 30902 | 95106 | 32557 | 94552 | 34202 | 93969 | 0 |
| ′ | Cosine. 74° | Sine. 74° | Cosine. 73° | Sine. 73° | Cosine. 72° | Sine. 72° | Cosine. 71° | Sine. 71° | Cosine. 70° | Sine. 70° | ′ |

| ′ | 20° | | 21° | | 22° | | 23° | | 24° | | ′ |
|---|---|---|---|---|---|---|---|---|---|---|---|
| | Sine. | Cosine. | Sine. | Cosine. | Sine. | Cosine. | Sine. | Cosine. | Sine. | Cosine. | |
| 0 | 34202 | 93969 | 35837 | 93358 | 37461 | 92718 | 39073 | 92050 | 40674 | 91355 | 60 |
| 1 | 34229 | 93959 | 35864 | 93348 | 37488 | 92707 | 39100 | 92039 | 40700 | 91343 | 59 |
| 2 | 34257 | 93949 | 35891 | 93337 | 37515 | 92697 | 39127 | 92028 | 40727 | 91331 | 58 |
| 3 | 34284 | 93939 | 35918 | 93327 | 37542 | 92686 | 39153 | 92016 | 40753 | 91319 | 57 |
| 4 | 34311 | 93929 | 35945 | 93316 | 37569 | 92675 | 39180 | 92005 | 40780 | 91307 | 56 |
| 5 | 34339 | 93919 | 35973 | 93306 | 37595 | 92664 | 39207 | 91994 | 40806 | 91295 | 55 |
| 6 | 34366 | 93909 | 36000 | 93295 | 37622 | 92653 | 39234 | 91982 | 40833 | 91283 | 54 |
| 7 | 34393 | 93899 | 36027 | 93285 | 37649 | 92642 | 39260 | 91971 | 40860 | 91272 | 53 |
| 8 | 34421 | 93889 | 36054 | 93274 | 37676 | 92631 | 39287 | 91959 | 40886 | 91260 | 52 |
| 9 | 34448 | 93879 | 36081 | 93264 | 37703 | 92620 | 39314 | 91948 | 40913 | 91248 | 51 |
| 10 | 34475 | 93869 | 36108 | 93253 | 37730 | 92609 | 39341 | 91936 | 40939 | 91236 | 50 |
| 11 | 34503 | 93859 | 36135 | 93243 | 37757 | 92598 | 39367 | 91925 | 40966 | 91224 | 49 |
| 12 | 34530 | 93849 | 36162 | 93232 | 37784 | 92587 | 39394 | 91914 | 40992 | 91212 | 48 |
| 13 | 34557 | 93839 | 36190 | 93222 | 37811 | 92576 | 39421 | 91902 | 41019 | 91200 | 47 |
| 14 | 34584 | 93829 | 36217 | 93211 | 37838 | 92565 | 39448 | 91891 | 41045 | 91188 | 46 |
| 15 | 34612 | 93819 | 36244 | 93201 | 37865 | 92554 | 39474 | 91879 | 41072 | 91176 | 45 |
| 16 | 34639 | 93809 | 36271 | 93190 | 37892 | 92543 | 39501 | 91868 | 41098 | 91164 | 44 |
| 17 | 34666 | 93799 | 36298 | 93180 | 37919 | 92532 | 39528 | 91856 | 41125 | 91152 | 43 |
| 18 | 34694 | 93789 | 36325 | 93169 | 37946 | 92521 | 39555 | 91845 | 41151 | 91140 | 42 |
| 19 | 34721 | 93779 | 36352 | 93159 | 37973 | 92510 | 39581 | 91833 | 41178 | 91128 | 41 |
| 20 | 34748 | 93769 | 36379 | 93148 | 37999 | 92499 | 39608 | 91822 | 41204 | 91116 | 40 |
| 21 | 34775 | 93759 | 36406 | 93137 | 38026 | 92488 | 39635 | 91810 | 41231 | 91104 | 39 |
| 22 | 34803 | 93748 | 36434 | 93127 | 38053 | 92477 | 39661 | 91799 | 41257 | 91092 | 38 |
| 23 | 34830 | 93738 | 36461 | 93116 | 38080 | 92466 | 39688 | 91787 | 41284 | 91080 | 37 |
| 24 | 34857 | 93728 | 36488 | 93106 | 38107 | 92455 | 39715 | 91775 | 41310 | 91068 | 36 |
| 25 | 34884 | 93718 | 36515 | 93095 | 38134 | 92444 | 39741 | 91764 | 41337 | 91056 | 35 |
| 26 | 34912 | 93708 | 36542 | 93084 | 38161 | 92432 | 39768 | 91752 | 41363 | 91044 | 34 |
| 27 | 34939 | 93698 | 36569 | 93074 | 38188 | 92421 | 39795 | 91741 | 41390 | 91032 | 33 |
| 28 | 34966 | 93688 | 36596 | 93063 | 38215 | 92410 | 39822 | 91729 | 41416 | 91020 | 32 |
| 29 | 34993 | 93677 | 36623 | 93052 | 38241 | 92399 | 39848 | 91718 | 41443 | 91008 | 31 |
| 30 | 35021 | 93667 | 36650 | 93042 | 38268 | 92388 | 39875 | 91706 | 41469 | 90996 | 30 |
| 31 | 35048 | 93657 | 36677 | 93031 | 38295 | 92377 | 39902 | 91694 | 41496 | 90984 | 29 |
| 32 | 35075 | 93647 | 36704 | 93020 | 38322 | 92366 | 39928 | 91683 | 41522 | 90972 | 28 |
| 33 | 35102 | 93637 | 36731 | 93010 | 38349 | 92355 | 39955 | 91671 | 41549 | 90960 | 27 |
| 34 | 35130 | 93626 | 36758 | 92999 | 38376 | 92343 | 39982 | 91660 | 41575 | 90948 | 26 |
| 35 | 35157 | 93616 | 36785 | 92988 | 38403 | 92332 | 40008 | 91648 | 41602 | 90936 | 25 |
| 36 | 35184 | 93606 | 36812 | 92978 | 38430 | 92321 | 40035 | 91636 | 41628 | 90924 | 24 |
| 37 | 35211 | 93596 | 36839 | 92967 | 38456 | 92310 | 40062 | 91625 | 41655 | 90911 | 23 |
| 38 | 35239 | 93585 | 36867 | 92956 | 38483 | 92299 | 40088 | 91613 | 41681 | 90899 | 22 |
| 39 | 35266 | 93575 | 36894 | 92945 | 38510 | 92287 | 40115 | 91601 | 41707 | 90887 | 21 |
| 40 | 35293 | 93565 | 36921 | 92935 | 38537 | 92276 | 40141 | 91590 | 41734 | 90875 | 20 |
| 41 | 35320 | 93555 | 36948 | 92924 | 38564 | 92265 | 40168 | 91578 | 41760 | 90863 | 19 |
| 42 | 35347 | 93544 | 36975 | 92913 | 38591 | 92254 | 40195 | 91566 | 41787 | 90851 | 18 |
| 43 | 35375 | 93534 | 37002 | 92902 | 38617 | 92243 | 40221 | 91555 | 41813 | 90839 | 17 |
| 44 | 35402 | 93524 | 37029 | 92892 | 38644 | 92231 | 40248 | 91543 | 41840 | 90826 | 16 |
| 45 | 35429 | 93514 | 37056 | 92881 | 38671 | 92220 | 40275 | 91531 | 41866 | 90814 | 15 |
| 46 | 35456 | 93503 | 37083 | 92870 | 38698 | 92209 | 40301 | 91519 | 41892 | 90802 | 14 |
| 47 | 35484 | 93493 | 37110 | 92859 | 38725 | 92198 | 40328 | 91508 | 41919 | 90790 | 13 |
| 48 | 35511 | 93483 | 37137 | 92849 | 38752 | 92186 | 40355 | 91496 | 41945 | 90778 | 12 |
| 49 | 35538 | 93472 | 37164 | 92838 | 38778 | 92175 | 40381 | 91484 | 41972 | 90766 | 11 |
| 50 | 35565 | 93462 | 37191 | 92827 | 38805 | 92164 | 40408 | 91472 | 41998 | 90753 | 10 |
| 51 | 35592 | 93452 | 37218 | 92816 | 38832 | 92152 | 40434 | 91461 | 42024 | 90741 | 9 |
| 52 | 35619 | 93441 | 37245 | 92805 | 38859 | 92141 | 40461 | 91449 | 42051 | 90729 | 8 |
| 53 | 35647 | 93431 | 37272 | 92794 | 38886 | 92130 | 40488 | 91437 | 42077 | 90717 | 7 |
| 54 | 35674 | 93420 | 37299 | 92784 | 38912 | 92119 | 40514 | 91425 | 42104 | 90704 | 6 |
| 55 | 35701 | 93410 | 37326 | 92773 | 38939 | 92107 | 40541 | 91414 | 42130 | 90692 | 5 |
| 56 | 35728 | 93400 | 37353 | 92762 | 38966 | 92096 | 40567 | 91402 | 42156 | 90680 | 4 |
| 57 | 35755 | 93389 | 37380 | 92751 | 38993 | 92085 | 40594 | 91390 | 42183 | 90668 | 3 |
| 58 | 35782 | 93379 | 37407 | 92740 | 39020 | 92073 | 40621 | 91378 | 42209 | 90655 | 2 |
| 59 | 35810 | 93368 | 37434 | 92729 | 39046 | 92062 | 40647 | 91366 | 42235 | 90643 | 1 |
| 60 | 35837 | 93358 | 37461 | 92718 | 39073 | 92050 | 40674 | 91355 | 42262 | 90631 | 0 |
| | Cosine. | Sine. | Cosine. | Sine. | Cosine. | Sine. | Cosine. | Sine. | Cosine. | Sine. | |
| ′ | 69° | | 68° | | 67° | | 66° | | 65° | | ′ |

| ′ | 25° Sine. | 25° Cosine. | 26° Sine. | 26° Cosine. | 27° Sine. | 27° Cosine. | 28° Sine. | 28° Cosine. | 29° Sine. | 29° Cosine. | ′ |
|---|---|---|---|---|---|---|---|---|---|---|---|
| 0 | 42262 | 90631 | 43837 | 89879 | 45399 | 89101 | 46947 | 88295 | 48481 | 87462 | 60 |
| 1 | 42288 | 90618 | 43863 | 89867 | 45425 | 89087 | 46973 | 88281 | 48506 | 87448 | 59 |
| 2 | 42315 | 90606 | 43889 | 89854 | 45451 | 89074 | 46999 | 88267 | 48532 | 87434 | 58 |
| 3 | 42341 | 90594 | 43916 | 89841 | 45477 | 89061 | 47024 | 88254 | 48557 | 87420 | 57 |
| 4 | 42367 | 90582 | 43942 | 89828 | 45503 | 89048 | 47050 | 88240 | 48583 | 87406 | 56 |
| 5 | 42394 | 90569 | 43968 | 89816 | 45529 | 89035 | 47076 | 88226 | 48608 | 87391 | 55 |
| 6 | 42420 | 90557 | 43994 | 89803 | 45554 | 89021 | 47101 | 88213 | 48634 | 87377 | 54 |
| 7 | 42446 | 90545 | 44020 | 89790 | 45580 | 89008 | 47127 | 88199 | 48659 | 87363 | 53 |
| 8 | 42473 | 90532 | 44046 | 89777 | 45606 | 88995 | 47153 | 88185 | 48684 | 87349 | 52 |
| 9 | 42499 | 90520 | 44072 | 89764 | 45632 | 88981 | 47178 | 88172 | 48710 | 87335 | 51 |
| 10 | 42525 | 90507 | 44098 | 89752 | 45658 | 88968 | 47204 | 88158 | 48735 | 87321 | 50 |
| 11 | 42552 | 90495 | 44124 | 89739 | 45684 | 88955 | 47229 | 88144 | 48761 | 87306 | 49 |
| 12 | 42578 | 90483 | 44151 | 89726 | 45710 | 88942 | 47255 | 88130 | 48786 | 87292 | 48 |
| 13 | 42604 | 90470 | 44177 | 89713 | 45736 | 88928 | 47281 | 88117 | 48811 | 87278 | 47 |
| 14 | 42631 | 90458 | 44203 | 89700 | 45762 | 88915 | 47306 | 88103 | 48837 | 87264 | 46 |
| 15 | 42657 | 90446 | 44229 | 89687 | 45787 | 88902 | 47332 | 88089 | 48862 | 87250 | 45 |
| 16 | 42683 | 90433 | 44255 | 89674 | 45813 | 88888 | 47358 | 88075 | 48888 | 87235 | 44 |
| 17 | 42709 | 90421 | 44281 | 89662 | 45839 | 88875 | 47383 | 88062 | 48913 | 87221 | 43 |
| 18 | 42736 | 90408 | 44307 | 89649 | 45865 | 88862 | 47409 | 88048 | 48938 | 87207 | 42 |
| 19 | 42762 | 90396 | 44333 | 89636 | 45891 | 88848 | 47434 | 88034 | 48964 | 87193 | 41 |
| 20 | 42788 | 90383 | 44359 | 89623 | 45917 | 88835 | 47460 | 88020 | 48989 | 87178 | 40 |
| 21 | 42815 | 90371 | 44385 | 89610 | 45942 | 88822 | 47486 | 88006 | 49014 | 87164 | 39 |
| 22 | 42841 | 90358 | 44411 | 89597 | 45968 | 88808 | 47511 | 87993 | 49040 | 87150 | 38 |
| 23 | 42867 | 90346 | 44437 | 89584 | 45994 | 88795 | 47537 | 87979 | 49065 | 87136 | 37 |
| 24 | 42894 | 90334 | 44464 | 89571 | 46020 | 88782 | 47562 | 87965 | 49090 | 87121 | 36 |
| 25 | 42920 | 90321 | 44490 | 89558 | 46046 | 88768 | 47588 | 87951 | 49116 | 87107 | 35 |
| 26 | 42946 | 90309 | 44516 | 89545 | 46072 | 88755 | 47614 | 87937 | 49141 | 87093 | 34 |
| 27 | 42972 | 90296 | 44542 | 89532 | 46097 | 88741 | 47639 | 87923 | 49166 | 87079 | 33 |
| 28 | 42999 | 90284 | 44568 | 89519 | 46123 | 88728 | 47665 | 87909 | 49192 | 87064 | 32 |
| 29 | 43025 | 90271 | 44594 | 89506 | 46149 | 88715 | 47690 | 87896 | 49217 | 87050 | 31 |
| 30 | 43051 | 90259 | 44620 | 89493 | 46175 | 88701 | 47716 | 87882 | 49242 | 87036 | 30 |
| 31 | 43077 | 90246 | 44646 | 89480 | 46201 | 88688 | 47741 | 87868 | 49268 | 87021 | 29 |
| 32 | 43104 | 90233 | 44672 | 89467 | 46226 | 88674 | 47767 | 87854 | 49293 | 87007 | 28 |
| 33 | 43130 | 90221 | 44698 | 89454 | 46252 | 88661 | 47793 | 87840 | 49318 | 86993 | 27 |
| 34 | 43156 | 90208 | 44724 | 89441 | 46278 | 88647 | 47818 | 87826 | 49344 | 86978 | 26 |
| 35 | 43182 | 90196 | 44750 | 89428 | 46304 | 88634 | 47844 | 87812 | 49369 | 86964 | 25 |
| 36 | 43209 | 90183 | 44776 | 89415 | 46330 | 88620 | 47869 | 87798 | 49394 | 86949 | 24 |
| 37 | 43235 | 90171 | 44802 | 89402 | 46355 | 88607 | 47895 | 87784 | 49419 | 86935 | 23 |
| 38 | 43261 | 90158 | 44828 | 89389 | 46381 | 88593 | 47920 | 87770 | 49445 | 86921 | 22 |
| 39 | 43287 | 90146 | 44854 | 89376 | 46407 | 88580 | 47946 | 87756 | 49470 | 86906 | 21 |
| 40 | 43313 | 90133 | 44880 | 89363 | 46433 | 88566 | 47971 | 87743 | 49495 | 86892 | 20 |
| 41 | 43340 | 90120 | 44906 | 89350 | 46458 | 88553 | 47997 | 87729 | 49521 | 86878 | 19 |
| 42 | 43366 | 90108 | 44932 | 89337 | 46484 | 88539 | 48022 | 87715 | 49546 | 86863 | 18 |
| 43 | 43392 | 90095 | 44958 | 89324 | 46510 | 88526 | 48048 | 87701 | 49571 | 86849 | 17 |
| 44 | 43418 | 90082 | 44984 | 89311 | 46536 | 88512 | 48073 | 87687 | 49596 | 86834 | 16 |
| 45 | 43445 | 90070 | 45010 | 89298 | 46561 | 88499 | 48099 | 87673 | 49622 | 86820 | 15 |
| 46 | 43471 | 90057 | 45036 | 89285 | 46587 | 88485 | 48124 | 87659 | 49647 | 86805 | 14 |
| 47 | 43497 | 90045 | 45062 | 89272 | 46613 | 88472 | 48150 | 87645 | 49672 | 86791 | 13 |
| 48 | 43523 | 90032 | 45088 | 89259 | 46639 | 88458 | 48175 | 87631 | 49697 | 86777 | 12 |
| 49 | 43549 | 90019 | 45114 | 89245 | 46664 | 88445 | 48201 | 87617 | 49723 | 86762 | 11 |
| 50 | 43575 | 90007 | 45140 | 89232 | 46690 | 88431 | 48226 | 87603 | 49748 | 86748 | 10 |
| 51 | 43602 | 89994 | 45166 | 89219 | 46716 | 88417 | 48252 | 87589 | 49773 | 86733 | 9 |
| 52 | 43628 | 89981 | 45192 | 89206 | 46742 | 88404 | 48277 | 87575 | 49798 | 86719 | 8 |
| 53 | 43654 | 89968 | 45218 | 89193 | 46767 | 88390 | 48303 | 87561 | 49824 | 86704 | 7 |
| 54 | 43680 | 89956 | 45243 | 89180 | 46793 | 88377 | 48328 | 87546 | 49849 | 86690 | 6 |
| 55 | 43706 | 89943 | 45269 | 89167 | 46819 | 88363 | 48354 | 87532 | 49874 | 86675 | 5 |
| 56 | 43733 | 89930 | 45295 | 89153 | 46844 | 88349 | 48379 | 87518 | 49899 | 86661 | 4 |
| 57 | 43759 | 89918 | 45321 | 89140 | 46870 | 88336 | 48405 | 87504 | 49924 | 86646 | 3 |
| 58 | 43785 | 89905 | 45347 | 89127 | 46896 | 88322 | 48430 | 87490 | 49950 | 86632 | 2 |
| 59 | 43811 | 89892 | 45373 | 89114 | 46921 | 88308 | 48456 | 87476 | 49975 | 86617 | 1 |
| 60 | 43837 | 89879 | 45399 | 89101 | 46947 | 88295 | 48481 | 87462 | 50000 | 86603 | 0 |
| ′ | Cosine. 64° | Sine. 64° | Cosine. 63° | Sine. 63° | Cosine. 62° | Sine. 62° | Cosine. 61° | Sine. 61° | Cosine. 60° | Sine. 60° | ′ |

| ′ | 30° | | 31° | | 32° | | 33° | | 34° | | ′ |
|---|---|---|---|---|---|---|---|---|---|---|---|
| | Sine. | Cosine. | Sine. | Cosine. | Sine. | Cosine. | Sine. | Cosine. | Sine. | Cosine. | |
| 0 | 50000 | 86603 | 51504 | 85717 | 52992 | 84805 | 54464 | 83867 | 55919 | 82904 | 60 |
| 1 | 50025 | 86588 | 51529 | 85702 | 53017 | 84789 | 54488 | 83851 | 55943 | 82887 | 59 |
| 2 | 50050 | 86573 | 51554 | 85687 | 53041 | 84774 | 54513 | 83835 | 55968 | 82871 | 58 |
| 3 | 50076 | 86559 | 51579 | 85672 | 53066 | 84759 | 54537 | 83819 | 55992 | 82855 | 57 |
| 4 | 50101 | 86544 | 51604 | 85657 | 53091 | 84743 | 54561 | 83804 | 56016 | 82839 | 56 |
| 5 | 50126 | 86530 | 51628 | 85642 | 53115 | 84728 | 54586 | 83788 | 56040 | 82822 | 55 |
| 6 | 50151 | 86515 | 51653 | 85627 | 53140 | 84712 | 54610 | 83772 | 56064 | 82806 | 54 |
| 7 | 50176 | 86501 | 51678 | 85612 | 53164 | 84697 | 54635 | 83756 | 56088 | 82790 | 53 |
| 8 | 50201 | 86486 | 51703 | 85597 | 53189 | 84681 | 54659 | 83740 | 56112 | 82773 | 52 |
| 9 | 50227 | 86471 | 51728 | 85582 | 53214 | 84666 | 54683 | 83724 | 56136 | 82757 | 51 |
| 10 | 50252 | 86457 | 51753 | 85567 | 53238 | 84650 | 54708 | 83708 | 56160 | 82741 | 50 |
| 11 | 50277 | 86442 | 51778 | 85551 | 53263 | 84635 | 54732 | 83692 | 56184 | 82724 | 49 |
| 12 | 50302 | 86427 | 51803 | 85536 | 53288 | 84619 | 54756 | 83676 | 56208 | 82708 | 48 |
| 13 | 50327 | 86413 | 51828 | 85521 | 53312 | 84604 | 54781 | 83660 | 56232 | 82692 | 47 |
| 14 | 50352 | 86398 | 51852 | 85506 | 53337 | 84588 | 54805 | 83645 | 56256 | 82675 | 46 |
| 15 | 50377 | 86384 | 51877 | 85491 | 53361 | 84573 | 54829 | 83629 | 56280 | 82659 | 45 |
| 16 | 50403 | 86369 | 51902 | 85476 | 53386 | 84557 | 54854 | 83613 | 56305 | 82643 | 44 |
| 17 | 50428 | 86354 | 51927 | 85461 | 53411 | 84542 | 54878 | 83597 | 56329 | 82626 | 43 |
| 18 | 50453 | 86340 | 51952 | 85446 | 53435 | 84526 | 54902 | 83581 | 56353 | 82610 | 42 |
| 19 | 50478 | 86325 | 51977 | 85431 | 53460 | 84511 | 54927 | 83565 | 56377 | 82593 | 41 |
| 20 | 50503 | 86310 | 52002 | 85416 | 53484 | 84495 | 54951 | 83549 | 56401 | 82577 | 40 |
| 21 | 50528 | 86295 | 52026 | 85401 | 53509 | 84480 | 54975 | 83533 | 56425 | 82561 | 39 |
| 22 | 50553 | 86281 | 52051 | 85385 | 53534 | 84464 | 54999 | 83517 | 56449 | 82544 | 38 |
| 23 | 50578 | 86266 | 52076 | 85370 | 53558 | 84448 | 55024 | 83501 | 56473 | 82528 | 37 |
| 24 | 50603 | 86251 | 52101 | 85355 | 53583 | 84433 | 55048 | 83485 | 56497 | 82511 | 36 |
| 25 | 50628 | 86237 | 52126 | 85340 | 53607 | 84417 | 55072 | 83469 | 56521 | 82495 | 35 |
| 26 | 50654 | 86222 | 52151 | 85325 | 53632 | 84402 | 55097 | 83453 | 56545 | 82478 | 34 |
| 27 | 50679 | 86207 | 52175 | 85310 | 53656 | 84386 | 55121 | 83437 | 56569 | 82462 | 33 |
| 28 | 50704 | 86192 | 52200 | 85294 | 53681 | 84370 | 55145 | 83421 | 56593 | 82446 | 32 |
| 29 | 50729 | 86178 | 52225 | 85279 | 53705 | 84355 | 55169 | 83405 | 56617 | 82429 | 31 |
| 30 | 50754 | 86163 | 52250 | 85264 | 53730 | 84339 | 55194 | 83389 | 56641 | 82413 | 30 |
| 31 | 50779 | 86148 | 52275 | 85249 | 53754 | 84324 | 55218 | 83373 | 56665 | 82396 | 29 |
| 32 | 50804 | 86133 | 52299 | 85234 | 53779 | 84308 | 55242 | 83356 | 56689 | 82380 | 28 |
| 33 | 50829 | 86119 | 52324 | 85218 | 53804 | 84292 | 55266 | 83340 | 56713 | 82363 | 27 |
| 34 | 50854 | 86104 | 52349 | 85203 | 53828 | 84277 | 55291 | 83324 | 56736 | 82347 | 26 |
| 35 | 50879 | 86089 | 52374 | 85188 | 53853 | 84261 | 55315 | 83308 | 56760 | 82330 | 25 |
| 36 | 50904 | 86074 | 52399 | 85173 | 53877 | 84245 | 55339 | 83292 | 56784 | 82314 | 24 |
| 37 | 50929 | 86059 | 52423 | 85157 | 53902 | 84230 | 55363 | 83276 | 56808 | 82297 | 23 |
| 38 | 50954 | 86045 | 52448 | 85142 | 53926 | 84214 | 55388 | 83260 | 56832 | 82281 | 22 |
| 39 | 50979 | 86030 | 52473 | 85127 | 53951 | 84198 | 55412 | 83244 | 56856 | 82264 | 21 |
| 40 | 51004 | 86015 | 52498 | 85112 | 53975 | 84182 | 55436 | 83228 | 56880 | 82248 | 20 |
| 41 | 51029 | 86000 | 52522 | 85096 | 54000 | 84167 | 55460 | 83212 | 56904 | 82231 | 19 |
| 42 | 51054 | 85985 | 52547 | 85081 | 54024 | 84151 | 55484 | 83195 | 56928 | 82214 | 18 |
| 43 | 51079 | 85970 | 52572 | 85066 | 54049 | 84135 | 55509 | 83179 | 56952 | 82198 | 17 |
| 44 | 51104 | 85956 | 52597 | 85051 | 54073 | 84120 | 55533 | 83163 | 56976 | 82181 | 16 |
| 45 | 51129 | 85941 | 52621 | 85035 | 54097 | 84104 | 55557 | 83147 | 57000 | 82165 | 15 |
| 46 | 51154 | 85926 | 52646 | 85020 | 54122 | 84088 | 55581 | 83131 | 57024 | 82148 | 14 |
| 47 | 51179 | 85911 | 52671 | 85005 | 54146 | 84072 | 55605 | 83115 | 57047 | 82132 | 13 |
| 48 | 51204 | 85896 | 52696 | 84989 | 54171 | 84057 | 55630 | 83098 | 57071 | 82115 | 12 |
| 49 | 51229 | 85881 | 52720 | 84974 | 54195 | 84041 | 55654 | 83082 | 57095 | 82098 | 11 |
| 50 | 51254 | 85866 | 52745 | 84959 | 54220 | 84025 | 55678 | 83066 | 57119 | 82082 | 10 |
| 51 | 51279 | 85851 | 52770 | 84943 | 54244 | 84009 | 55702 | 83050 | 57143 | 82065 | 9 |
| 52 | 51304 | 85836 | 52794 | 84928 | 54269 | 83994 | 55726 | 83034 | 57167 | 82048 | 8 |
| 53 | 51329 | 85821 | 52819 | 84913 | 54293 | 83978 | 55750 | 83017 | 57191 | 82032 | 7 |
| 54 | 51354 | 85806 | 52844 | 84897 | 54317 | 83962 | 55775 | 83001 | 57215 | 82015 | 6 |
| 55 | 51379 | 85792 | 52869 | 84882 | 54342 | 83946 | 55799 | 82985 | 57238 | 81999 | 5 |
| 56 | 51404 | 85777 | 52893 | 84866 | 54366 | 83930 | 55823 | 82969 | 57262 | 81982 | 4 |
| 57 | 51429 | 85762 | 52918 | 84851 | 54391 | 83915 | 55847 | 82953 | 57286 | 81965 | 3 |
| 58 | 51454 | 85747 | 52943 | 84836 | 54415 | 83899 | 55871 | 82936 | 57310 | 81949 | 2 |
| 59 | 51479 | 85732 | 52967 | 84820 | 54440 | 83883 | 55895 | 82920 | 57334 | 81932 | 1 |
| 60 | 51504 | 85717 | 52992 | 84805 | 54464 | 83867 | 55919 | 82904 | 57358 | 81915 | 0 |
| | Cosine. | Sine. | Cosine. | Sine. | Cosine. | Sine. | Cosine. | Sine. | Cosine. | Sine. | |
| ′ | 59° | | 58° | | 57° | | 56° | | 55° | | ′ |

| ′ | 35° | | 36° | | 37° | | 38° | | 39° | | ′ |
|---|---|---|---|---|---|---|---|---|---|---|---|
| | Sine. | Cosine. | Sine. | Cosine. | Sine. | Cosine. | Sine. | Cosine. | Sine. | Cosine. | |
| 0 | 57358 | 81915 | 58779 | 80902 | 60182 | 79864 | 61566 | 78801 | 62932 | 77715 | 60 |
| 1 | 57381 | 81899 | 58802 | 80885 | 60205 | 79846 | 61589 | 78783 | 62955 | 77696 | 59 |
| 2 | 57405 | 81882 | 58826 | 80867 | 60228 | 79829 | 61612 | 78765 | 62977 | 77678 | 58 |
| 3 | 57429 | 81865 | 58849 | 80850 | 60251 | 79811 | 61635 | 78747 | 63000 | 77660 | 57 |
| 4 | 57453 | 81848 | 58873 | 80833 | 60274 | 79793 | 61658 | 78729 | 63022 | 77641 | 56 |
| 5 | 57477 | 81832 | 58896 | 80816 | 60298 | 79776 | 61681 | 78711 | 63045 | 77623 | 55 |
| 6 | 57501 | 81815 | 58920 | 80799 | 60321 | 79758 | 61704 | 78694 | 63068 | 77605 | 54 |
| 7 | 57524 | 81798 | 58943 | 80782 | 60344 | 79741 | 61726 | 78676 | 63090 | 77586 | 53 |
| 8 | 57548 | 81782 | 58967 | 80765 | 60367 | 79723 | 61749 | 78658 | 63113 | 77568 | 52 |
| 9 | 57572 | 81765 | 58990 | 80748 | 60390 | 79706 | 61772 | 78640 | 63135 | 77550 | 51 |
| 10 | 57596 | 81748 | 59014 | 80730 | 60414 | 79688 | 61795 | 78622 | 63158 | 77531 | 50 |
| 11 | 57619 | 81731 | 59037 | 80713 | 60437 | 79671 | 61818 | 78604 | 63180 | 77513 | 49 |
| 12 | 57643 | 81714 | 59061 | 80696 | 60460 | 79653 | 61841 | 78586 | 63203 | 77494 | 48 |
| 13 | 57667 | 81698 | 59084 | 80679 | 60483 | 79635 | 61864 | 78568 | 63225 | 77476 | 47 |
| 14 | 57691 | 81681 | 59108 | 80662 | 60506 | 79618 | 61887 | 78550 | 63248 | 77458 | 46 |
| 15 | 57715 | 81664 | 59131 | 80644 | 60529 | 79600 | 61909 | 78532 | 63271 | 77439 | 45 |
| 16 | 57738 | 81647 | 59154 | 80627 | 60553 | 79583 | 61932 | 78514 | 63293 | 77421 | 44 |
| 17 | 57762 | 81631 | 59178 | 80610 | 60576 | 79565 | 61955 | 78496 | 63316 | 77402 | 43 |
| 18 | 57786 | 81614 | 59201 | 80593 | 60599 | 79547 | 61978 | 78478 | 63338 | 77384 | 42 |
| 19 | 57810 | 81597 | 59225 | 80576 | 60622 | 79530 | 62001 | 78460 | 63361 | 77366 | 41 |
| 20 | 57833 | 81580 | 59248 | 80558 | 60645 | 79512 | 62024 | 78442 | 63383 | 77347 | 40 |
| 21 | 57857 | 81563 | 59272 | 80541 | 60668 | 79494 | 62046 | 78424 | 63406 | 77329 | 39 |
| 22 | 57881 | 81546 | 59295 | 80524 | 60691 | 79477 | 62069 | 78405 | 63428 | 77310 | 38 |
| 23 | 57904 | 81530 | 59318 | 80507 | 60714 | 79459 | 62092 | 78387 | 63451 | 77292 | 37 |
| 24 | 57928 | 81513 | 59342 | 80489 | 60738 | 79441 | 62115 | 78369 | 63473 | 77273 | 36 |
| 25 | 57952 | 81496 | 59365 | 80472 | 60761 | 79424 | 62138 | 78351 | 63496 | 77255 | 35 |
| 26 | 57976 | 81479 | 59389 | 80455 | 60784 | 79406 | 62160 | 78333 | 63518 | 77236 | 34 |
| 27 | 57999 | 81462 | 59412 | 80438 | 60807 | 79388 | 62183 | 78315 | 63540 | 77218 | 33 |
| 28 | 58023 | 81445 | 59436 | 80420 | 60830 | 79371 | 62206 | 78297 | 63563 | 77199 | 32 |
| 29 | 58047 | 81428 | 59459 | 80403 | 60853 | 79353 | 62229 | 78279 | 63585 | 77181 | 31 |
| 30 | 58070 | 81412 | 59482 | 80386 | 60876 | 79335 | 62251 | 78261 | 63608 | 77162 | 30 |
| 31 | 58094 | 81395 | 59506 | 80368 | 60899 | 79318 | 62274 | 78243 | 63630 | 77144 | 29 |
| 32 | 58118 | 81378 | 59529 | 80351 | 60922 | 79300 | 62297 | 78225 | 63653 | 77125 | 28 |
| 33 | 58141 | 81361 | 59552 | 80334 | 60945 | 79282 | 62320 | 78206 | 63675 | 77107 | 27 |
| 34 | 58165 | 81344 | 59576 | 80316 | 60968 | 79264 | 62342 | 78188 | 63698 | 77088 | 26 |
| 35 | 58189 | 81327 | 59599 | 80299 | 60991 | 79247 | 62365 | 78170 | 63720 | 77070 | 25 |
| 36 | 58212 | 81310 | 59622 | 80282 | 61015 | 79229 | 62388 | 78152 | 63742 | 77051 | 24 |
| 37 | 58236 | 81293 | 59646 | 80264 | 61038 | 79211 | 62411 | 78134 | 63765 | 77033 | 23 |
| 38 | 58260 | 81276 | 59669 | 80247 | 61061 | 79193 | 62433 | 78116 | 63787 | 77014 | 22 |
| 39 | 58283 | 81259 | 59693 | 80230 | 61084 | 79176 | 62456 | 78098 | 63810 | 76996 | 21 |
| 40 | 58307 | 81242 | 59716 | 80212 | 61107 | 79158 | 62479 | 78079 | 63832 | 76977 | 20 |
| 41 | 58330 | 81225 | 59739 | 80195 | 61130 | 79140 | 62502 | 78061 | 63854 | 76959 | 19 |
| 42 | 58354 | 81208 | 59763 | 80178 | 61153 | 79122 | 62524 | 78043 | 63877 | 76940 | 18 |
| 43 | 58378 | 81191 | 59786 | 80160 | 61176 | 79105 | 62547 | 78025 | 63899 | 76921 | 17 |
| 44 | 58401 | 81174 | 59809 | 80143 | 61199 | 79087 | 62570 | 78007 | 63922 | 76903 | 16 |
| 45 | 58425 | 81157 | 59832 | 80125 | 61222 | 79069 | 62592 | 77988 | 63944 | 76884 | 15 |
| 46 | 58449 | 81140 | 59856 | 80108 | 61245 | 79051 | 62615 | 77970 | 63966 | 76866 | 14 |
| 47 | 58472 | 81123 | 59879 | 80091 | 61268 | 79033 | 62638 | 77952 | 63989 | 76847 | 13 |
| 48 | 58496 | 81106 | 59902 | 80073 | 61291 | 79016 | 62660 | 77934 | 64011 | 76828 | 12 |
| 49 | 58519 | 81089 | 59926 | 80056 | 61314 | 78998 | 62683 | 77916 | 64033 | 76810 | 11 |
| 50 | 58543 | 81072 | 59949 | 80038 | 61337 | 78980 | 62706 | 77897 | 64056 | 76791 | 10 |
| 51 | 58567 | 81055 | 59972 | 80021 | 61360 | 78962 | 62728 | 77879 | 64078 | 76772 | 9 |
| 52 | 58590 | 81038 | 59995 | 80003 | 61383 | 78944 | 62751 | 77861 | 64100 | 76754 | 8 |
| 53 | 58614 | 81021 | 60019 | 79986 | 61406 | 78926 | 62774 | 77843 | 64123 | 76735 | 7 |
| 54 | 58637 | 81004 | 60042 | 79968 | 61429 | 78908 | 62796 | 77824 | 64145 | 76717 | 6 |
| 55 | 58661 | 80987 | 60065 | 79951 | 61451 | 78891 | 62819 | 77806 | 64167 | 76698 | 5 |
| 56 | 58684 | 80970 | 60089 | 79934 | 61474 | 78873 | 62842 | 77788 | 64190 | 76679 | 4 |
| 57 | 58708 | 80953 | 60112 | 79916 | 61497 | 78855 | 62864 | 77769 | 64212 | 76661 | 3 |
| 58 | 58731 | 80936 | 60135 | 79899 | 61520 | 78837 | 62887 | 77751 | 64234 | 76642 | 2 |
| 59 | 58755 | 80919 | 60158 | 79881 | 61543 | 78819 | 62909 | 77733 | 64256 | 76623 | 1 |
| 60 | 58779 | 80902 | 60182 | 79864 | 61566 | 78801 | 62932 | 77715 | 64279 | 76604 | 0 |
| ′ | Cosine. | Sine. | Cosine. | Sine. | Cosine. | Sine. | Cosine. | Sine. | Cosine. | Sine. | ′ |
| | 54° | | 53° | | 52° | | 51° | | 50° | | |

| ′ | 40° | | 41° | | 42° | | 43° | | 44° | | ′ |
|---|---|---|---|---|---|---|---|---|---|---|---|
| | Sine. | Cosine. | Sine. | Cosine. | Sine. | Cosine. | Sine. | Cosine. | Sine. | Cosine. | |
| 0 | 64279 | 76604 | 65606 | 75471 | 66913 | 74314 | 68200 | 73135 | 69466 | 71934 | 60 |
| 1 | 64301 | 76586 | 65628 | 75452 | 66935 | 74295 | 68221 | 73116 | 69487 | 71914 | 59 |
| 2 | 64323 | 76567 | 65650 | 75433 | 66956 | 74276 | 68242 | 73096 | 69508 | 71894 | 58 |
| 3 | 64346 | 76548 | 65672 | 75414 | 66978 | 74256 | 68264 | 73076 | 69529 | 71873 | 57 |
| 4 | 64368 | 76530 | 65694 | 75395 | 66999 | 74237 | 68285 | 73056 | 69549 | 71853 | 56 |
| 5 | 64390 | 76511 | 65716 | 75375 | 67021 | 74217 | 68306 | 73036 | 69570 | 71833 | 55 |
| 6 | 64412 | 76492 | 65738 | 75356 | 67043 | 74198 | 68327 | 73016 | 69591 | 71813 | 54 |
| 7 | 64435 | 76473 | 65759 | 75337 | 67064 | 74178 | 68349 | 72996 | 69612 | 71792 | 53 |
| 8 | 64457 | 76455 | 65781 | 75318 | 67086 | 74159 | 68370 | 72976 | 69633 | 71772 | 52 |
| 9 | 64479 | 76436 | 65803 | 75299 | 67107 | 74139 | 68391 | 72957 | 69654 | 71752 | 51 |
| 10 | 64501 | 76417 | 65825 | 75280 | 67129 | 74120 | 68412 | 72937 | 69675 | 71732 | 50 |
| 11 | 64524 | 76398 | 65847 | 75261 | 67151 | 74100 | 68434 | 72917 | 69696 | 71711 | 49 |
| 12 | 64546 | 76380 | 65869 | 75241 | 67172 | 74080 | 68455 | 72897 | 69717 | 71691 | 48 |
| 13 | 64568 | 76361 | 65891 | 75222 | 67194 | 74061 | 68476 | 72877 | 69737 | 71671 | 47 |
| 14 | 64590 | 76342 | 65913 | 75203 | 67215 | 74041 | 68497 | 72857 | 69758 | 71650 | 46 |
| 15 | 64612 | 76323 | 65935 | 75184 | 67237 | 74022 | 68518 | 72837 | 69779 | 71630 | 45 |
| 16 | 64635 | 76304 | 65956 | 75165 | 67258 | 74002 | 68539 | 72817 | 69800 | 71610 | 44 |
| 17 | 64657 | 76286 | 65978 | 75146 | 67280 | 73983 | 68561 | 72797 | 69821 | 71590 | 43 |
| 18 | 64679 | 76267 | 66000 | 75126 | 67301 | 73963 | 68582 | 72777 | 69842 | 71569 | 42 |
| 19 | 64701 | 76248 | 66022 | 75107 | 67323 | 73944 | 68603 | 72757 | 69862 | 71549 | 41 |
| 20 | 64723 | 76229 | 66044 | 75088 | 67344 | 73924 | 68624 | 72737 | 69883 | 71529 | 40 |
| 21 | 64746 | 76210 | 66066 | 75069 | 67366 | 73904 | 68645 | 72717 | 69904 | 71508 | 39 |
| 22 | 64768 | 76192 | 66088 | 75050 | 67387 | 73885 | 68666 | 72697 | 69925 | 71488 | 38 |
| 23 | 64790 | 76173 | 66109 | 75030 | 67409 | 73865 | 68688 | 72677 | 69946 | 71468 | 37 |
| 24 | 64812 | 76154 | 66131 | 75011 | 67430 | 73846 | 68709 | 72657 | 69966 | 71447 | 36 |
| 25 | 64834 | 76135 | 66153 | 74992 | 67452 | 73826 | 68730 | 72637 | 69987 | 71427 | 35 |
| 26 | 64856 | 76116 | 66175 | 74973 | 67473 | 73806 | 68751 | 72617 | 70008 | 71407 | 34 |
| 27 | 64878 | 76097 | 66197 | 74953 | 67495 | 73787 | 68772 | 72597 | 70029 | 71386 | 33 |
| 28 | 64901 | 76078 | 66218 | 74934 | 67516 | 73767 | 68793 | 72577 | 70049 | 71366 | 32 |
| 29 | 64923 | 76059 | 66240 | 74915 | 67538 | 73747 | 68814 | 72557 | 70070 | 71345 | 31 |
| 30 | 64945 | 76041 | 66262 | 74896 | 67559 | 73728 | 68835 | 72537 | 70091 | 71325 | 30 |
| 31 | 64967 | 76022 | 66284 | 74876 | 67580 | 73708 | 68857 | 72517 | 70112 | 71305 | 29 |
| 32 | 64989 | 76003 | 66306 | 74857 | 67602 | 73688 | 68878 | 72497 | 70132 | 71284 | 28 |
| 33 | 65011 | 75984 | 66327 | 74838 | 67623 | 73669 | 68899 | 72477 | 70153 | 71264 | 27 |
| 34 | 65033 | 75965 | 66349 | 74818 | 67645 | 73649 | 68920 | 72457 | 70174 | 71243 | 26 |
| 35 | 65055 | 75946 | 66371 | 74799 | 67666 | 73629 | 68941 | 72437 | 70195 | 71223 | 25 |
| 36 | 65077 | 75927 | 66393 | 74780 | 67688 | 73610 | 68962 | 72417 | 70215 | 71203 | 24 |
| 37 | 65100 | 75908 | 66414 | 74760 | 67709 | 73590 | 68983 | 72397 | 70236 | 71182 | 23 |
| 38 | 65122 | 75889 | 66436 | 74741 | 67730 | 73570 | 69004 | 72377 | 70257 | 71162 | 22 |
| 39 | 65144 | 75870 | 66458 | 74722 | 67752 | 73551 | 69025 | 72357 | 70277 | 71141 | 21 |
| 40 | 65166 | 75851 | 66480 | 74703 | 67773 | 73531 | 69046 | 72337 | 70298 | 71121 | 20 |
| 41 | 65188 | 75832 | 66501 | 74683 | 67795 | 73511 | 69067 | 72317 | 70319 | 71100 | 19 |
| 42 | 65210 | 75813 | 66523 | 74664 | 67816 | 73491 | 69088 | 72297 | 70339 | 71080 | 18 |
| 43 | 65232 | 75794 | 66545 | 74644 | 67837 | 73472 | 69109 | 72277 | 70360 | 71059 | 17 |
| 44 | 65254 | 75775 | 66566 | 74625 | 67859 | 73452 | 69130 | 72257 | 70381 | 71039 | 16 |
| 45 | 65276 | 75756 | 66588 | 74606 | 67880 | 73432 | 69151 | 72236 | 70401 | 71019 | 15 |
| 46 | 65298 | 75738 | 66610 | 74586 | 67901 | 73413 | 69172 | 72216 | 70422 | 70998 | 14 |
| 47 | 65320 | 75719 | 66632 | 74567 | 67923 | 73393 | 69193 | 72196 | 70443 | 70978 | 13 |
| 48 | 65342 | 75700 | 66653 | 74548 | 67944 | 73373 | 69214 | 72176 | 70463 | 70957 | 12 |
| 49 | 65364 | 75680 | 66675 | 74528 | 67965 | 73353 | 69235 | 72156 | 70484 | 70937 | 11 |
| 50 | 65386 | 75661 | 66697 | 74509 | 67987 | 73333 | 69256 | 72136 | 70505 | 70916 | 10 |
| 51 | 65408 | 75642 | 66718 | 74489 | 68008 | 73314 | 69277 | 72116 | 70525 | 70896 | 9 |
| 52 | 65430 | 75623 | 66740 | 74470 | 68029 | 73294 | 69298 | 72095 | 70546 | 70875 | 8 |
| 53 | 65452 | 75604 | 66762 | 74451 | 68051 | 73274 | 69319 | 72075 | 70567 | 70855 | 7 |
| 54 | 65474 | 75585 | 66783 | 74431 | 68072 | 73254 | 69340 | 72055 | 70587 | 70834 | 6 |
| 55 | 65496 | 75566 | 66805 | 74412 | 68093 | 73234 | 69361 | 72035 | 70608 | 70813 | 5 |
| 56 | 65518 | 75547 | 66827 | 74392 | 68115 | 73215 | 69382 | 72015 | 70628 | 70793 | 4 |
| 57 | 65540 | 75528 | 66848 | 74373 | 68136 | 73195 | 69403 | 71995 | 70649 | 70772 | 3 |
| 58 | 65562 | 75509 | 66870 | 74353 | 68157 | 73175 | 69424 | 71974 | 70670 | 70752 | 2 |
| 59 | 65584 | 75490 | 66891 | 74334 | 68179 | 73155 | 69445 | 71954 | 70690 | 70731 | 1 |
| 60 | 65606 | 75471 | 66913 | 74314 | 68200 | 73135 | 69466 | 71934 | 70711 | 70711 | 0 |
| | Cosine. | Sine. | Cosine. | Sine. | Cosine. | Sine. | Cosine. | Sine. | Cosine. | Sine. | |
| ′ | 49° | | 48° | | 47° | | 46° | | 45° | | ′ |

| ′ | 0° | | 1° | | 2° | | 3° | | ′ |
|---|---|---|---|---|---|---|---|---|---|
| | Tangent. | Cotang. | Tangent. | Cotang. | Tangent. | Cotang. | Tangent. | Cotang. | |
| 0 | 00000 | Infinite. | 01746 | 57·2900 | 03492 | 28·6363 | 05241 | 19·0811 | 60 |
| 1 | 00029 | 3437·75 | 01775 | 56·3506 | 03521 | 28·3994 | 05270 | 18·9755 | 59 |
| 2 | 00058 | 1718·87 | 01804 | 55·4415 | 03550 | 28·1664 | 05299 | 18·8711 | 58 |
| 3 | 00087 | 1145·92 | 01833 | 54·5613 | 03579 | 27·9372 | 05328 | 18·7678 | 57 |
| 4 | 00116 | 859·436 | 01862 | 53·7086 | 03609 | 27·7117 | 05357 | 18·6656 | 56 |
| 5 | 00145 | 687·549 | 01891 | 52·8821 | 03638 | 27·4899 | 05387 | 18·5645 | 55 |
| 6 | 00175 | 572·957 | 01920 | 52·0807 | 03667 | 27·2715 | 05416 | 18·4645 | 54 |
| 7 | 00204 | 491·106 | 01949 | 51·3032 | 03696 | 27·0566 | 05445 | 18·3655 | 53 |
| 8 | 00233 | 429·718 | 01978 | 50·5485 | 03725 | 26·8450 | 05474 | 18·2677 | 52 |
| 9 | 00262 | 381·971 | 02007 | 49·8157 | 03754 | 26·6367 | 05503 | 18·1708 | 51 |
| 10 | 00291 | 343·774 | 02036 | 49·1039 | 03783 | 26·4316 | 05533 | 18·0750 | 50 |
| 11 | 00320 | 312·521 | 02066 | 48·4121 | 03812 | 26·2296 | 05562 | 17·9802 | 49 |
| 12 | 00349 | 286·478 | 02095 | 47·7395 | 03842 | 26·0307 | 05591 | 17·8863 | 48 |
| 13 | 00378 | 264·441 | 02124 | 47·0853 | 03871 | 25·8348 | 05620 | 17·7934 | 47 |
| 14 | 00407 | 245·552 | 02153 | 46·4489 | 03900 | 25·6418 | 05649 | 17·7015 | 46 |
| 15 | 00436 | 229·182 | 02182 | 45·8294 | 03929 | 25·4517 | 05678 | 17·6106 | 45 |
| 16 | 00465 | 214·858 | 02211 | 45·2261 | 03958 | 25·2644 | 05708 | 17·5205 | 44 |
| 17 | 00495 | 202·219 | 02240 | 44·6386 | 03987 | 25·0798 | 05737 | 17·4314 | 43 |
| 18 | 00524 | 190·984 | 02269 | 44·0661 | 04016 | 24·8978 | 05766 | 17·3432 | 42 |
| 19 | 00553 | 180·932 | 02298 | 43·5081 | 04046 | 24·7185 | 05795 | 17·2558 | 41 |
| 20 | 00582 | 171·885 | 02328 | 42·9641 | 04075 | 24·5418 | 05824 | 17·1693 | 40 |
| 21 | 00611 | 163·700 | 02357 | 42·4335 | 04104 | 24·3675 | 05854 | 17·0837 | 39 |
| 22 | 00640 | 156·259 | 02386 | 41·9158 | 04133 | 24·1957 | 05883 | 16·9990 | 38 |
| 23 | 00669 | 149·465 | 02415 | 41·4106 | 04162 | 24·0263 | 05912 | 16·9150 | 37 |
| 24 | 00698 | 143·237 | 02444 | 40·9174 | 04191 | 23·8593 | 05941 | 16·8319 | 36 |
| 25 | 00727 | 137·507 | 02473 | 40·4358 | 04220 | 23·6945 | 05970 | 16·7496 | 35 |
| 26 | 00756 | 132·219 | 02502 | 39·9655 | 04250 | 23·5321 | 05999 | 16·6681 | 34 |
| 27 | 00785 | 127·321 | 02531 | 39·5059 | 04279 | 23·3718 | 06029 | 16·5874 | 33 |
| 28 | 00814 | 122·774 | 02560 | 39·0568 | 04308 | 23·2137 | 06058 | 16·5075 | 32 |
| 29 | 00844 | 118·540 | 02589 | 38·6177 | 04337 | 23·0577 | 06087 | 16·4283 | 31 |
| 30 | 00873 | 114·589 | 02619 | 38·1885 | 04366 | 22·9038 | 06116 | 16·3499 | 30 |
| 31 | 00902 | 110·892 | 02648 | 37·7686 | 04395 | 22·7519 | 06145 | 16·2722 | 29 |
| 32 | 00931 | 107·426 | 02677 | 37·3579 | 04424 | 22·6020 | 06175 | 16·1952 | 28 |
| 33 | 00960 | 104·171 | 02706 | 36·9560 | 04454 | 22·4541 | 06204 | 16·1190 | 27 |
| 34 | 00989 | 101·107 | 02735 | 36·5627 | 04483 | 22·3081 | 06233 | 16·0435 | 26 |
| 35 | 01018 | 98·2179 | 02764 | 36·1776 | 04512 | 22·1640 | 06262 | 15·9687 | 25 |
| 36 | 01047 | 95·4895 | 02793 | 35·8006 | 04541 | 22·0217 | 06291 | 15·8945 | 24 |
| 37 | 01076 | 92·9085 | 02822 | 35·4313 | 04570 | 21·8813 | 06321 | 15·8211 | 23 |
| 38 | 01105 | 90·4633 | 02851 | 35·0695 | 04599 | 21·7426 | 06350 | 15·7483 | 22 |
| 39 | 01135 | 88·1436 | 02881 | 34·7151 | 04628 | 21·6056 | 06379 | 15·6762 | 21 |
| 40 | 01164 | 85·9398 | 02910 | 34·3678 | 04658 | 21·4704 | 06408 | 15·6048 | 20 |
| 41 | 01193 | 83·8435 | 02939 | 34·0273 | 04687 | 21·3369 | 06437 | 15·5340 | 19 |
| 42 | 01222 | 81·8470 | 02968 | 33·6935 | 04716 | 21·2049 | 06467 | 15·4638 | 18 |
| 43 | 01251 | 79·9434 | 02997 | 33·3662 | 04745 | 21·0747 | 06496 | 15·3943 | 17 |
| 44 | 01280 | 78·1263 | 03026 | 33·0452 | 04774 | 20·9460 | 06525 | 15·3254 | 16 |
| 45 | 01309 | 76·3900 | 03055 | 32·7303 | 04803 | 20·8188 | 06554 | 15·2571 | 15 |
| 46 | 01338 | 74·7292 | 03084 | 32·4213 | 04832 | 20·6932 | 06584 | 15·1893 | 14 |
| 47 | 01367 | 73·1390 | 03114 | 32·1181 | 04862 | 20·5691 | 06613 | 15·1222 | 13 |
| 48 | 01396 | 71·6151 | 03143 | 31·8205 | 04891 | 20·4465 | 06642 | 15·0557 | 12 |
| 49 | 01425 | 70·1533 | 03172 | 31·5284 | 04920 | 20·3253 | 06671 | 14·9898 | 11 |
| 50 | 01455 | 68·7501 | 03201 | 31·2416 | 04949 | 20·2056 | 06700 | 14·9244 | 10 |
| 51 | 01484 | 67·4019 | 03230 | 30·9599 | 04978 | 20·0872 | 06730 | 14·8596 | 9 |
| 52 | 01513 | 66·1055 | 03259 | 30·6833 | 05007 | 19·9702 | 06759 | 14·7954 | 8 |
| 53 | 01542 | 64·0500 | 03288 | 30·4116 | 05037 | 19·8046 | 06788 | 14·7317 | 7 |
| 54 | 01571 | 63·6567 | 03317 | 30·1446 | 05066 | 19·7403 | 06817 | 14·6685 | 6 |
| 55 | 01600 | 62·4992 | 03346 | 29·8823 | 05095 | 19·6273 | 06847 | 14·6059 | 5 |
| 56 | 01629 | 61·3829 | 03376 | 29·6245 | 05124 | 19·5156 | 06876 | 14·5438 | 4 |
| 57 | 01658 | 60·3058 | 03405 | 29·3711 | 05153 | 19·4051 | 06905 | 14·4823 | 3 |
| 58 | 01687 | 59·2659 | 03434 | 29·1220 | 05182 | 19·2959 | 06934 | 14·4212 | 2 |
| 59 | 01716 | 58·2612 | 03463 | 28·8771 | 05212 | 19·1879 | 06963 | 14·3607 | 1 |
| 60 | 01746 | 57·2900 | 03492 | 28·6363 | 05241 | 19·0811 | 06993 | 14·3007 | 0 |
| | Cotang. | Tangent. | Cotang. | Tangent. | Cotang. | Tangent. | Cotang. | Tangent. | |
| ′ | 89° | | 88° | | 87° | | 86° | | ′ |

| ′ | 4° | | 5° | | 6° | | 7° | | ′ |
|---|---|---|---|---|---|---|---|---|---|
| | Tangent. | Cotang. | Tangent. | Cotang. | Tangent. | Cotang. | Tangent. | Cotang. | |
| 0 | 06993 | 14·3007 | 08749 | 11·4301 | 10510 | 9·51436 | 12278 | 8·14435 | 60 |
| 1 | 07022 | 14·2411 | 08778 | 11·3919 | 10540 | 9·48781 | 12308 | 8·12481 | 59 |
| 2 | 07051 | 14·1821 | 08807 | 11·3540 | 10569 | 9·46141 | 12338 | 8·10536 | 58 |
| 3 | 07080 | 14·1235 | 08837 | 11·3163 | 10599 | 9·43515 | 12367 | 8·08600 | 57 |
| 4 | 07110 | 14·0655 | 08866 | 11·2789 | 10628 | 9·40904 | 12397 | 8·06674 | 56 |
| 5 | 07139 | 14·0079 | 08895 | 11·2417 | 10657 | 9·38307 | 12426 | 8·04756 | 55 |
| 6 | 07168 | 13·9507 | 08925 | 11·2048 | 10687 | 9·35724 | 12456 | 8·02848 | 54 |
| 7 | 07197 | 13·8940 | 08954 | 11·1681 | 10716 | 9·33154 | 12485 | 8·00948 | 53 |
| 8 | 07227 | 13·8378 | 08983 | 11·1316 | 10746 | 9·30599 | 12515 | 7·99058 | 52 |
| 9 | 07256 | 13·7821 | 09013 | 11·0954 | 10775 | 9·28058 | 12544 | 7·97176 | 51 |
| 10 | 07285 | 13·7267 | 09042 | 11·0594 | 10805 | 9·25530 | 12574 | 7·95302 | 50 |
| 11 | 07314 | 13·6719 | 09071 | 11·0237 | 10834 | 9·23016 | 12603 | 7·93438 | 49 |
| 12 | 07344 | 13·6174 | 09101 | 10·9882 | 10863 | 9·20516 | 12633 | 7·91582 | 48 |
| 13 | 07373 | 13·5634 | 09130 | 10·9529 | 10893 | 9·18028 | 12662 | 7·89734 | 47 |
| 14 | 07402 | 13·5098 | 09159 | 10·9178 | 10922 | 9·15554 | 12692 | 7·87895 | 46 |
| 15 | 07431 | 13·4566 | 09189 | 10·8829 | 10952 | 9·13093 | 12722 | 7·86064 | 45 |
| 16 | 07461 | 13·4039 | 09218 | 10·8483 | 10981 | 9·10646 | 12751 | 7·84242 | 44 |
| 17 | 07490 | 13·3515 | 09247 | 10·8139 | 11011 | 9·08211 | 12781 | 7·82428 | 43 |
| 18 | 07519 | 13·2996 | 09277 | 10·7797 | 11040 | 9·05789 | 12810 | 7·80622 | 42 |
| 19 | 07548 | 13·2480 | 09306 | 10·7457 | 11070 | 9·03379 | 12840 | 7·78825 | 41 |
| 20 | 07578 | 13·1969 | 09335 | 10·7119 | 11099 | 9·00983 | 12869 | 7·77035 | 40 |
| 21 | 07607 | 13·1461 | 09365 | 10·6783 | 11128 | 8·98598 | 12899 | 7·75254 | 39 |
| 22 | 07636 | 13·0958 | 09394 | 10·6450 | 11158 | 8·96227 | 12929 | 7·73480 | 38 |
| 23 | 07665 | 13·0458 | 09423 | 10·6118 | 11187 | 8·93867 | 12958 | 7·71715 | 37 |
| 24 | 07695 | 12·9962 | 09453 | 10·5789 | 11217 | 8·91520 | 12988 | 7·69957 | 36 |
| 25 | 07724 | 12·9469 | 09482 | 10·5462 | 11246 | 8·89185 | 13017 | 7·68208 | 35 |
| 26 | 07753 | 12·8981 | 09511 | 10·5136 | 11276 | 8·86862 | 13047 | 7·66466 | 34 |
| 27 | 07782 | 12·8496 | 09541 | 10·4813 | 11305 | 8·84551 | 13076 | 7·64732 | 33 |
| 28 | 07812 | 12·8014 | 09570 | 10·4491 | 11335 | 8·82252 | 13106 | 7·63005 | 32 |
| 29 | 07841 | 12·7536 | 09600 | 10·4172 | 11364 | 8·79964 | 13136 | 7·61287 | 31 |
| 30 | 07870 | 12·7062 | 09629 | 10·3854 | 11394 | 8·77689 | 13165 | 7·59575 | 30 |
| 31 | 07899 | 12·6591 | 09658 | 10·3538 | 11423 | 8·75425 | 13195 | 7·57872 | 29 |
| 32 | 07929 | 12·6124 | 09688 | 10·3224 | 11452 | 8·73172 | 13224 | 7·56176 | 28 |
| 33 | 07958 | 12·5660 | 09717 | 10·2913 | 11482 | 8·70931 | 13254 | 7·54487 | 27 |
| 34 | 07987 | 12·5199 | 09746 | 10·2602 | 11511 | 8·68701 | 13284 | 7·52806 | 26 |
| 35 | 08017 | 12·4742 | 09776 | 10·2294 | 11541 | 8·66482 | 13313 | 7·51132 | 25 |
| 36 | 08046 | 12·4288 | 09805 | 10·1988 | 11570 | 8·64275 | 13343 | 7·49465 | 24 |
| 37 | 08075 | 12·3838 | 09834 | 10·1683 | 11600 | 8·62078 | 13372 | 7·47806 | 23 |
| 38 | 08104 | 12·3390 | 09864 | 10·1381 | 11629 | 8·59893 | 13402 | 7·46154 | 22 |
| 39 | 08134 | 12·2946 | 09893 | 10·1080 | 11659 | 8·57718 | 13432 | 7·44509 | 21 |
| 40 | 08163 | 12·2505 | 09923 | 10·0780 | 11688 | 8·55555 | 13461 | 7·42871 | 20 |
| 41 | 08192 | 12·2067 | 09952 | 10·0483 | 11718 | 8·53402 | 13491 | 7·41240 | 19 |
| 42 | 08221 | 12·1632 | 09981 | 10·0187 | 11747 | 8·51259 | 13521 | 7·39616 | 18 |
| 43 | 08251 | 12·1201 | 10011 | 9·98930 | 11777 | 8·49128 | 13550 | 7·37999 | 17 |
| 44 | 08280 | 12·0772 | 10040 | 9·96007 | 11806 | 8·47007 | 13580 | 7·36389 | 16 |
| 45 | 08309 | 12·0346 | 10069 | 9·93101 | 11836 | 8·44896 | 13609 | 7·34786 | 15 |
| 46 | 08339 | 11·9923 | 10099 | 9·90211 | 11865 | 8·42795 | 13639 | 7·33190 | 14 |
| 47 | 08368 | 11·9504 | 10128 | 9·87338 | 11895 | 8·40705 | 13669 | 7·31600 | 13 |
| 48 | 08397 | 11·9087 | 10158 | 9·84482 | 11924 | 8·38625 | 13698 | 7·30018 | 12 |
| 49 | 08427 | 11·8673 | 10187 | 9·81641 | 11954 | 8·36555 | 13728 | 7·28442 | 11 |
| 50 | 08456 | 11·8262 | 10216 | 9·78817 | 11983 | 8·34496 | 13758 | 7·26873 | 10 |
| 51 | 08485 | 11·7853 | 10246 | 9·76009 | 12013 | 8·32446 | 13787 | 7·25310 | 9 |
| 52 | 08514 | 11·7448 | 10275 | 9·73217 | 12042 | 8·30406 | 13817 | 7·23754 | 8 |
| 53 | 08544 | 11·7045 | 10305 | 9·70441 | 12072 | 8·28376 | 13846 | 7·22204 | 7 |
| 54 | 08573 | 11·6645 | 10334 | 9·67680 | 12101 | 8·26355 | 13876 | 7·20661 | 6 |
| 55 | 08602 | 11·6248 | 10363 | 9·64935 | 12131 | 8·24345 | 13906 | 7·19125 | 5 |
| 56 | 08632 | 11·5853 | 10393 | 9·62205 | 12160 | 8·22344 | 13935 | 7·17594 | 4 |
| 57 | 08661 | 11·5461 | 10422 | 9·59490 | 12190 | 8·20352 | 13965 | 7·16071 | 3 |
| 58 | 08690 | 11·5072 | 10452 | 9·56791 | 12219 | 8·18370 | 13995 | 7·14553 | 2 |
| 59 | 08720 | 11·4685 | 10481 | 9·54106 | 12249 | 8·16398 | 14024 | 7·13042 | 1 |
| 60 | 08749 | 11·4301 | 10510 | 9·51436 | 12278 | 8·14435 | 14054 | 7·11537 | 0 |
| | Cotang. | Tangent. | Cotang. | Tangent. | Cotang. | Tangent. | Cotang. | Tangent. | |
| ′ | 85° | | 84° | | 83° | | 82° | | ′ |

| ′ | 8° | | 9° | | 10° | | 11° | | ′ |
|---|---|---|---|---|---|---|---|---|---|
| | Tangent. | Cotang. | Tangent. | Cotang. | Tangent. | Cotang. | Tangent. | Cotang. | |
| 0 | 14054 | 7·11537 | 15838 | 6·31375 | 17633 | 5·67128 | 19438 | 5·14455 | 60 |
| 1 | 14084 | 7·10038 | 15868 | 6·30189 | 17663 | 5·66165 | 19468 | 5·13658 | 59 |
| 2 | 14113 | 7·08546 | 15898 | 6·29007 | 17693 | 5·65205 | 19498 | 5·12862 | 58 |
| 3 | 14143 | 7·07059 | 15928 | 6·27829 | 17723 | 5·64248 | 19529 | 5·12069 | 57 |
| 4 | 14173 | 7·05579 | 15958 | 6·26655 | 17753 | 5·63295 | 19559 | 5·11279 | 56 |
| 5 | 14202 | 7·04105 | 15988 | 6·25486 | 17783 | 5·62344 | 19589 | 5·10490 | 55 |
| 6 | 14232 | 7·02637 | 16017 | 6·24321 | 17813 | 5·61397 | 19619 | 5·09704 | 54 |
| 7 | 14262 | 7·01174 | 16047 | 6·23160 | 17843 | 5·60452 | 19649 | 5·08921 | 53 |
| 8 | 14291 | 6·99718 | 16077 | 6·22003 | 17873 | 5·59511 | 19680 | 5·08139 | 52 |
| 9 | 14321 | 6·98268 | 16107 | 6·20851 | 17903 | 5·58573 | 19710 | 5·07360 | 51 |
| 10 | 14351 | 6·96823 | 16137 | 6·19703 | 17933 | 5·57638 | 19740 | 5·06584 | 50 |
| 11 | 14381 | 6·95385 | 16167 | 6·18559 | 17963 | 5·56706 | 19770 | 5·05809 | 49 |
| 12 | 14410 | 6·93952 | 16196 | 6·17419 | 17993 | 5·55777 | 19801 | 5·05037 | 48 |
| 13 | 14440 | 6·92525 | 16226 | 6·16283 | 18023 | 5·54851 | 19831 | 5·04267 | 47 |
| 14 | 14470 | 6·91104 | 16256 | 6·15151 | 18053 | 5·53927 | 19861 | 5·03499 | 46 |
| 15 | 14499 | 6·89688 | 16286 | 6·14023 | 18083 | 5·53007 | 19891 | 5·02734 | 45 |
| 16 | 14529 | 6·88278 | 16316 | 6·12899 | 18113 | 5·52090 | 19921 | 5·01971 | 44 |
| 17 | 14559 | 6·86874 | 16346 | 6·11779 | 18143 | 5·51176 | 19952 | 5·01210 | 43 |
| 18 | 14588 | 6·85475 | 16376 | 6·10664 | 18173 | 5·50264 | 19982 | 5·00451 | 42 |
| 19 | 14618 | 6·84082 | 16405 | 6·09552 | 18203 | 5·49356 | 20012 | 4·99695 | 41 |
| 20 | 14648 | 6·82694 | 16435 | 6·08444 | 18233 | 5·48451 | 20042 | 4·98940 | 40 |
| 21 | 14678 | 6·81312 | 16465 | 6·07340 | 18263 | 5·47548 | 20073 | 4·98188 | 39 |
| 22 | 14707 | 6·79936 | 16495 | 6·06240 | 18293 | 5·46648 | 20103 | 4·97438 | 38 |
| 23 | 14737 | 6·78564 | 16525 | 6·05143 | 18323 | 5·45751 | 20133 | 4·96690 | 37 |
| 24 | 14767 | 6·77199 | 16555 | 6·04051 | 18353 | 5·44857 | 20164 | 4·95945 | 36 |
| 25 | 14796 | 6·75838 | 16585 | 6·02962 | 18383 | 5·43966 | 20194 | 4·95201 | 35 |
| 26 | 14826 | 6·74483 | 16615 | 6·01878 | 18414 | 5·43077 | 20224 | 4·94460 | 34 |
| 27 | 14856 | 6·73133 | 16645 | 6·00797 | 18444 | 5·42192 | 20254 | 4·93721 | 33 |
| 28 | 14886 | 6·71789 | 16674 | 5·99720 | 18474 | 5·41309 | 20285 | 4·92984 | 32 |
| 29 | 14915 | 6·70450 | 16704 | 5·98646 | 18504 | 5·40429 | 20315 | 4·92249 | 31 |
| 30 | 14945 | 6·69116 | 16734 | 5·97576 | 18534 | 5·39552 | 20345 | 4·91516 | 30 |
| 31 | 14975 | 6·67787 | 16764 | 5·96510 | 18564 | 5·38677 | 20376 | 4·90785 | 29 |
| 32 | 15005 | 6·66463 | 16794 | 5·95448 | 18594 | 5·37805 | 20406 | 4·90056 | 28 |
| 33 | 15034 | 6·65144 | 16824 | 5·94390 | 18624 | 5·36936 | 20436 | 4·89330 | 27 |
| 34 | 15064 | 6·63831 | 16854 | 5·93335 | 18654 | 5·36070 | 20466 | 4·88605 | 26 |
| 35 | 15094 | 6·62523 | 16884 | 5·92283 | 18684 | 5·35206 | 20497 | 4·87882 | 25 |
| 36 | 15124 | 6·61219 | 16914 | 5·91235 | 18714 | 5·34345 | 20527 | 4·87162 | 24 |
| 37 | 15153 | 6·59921 | 16944 | 5·90191 | 18745 | 5·33487 | 20557 | 4·86444 | 23 |
| 38 | 15183 | 6·58627 | 16974 | 5·89151 | 18775 | 5·32631 | 20588 | 4·85727 | 22 |
| 39 | 15213 | 6·57339 | 17004 | 5·88114 | 18805 | 5·31778 | 20618 | 4·85013 | 21 |
| 40 | 15243 | 6·56055 | 17033 | 5·87080 | 18835 | 5·30928 | 20648 | 4·84300 | 20 |
| 41 | 15272 | 6·54777 | 17063 | 5·86051 | 18865 | 5·30080 | 20679 | 4·83590 | 19 |
| 42 | 15302 | 6·53503 | 17093 | 5·85024 | 18895 | 5·29235 | 20709 | 4·82882 | 18 |
| 43 | 15332 | 6·52234 | 17123 | 5·84001 | 18925 | 5·28393 | 20739 | 4·82175 | 17 |
| 44 | 15362 | 6·50970 | 17153 | 5·82982 | 18955 | 5·27553 | 20770 | 4·81471 | 16 |
| 45 | 15391 | 6·49710 | 17183 | 5·81966 | 18986 | 5·26715 | 20800 | 4·80769 | 15 |
| 46 | 15421 | 6·48456 | 17213 | 5·80953 | 19016 | 5·25880 | 20830 | 4·80068 | 14 |
| 47 | 15451 | 6·47206 | 17243 | 5·79944 | 19046 | 5·25048 | 20861 | 4·79370 | 13 |
| 48 | 15481 | 6·45961 | 17273 | 5·78938 | 19076 | 5·24218 | 20891 | 4·78673 | 12 |
| 49 | 15511 | 6·44720 | 17303 | 5·77936 | 19106 | 5·23391 | 20921 | 4·77978 | 11 |
| 50 | 15540 | 6·43484 | 17333 | 5·76937 | 19136 | 5·22566 | 20952 | 4·77286 | 10 |
| 51 | 15570 | 6·42253 | 17363 | 5·75941 | 19166 | 5·21744 | 20982 | 4·76595 | 9 |
| 52 | 15600 | 6·41026 | 17393 | 5·74949 | 19197 | 5·20925 | 21013 | 4·75906 | 8 |
| 53 | 15630 | 6·39804 | 17423 | 5·73960 | 19227 | 5·20107 | 21043 | 4·75219 | 7 |
| 54 | 15660 | 6·38587 | 17453 | 5·72974 | 19257 | 5·19293 | 21073 | 4·74534 | 6 |
| 55 | 15689 | 6·37374 | 17483 | 5·71992 | 19287 | 5·18480 | 21104 | 4·73851 | 5 |
| 56 | 15719 | 6·36165 | 17513 | 5·71013 | 19317 | 5·17671 | 21134 | 4·73170 | 4 |
| 57 | 15749 | 6·34961 | 17543 | 5·70037 | 19347 | 5·16863 | 21164 | 4·72490 | 3 |
| 58 | 15779 | 6·33761 | 17573 | 5·69064 | 19378 | 5·16058 | 21195 | 4·71813 | 2 |
| 59 | 15809 | 6·32566 | 17603 | 5·68094 | 19408 | 5·15256 | 21225 | 4·71137 | 1 |
| 60 | 15838 | 6·31375 | 17633 | 5·67128 | 19438 | 5·14455 | 21256 | 4·70463 | 0 |
| | Cotang. | Tangent. | Cotang. | Tangent. | Cotang. | Tangent. | Cotang. | Tangent. | |
| ′ | 81° | | 80° | | 79° | | 78° | | ′ |

| ′ | 12° | | 13° | | 14° | | 15° | | ′ |
|---|---|---|---|---|---|---|---|---|---|
| | Tangent. | Cotang. | Tangent. | Cotang. | Tangent. | Cotang. | Tangent. | Cotang. | |
| 0 | 21256 | 4·70463 | 23087 | 4·33148 | 24933 | 4·01078 | 26795 | 3·73205 | 60 |
| 1 | 21286 | 4·69791 | 23117 | 4·32573 | 24964 | 4·00582 | 26826 | 3·72771 | 59 |
| 2 | 21316 | 4·69121 | 23148 | 4·32001 | 24995 | 4·00086 | 26857 | 3·72338 | 58 |
| 3 | 21347 | 4·68452 | 23179 | 4·31430 | 25026 | 3·99592 | 26888 | 3·71907 | 57 |
| 4 | 21377 | 4·67786 | 23209 | 4·30860 | 25056 | 3·99099 | 26920 | 3·71476 | 56 |
| 5 | 21408 | 4·67121 | 23240 | 4·30291 | 25087 | 3·98607 | 26951 | 3·71046 | 55 |
| 6 | 21438 | 4·66458 | 23271 | 4·29724 | 25118 | 3·98117 | 26982 | 3·70616 | 54 |
| 7 | 21469 | 4·65797 | 23301 | 4·29159 | 25149 | 3·97627 | 27013 | 3·70188 | 53 |
| 8 | 21499 | 4·65138 | 23332 | 4·28595 | 25180 | 3·97139 | 27044 | 3·69761 | 52 |
| 9 | 21529 | 4·64480 | 23363 | 4·28032 | 25211 | 3·96651 | 27076 | 3·69335 | 51 |
| 10 | 21560 | 4·63825 | 23393 | 4·27471 | 25242 | 3·96165 | 27107 | 3·68909 | 50 |
| 11 | 21590 | 4·63171 | 23424 | 4·26911 | 25273 | 3·95680 | 27138 | 3·68485 | 49 |
| 12 | 21621 | 4·62518 | 23455 | 4·26352 | 25304 | 3·95196 | 27169 | 3·68061 | 48 |
| 13 | 21651 | 4·61868 | 23485 | 4·25795 | 25335 | 3·94713 | 27201 | 3·67638 | 47 |
| 14 | 21682 | 4·61219 | 23516 | 4·25239 | 25366 | 3·94232 | 27232 | 3·67217 | 46 |
| 15 | 21712 | 4·60572 | 23547 | 4·24685 | 25397 | 3·93751 | 27263 | 3·66796 | 45 |
| 16 | 21743 | 4·59927 | 23578 | 4·24132 | 25428 | 3·93271 | 27294 | 3·66376 | 44 |
| 17 | 21773 | 4·59283 | 23608 | 4·23580 | 25459 | 3·92793 | 27326 | 3·65957 | 43 |
| 18 | 21804 | 4·58641 | 23639 | 4·23030 | 25490 | 3·92316 | 27357 | 3·65538 | 42 |
| 19 | 21834 | 4·58001 | 23670 | 4·22481 | 25521 | 3·91839 | 27388 | 3·65121 | 41 |
| 20 | 21864 | 4·57363 | 23700 | 4·21933 | 25552 | 3·91364 | 27419 | 3·64705 | 40 |
| 21 | 21895 | 4·56726 | 23731 | 4·21387 | 25583 | 3·90890 | 27451 | 3·64289 | 39 |
| 22 | 21925 | 4·56091 | 23762 | 4·20842 | 25614 | 3·90417 | 27482 | 3·63874 | 38 |
| 23 | 21956 | 4·55458 | 23793 | 4·20298 | 25645 | 3·89945 | 27513 | 3·63461 | 37 |
| 24 | 21986 | 4·54826 | 23823 | 4·19756 | 25676 | 3·89474 | 27545 | 3·63048 | 36 |
| 25 | 22017 | 4·54196 | 23854 | 4·19215 | 25707 | 3·89004 | 27576 | 3·62636 | 35 |
| 26 | 22047 | 4·53568 | 23885 | 4·18675 | 25738 | 3·88536 | 27607 | 3·62224 | 34 |
| 27 | 22078 | 4·52941 | 23916 | 4·18137 | 25769 | 3·88068 | 27638 | 3·61814 | 33 |
| 28 | 22108 | 4·52316 | 23946 | 4·17600 | 25800 | 3·87601 | 27670 | 3·61405 | 32 |
| 29 | 22139 | 4·51693 | 23977 | 4·17064 | 25831 | 3·87136 | 27701 | 3·60996 | 31 |
| 30 | 22169 | 4·51071 | 24008 | 4·16530 | 25862 | 3·86671 | 27732 | 3·60588 | 30 |
| 31 | 22200 | 4·50451 | 24039 | 4·15997 | 25893 | 3·86208 | 27764 | 3·60181 | 29 |
| 32 | 22231 | 4·49832 | 24069 | 4·15465 | 25924 | 3·85745 | 27795 | 3·59775 | 28 |
| 33 | 22261 | 4·49215 | 24100 | 4·14934 | 25955 | 3·85284 | 27826 | 3·59370 | 27 |
| 34 | 22292 | 4·48600 | 24131 | 4·14405 | 25986 | 3·84824 | 27858 | 3·58966 | 26 |
| 35 | 22322 | 4·47986 | 24162 | 4·13877 | 26017 | 3·84364 | 27889 | 3·58562 | 25 |
| 36 | 22353 | 4·47374 | 24193 | 4·13350 | 26048 | 3·83906 | 27920 | 3·58160 | 24 |
| 37 | 22383 | 4·46764 | 24223 | 4·12825 | 26079 | 3·83449 | 27952 | 3·57758 | 23 |
| 38 | 22414 | 4·46155 | 24254 | 4·12301 | 26110 | 3·82992 | 27983 | 3·57357 | 22 |
| 39 | 22444 | 4·45548 | 24285 | 4·11778 | 26141 | 3·82537 | 28015 | 3·56957 | 21 |
| 40 | 22475 | 4·44942 | 24316 | 4·11256 | 26172 | 3·82083 | 28046 | 3·56557 | 20 |
| 41 | 22505 | 4·44338 | 24347 | 4·10736 | 26203 | 3·81630 | 28077 | 3·56159 | 19 |
| 42 | 22536 | 4·43735 | 24377 | 4·10216 | 26235 | 3·81177 | 28109 | 3·55761 | 18 |
| 43 | 22567 | 4·43134 | 24408 | 4·09699 | 26266 | 3·80726 | 28140 | 3·55364 | 17 |
| 44 | 22597 | 4·42534 | 24439 | 4·09182 | 26297 | 3·80276 | 28172 | 3·54968 | 16 |
| 45 | 22628 | 4·41936 | 24470 | 4·08666 | 26328 | 3·79827 | 28203 | 3·54573 | 15 |
| 46 | 22658 | 4·41340 | 24501 | 4·08152 | 26359 | 3·79378 | 28234 | 3·54179 | 14 |
| 47 | 22689 | 4·40745 | 24532 | 4·07639 | 26390 | 3·78931 | 28266 | 3·53785 | 13 |
| 48 | 22719 | 4·40152 | 24562 | 4·07127 | 26421 | 3·78485 | 28297 | 3·53393 | 12 |
| 49 | 22750 | 4·39560 | 24593 | 4·06616 | 26452 | 3·78040 | 28329 | 3·53001 | 11 |
| 50 | 22781 | 4·38969 | 24624 | 4·06107 | 26483 | 3·77595 | 28360 | 3·52609 | 10 |
| 51 | 22811 | 4·38381 | 24655 | 4·05599 | 26515 | 3·77152 | 28391 | 3·52219 | 9 |
| 52 | 22842 | 4·37793 | 24686 | 4·05092 | 26546 | 3·76709 | 28423 | 3·51829 | 8 |
| 53 | 22872 | 4·37207 | 24717 | 4·04586 | 26577 | 3·76268 | 28454 | 3·51441 | 7 |
| 54 | 22903 | 4·36623 | 24747 | 4·04081 | 26608 | 3·75828 | 28486 | 3·51053 | 6 |
| 55 | 22934 | 4·36040 | 24778 | 4·03578 | 26639 | 3·75388 | 28517 | 3·50666 | 5 |
| 56 | 22964 | 4·35459 | 24809 | 4·03075 | 26670 | 3·74950 | 28549 | 3·50279 | 4 |
| 57 | 22995 | 4·34879 | 24840 | 4·02574 | 26701 | 3·74512 | 28580 | 3·49894 | 3 |
| 58 | 23026 | 4·34300 | 24871 | 4·02074 | 26733 | 3·74075 | 28612 | 3·49509 | 2 |
| 59 | 23056 | 4·33723 | 24902 | 4·01576 | 26764 | 3·73640 | 28643 | 3·49125 | 1 |
| 60 | 23087 | 4·33148 | 24933 | 4·01078 | 26795 | 3·73205 | 28675 | 3·48741 | 0 |
| ′ | Cotang. | Tangent. | Cotang. | Tangent. | Cotang. | Tangent. | Cotang. | Tangent. | ′ |
| | 77° | | 76° | | 75° | | 74° | | |

| ′ | 16° | | 17° | | 18° | | 19° | | ′ |
|---|---|---|---|---|---|---|---|---|---|
| | Tangent. | Cotang. | Tangent. | Cotang. | Tangent. | Cotang. | Tangent. | Cotang. | |
| 0 | 28675 | 3·48741 | 30573 | 3·27085 | 32492 | 3·07768 | 34433 | 2·90421 | 60 |
| 1 | 28706 | 3·48359 | 30605 | 3·26745 | 32524 | 3·07464 | 34465 | 2·90147 | 59 |
| 2 | 28738 | 3·47977 | 30637 | 3·26406 | 32556 | 3·07160 | 34498 | 2·89873 | 58 |
| 3 | 28769 | 3·47596 | 30669 | 3·26067 | 32588 | 3·06857 | 34530 | 2·89600 | 57 |
| 4 | 28800 | 3·47216 | 30700 | 3·25729 | 32621 | 3·06554 | 34563 | 2·89327 | 56 |
| 5 | 28832 | 3·46837 | 30732 | 3·25392 | 32653 | 3·06252 | 34596 | 2·89055 | 55 |
| 6 | 28864 | 3·46458 | 30764 | 3·25055 | 32685 | 3·05950 | 34628 | 2·88783 | 54 |
| 7 | 28895 | 3·46080 | 30796 | 3·24719 | 32717 | 3·05649 | 34661 | 2·88511 | 53 |
| 8 | 28927 | 3·45703 | 30828 | 3·24383 | 32749 | 3·05349 | 34693 | 2·88240 | 52 |
| 9 | 28958 | 3·45327 | 30860 | 3·24049 | 32782 | 3·05049 | 34726 | 2·87970 | 51 |
| 10 | 28990 | 3·44951 | 30891 | 3·23714 | 32814 | 3·04749 | 34758 | 2·87700 | 50 |
| 11 | 29021 | 3·44576 | 30923 | 3·23381 | 32846 | 3·04450 | 34791 | 2·87430 | 49 |
| 12 | 29053 | 3·44202 | 30955 | 3·23048 | 32878 | 3·04152 | 34824 | 2·87161 | 48 |
| 13 | 29084 | 3·43829 | 30987 | 3·22715 | 32911 | 3·03854 | 34856 | 2·86892 | 47 |
| 14 | 29116 | 3·43456 | 31019 | 3·22384 | 32943 | 3·03556 | 34889 | 2·86624 | 46 |
| 15 | 29147 | 3·43084 | 31051 | 3·22053 | 32975 | 3·03260 | 34922 | 2·86356 | 45 |
| 16 | 29179 | 3·42713 | 31083 | 3·21722 | 33007 | 3·02963 | 34954 | 2·86089 | 44 |
| 17 | 29210 | 3·42343 | 31115 | 3·21392 | 33040 | 3·02667 | 34987 | 2·85822 | 43 |
| 18 | 29242 | 3·41973 | 31147 | 3·21063 | 33072 | 3·02372 | 35019 | 2·85555 | 42 |
| 19 | 29274 | 3·41604 | 31178 | 3·20734 | 33104 | 3·02077 | 35052 | 2·85289 | 41 |
| 20 | 29305 | 3·41236 | 31210 | 3·20406 | 33136 | 3·01783 | 35085 | 2·85023 | 40 |
| 21 | 29337 | 3·40869 | 31242 | 3·20079 | 33169 | 3·01489 | 35117 | 2·84758 | 39 |
| 22 | 29368 | 3·40502 | 31274 | 3·19752 | 33201 | 3·01196 | 35150 | 2·84494 | 38 |
| 23 | 29400 | 3·40136 | 31306 | 3·19426 | 33233 | 3·00903 | 35183 | 2·84229 | 37 |
| 24 | 29432 | 3·39771 | 31338 | 3·19100 | 33266 | 3·00611 | 35216 | 2·83965 | 36 |
| 25 | 29463 | 3·39406 | 31370 | 3·18775 | 33298 | 3·00319 | 35248 | 2·83702 | 35 |
| 26 | 29495 | 3·39042 | 31402 | 3·18451 | 33330 | 3·00028 | 35281 | 2·83439 | 34 |
| 27 | 29526 | 3·38679 | 31434 | 3·18127 | 33363 | 2·99738 | 35314 | 2·83176 | 33 |
| 28 | 29558 | 3·38317 | 31466 | 3·17804 | 33395 | 2·99447 | 35346 | 2·82914 | 32 |
| 29 | 29590 | 3·37955 | 31498 | 3·17481 | 33427 | 2·99158 | 35379 | 2·82653 | 31 |
| 30 | 29621 | 3·37594 | 31530 | 3·17159 | 33460 | 2·98868 | 35412 | 2·82391 | 30 |
| 31 | 29653 | 3·37234 | 31562 | 3·16838 | 33492 | 2·98580 | 35445 | 2·82130 | 29 |
| 32 | 29685 | 3·36875 | 31594 | 3·16517 | 33524 | 2·98292 | 35477 | 2·81870 | 28 |
| 33 | 29716 | 3·36516 | 31626 | 3·16197 | 33557 | 2·98004 | 35510 | 2·81610 | 27 |
| 34 | 29748 | 3·36158 | 31658 | 3·15877 | 33589 | 2·97717 | 35543 | 2·81350 | 26 |
| 35 | 29780 | 3·35800 | 31690 | 3·15558 | 33621 | 2·97430 | 35576 | 2·81091 | 25 |
| 36 | 29811 | 3·35443 | 31722 | 3·15240 | 33654 | 2·97144 | 35608 | 2·80833 | 24 |
| 37 | 29843 | 3·35087 | 31754 | 3·14922 | 33686 | 2·96858 | 35641 | 2·80574 | 23 |
| 38 | 29875 | 3·34732 | 31786 | 3·14605 | 33718 | 2·96573 | 35674 | 2·80316 | 22 |
| 39 | 29906 | 3·34377 | 31818 | 3·14288 | 33751 | 2·96288 | 35707 | 2·80059 | 21 |
| 40 | 29938 | 3·34023 | 31850 | 3·13972 | 33783 | 2·96004 | 35740 | 2·79802 | 20 |
| 41 | 29970 | 3·33670 | 31882 | 3·13656 | 33816 | 2·95721 | 35772 | 2·79545 | 19 |
| 42 | 30001 | 3·33317 | 31914 | 3·13341 | 33848 | 2·95437 | 35805 | 2·79289 | 18 |
| 43 | 30033 | 3·32965 | 31946 | 3·13027 | 33881 | 2·95155 | 35838 | 2·79033 | 17 |
| 44 | 30065 | 3·32614 | 31978 | 3·12713 | 33913 | 2·94872 | 35871 | 2·78778 | 16 |
| 45 | 30097 | 3·32264 | 32010 | 3·12400 | 33945 | 2·94590 | 35904 | 2·78523 | 15 |
| 46 | 30128 | 3·31914 | 32042 | 3·12087 | 33978 | 2·94309 | 35937 | 2·78269 | 14 |
| 47 | 30160 | 3·31565 | 32074 | 3·11775 | 34010 | 2·94028 | 35969 | 2·78014 | 13 |
| 48 | 30192 | 3·31216 | 32106 | 3·11464 | 34043 | 2·93748 | 36002 | 2·77761 | 12 |
| 49 | 30224 | 3·30868 | 32139 | 3·11153 | 34075 | 2·93468 | 36035 | 2·77507 | 11 |
| 50 | 30255 | 3·30521 | 32171 | 3·10842 | 34108 | 2·93189 | 36068 | 2·77254 | 10 |
| 51 | 30287 | 3·30174 | 32203 | 3·10532 | 34140 | 2·92910 | 36101 | 2·77002 | 9 |
| 52 | 30319 | 3·29829 | 32235 | 3·10223 | 34173 | 2·92632 | 36134 | 2·76750 | 8 |
| 53 | 30351 | 3·29483 | 32267 | 3·09914 | 34205 | 2·92354 | 36167 | 2·76498 | 7 |
| 54 | 30382 | 3·29139 | 32299 | 3·09606 | 34238 | 2·92076 | 36199 | 2·76247 | 6 |
| 55 | 30414 | 3·28795 | 32331 | 3·09298 | 34270 | 2·91799 | 36232 | 2·75996 | 5 |
| 56 | 30446 | 3·28452 | 32363 | 3·08991 | 34303 | 2·91523 | 36265 | 2·75746 | 4 |
| 57 | 30478 | 3·28109 | 32396 | 3·08685 | 34335 | 2·91246 | 36298 | 2·75496 | 3 |
| 58 | 30509 | 3·27767 | 32428 | 3·08379 | 34368 | 2·90971 | 36331 | 2·75246 | 2 |
| 59 | 30541 | 3·27426 | 32460 | 3·08073 | 34400 | 2·90696 | 36364 | 2·74997 | 1 |
| 60 | 30573 | 3·27085 | 32492 | 3·07768 | 34433 | 2·90421 | 36397 | 2·74748 | 0 |
| | Cotang. | Tangent. | Cotang. | Tangent. | Cotang. | Tangent. | Cotang. | Tangent. | |
| ′ | 73° | | 72° | | 71° | | 70° | | ′ |

| ′ | 20° | | 21° | | 22° | | 23° | | ′ |
|---|---|---|---|---|---|---|---|---|---|
| | Tangent. | Cotang. | Tangent. | Cotang. | Tangent. | Cotang. | Tangent. | Cotang. | |
| 0 | 36397 | 2·74748 | 38386 | 2·60509 | 40403 | 2·47509 | 42447 | 2·35585 | 60 |
| 1 | 36430 | 2·74499 | 38420 | 2·60283 | 40436 | 2·47302 | 42482 | 2·35395 | 59 |
| 2 | 36463 | 2·74251 | 38453 | 2·60057 | 40470 | 2·47095 | 42516 | 2·35205 | 58 |
| 3 | 36496 | 2·74004 | 38487 | 2·59831 | 40504 | 2·46888 | 42551 | 2·35015 | 57 |
| 4 | 36529 | 2·73756 | 38520 | 2·59606 | 40538 | 2·46682 | 42585 | 2·34825 | 56 |
| 5 | 36562 | 2·73509 | 38553 | 2·59381 | 40572 | 2·46476 | 42619 | 2·34636 | 55 |
| 6 | 36595 | 2·73263 | 38587 | 2·59156 | 40606 | 2·46270 | 42654 | 2·34447 | 54 |
| 7 | 36628 | 2·73017 | 38620 | 2·58932 | 40640 | 2·46065 | 42688 | 2·34258 | 53 |
| 8 | 36661 | 2·72771 | 38654 | 2·58708 | 40674 | 2·45860 | 42722 | 2·34069 | 52 |
| 9 | 36694 | 2·72526 | 38687 | 2·58484 | 40707 | 2·45655 | 42757 | 2·33881 | 51 |
| 10 | 36727 | 2·72281 | 38721 | 2·58261 | 40741 | 2·45451 | 42791 | 2·33693 | 50 |
| 11 | 36760 | 2·72036 | 38754 | 2·58038 | 40775 | 2·45246 | 42826 | 2·33505 | 49 |
| 12 | 36793 | 2·71792 | 38787 | 2·57815 | 40809 | 2·45043 | 42860 | 2·33317 | 48 |
| 13 | 36826 | 2·71548 | 38821 | 2·57593 | 40843 | 2·44839 | 42894 | 2·33130 | 47 |
| 14 | 36859 | 2·71305 | 38854 | 2·57371 | 40877 | 2·44636 | 42929 | 2·32943 | 46 |
| 15 | 36892 | 2·71062 | 38888 | 2·57150 | 40911 | 2·44433 | 42963 | 2·32756 | 45 |
| 16 | 36925 | 2·70819 | 38921 | 2·56928 | 40945 | 2·44230 | 42998 | 2·32570 | 44 |
| 17 | 36958 | 2·70577 | 38955 | 2·56707 | 40979 | 2·44027 | 43032 | 2·32383 | 43 |
| 18 | 36991 | 2·70335 | 38988 | 2·56487 | 41013 | 2·43825 | 43067 | 2·32197 | 42 |
| 19 | 37024 | 2·70094 | 39022 | 2·56266 | 41047 | 2·43623 | 43101 | 2·32012 | 41 |
| 20 | 37057 | 2·69853 | 39055 | 2·56046 | 41081 | 2·43422 | 43136 | 2·31826 | 40 |
| 21 | 37090 | 2·69612 | 39089 | 2·55827 | 41115 | 2·43220 | 43170 | 2·31641 | 39 |
| 22 | 37124 | 2·69371 | 39122 | 2·55608 | 41149 | 2·43019 | 43205 | 2·31456 | 38 |
| 23 | 37157 | 2·69131 | 39156 | 2·55389 | 41183 | 2·42819 | 43239 | 2·31271 | 37 |
| 24 | 37190 | 2·68892 | 39190 | 2·55170 | 41217 | 2·42618 | 43274 | 2·31086 | 36 |
| 25 | 37223 | 2·68653 | 39223 | 2·54952 | 41251 | 2·42418 | 43308 | 2·30902 | 35 |
| 26 | 37256 | 2·68414 | 39257 | 2·54734 | 41285 | 2·42218 | 43343 | 2·30718 | 34 |
| 27 | 37289 | 2·68175 | 39290 | 2·54516 | 41319 | 2·42019 | 43378 | 2·30534 | 33 |
| 28 | 37322 | 2·67937 | 39324 | 2·54299 | 41353 | 2·41819 | 43412 | 2·30351 | 32 |
| 29 | 37355 | 2·67700 | 39357 | 2·54082 | 41387 | 2·41620 | 43447 | 2·30167 | 31 |
| 30 | 37388 | 2·67462 | 39391 | 2·53865 | 41421 | 2·41421 | 43481 | 2·29984 | 30 |
| 31 | 37422 | 2·67225 | 39425 | 2·53648 | 41455 | 2·41223 | 43516 | 2·29801 | 29 |
| 32 | 37455 | 2·66989 | 39458 | 2·53432 | 41490 | 2·41025 | 43550 | 2·29619 | 28 |
| 33 | 37488 | 2·66752 | 39492 | 2·53217 | 41524 | 2·40827 | 43585 | 2·29437 | 27 |
| 34 | 37521 | 2·66516 | 39526 | 2·53001 | 41558 | 2·40629 | 43620 | 2·29254 | 26 |
| 35 | 37554 | 2·66281 | 39559 | 2·52786 | 41592 | 2·40432 | 43654 | 2·29073 | 25 |
| 36 | 37588 | 2·66046 | 39593 | 2·52571 | 41626 | 2·40235 | 43689 | 2·28891 | 24 |
| 37 | 37621 | 2·65811 | 39626 | 2·52357 | 41660 | 2·40038 | 43724 | 2·28710 | 23 |
| 38 | 37654 | 2·65576 | 39660 | 2·52142 | 41694 | 2·39841 | 43758 | 2·28528 | 22 |
| 39 | 37687 | 2·65342 | 39694 | 2·51929 | 41728 | 2·39645 | 43793 | 2·28348 | 21 |
| 40 | 37720 | 2·65109 | 39727 | 2·51715 | 41763 | 2·39449 | 43828 | 2·28167 | 20 |
| 41 | 37754 | 2·64875 | 39761 | 2·51502 | 41797 | 2·39253 | 43862 | 2·27987 | 19 |
| 42 | 37787 | 2·64642 | 39795 | 2·51289 | 41831 | 2·39058 | 43897 | 2·27806 | 18 |
| 43 | 37820 | 2·64410 | 39829 | 2·51076 | 41865 | 2·38862 | 43932 | 2·27626 | 17 |
| 44 | 37853 | 2·64177 | 39862 | 2·50864 | 41899 | 2·38668 | 43966 | 2·27447 | 16 |
| 45 | 37887 | 2·63945 | 39896 | 2·50652 | 41933 | 2·38473 | 44001 | 2·27267 | 15 |
| 46 | 37920 | 2·63714 | 39930 | 2·50440 | 41968 | 2·38279 | 44036 | 2·27088 | 14 |
| 47 | 37953 | 2·63483 | 39963 | 2·50229 | 42002 | 2·38084 | 44071 | 2·26909 | 13 |
| 48 | 37986 | 2·63252 | 39997 | 2·50018 | 42036 | 2·37891 | 44105 | 2·26730 | 12 |
| 49 | 38020 | 2·63021 | 40031 | 2·49807 | 42070 | 2·37697 | 44140 | 2·26552 | 11 |
| 50 | 38053 | 2·62791 | 40065 | 2·49597 | 42105 | 2·37504 | 44175 | 2·26374 | 10 |
| 51 | 38086 | 2·62561 | 40098 | 2·49386 | 42139 | 2·37311 | 44210 | 2·26196 | 9 |
| 52 | 38120 | 2·62332 | 40132 | 2·49177 | 42173 | 2·37118 | 44244 | 2·26018 | 8 |
| 53 | 38153 | 2·62103 | 40166 | 2·48967 | 42207 | 2·36925 | 44279 | 2·25840 | 7 |
| 54 | 38186 | 2·61874 | 40200 | 2·48758 | 42242 | 2·36733 | 44314 | 2·25663 | 6 |
| 55 | 38220 | 2·61646 | 40234 | 2·48549 | 42276 | 2·36541 | 44349 | 2·25486 | 5 |
| 56 | 38253 | 2·61418 | 40267 | 2·48340 | 42310 | 2·36349 | 44384 | 2·25309 | 4 |
| 57 | 38286 | 2·61190 | 40301 | 2·48132 | 42345 | 2·36158 | 44418 | 2·25132 | 3 |
| 58 | 38320 | 2·60963 | 40335 | 2·47924 | 42379 | 2·35967 | 44453 | 2·24956 | 2 |
| 59 | 38353 | 2·60736 | 40369 | 2·47716 | 42413 | 2·35776 | 44488 | 2·24780 | 1 |
| 60 | 38386 | 2·60509 | 40403 | 2·47509 | 42447 | 2·35585 | 44523 | 2·24604 | 0 |
| ′ | Cotang. | Tangent. | Cotang. | Tangent. | Cotang. | Tangent. | Cotang. | Tangent. | ′ |
| | 69° | | 68° | | 67° | | 66° | | |

| ′ | 24° | | 25° | | 26° | | 27° | | ′ |
|---|---|---|---|---|---|---|---|---|---|
| | Tangent. | Cotang. | Tangent. | Cotang. | Tangent. | Cotang. | Tangent. | Cotang. | |
| 0 | 44523 | 2·24604 | 46631 | 2·14451 | 48773 | 2·05030 | 50953 | 1·96261 | 60 |
| 1 | 44558 | 2·24428 | 46666 | 2·14288 | 48809 | 2·04879 | 50989 | 1·96120 | 59 |
| 2 | 44593 | 2·24252 | 46702 | 2·14125 | 48845 | 2·04728 | 51026 | 1·95979 | 58 |
| 3 | 44627 | 2·24077 | 46737 | 2·13963 | 48881 | 2·04577 | 51063 | 1·95838 | 57 |
| 4 | 44662 | 2·23902 | 46772 | 2·13801 | 48917 | 2·04426 | 51099 | 1·95698 | 56 |
| 5 | 44697 | 2·23727 | 46808 | 2·13639 | 48953 | 2·04276 | 51136 | 1·95557 | 55 |
| 6 | 44732 | 2·23553 | 46843 | 2·13477 | 48989 | 2·04125 | 51173 | 1·95417 | 54 |
| 7 | 44767 | 2·23378 | 46879 | 2·13316 | 49026 | 2·03975 | 51209 | 1·95277 | 53 |
| 8 | 44802 | 2·23204 | 46914 | 2·13154 | 49062 | 2·03825 | 51246 | 1·95137 | 52 |
| 9 | 44837 | 2·23030 | 46950 | 2·12993 | 49098 | 2·03675 | 51283 | 1·94997 | 51 |
| 10 | 44872 | 2·22857 | 46985 | 2·12832 | 49134 | 2·03526 | 51319 | 1·94858 | 50 |
| 11 | 44907 | 2·22683 | 47021 | 2·12671 | 49170 | 2·03376 | 51356 | 1·94718 | 49 |
| 12 | 44942 | 2·22510 | 47056 | 2·12511 | 49206 | 2·03227 | 51393 | 1·94579 | 48 |
| 13 | 44977 | 2·22337 | 47092 | 2·12350 | 49242 | 2·03078 | 51430 | 1·94440 | 47 |
| 14 | 45012 | 2·22164 | 47128 | 2·12190 | 49278 | 2·02929 | 51467 | 1·94301 | 46 |
| 15 | 45047 | 2·21992 | 47163 | 2·12030 | 49315 | 2·02780 | 51503 | 1·94162 | 45 |
| 16 | 45082 | 2·21819 | 47199 | 2·11871 | 49351 | 2·02631 | 51540 | 1·94023 | 44 |
| 17 | 45117 | 2·21647 | 47234 | 2·11711 | 49387 | 2·02483 | 51577 | 1·93885 | 43 |
| 18 | 45152 | 2·21475 | 47270 | 2·11552 | 49423 | 2·02335 | 51614 | 1·93746 | 42 |
| 19 | 45187 | 2·21304 | 47305 | 2·11392 | 49459 | 2·02187 | 51651 | 1·93608 | 41 |
| 20 | 45222 | 2·21132 | 47341 | 2·11233 | 49495 | 2·02039 | 51688 | 1·93470 | 40 |
| 21 | 45257 | 2·20961 | 47377 | 2·11075 | 49532 | 2·01891 | 51724 | 1·93332 | 39 |
| 22 | 45292 | 2·20790 | 47412 | 2·10916 | 49568 | 2·01743 | 51761 | 1·93195 | 38 |
| 23 | 45327 | 2·20619 | 47448 | 2·10758 | 49604 | 2·01596 | 51798 | 1·93057 | 37 |
| 24 | 45362 | 2·20449 | 47483 | 2·10600 | 49640 | 2·01449 | 51835 | 1·92920 | 36 |
| 25 | 45397 | 2·20278 | 47519 | 2·10442 | 49677 | 2·01302 | 51872 | 1·92782 | 35 |
| 26 | 45432 | 2·20108 | 47555 | 2·10284 | 49713 | 2·01155 | 51909 | 1·92645 | 34 |
| 27 | 45467 | 2·19938 | 47590 | 2·10126 | 49749 | 2·01008 | 51946 | 1·92508 | 33 |
| 28 | 45502 | 2·19769 | 47626 | 2·09969 | 49786 | 2·00862 | 51983 | 1·92371 | 32 |
| 29 | 45537 | 2·19599 | 47662 | 2·09811 | 49822 | 2·00715 | 52020 | 1·92235 | 31 |
| 30 | 45573 | 2·19430 | 47698 | 2·09654 | 49858 | 2·00569 | 52057 | 1·92098 | 30 |
| 31 | 45608 | 2·19261 | 47733 | 2·09498 | 49894 | 2·00423 | 52094 | 1·91962 | 29 |
| 32 | 45643 | 2·19092 | 47769 | 2·09341 | 49931 | 2·00277 | 52131 | 1·91826 | 28 |
| 33 | 45678 | 2·18923 | 47805 | 2·09184 | 49967 | 2·00131 | 52168 | 1·91690 | 27 |
| 34 | 45713 | 2·18755 | 47840 | 2·09028 | 50004 | 1·99986 | 52205 | 1·91554 | 26 |
| 35 | 45748 | 2·18587 | 47876 | 2·08872 | 50040 | 1·99841 | 52242 | 1·91418 | 25 |
| 36 | 45784 | 2·18419 | 47912 | 2·08716 | 50076 | 1·99695 | 52279 | 1·91282 | 24 |
| 37 | 45819 | 2·18251 | 47948 | 2·08560 | 50113 | 1·99550 | 52316 | 1·91147 | 23 |
| 38 | 45854 | 2·18084 | 47984 | 2·08405 | 50149 | 1·99406 | 52353 | 1·91012 | 22 |
| 39 | 45889 | 2·17916 | 48019 | 2·08250 | 50185 | 1·99261 | 52390 | 1·90876 | 21 |
| 40 | 45924 | 2·17749 | 48055 | 2·08094 | 50222 | 1·99116 | 52427 | 1·90741 | 20 |
| 41 | 45960 | 2·17582 | 48091 | 2·07939 | 50258 | 1·98972 | 52464 | 1·90607 | 19 |
| 42 | 45995 | 2·17416 | 48127 | 2·07785 | 50295 | 1·98828 | 52501 | 1·90472 | 18 |
| 43 | 46030 | 2·17249 | 48163 | 2·07630 | 50331 | 1·98684 | 52538 | 1·90337 | 17 |
| 44 | 46065 | 2·17083 | 48198 | 2·07476 | 50368 | 1·98540 | 52575 | 1·90203 | 16 |
| 45 | 46101 | 2·16917 | 48234 | 2·07321 | 50404 | 1·98396 | 52613 | 1·90069 | 15 |
| 46 | 46136 | 2·16751 | 48270 | 2·07167 | 50441 | 1·98253 | 52650 | 1·89935 | 14 |
| 47 | 46171 | 2·16585 | 48306 | 2·07014 | 50477 | 1·98110 | 52687 | 1·89801 | 13 |
| 48 | 46206 | 2·16420 | 48342 | 2·06860 | 50514 | 1·97966 | 52724 | 1·89667 | 12 |
| 49 | 46242 | 2·16255 | 48378 | 2·06706 | 50550 | 1·97823 | 52761 | 1·89533 | 11 |
| 50 | 46277 | 2·16090 | 48414 | 2·06553 | 50587 | 1·97680 | 52798 | 1·89400 | 10 |
| 51 | 46312 | 2·15925 | 48450 | 2·06400 | 50623 | 1·97538 | 52836 | 1·89266 | 9 |
| 52 | 46348 | 2·15760 | 48486 | 2·06247 | 50660 | 1·97395 | 52873 | 1·89133 | 8 |
| 53 | 46383 | 2·15596 | 48521 | 2·06094 | 50696 | 1·97253 | 52910 | 1·89000 | 7 |
| 54 | 46418 | 2·15432 | 48557 | 2·05942 | 50733 | 1·97111 | 52947 | 1·88867 | 6 |
| 55 | 46454 | 2·15268 | 48593 | 2·05790 | 50769 | 1·96969 | 52984 | 1·88734 | 5 |
| 56 | 46489 | 2·15104 | 48629 | 2·05637 | 50806 | 1·96827 | 53022 | 1·88602 | 4 |
| 57 | 46525 | 2·14940 | 48665 | 2·05485 | 50843 | 1·96685 | 53059 | 1·88469 | 3 |
| 58 | 46560 | 2·14777 | 48701 | 2·05333 | 50879 | 1·96544 | 53096 | 1·88337 | 2 |
| 59 | 46595 | 2·14614 | 48737 | 2·05182 | 50916 | 1·96402 | 53134 | 1·88205 | 1 |
| 60 | 46631 | 2·14451 | 48773 | 2·05030 | 50953 | 1·96261 | 53171 | 1·88073 | 0 |
| ′ | Cotang. | Tangent. | Cotang. | Tangent. | Cotang. | Tangent. | Cotang. | Tangent. | ′ |
| | 65° | | 64° | | 63° | | 62° | | |

| ′ | 28° | | 29° | | 30° | | 31° | | ′ |
|---|---|---|---|---|---|---|---|---|---|
| | Tangent. | Cotang. | Tangent. | Cotang. | Tangent. | Cotang. | Tangent. | Cotang. | |
| 0 | 53171 | 1·88073 | 55431 | 1·80405 | 57735 | 1·73205 | 60086 | 1·66428 | 60 |
| 1 | 53208 | 1·87941 | 55469 | 1·80281 | 57774 | 1·73089 | 60126 | 1·66318 | 59 |
| 2 | 53246 | 1·87809 | 55507 | 1·80158 | 57813 | 1·72973 | 60165 | 1·66209 | 58 |
| 3 | 53283 | 1·87677 | 55545 | 1·80034 | 57851 | 1·72857 | 60205 | 1·66099 | 57 |
| 4 | 53320 | 1·87546 | 55583 | 1·79911 | 57890 | 1·72741 | 60245 | 1·65990 | 56 |
| 5 | 53358 | 1·87415 | 55621 | 1·79788 | 57929 | 1·72625 | 60284 | 1·65881 | 55 |
| 6 | 53395 | 1·87283 | 55659 | 1·79665 | 57968 | 1·72509 | 60324 | 1·65772 | 54 |
| 7 | 53432 | 1·87152 | 55697 | 1·79542 | 58007 | 1·72393 | 60364 | 1·65663 | 53 |
| 8 | 53470 | 1·87021 | 55736 | 1·79419 | 58046 | 1·72278 | 60403 | 1·65554 | 52 |
| 9 | 53507 | 1·86891 | 55774 | 1·79296 | 58085 | 1·72163 | 60443 | 1·65445 | 51 |
| 10 | 53545 | 1·86760 | 55812 | 1·79174 | 58124 | 1·72047 | 60483 | 1·65337 | 50 |
| 11 | 53582 | 1·86630 | 55850 | 1·79051 | 58162 | 1·71932 | 60522 | 1·65228 | 49 |
| 12 | 53620 | 1·86499 | 55888 | 1·78929 | 58201 | 1·71817 | 60562 | 1·65120 | 48 |
| 13 | 53657 | 1·86369 | 55926 | 1·78807 | 58240 | 1·71702 | 60602 | 1·65011 | 47 |
| 14 | 53694 | 1·86239 | 55964 | 1·78685 | 58279 | 1·71588 | 60642 | 1·64903 | 46 |
| 15 | 53732 | 1·86109 | 56003 | 1·78563 | 58318 | 1·71473 | 60681 | 1·64795 | 45 |
| 16 | 53769 | 1·85979 | 56041 | 1·78441 | 58357 | 1·71358 | 60721 | 1·64687 | 44 |
| 17 | 53807 | 1·85850 | 56079 | 1·78319 | 58396 | 1·71244 | 60761 | 1·64579 | 43 |
| 18 | 53844 | 1·85720 | 56117 | 1·78198 | 58435 | 1·71129 | 60801 | 1·64471 | 42 |
| 19 | 53882 | 1·85591 | 56156 | 1·78077 | 58474 | 1·71015 | 60841 | 1·64363 | 41 |
| 20 | 53920 | 1·85462 | 56194 | 1·77955 | 58513 | 1·70901 | 60881 | 1·64256 | 40 |
| 21 | 53957 | 1·85333 | 56232 | 1·77834 | 58552 | 1·70787 | 60921 | 1·64148 | 39 |
| 22 | 53995 | 1·85204 | 56270 | 1·77713 | 58591 | 1·70673 | 60960 | 1·64041 | 38 |
| 23 | 54032 | 1·85075 | 56309 | 1·77592 | 58631 | 1·70560 | 61000 | 1·63934 | 37 |
| 24 | 54070 | 1·84946 | 56347 | 1·77471 | 58670 | 1·70446 | 61040 | 1·63826 | 36 |
| 25 | 54107 | 1·84818 | 56385 | 1·77351 | 58709 | 1·70332 | 61080 | 1·63719 | 35 |
| 26 | 54145 | 1·84689 | 56424 | 1·77230 | 58748 | 1·70219 | 61120 | 1·63612 | 34 |
| 27 | 54183 | 1·84561 | 56462 | 1·77110 | 58787 | 1·70106 | 61160 | 1·63505 | 33 |
| 28 | 54220 | 1·84433 | 56500 | 1·76990 | 58826 | 1·69992 | 61200 | 1·63398 | 32 |
| 29 | 54258 | 1·84305 | 56539 | 1·76869 | 58865 | 1·69879 | 61240 | 1·63292 | 31 |
| 30 | 54296 | 1·84177 | 56577 | 1·76749 | 58904 | 1·69766 | 61280 | 1·63185 | 30 |
| 31 | 54333 | 1·84049 | 56616 | 1·76630 | 58944 | 1·69653 | 61320 | 1·63079 | 29 |
| 32 | 54371 | 1·83922 | 56654 | 1·76510 | 58983 | 1·69541 | 61360 | 1·62972 | 28 |
| 33 | 54409 | 1·83794 | 56693 | 1·76390 | 59022 | 1·69428 | 61400 | 1·62866 | 27 |
| 34 | 54446 | 1·83667 | 56731 | 1·76271 | 59061 | 1·69316 | 61440 | 1·62760 | 26 |
| 35 | 54484 | 1·83540 | 56769 | 1·76151 | 59101 | 1·69203 | 61480 | 1·62654 | 25 |
| 36 | 54522 | 1·83413 | 56808 | 1·76032 | 59140 | 1·69091 | 61520 | 1·62548 | 24 |
| 37 | 54560 | 1·83286 | 56846 | 1·75913 | 59179 | 1·68979 | 61561 | 1·62442 | 23 |
| 38 | 54597 | 1·83159 | 56885 | 1·75794 | 59218 | 1·68866 | 61601 | 1·62336 | 22 |
| 39 | 54635 | 1·83033 | 56923 | 1·75675 | 59258 | 1·68754 | 61641 | 1·62230 | 21 |
| 40 | 54673 | 1·82906 | 56962 | 1·75556 | 59297 | 1·68643 | 61681 | 1·62125 | 20 |
| 41 | 54711 | 1·82780 | 57000 | 1·75437 | 59336 | 1·68531 | 61721 | 1·62019 | 19 |
| 42 | 54748 | 1·82654 | 57039 | 1·75319 | 59376 | 1·68419 | 61761 | 1·61914 | 18 |
| 43 | 54786 | 1·82528 | 57078 | 1·75200 | 59415 | 1·68308 | 61801 | 1·61808 | 17 |
| 44 | 54824 | 1·82402 | 57116 | 1·75082 | 59454 | 1·68196 | 61842 | 1·61703 | 16 |
| 45 | 54862 | 1·82276 | 57155 | 1·74964 | 59494 | 1·68085 | 61882 | 1·61598 | 15 |
| 46 | 54900 | 1·82150 | 57193 | 1·74846 | 59533 | 1·67974 | 61922 | 1·61493 | 14 |
| 47 | 54938 | 1·82025 | 57232 | 1·74728 | 59573 | 1·67863 | 61962 | 1·61388 | 13 |
| 48 | 54975 | 1·81899 | 57271 | 1·74610 | 59612 | 1·67752 | 62003 | 1·61283 | 12 |
| 49 | 55013 | 1·81774 | 57309 | 1·74492 | 59651 | 1·67641 | 62043 | 1·61179 | 11 |
| 50 | 55051 | 1·81649 | 57348 | 1·74375 | 59691 | 1·67530 | 62083 | 1·61074 | 10 |
| 51 | 55089 | 1·81524 | 57386 | 1·74257 | 59730 | 1·67419 | 62124 | 1·60970 | 9 |
| 52 | 55127 | 1·81399 | 57425 | 1·74140 | 59770 | 1·67309 | 62164 | 1·60865 | 8 |
| 53 | 55165 | 1·81274 | 57464 | 1·74022 | 59809 | 1·67198 | 62204 | 1·60761 | 7 |
| 54 | 55203 | 1·81150 | 57503 | 1·73905 | 59849 | 1·67088 | 62245 | 1·60657 | 6 |
| 55 | 55241 | 1·81025 | 57541 | 1·73788 | 59888 | 1·66978 | 62285 | 1·60553 | 5 |
| 56 | 55279 | 1·80901 | 57580 | 1·73671 | 59928 | 1·66867 | 62325 | 1·60449 | 4 |
| 57 | 55317 | 1·80777 | 57619 | 1·73555 | 59967 | 1·66757 | 62366 | 1·60345 | 3 |
| 58 | 55355 | 1·80653 | 57657 | 1·73438 | 60007 | 1·66647 | 62406 | 1·60241 | 2 |
| 59 | 55393 | 1·80529 | 57696 | 1·73321 | 60046 | 1·66538 | 62446 | 1·60137 | 1 |
| 60 | 55431 | 1·80405 | 57735 | 1·73205 | 60086 | 1·66428 | 62487 | 1·60033 | 0 |
| ′ | Cotang. | Tangent. | Cotang. | Tangent. | Cotang. | Tangent. | Cotang. | Tangent. | ′ |
| | 61° | | 60° | | 59° | | 58° | | |

| ′ | 32° | | 33° | | 34° | | 35° | | ′ |
|---|---|---|---|---|---|---|---|---|---|
| | Tangent. | Cotang. | Tangent. | Cotang. | Tangent. | Cotang. | Tangent. | Cotang. | |
| 0 | 62487 | 1·60033 | 64941 | 1·53986 | 67451 | 1·48256 | 70021 | 1·42815 | 60 |
| 1 | 62527 | 1·59930 | 64982 | 1·53888 | 67493 | 1·48163 | 70064 | 1·42726 | 59 |
| 2 | 62568 | 1·59826 | 65023 | 1·53791 | 67536 | 1·48070 | 70107 | 1·42638 | 58 |
| 3 | 62608 | 1·59723 | 65065 | 1·53693 | 67578 | 1·47977 | 70151 | 1·42550 | 57 |
| 4 | 62649 | 1·59620 | 65106 | 1·53595 | 67620 | 1·47885 | 70194 | 1·42462 | 56 |
| 5 | 62689 | 1·59517 | 65148 | 1·53497 | 67663 | 1·47792 | 70238 | 1·42374 | 55 |
| 6 | 62730 | 1·59414 | 65189 | 1·53400 | 67705 | 1·47699 | 70281 | 1·42286 | 54 |
| 7 | 62770 | 1·59311 | 65231 | 1·53302 | 67748 | 1·47607 | 70325 | 1·42198 | 53 |
| 8 | 62811 | 1·59208 | 65272 | 1·53205 | 67790 | 1·47514 | 70368 | 1·42110 | 52 |
| 9 | 62852 | 1·59105 | 65314 | 1·53107 | 67832 | 1·47422 | 70412 | 1·42022 | 51 |
| 10 | 62892 | 1·59002 | 65355 | 1·53010 | 67875 | 1·47330 | 70455 | 1·41934 | 50 |
| 11 | 62933 | 1·58900 | 65397 | 1·52913 | 67917 | 1·47238 | 70499 | 1·41847 | 49 |
| 12 | 62973 | 1·58797 | 65438 | 1·52816 | 67960 | 1·47146 | 70542 | 1·41759 | 48 |
| 13 | 63014 | 1·58695 | 65480 | 1·52719 | 68002 | 1·47053 | 70586 | 1·41672 | 47 |
| 14 | 63055 | 1·58593 | 65521 | 1·52622 | 68045 | 1·46962 | 70629 | 1·41584 | 46 |
| 15 | 63095 | 1·58490 | 65563 | 1·52525 | 68088 | 1·46870 | 70673 | 1·41497 | 45 |
| 16 | 63136 | 1·58388 | 65604 | 1·52429 | 68130 | 1·46778 | 70717 | 1·41409 | 44 |
| 17 | 63177 | 1·58286 | 65646 | 1·52332 | 68173 | 1·46686 | 70760 | 1·41322 | 43 |
| 18 | 63217 | 1·58184 | 65688 | 1·52235 | 68215 | 1·46595 | 70804 | 1·41235 | 42 |
| 19 | 63258 | 1·58083 | 65729 | 1·52139 | 68258 | 1·46503 | 70848 | 1·41148 | 41 |
| 20 | 63299 | 1·57981 | 65771 | 1·52043 | 68301 | 1·46411 | 70891 | 1·41061 | 40 |
| 21 | 63340 | 1·57879 | 65813 | 1·51946 | 68343 | 1·46320 | 70935 | 1·40974 | 39 |
| 22 | 63380 | 1·57778 | 65854 | 1·51850 | 68386 | 1·46229 | 70979 | 1·40887 | 38 |
| 23 | 63421 | 1·57676 | 65896 | 1·51754 | 68429 | 1·46137 | 71023 | 1·40800 | 37 |
| 24 | 63462 | 1·57575 | 65938 | 1·51658 | 68471 | 1·46046 | 71066 | 1·40714 | 36 |
| 25 | 63503 | 1·57474 | 65980 | 1·51562 | 68514 | 1·45955 | 71110 | 1·40627 | 35 |
| 26 | 63544 | 1·57372 | 66021 | 1·51466 | 98557 | 1·45864 | 71154 | 1·40540 | 34 |
| 27 | 63584 | 1·57271 | 66063 | 1·51370 | 68600 | 1·45773 | 71198 | 1·40454 | 33 |
| 28 | 63625 | 1·57170 | 66105 | 1·51275 | 68642 | 1·45682 | 71242 | 1·40367 | 32 |
| 29 | 63666 | 1·57069 | 66147 | 1·51179 | 68685 | 1·45592 | 71285 | 1·40281 | 31 |
| 30 | 63707 | 1·56969 | 66189 | 1·51084 | 68728 | 1·45501 | 71329 | 1·40195 | 30 |
| 31 | 63748 | 1·56868 | 66230 | 1·50988 | 68771 | 1·45410 | 71373 | 1·40109 | 29 |
| 32 | 63789 | 1·56767 | 66272 | 1·50893 | 68814 | 1·45320 | 71417 | 1·40022 | 28 |
| 33 | 63830 | 1·56667 | 66314 | 1·50797 | 68857 | 1·45229 | 71461 | 1·39936 | 27 |
| 34 | 63871 | 1·56566 | 66356 | 1·50702 | 68900 | 1·45139 | 71505 | 1·39850 | 26 |
| 35 | 63912 | 1·56466 | 66398 | 1·50607 | 68942 | 1·45049 | 71549 | 1·39764 | 25 |
| 36 | 63953 | 1·56366 | 66440 | 1·50512 | 68985 | 1·44958 | 71593 | 1·39679 | 24 |
| 37 | 63994 | 1·56265 | 66482 | 1·50417 | 69028 | 1·44868 | 71637 | 1·39593 | 23 |
| 38 | 64035 | 1·56165 | 66524 | 1·50322 | 69071 | 1·44778 | 71681 | 1·39507 | 22 |
| 39 | 64076 | 1·56065 | 66566 | 1·50228 | 69114 | 1·44688 | 71725 | 1·39421 | 21 |
| 40 | 64117 | 1·55966 | 66608 | 1·50133 | 69157 | 1·44598 | 71769 | 1·39336 | 20 |
| 41 | 64158 | 1·55866 | 66650 | 1·50038 | 69200 | 1·44508 | 71813 | 1·39250 | 19 |
| 42 | 64199 | 1·55766 | 66692 | 1·49944 | 69243 | 1·44418 | 71857 | 1·39165 | 18 |
| 43 | 64240 | 1·55666 | 66734 | 1·49849 | 69286 | 1·44329 | 71901 | 1·39079 | 17 |
| 44 | 64281 | 1·55567 | 66776 | 1·49755 | 69329 | 1·44239 | 71946 | 1·38994 | 16 |
| 45 | 64322 | 1·55467 | 66818 | 1·49661 | 69372 | 1·44149 | 71990 | 1·38909 | 15 |
| 46 | 64363 | 1·55368 | 66860 | 1·49566 | 69416 | 1·44060 | 72034 | 1·38824 | 14 |
| 47 | 64404 | 1·55269 | 66902 | 1·49472 | 69459 | 1·43970 | 72078 | 1·38738 | 13 |
| 48 | 64446 | 1·55170 | 66944 | 1·49378 | 69502 | 1·43881 | 72122 | 1·38653 | 12 |
| 49 | 64487 | 1·55071 | 66986 | 1·49284 | 69545 | 1·43792 | 72166 | 1·38568 | 11 |
| 50 | 64528 | 1·54972 | 67028 | 1·49190 | 69588 | 1·43703 | 72211 | 1·38484 | 10 |
| 51 | 64569 | 1·54873 | 67071 | 1·49097 | 69631 | 1·43614 | 72255 | 1·38399 | 9 |
| 52 | 64610 | 1·54774 | 67113 | 1·49003 | 69675 | 1·43525 | 72299 | 1·38314 | 8 |
| 53 | 64652 | 1·54675 | 67155 | 1·48909 | 69718 | 1·43436 | 72344 | 1·38229 | 7 |
| 54 | 64693 | 1·54576 | 67197 | 1·48816 | 69761 | 1·43347 | 72388 | 1·38145 | 6 |
| 55 | 64734 | 1·54478 | 67239 | 1·48722 | 69804 | 1·43258 | 72432 | 1·38060 | 5 |
| 56 | 64775 | 1·54379 | 67282 | 1·48629 | 69847 | 1·43169 | 72477 | 1·37976 | 4 |
| 57 | 64817 | 1·54281 | 67324 | 1·48536 | 69891 | 1·43080 | 72521 | 1·37891 | 3 |
| 58 | 64858 | 1·54183 | 67366 | 1·48442 | 69934 | 1·42992 | 72565 | 1·37807 | 2 |
| 59 | 64899 | 1·54085 | 67409 | 1·48349 | 69977 | 1·42903 | 72610 | 1·37722 | 1 |
| 60 | 64941 | 1·53986 | 67451 | 1·48256 | 70021 | 1·42815 | 72654 | 1·37638 | 0 |
| | Cotang. | Tangent. | Cotang. | Tangent. | Cotang. | Tangent. | Cotang. | Tangent. | |
| ′ | 57° | | 56° | | 55° | | 54° | | ′ |

| ′ | 36° Tangent. | 36° Cotang. | 37° Tangent. | 37° Cotang. | 38° Tangent. | 38° Cotang. | 39° Tangent. | 39° Cotang. | ′ |
|---|---|---|---|---|---|---|---|---|---|
| 0 | 72654 | 1·37638 | 75355 | 1·32704 | 78129 | 1·27994 | 80978 | 1·23490 | 60 |
| 1 | 72699 | 1·37554 | 75401 | 1·32624 | 78175 | 1·27917 | 81027 | 1·23416 | 59 |
| 2 | 72743 | 1·37470 | 75447 | 1·32544 | 78222 | 1·27841 | 81075 | 1·23343 | 58 |
| 3 | 72788 | 1·37386 | 75492 | 1·32464 | 78269 | 1·27764 | 81123 | 1·23270 | 57 |
| 4 | 72832 | 1·37302 | 75538 | 1·32384 | 78316 | 1·27688 | 81171 | 1·23196 | 56 |
| 5 | 72877 | 1·37218 | 75584 | 1·32304 | 78363 | 1·27611 | 81220 | 1·23123 | 55 |
| 6 | 72921 | 1·37134 | 75629 | 1·32224 | 78410 | 1·27535 | 81268 | 1·23050 | 54 |
| 7 | 72966 | 1·37050 | 75675 | 1·32144 | 78457 | 1·27458 | 81316 | 1·22977 | 53 |
| 8 | 73010 | 1·36967 | 75721 | 1·32064 | 78504 | 1·27382 | 81364 | 1·22904 | 52 |
| 9 | 73055 | 1·36883 | 75767 | 1·31984 | 78551 | 1·27306 | 81413 | 1·22831 | 51 |
| 10 | 73100 | 1·36800 | 75812 | 1·31904 | 78598 | 1·27230 | 81461 | 1·22758 | 50 |
| 11 | 73144 | 1·36716 | 75858 | 1·31825 | 78645 | 1·27153 | 81510 | 1·22685 | 49 |
| 12 | 73189 | 1·36633 | 75904 | 1·31745 | 78692 | 1·27077 | 81558 | 1·22612 | 48 |
| 13 | 73234 | 1·36549 | 75950 | 1·31666 | 78739 | 1·27001 | 81606 | 1·22539 | 47 |
| 14 | 73278 | 1·36466 | 75996 | 1·31586 | 78786 | 1·26925 | 81655 | 1·22467 | 46 |
| 15 | 73323 | 1·36383 | 76042 | 1·31507 | 78834 | 1·26849 | 81703 | 1·22394 | 45 |
| 16 | 73368 | 1·36300 | 76088 | 1·31427 | 78881 | 1·26774 | 81752 | 1·22321 | 44 |
| 17 | 73413 | 1·36217 | 76134 | 1·31348 | 78928 | 1·26698 | 81800 | 1·22249 | 43 |
| 18 | 73457 | 1·36133 | 76180 | 1·31269 | 78975 | 1·26622 | 81849 | 1·22176 | 42 |
| 19 | 73502 | 1·36051 | 76226 | 1·31190 | 79022 | 1·26546 | 81898 | 1·22104 | 41 |
| 20 | 73547 | 1·35968 | 76272 | 1·31110 | 79070 | 1·26471 | 81946 | 1·22031 | 40 |
| 21 | 73592 | 1·35885 | 76318 | 1·31031 | 79117 | 1·26395 | 81995 | 1·21959 | 39 |
| 22 | 73637 | 1·35802 | 76364 | 1·30952 | 79164 | 1·26319 | 82044 | 1·21886 | 38 |
| 23 | 73681 | 1·35719 | 76410 | 1·30873 | 79212 | 1·26244 | 82092 | 1·21814 | 37 |
| 24 | 73726 | 1·35637 | 76456 | 1·30795 | 79259 | 1·26169 | 82141 | 1·21742 | 36 |
| 25 | 73771 | 1·35554 | 76502 | 1·30716 | 79306 | 1·26093 | 82190 | 1·21670 | 35 |
| 26 | 73816 | 1·35472 | 76548 | 1·30637 | 79354 | 1·26018 | 82238 | 1·21598 | 34 |
| 27 | 73861 | 1·35389 | 76594 | 1·30558 | 79401 | 1·25943 | 82287 | 1·21526 | 33 |
| 28 | 73906 | 1·35307 | 76640 | 1·30480 | 79449 | 1·25867 | 82336 | 1·21454 | 32 |
| 29 | 73951 | 1·35224 | 76686 | 1·30401 | 79496 | 1·25792 | 82385 | 1·21382 | 31 |
| 30 | 73996 | 1·35142 | 76733 | 1·30323 | 79544 | 1·25717 | 82434 | 1·21310 | 30 |
| 31 | 74041 | 1·35060 | 76779 | 1·30244 | 79591 | 1·25642 | 82483 | 1·21238 | 29 |
| 32 | 74086 | 1·34978 | 76825 | 1·30166 | 79639 | 1·25567 | 82531 | 1·21166 | 28 |
| 33 | 74131 | 1·34896 | 76871 | 1·30087 | 79686 | 1·25492 | 82580 | 1·21094 | 27 |
| 34 | 74176 | 1·34814 | 76918 | 1·30009 | 79734 | 1·25417 | 82629 | 1·21023 | 26 |
| 35 | 74221 | 1·34732 | 76964 | 1·29931 | 79781 | 1·25343 | 82678 | 1·20951 | 25 |
| 36 | 74267 | 1·34650 | 77010 | 1·29853 | 79829 | 1·25268 | 82727 | 1·20879 | 24 |
| 37 | 74312 | 1·34568 | 77057 | 1·29775 | 79877 | 1·25193 | 82776 | 1·20808 | 23 |
| 38 | 74357 | 1·34487 | 77103 | 1·29696 | 79924 | 1·25118 | 82825 | 1·20736 | 22 |
| 39 | 74402 | 1·34405 | 77149 | 1·29618 | 79972 | 1·25044 | 82874 | 1·20665 | 21 |
| 40 | 74447 | 1·34323 | 77196 | 1·29541 | 80020 | 1·24969 | 82923 | 1·20593 | 20 |
| 41 | 74492 | 1·34242 | 77242 | 1·29463 | 80067 | 1·24895 | 82972 | 1·20522 | 19 |
| 42 | 74538 | 1·34160 | 77289 | 1·29385 | 80115 | 1·24820 | 83022 | 1·20451 | 18 |
| 43 | 74583 | 1·34079 | 77335 | 1·29307 | 80163 | 1·24746 | 83071 | 1·20379 | 17 |
| 44 | 74628 | 1·33998 | 77382 | 1·29229 | 80211 | 1·24672 | 83120 | 1·20308 | 16 |
| 45 | 74674 | 1·33916 | 77428 | 1·29152 | 80258 | 1·24597 | 83169 | 1·20237 | 15 |
| 46 | 74719 | 1·33835 | 77475 | 1·29074 | 80306 | 1·24523 | 83218 | 1·20166 | 14 |
| 47 | 74764 | 1·33754 | 77521 | 1·28997 | 80354 | 1·24449 | 83268 | 1·20095 | 13 |
| 48 | 74810 | 1·33673 | 77568 | 1·28919 | 80402 | 1·24375 | 83317 | 1·20024 | 12 |
| 49 | 74855 | 1·33592 | 77615 | 1·28842 | 80450 | 1·24301 | 83366 | 1·19953 | 11 |
| 50 | 74900 | 1·33511 | 77661 | 1·28764 | 80498 | 1·24227 | 83415 | 1·19882 | 10 |
| 51 | 74946 | 1·33430 | 77708 | 1·28687 | 80546 | 1·24153 | 83465 | 1·19811 | 9 |
| 52 | 74991 | 1·33349 | 77754 | 1·28610 | 80594 | 1·24079 | 83514 | 1·19740 | 8 |
| 53 | 75037 | 1·33268 | 77801 | 1·28533 | 80642 | 1·24005 | 83564 | 1·19669 | 7 |
| 54 | 75082 | 1·33187 | 77848 | 1·28456 | 80690 | 1·23931 | 83613 | 1·19599 | 6 |
| 55 | 75128 | 1·33107 | 77895 | 1·28379 | 80738 | 1·23858 | 83662 | 1·19528 | 5 |
| 56 | 75173 | 1·33026 | 77941 | 1·28302 | 80786 | 1·23784 | 83712 | 1·19457 | 4 |
| 57 | 75219 | 1·32946 | 77988 | 1·28225 | 80834 | 1·23710 | 83761 | 1·19387 | 3 |
| 58 | 75264 | 1·32865 | 78035 | 1·28148 | 80882 | 1·23637 | 83811 | 1·19316 | 2 |
| 59 | 75310 | 1·32785 | 78082 | 1·28071 | 80930 | 1·23563 | 83860 | 1·19246 | 1 |
| 60 | 75355 | 1·32704 | 78129 | 1·27994 | 80978 | 1·23490 | 83910 | 1·19175 | 0 |
| ′ | Cotang. 53° | Tangent. 53° | Cotang. 52° | Tangent. 52° | Cotang. 51° | Tangent. 51° | Cotang. 50° | Tangent. 50° | ′ |

| ′ | 40° | | 41° | | 42° | | 43° | | ′ |
|---|---|---|---|---|---|---|---|---|---|
| | Tangent. | Cotang. | Tangent. | Cotang. | Tangent. | Cotang. | Tangent. | Cotang. | |
| 0 | 83910 | 1·19175 | 86929 | 1·15037 | 90040 | 1·11061 | 93252 | 1·07237 | 60 |
| 1 | 83960 | 1·19105 | 86980 | 1·14969 | 90093 | 1·10996 | 93306 | 1·07174 | 59 |
| 2 | 84009 | 1·19035 | 87031 | 1·14902 | 90146 | 1·10931 | 93360 | 1·07112 | 58 |
| 3 | 84059 | 1·18964 | 87082 | 1·14834 | 90199 | 1·10867 | 93415 | 1·07049 | 57 |
| 4 | 84108 | 1·18894 | 87133 | 1·14767 | 90251 | 1·10802 | 93469 | 1·06987 | 56 |
| 5 | 84158 | 1·18824 | 87184 | 1·14699 | 90304 | 1·10737 | 93524 | 1·06925 | 55 |
| 6 | 84208 | 1·18754 | 87236 | 1·14632 | 90357 | 1·10672 | 93578 | 1·06862 | 54 |
| 7 | 84258 | 1·18684 | 87287 | 1·14565 | 90410 | 1·10607 | 93633 | 1·06800 | 53 |
| 8 | 84307 | 1·18614 | 87338 | 1·14498 | 90463 | 1·10543 | 93688 | 1·06738 | 52 |
| 9 | 84357 | 1·18544 | 87389 | 1·14430 | 90516 | 1·10478 | 93742 | 1·06676 | 51 |
| 10 | 84407 | 1·18474 | 87441 | 1·14363 | 90569 | 1·10414 | 93797 | 1·06613 | 50 |
| 11 | 84457 | 1·18404 | 87492 | 1·14296 | 90621 | 1·10349 | 93852 | 1·06551 | 49 |
| 12 | 84507 | 1·18334 | 87543 | 1·14229 | 90674 | 1·10285 | 93906 | 1·06489 | 48 |
| 13 | 84556 | 1·18264 | 87595 | 1·14162 | 90727 | 1·10220 | 93961 | 1·06427 | 47 |
| 14 | 84606 | 1·18194 | 87646 | 1·14095 | 90781 | 1·10156 | 94016 | 1·06365 | 46 |
| 15 | 84656 | 1·18125 | 87698 | 1·14028 | 90834 | 1·10091 | 94071 | 1·06303 | 45 |
| 16 | 84706 | 1·18055 | 87749 | 1·13961 | 90887 | 1·10027 | 94125 | 1·06241 | 44 |
| 17 | 84756 | 1·17986 | 87801 | 1·13894 | 90940 | 1·09963 | 94180 | 1·06179 | 43 |
| 18 | 84806 | 1·17916 | 87852 | 1·13828 | 90993 | 1·09899 | 94235 | 1·06117 | 42 |
| 19 | 84856 | 1·17846 | 87904 | 1·13761 | 91046 | 1·09834 | 94290 | 1·06056 | 41 |
| 20 | 84906 | 1·17777 | 87955 | 1·13694 | 91099 | 1·09770 | 94345 | 1·05994 | 40 |
| 21 | 84956 | 1·17708 | 88007 | 1·13627 | 91153 | 1·09706 | 94400 | 1·05932 | 39 |
| 22 | 85006 | 1·17638 | 88059 | 1·13561 | 91206 | 1·09642 | 94455 | 1·05870 | 38 |
| 23 | 85057 | 1·17569 | 88110 | 1·13494 | 91259 | 1·09578 | 94510 | 1·05809 | 37 |
| 24 | 85107 | 1·17500 | 88162 | 1·13428 | 91313 | 1·09514 | 94565 | 1·05747 | 36 |
| 25 | 85157 | 1·17430 | 88214 | 1·13361 | 91366 | 1·09450 | 94620 | 1·05685 | 35 |
| 26 | 85207 | 1·17361 | 88265 | 1·13295 | 91419 | 1·09386 | 94676 | 1·05624 | 34 |
| 27 | 85257 | 1·17292 | 88317 | 1·13228 | 91473 | 1·09322 | 94731 | 1·05562 | 33 |
| 28 | 85307 | 1·17223 | 88369 | 1·13162 | 91526 | 1·09258 | 94786 | 1·05501 | 32 |
| 29 | 85358 | 1·17154 | 88421 | 1·13096 | 91580 | 1·09195 | 94841 | 1·05439 | 31 |
| 30 | 85408 | 1·17085 | 88473 | 1·13029 | 91633 | 1·09131 | 94896 | 1·05378 | 30 |
| 31 | 85458 | 1·17016 | 88524 | 1·12963 | 91687 | 1·09067 | 94952 | 1·05317 | 29 |
| 32 | 85509 | 1·16947 | 88576 | 1·12897 | 91740 | 1·09003 | 95007 | 1·05255 | 28 |
| 33 | 85559 | 1·16878 | 88628 | 1·12831 | 91794 | 1·08940 | 95062 | 1·05194 | 27 |
| 34 | 85609 | 1·16809 | 88680 | 1·12765 | 91847 | 1·08876 | 95118 | 1·05133 | 26 |
| 35 | 85660 | 1·16741 | 88732 | 1·12699 | 91901 | 1·08813 | 95173 | 1·05072 | 25 |
| 36 | 85710 | 1·16672 | 88784 | 1·12633 | 91955 | 1·08749 | 95229 | 1·05010 | 24 |
| 37 | 85761 | 1·16603 | 88836 | 1·12567 | 92008 | 1·08686 | 95284 | 1·04949 | 23 |
| 38 | 85811 | 1·16535 | 88888 | 1·12501 | 92062 | 1·08622 | 95340 | 1·04888 | 22 |
| 39 | 85862 | 1·16466 | 88940 | 1·12435 | 92116 | 1·08559 | 95395 | 1·04827 | 21 |
| 40 | 85912 | 1·16398 | 88992 | 1·12369 | 92170 | 1·08496 | 95451 | 1·04766 | 20 |
| 41 | 85963 | 1·16329 | 89045 | 1·12303 | 92223 | 1·08432 | 95506 | 1·04705 | 19 |
| 42 | 86014 | 1·16261 | 89097 | 1·12238 | 92277 | 1·08369 | 95562 | 1·04644 | 18 |
| 43 | 86064 | 1·16192 | 89149 | 1·12172 | 92331 | 1·08306 | 95618 | 1·04583 | 17 |
| 44 | 86115 | 1·16124 | 89201 | 1·12106 | 92385 | 1·08243 | 95673 | 1·04522 | 16 |
| 45 | 86166 | 1·16056 | 89253 | 1·12041 | 92439 | 1·08179 | 95729 | 1·04461 | 15 |
| 46 | 86216 | 1·15987 | 89306 | 1·11975 | 92493 | 1·08116 | 95785 | 1·04401 | 14 |
| 47 | 86267 | 1·15919 | 89358 | 1·11909 | 92547 | 1·08053 | 95841 | 1·04340 | 13 |
| 48 | 86318 | 1·15851 | 89410 | 1·11844 | 92601 | 1·07990 | 95897 | 1·04279 | 12 |
| 49 | 86368 | 1·15783 | 89463 | 1·11778 | 92655 | 1·07927 | 95952 | 1·04218 | 11 |
| 50 | 86419 | 1·15715 | 89515 | 1·11713 | 92709 | 1·07864 | 96008 | 1·04158 | 10 |
| 51 | 86470 | 1·15647 | 89567 | 1·11648 | 92763 | 1·07801 | 96064 | 1·04097 | 9 |
| 52 | 86521 | 1·15579 | 89620 | 1·11582 | 92817 | 1·07738 | 96120 | 1·04036 | 8 |
| 53 | 86572 | 1·15511 | 89672 | 1·11517 | 92872 | 1·07676 | 96176 | 1·03976 | 7 |
| 54 | 86623 | 1·15443 | 89725 | 1·11452 | 92926 | 1·07613 | 96232 | 1·03915 | 6 |
| 55 | 86674 | 1·15375 | 89777 | 1·11387 | 92980 | 1·07550 | 96288 | 1·03855 | 5 |
| 56 | 86725 | 1·15308 | 89830 | 1·11321 | 93034 | 1·07487 | 96344 | 1·03794 | 4 |
| 57 | 86776 | 1·15240 | 89883 | 1·11256 | 93088 | 1·07425 | 96400 | 1·03734 | 3 |
| 58 | 86827 | 1·15172 | 89935 | 1·11191 | 93143 | 1·07362 | 96457 | 1·03674 | 2 |
| 59 | 86878 | 1·15104 | 89988 | 1·11126 | 93197 | 1·07299 | 96513 | 1·03613 | 1 |
| 60 | 86929 | 1·15037 | 90040 | 1·11061 | 93252 | 1·07237 | 96569 | 1·03553 | 0 |
| | Cotang. | Tangent. | Cotang. | Tangent. | Cotang. | Tangent. | Cotang. | Tangent. | |
| ′ | 49° | | 48° | | 47° | | 46° | | ′ |

| ′ | 44° Tangent. | 44° Cotang. | ′ | ′ | 44° Tangent. | 44° Cotang. | ′ |
|---|---|---|---|---|---|---|---|
| 0 | 96569 | 1·03553 | 60 | 31 | 98327 | 1·01702 | 29 |
| 1 | 96625 | 1·03493 | 59 | 32 | 98384 | 1·01642 | 28 |
| 2 | 96681 | 1·03433 | 58 | 33 | 98441 | 1·01583 | 27 |
| 3 | 96738 | 1·03372 | 57 | 34 | 98499 | 1·01524 | 26 |
| 4 | 96794 | 1·03312 | 56 | 35 | 98556 | 1·01465 | 25 |
| 5 | 96850 | 1·03252 | 55 | 36 | 98613 | 1·01406 | 24 |
| 6 | 96907 | 1·03192 | 54 | 37 | 98671 | 1·01347 | 23 |
| 7 | 96963 | 1·03132 | 53 | 38 | 98728 | 1·01288 | 22 |
| 8 | 97020 | 1·03072 | 52 | 39 | 98786 | 1·01229 | 21 |
| 9 | 97076 | 1·03012 | 51 | 40 | 98843 | 1·01170 | 20 |
| 10 | 97133 | 1·02952 | 50 | 41 | 98901 | 1·01112 | 19 |
| 11 | 97189 | 1·02892 | 49 | 42 | 98958 | 1·01053 | 18 |
| 12 | 97246 | 1·02832 | 48 | 43 | 99016 | 1·00994 | 17 |
| 13 | 97302 | 1·02772 | 47 | 44 | 99073 | 1·00935 | 16 |
| 14 | 97359 | 1·02713 | 46 | 45 | 99131 | 1·00876 | 15 |
| 15 | 97416 | 1·02653 | 45 | 46 | 99189 | 1·00818 | 14 |
| 16 | 97472 | 1·02593 | 44 | 47 | 99247 | 1·00759 | 13 |
| 17 | 97529 | 1·02533 | 43 | 48 | 99304 | 1·00701 | 12 |
| 18 | 97586 | 1·02474 | 42 | 49 | 99362 | 1·00642 | 11 |
| 19 | 97643 | 1·02414 | 41 | 50 | 99420 | 1·00583 | 10 |
| 20 | 97700 | 1·02355 | 40 | 51 | 99478 | 1·00525 | 9 |
| 21 | 97756 | 1·02295 | 39 | 52 | 99536 | 1·00467 | 8 |
| 22 | 97813 | 1·02236 | 38 | 53 | 99594 | 1·00408 | 7 |
| 23 | 97870 | 1·02176 | 37 | 54 | 99652 | 1·00350 | 6 |
| 24 | 97927 | 1·02117 | 36 | 55 | 99710 | 1·00291 | 5 |
| 25 | 97984 | 1·02057 | 35 | 56 | 99768 | 1·00233 | 4 |
| 26 | 98041 | 1·01998 | 34 | 57 | 99826 | 1·00175 | 3 |
| 27 | 98098 | 1 01939 | 33 | 58 | 99884 | 1·00116 | 2 |
| 28 | 98155 | 1·01879 | 32 | 59 | 99942 | 1·00058 | 1 |
| 29 | 98213 | 1·01820 | 31 | 60 | Unit. | Unit. | 0 |
| 30 | 98270 | 1·01761 | 30 | | | | |
| ′ | 45° Cotang. | 45° Tangent. | ′ | ′ | 45° Cotang. | 45° Tangent. | ′ |

## TABLE OF CONSTANTS.

Base of Napier's system of logarithms = ................ $\varepsilon$ = 2·718281828459

Mod. of common syst. of logarithms = .... com. log. $\varepsilon$ = M = 0·434294481903

Ratio of circumference to diameter of a circle = .......... $\pi$ = 3·14159265359 0

log. $\pi$ = 0·497149872694

$\pi^2$ = 9·869604401089 ............ $\sqrt{\pi}$ = 1·772453850906

Arc of same length as radius = .......... 180° ÷ $\pi$ = 10800′ ÷ $\pi$ = 648000″ ÷ $\pi$

180° ÷ $\pi$ = 57°·2957795130, ............................ log. = 1·758122632409

10800′ ÷ $\pi$ = 3437′·7467707849, ........................ log. = 3·536273882793

648000″ ÷ $\pi$ = 206264″·8062470964, .................... log. = 5·314425133176

Tropical year = 365d. 5h. 48m. 47s. ·588 = 365d. ·242217456, log. = 2·5625810

Sidereal year = 365d. 6h. 9m. 10s. ·742 = 365d. ·256374332, log. = 2·5625978

24h. sol. t. = 24h. 3m. 56s. ·555335 sid. t. = 24h. × 1·00273791, log. 1·002 = 0·0011874

24h. sid. t. = 24h. − (3m. 55s. ·90944) sol. t. = 24h. × 0·9972696, log. 0·997 = 9·9988126

British imperial gallon = 277·274 cubic inches, ................ log. = 2·4429091

Length of sec. pend., in inches, at London, 39·13929; Paris, 39·1285; New York, 39·1285.

French *metre* = 3·2808992 English *feet* = 39·3707904 *inches*.

1 cubic inch of water (bar. 30 inches, Fahr. therm. 62°) = 252·458 Troy grains.

# MATHEMATICAL WORKS.

**BY GEO. R. PERKINS, LL. D.**

ELEMENTS OF ALGEBRA. 12mo. sheep. Price 75 Cents.

Designed as an introduction to the author's "Treatise on Algebra."

TREATISE ON ALGEBRA. 8vo. sheep, 420 pp. Price, $1 50.

Adapted to the wants of advanced schools and colleges. It will be found to contain a full and complete development of all the various subjects usually taught in our colleges, including a demonstration and application of the THEOREM OF STURM.

The present edition has been carefully revised and considerably enlarged. One entire chapter on the subject of Continued Fractions, treated in a general manner, has been added. The subject of RECURRING SERIES has been rewritten and simplified, and many other important changes have been made.

ELEMENTS OF GEOMETRY, WITH PRACTICAL APPLICATIONS. 12mo Sheep. Price $1.

The following from the author's preface will give the teacher an idea of the design of the work:

"We have found from experience in teaching, that, as a general thing, beginners in the study of Geometry consider it as *a dry, uninteresting science.* They have but little difficulty in following the demonstration, and arriving at a full conviction of its truth; but they ask, what if the proposition is true? what use can be made of it?

"Now, to meet these difficulties we have, in the body of the work, added in a smaller sized type, such remarks, suggestions, and *practical applications* as we have found to interest the pupil."

PLANE TRIGONOMETRY, AND ITS APPLICATION TO MENSURATION AND LAND SURVEYING, ACCOMPANIED WITH ALL THE NECESSARY LOGARITHMIC AND TRIGONOMETRIC TABLES. 8vo. sheep. Price $1 50.

The work will be found as simple and practical as the nature of the different subjects will admit, and differs in many respects from other similar works. The chapter upon the division of land is new and highly interesting to all lovers of the science. Great care has been taken to have the tables accurate; they are from the old style of figures, which are less fatiguing to the eye than those in common use

## RECOMMENDATIONS OF PERKINS' MATHEMATICAL SERIES.

*From* G. B. Docharty, *Prof. of Math., Free Academy, New York.*

"I have examined with more than ordinary interest a 'Treatise on Algebra' by George R. Perkins, L.L. D., and also a 'Treatise on Plane Trigonometry,' by the same author. I consider them both works of great merit. The 'Treatise on Algebra, as a text-book, has no superior. Its arrangement is excellent; the rules explicit, the explanations lucid, and the style in which it is written is well adapted to lead the tyro rapidly to a skilful knowledge of the higher analysis.

"The Treatise on Plane Trigonometry is just such a manual as the pupil requires It contains every thing necessary for a correct understanding of the principles of the science, and its practical application to the division of land. It will give me great pleasure at all times to recommend either of the above-named works to the teachers and students of my acquaintance."

*From Professors* Jackson *and* Potter, *of Union College.*

"I have examined Prof. Perkins' Elements of Algebra. It is a work in which the peculiar merits of the French and English systems are combined; the practical and theoretical being made to illustrate each other. It is consequently better adapted to elementary instruction in our seminaries than any foreign work I have seen. Indeed, it is equally fitted for the common school and the college, as the elementary principles are exhibited sufficiently in detail and with admirable clearness, and the higher parts of the science are fully and ably discussed."

*From* B. F. Joslin, *Prof. of Math., and Nat. Philosophy, in the City of Boston.*

"The treatise on Algebra by George R. Perkins, is evidently the work of an experienced and judicious teacher. The illustrations possess in a good degree the merits of simplicity and conciseness, whilst the prominence given to important theorems and rules, facilitates both recollection and reference."

*From* B. W. W. Redfield, *Asst. Prof. of Mathematics, N. Y. University.*

"I have examined, with much satisfaction, Perkins' Mathematical Series, and believe the treatise is inferior to none in point of clearness, accuracy and logical development. For his Elementary work on Algebra, the author deserves the thanks of teachers. He seems to have been the first to discover that, even in the education of children, a course of conclusive reasoning may with profit supersede the old system of dogmatical dictation."

*From* J. Zehner, *Prof. of Mathematics, College of St. Michael and All Angels.*

"It contains in a condensed form all the principles of Plane Trigonometry, that will most likely be needed in its application to the practical purposes of life. They are explained and demonstrated in a plain and concise manner, and the various examples and problems solved and unsolved are well calculated to make the student familiar with those principles, and expert in the application of them. The practical bearing of the book will, no doubt, make it a useful manual to the surveyor and engineer, particularly to those who have no opportunity to consult more extensive works on that subject, and wish to obtain the most necessary knowledge of it in a short time."

*From* Wm. J. Rolfe, *recent Principal of Day Academy, Wrentham, Mass.*

"I have put two classes through 'Perkins' Trigonometry,' and do not hesitate to place it very far above any similar work that I have ever taught. I do not believe there *is* a book before the public, which contains half as much practical matter on the subject of surveying. I have taken every opportunity to recommend it to my fellow teachers, and have no doubt that some of them will adopt it."

www.ingramcontent.com/pod-product-compliance
Lightning Source LLC
LaVergne TN
LVHW021124110826
845150LV00005B/936

* 9 7 8 1 4 2 5 5 5 0 8 6 8 *